AERODYNAMICS FO[R]

AVIATORS

BY

H. H. HURT, JR.

UNIVERSITY OF SOUTHERN CALIFORNIA

ISSUED BY
THE OFFICE OF THE CHIEF OF NAVAL OPERATIONS
AVIATION TRAINING DIVISION

U.S. NAVY, 1960 NAVWEPS 00-80T-80

REVISED JANUARY 1965

HBC0894 Printed in the USA

PREFACE

The purpose of this textbook is to present the elements of applied aerodynamics and aeronautical engineering which relate directly to the problems of flying operations. All Naval Aviators possess a natural interest in the basic aerodynamic factors which affect the performance of all aircraft. Due to the increasing complexity of modern aircraft, this natural interest must be applied to develop a sound understanding of basic engineering principles and an appreciation of some of the more advanced problems of aerodynamics and engineering. The safety and effectiveness of flying operations will depend greatly on the understanding and appreciation of how and why an airplane flies. The principles of aerodynamics will provide the foundations for developing exacting and precise flying techniques and operational procedures.

The content of this textbook has been arranged to provide as complete as possible a reference for all phases of flying in Naval Aviation. Hence, the text material is applicable to the problems of flight training, transition training, and general flying operations. The manner of presentation throughout the text has been designed to provide the elements of both theory and application and will allow either directed or unassisted study. As a result, the text material will be applicable to supplement formal class lectures and briefings and provide reading material as a background for training and flying operations.

Much of the specialized mathematical detail of aerodynamics has been omitted wherever it was considered unnecessary in the field of flying operations. Also, many of the basic assumptions and limitations of certain parts of aerodynamic theory have been omitted for the sake of simplicity and clarity of presentation. In order to contend with these specific shortcomings, the Naval Aviator should rely on the assistance of certain specially qualified individuals within Naval Aviation. For example, graduate aeronautical engineers, graduates of the Test Pilot Training School at the Naval Air Test Center, graduates of the Naval Aviation Safety Officers Course, and technical representatives of the manufacturers are qualified to assist in interpreting and applying the more difficult parts of aerodynamics and aeronautical engineering. To be sure, the specialized qualifications of these individuals should be utilized wherever possible.

The majority of aircraft accidents are due to some type of error of the pilot. This fact has been true in the past and, unfortunately, most probably will be true in the future. Each Naval Aviator should strive to arm himself with knowledge, training, and exacting, professional attitudes and techniques. The fundamentals of aerodynamics as presented in this text will provide the knowledge and background for safe and effective flying operations. The flight handbooks for the aircraft will provide the particular techniques, procedures, and operating data which are necessary for each aircraft. Diligent study and continuous training are necessary to develop the professional skills and techniques for successful flying operations.

The author takes this opportunity to express appreciation to those who have assisted in the preparation of the manuscript. In particular, thanks are due to Mr. J. E. Fairchild for his assistance with the portions dealing with helicopter aerodynamics and roll coupling phenomena. Also, thanks are due to Mr. J. F. Detwiler and Mr. E. Dimitruk for their review of the text material.

HUGH HARRISON HURT, Jr.

August 1959
University of Southern California
Los Angeles, Calif.

TABLE OF CONTENTS

PLANFORM EFFECTS AND AIRPLANE DRAG

CHAPTER 5. OPERATING STRENGTH LIMITATIONS

GENERAL DEFINITIONS AND STRUCTURAL REQUIREMENTS

CHAPTER 6. APPLICATION OF AERODYNAMICS TO SPECIFIC PROBLEMS OF FLYING

Chapter 1

BASIC AERODYNAMICS

In order to understand the characteristics of his aircraft and develop precision flying techniques, the Naval Aviator must be familiar with the fundamentals of aerodynamics. There are certain physical laws which describe the behavior of airflow and define the various aerodynamic forces and moments acting on a surface. These principles of aerodynamics provide the foundations for good, precise flying techniques.

WING AND AIRFOIL FORCES

PROPERTIES OF THE ATMOSPHERE

The aerodynamic forces and moments acting on a surface are due in great part to the properties of the air mass in which the surface is operating. The composition of the earth's atmosphere by volume is approximately 78 percent nitrogen, 21 percent oxygen, and 1 percent water vapor, argon, carbon dioxide, etc. For the majority of all aerodynamic considerations air is considered as a uniform mixture of these gases. The usual quantities used to define the properties of an air mass are as follows:

STATIC PRESSURE. The absolute static pressure of the air is a property of primary importance. The static pressure of the air at any altitude results from the mass of air supported above that level. At standard sea level conditions the static pressure of the air is 2,116 psf (or 14.7 psi, 29.92 in. Hg, etc.) and at 40,000 feet altitude this static pressure decreases to approximately 19 percent of the sea level value. The shorthand notation for the ambient static pressure is "p" and the standard sea level static pressure is given the subscript "o" for zero altitude, p_0. A more usual reference in aerodynamics and performance is the proportion of the ambient static pressure and the standard sea level static pressure. This static pressure ratio is assigned the shorthand notation of δ (delta).

$$\text{Altitude pressure ratio} = \frac{\text{Ambient static pressure}}{\text{Standard sea level static pressure}}$$

$$\delta = p/p_0$$

Many items of gas turbine engine performance are directly related to some parameter involving the altitude pressure ratio.

TEMPERATURE. The absolute temperature of the air is another important property. The ordinary temperature measurement by the Centigrade scale has a datum at the freezing point of water but absolute zero temperature is obtained at a temperature of $-273°$ Centigrade. Thus, the standard sea level temperature of 15° C. is an absolute temperature of 288°. This scale of absolute temperature using the Centigrade increments is the Kelvin scale, e.g., ° K. The shorthand notation for the ambient air temperature is "T" and the standard sea level air temperature of 288° K. is signified by T_0. The more usual reference is the proportion of the ambient air temperature and the standard sea level air temperature. This temperature ratio is assigned the shorthand notation of θ (theta).

$$\text{Temperature ratio} = \frac{\text{Ambient air temperature}}{\text{Standard sea level air temperature}}$$

$$\theta = T/T_0$$

$$\theta = \frac{C^0 + 273}{288}$$

Many items of compressibility effects and jet engine performance involve consideration of the temperature ratio.

DENSITY. The density of the air is a property of greatest importance in the study of aerodynamics. The density of air is simply the mass of air per cubic foot of volume and is a direct measure of the quantity of matter in each cubic foot of air. Air at standard sea level conditions weighs 0.0765 pounds per cubic foot and has a density of 0.002378 slugs per cubic foot. At an altitude of 40,000 feet the air density is approximately 25 percent of the sea level value.

The shorthand notation used for air density is ρ (rho) and the standard sea level air density is then ρ_0. In many parts of aerodynamics it is very convenient to consider the proportion of the ambient air density and standard sea level air density. This density ratio is assigned the shorthand notation of σ (sigma).

$$\text{density ratio} = \frac{\text{ambient air density}}{\text{standard sea level air density}}$$

$$\sigma = \rho/\rho_0$$

A general gas law defines the relationship of pressure temperature, and density when there is no change of state or heat transfer. Simply stated this would be "density varies directly with pressure, inversely with temperature." Using the properties previously defined,

$$\text{density ratio} = \frac{\text{pressure ratio}}{\text{temperature ratio}}$$

$$\frac{\rho}{\rho_0} = \left(\frac{P}{P_0}\right)\left(\frac{T_0}{T}\right)$$

$$\sigma = \delta/\theta$$

NAVY

This relationship has great application in aerodynamics and is quite fundamental and necessary in certain parts of airplane performance.

VISCOSITY. The viscosity of the air is important in scale and friction effects. The coefficient of absolute viscosity is the proportion between the shearing stress and velocity gradient for a fluid flow. The viscosity of gases is unusual in that the viscosity is generally a function of temperature alone and an increase in temperature increases the viscosity. The coefficient of absolute viscosity is assigned the shorthand notation μ (mu). Since many parts of aerodynamics involve consideration of viscosity and density, a more usual form of viscosity measure is the proportion of the coefficient of absolute viscosity and density. This combination is termed the "kinematic viscosity" and is noted by ν (nu).

$$\text{kinematic viscosity} = \frac{\text{coefficient of absolute viscosity}}{\text{density}}$$

$$\nu = \mu / \rho$$

The kinematic viscosity of air at standard sea level conditions is 0.0001576 square feet per second. At an altitude of 40,000 feet the kinematic viscosity is increased to 0.0005059 square foot per second.

In order to provide a common denominator for comparison of various aircraft, a standard atmosphere has been adopted. The standard atmosphere actually represents the mean or average properties of the atmosphere. Figure 1.1 illustrates the variation of the most important properties of the air throughout the standard atmosphere. Notice that the lapse rate is constant in the troposphere and the stratosphere begins with the isothermal region.

Since all aircraft performance is compared and evaluated in the environment of the standard atmosphere, all of the aircraft instrumentation is calibrated for the standard atmosphere. Thus, certain corrections must apply to the instrumentation as well as the aircraft performance if the operating conditions do not fit the standard atmosphere. In order to properly account for the nonstandard atmosphere certain terms must be defined. *Pressure altitude* is the altitude in the standard atmosphere corresponding to a particular pressure. The aircraft altimeter is essentially a sensitive barometer calibrated to indicate altitude in the standard atmosphere. If the altimeter is set for 29.92 in. Hg the altitude indicated is the pressure altitude—the altitude in the standard atmosphere corresponding to the sensed pressure. Of course, this indicated pressure altitude may not be the actual height above sea level due to variations in temperature, lapse rate, atmospheric pressure, and possible errors in the sensed pressure.

The more appropriate term for correlating aerodynamic performance in the nonstandard atmosphere is *density altitude*—the altitude in the standard atmosphere corresponding to a particular value of air density. The computation of density altitude must certainly involve consideration of pressure (pressure altitude) and temperature. Figure 1.6 illustrates the manner in which pressure altitude and temperature combine to produce a certain density altitude. This chart is quite standard in use and is usually included in the performance section of the flight handbook. Many subject areas of aerodynamics and aircraft performance will emphasize *density altitude* and *temperature* as the most important factors requiring consideration.

BERNOULLI'S PRINCIPLE AND SUBSONIC AIRFLOW

All of the external aerodynamic forces on a surface are the result of air *pressure* or air *friction*. Friction effects are generally confined to a thin layer of air in the immediate vicinity of the surface and friction forces are not the predominating aerodynamic forces. Therefore,

ICAO STANDARD ATMOSPHERE

ALTITUDE FT.	DENSITY RATIO σ	$\sqrt{\sigma}$	PRESSURE RATIO δ	TEMPER-ATURE °F	TEMPER-ATURE RATIO θ	SPEED OF SOUND a KNOTS	KINEMATIC VISCOSITY ν FT²/SEC
0	1.0000	1.0000	1.0000	59.00	1.0000	661.7	.000158
1000	0.9711	0.9854	0.9644	55.43	0.9931	659.5	.000161
2000	0.9428	0.9710	0.9298	51.87	0.9862	657.2	.000165
3000	0.9151	0.9566	0.8962	48.30	0.9794	654.9	.000169
4000	0.8881	0.9424	0.8637	44.74	0.9725	652.6	.000174
5000	0.8617	0.9283	0.8320	41.17	0.9656	650.3	.000178
6000	0.8359	0.9143	0.8014	37.60	0.9587	647.9	.000182
7000	0.8106	0.9004	0.7716	34.04	0.9519	645.6	.000187
8000	0.7860	0.8866	0.7428	30.47	0.9450	643.3	.000192
9000	0.7620	0.8729	0.7148	26.90	0.9381	640.9	.000197
10000	0.7385	0.8593	0.6877	23.34	0.9312	638.6	.000202
15000	0.6292	0.7932	0.5643	5.51	0.8969	626.7	.000229
20000	0.5328	0.7299	0.4595	−12.32	0.8625	614.6	.000262
25000	0.4481	0.6694	0.3711	−30.15	0.8281	602.2	.000302
30000	0.3741	0.6117	0.2970	−47.98	0.7937	589.5	.000349
35000	0.3099	0.5567	0.2353	−65.82	0.7594	576.6	.000405
* 36089	0.2971	0.5450	0.2234	−69.70	0.7519	573.8	.000419
40000	0.2462	0.4962	0.1851	−69.70	0.7519	573.8	.000506
45000	0.1936	0.4400	0.1455	−69.70	0.7519	573.8	.000643
50000	0.1522	0.3902	0.1145	−69.70	0.7519	573.8	.000818
55000	0.1197	0.3460	0.0900	−69.70	0.7519	573.8	.001040
60000	0.0941	0.3068	0.0708	−69.70	0.7519	573.8	.001323
65000	0.0740	0.2721	0.0557	−69.70	0.7519	573.8	.001682
70000	0.0582	0.2413	0.0438	−69.70	0.7519	573.8	.002139
75000	0.0458	0.2140	0.0344	−69.70	0.7519	573.8	.002721
80000	0.0360	0.1897	0.0271	−69.70	0.7519	573.8	.003460
85000	0.0280	0.1673	0.0213	−64.80	0.7613	577.4	.004499
90000	0.0217	0.1472	0.0168	−56.57	0.7772	583.4	.00591
95000	0.0169	0.1299	0.0134	−48.34	0.7931	589.3	.00772
100000	0.0132	0.1149	0.0107	−40.11	0.8089	595.2	.01004

* GEOPOTENTIAL OF THE TROPOPAUSE

Figure 1.1. Standard Altitude Table

the pressure forces created on an aerodynamic surface can be studied in a simple form which at first neglects the effect of friction and viscosity of the airflow. The most appropriate means of visualizing the effect of airflow and the resulting aerodynamic pressures is to study the fluid flow within a closed tube.

Suppose a stream of air is flowing through the tube shown in figure 1.2. The airflow at station 1 in the tube has a certain velocity, static pressure, and density. As the airstream approaches the constriction at station 2 certain changes must take place. Since the airflow is enclosed within the tube, the mass flow at any point along the tube must be the same and the velocity, pressure, or density must change to accommodate this continuity of flow.

BERNOULLI'S EQUATION. A distinguishing feature of *subsonic* airflow is that changes in pressure and velocity take place with small and negligible changes in density. For this reason the study of subsonic airflow can be simplified by neglecting the variation of density in the flow and assuming the flow to be *incompressible*. Of course, at high flow speeds which approach the speed of sound, the flow must be considered as compressible and "compressibility effects" taken into account. However, if the flow through the tube of figure 1.2 is considered subsonic, the density of the airstream is essentially constant at all stations along the length.

If the density of the flow remains constant, static pressure and velocity are the variable quantities. As the flow approaches the constriction of station 2 the velocity must increase to maintain the same mass flow. As the velocity increases the static pressure will decrease and the decrease in static pressure which accompanies the increase in velocity can be verified in two ways:

(1) Newton's laws of motion state the requirement of an unbalanced force to produce an acceleration (velocity change). If the airstream experiences an increase in velocity approaching the constriction, there must be an unbalance of force to provide the acceleration. Since there is only air within the tube, the unbalance of force is provided by the static pressure at station 1 being greater than the static pressure at the constriction, station 2.

(2) The total energy of the air stream in the tube is unchanged. However, the airstream energy may be in two forms. The airstream may have a *potential* energy which is related by the static pressure and a *kinetic* energy by virtue of mass and motion. As the total energy is unchanged, an increase in velocity (kinetic energy) will be accompanied by a decrease in static pressure (potential energy). This situation is analagous to a ball rolling along a smooth surface. As the ball rolls downhill, the potential energy due to position is exchanged for kinetic energy of motion. If friction were negligible, the change of potential energy would equal the change in kinetic energy. This is also the case for the airflow within the tube.

The relationship of static pressure and velocity is maintained throughout the length of the tube. As the flow moves past the constriction toward station 3, the velocity decreases and the static pressure increases.

The Bernoulli equation for incompressible flow is most readily explained by accounting for the energy of the airflow within the tube. As the airstream has no energy added or subtracted at any point, the sum of the potential and kinetic energy must be constant. The kinetic energy of an object is found by:

$$K.E.=\frac{1}{2}MV^2$$

where $K.E.$ = kinetic energy, ft.-lbs.
M = mass, slugs
V = velocity, ft./sec.

The kinetic energy of a cubic foot of air is:

$$\frac{K.E.}{ft.^3}=\frac{1}{2}\rho V^2$$

where $\frac{K.E.}{ft.^3}$ = kinetic energy per cu. ft., psf
ρ = air density, slugs per cu. ft.
V = air velocity, ft./sec.

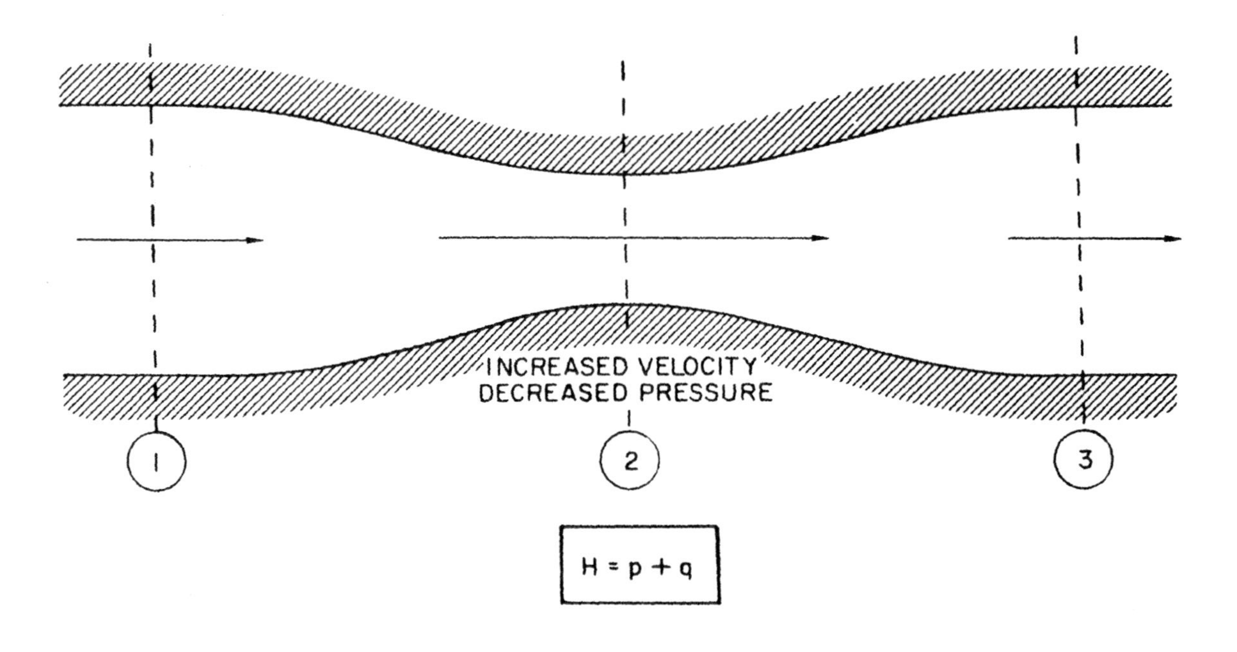

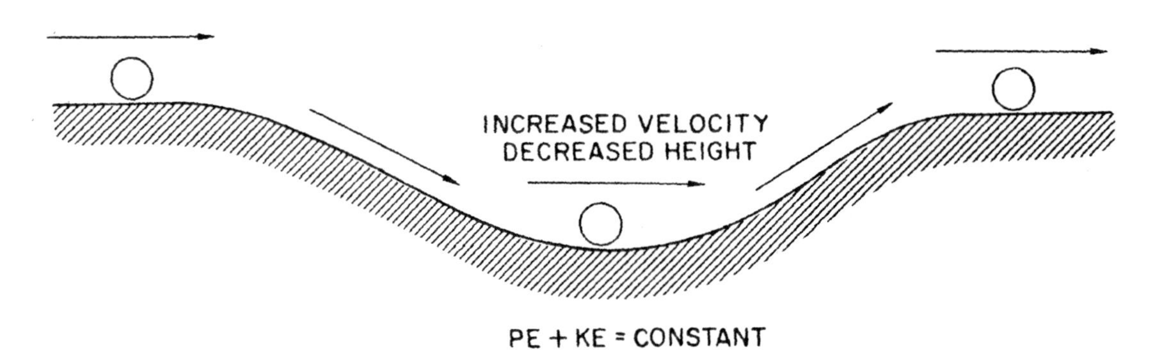

Figure 1.2. Airflow Within a Tube

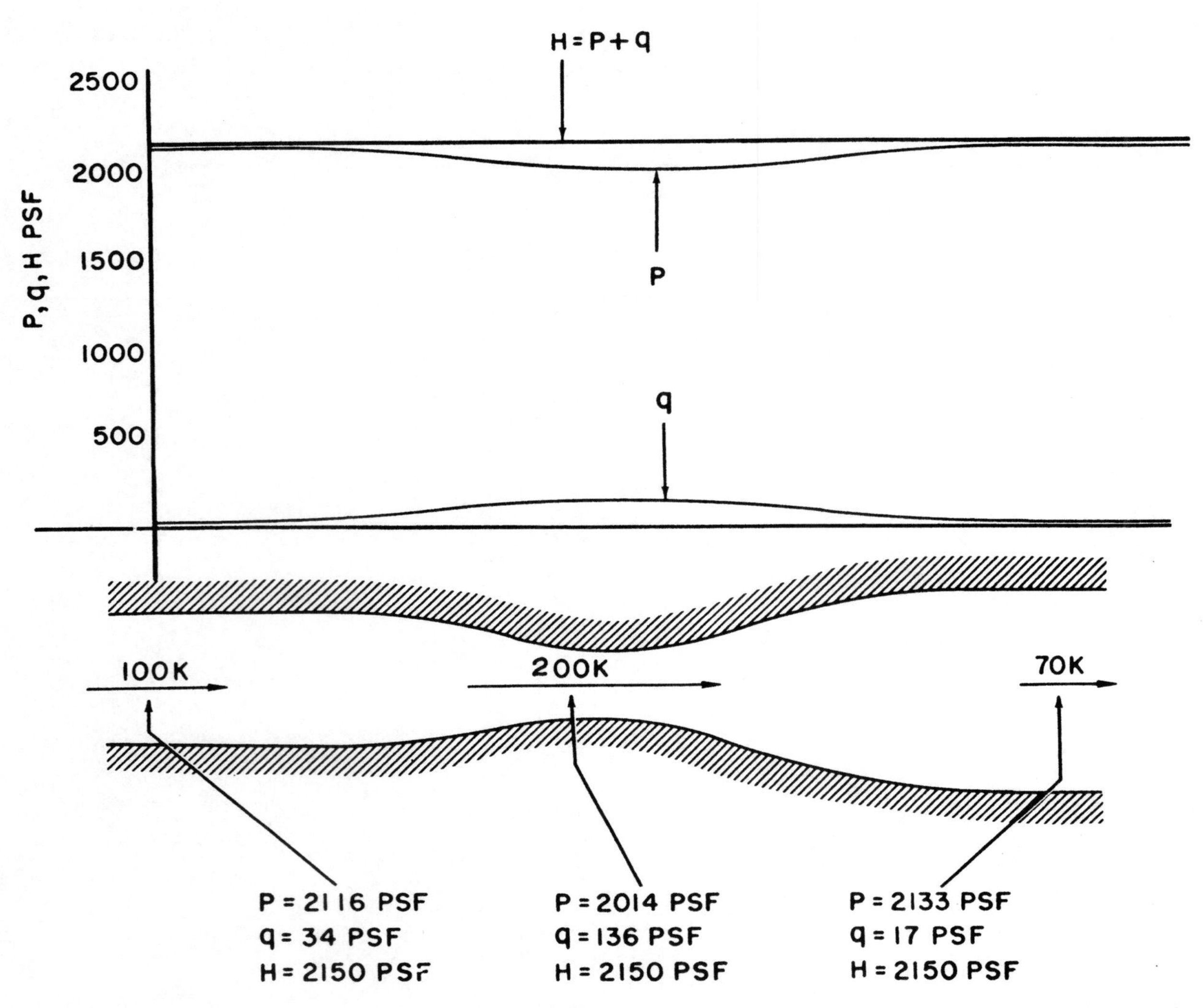

Figure 1.3. Variation of Pressure in Tube

If the potential energy is represented by the static pressure, p, the sum of the potential and kinetic energy is the total pressure of the airstream.

$$H=p+\tfrac{1}{2}\rho V^2$$

where H=total pressure, psf (sometimes referred to as "head" pressure)
p=static pressure, psf.
ρ=density, slugs per cu. ft.
V=velocity, ft./sec.

This equation is the Bernoulli equation for incompressible flow. It is important to appreciate that the term $\tfrac{1}{2}\rho V^2$ has the units of pressure, psf. This term is one of the most important in all aerodynamics and appears so frequently that it is given the name "dynamic pressure" and the shorthand notation "q".

$$q=\text{dynamic pressure, psf}$$
$$=\tfrac{1}{2}\rho V^2$$

With this definition it could be said that the sum of static and dynamic pressure in the flow tube remains constant.

Figure 1.3 illustrates the variation of static, dynamic, and total pressure of air flowing through a closed tube. Note that the total pressure is constant throughout the length and any change in dynamic pressure produces the same magnitude change in static pressure.

The dynamic pressure of a free airstream is the one common denominator of all aerodynamic forces and moments. Dynamic pressure represents the kinetic energy of the free airstream and is a factor relating the capability for producing changes in static pressure on a surface. As defined, the dynamic pressure varies directly as the density and the square of the velocity. Typical values of dynamic pressure, q, are shown in table 1–1 for various true airspeeds in the standard atmosphere. Notice that the dynamic pressure at some fixed velocity varies directly with the density ratio at any altitude. Also, appreciate the fact that at an altitude of 40,000 feet (where the density ratio, σ, is 0.2462) it is necessary to have a true air velocity twice that at sea level in order to product the same dynamic pressure.

TABLE 1–1. Effect of Speed and Altitude on Dynamic Pressure

Velocity (knots)	True air speed (ft./sec.)	Dynamic pressure, q, psf				
		Sea level	10,000 ft.	20,000 ft.	30,000 ft.	40,000 ft.
	$\sigma=$	1.000	0.7385	0.5328	0.3741	0.2462)
100	169	33.9	25.0	18.1	12.7	8.4
200	338	135.6	100.2	72.3	50.7	33.4
300	507	305	225	163	114	75.0
400	676	542	400	289	203	133
500	845	847	625	451	317	208
600	1,013	1,221	902	651	457	300

$q=\tfrac{1}{2}\rho V^2$
where q=dynamic pressure, psf
ρ=air density, slugs per cu. ft.
V=air velocity, ft. per sec.
or $q=.00339\sigma V^2$
where σ=density ratio
V=true velocity, *knots*
0.00339=constant which allows use of knots as velocity units and the altitude density ratio
an alternate form is

$$q=\frac{\sigma V^2}{295}\quad\left(0.00339=\frac{1}{295}\right)$$

AIRSPEED MEASUREMENT. If a symmetrically shaped object were placed in a moving airstream, the flow pattern typical of figure 1.4 would result. The airstream at the very nose of the object would stagnate and the relative flow velocity at this point would be zero. The airflow ahead of the object possesses some certain dynamic pressure and ambient static pressure. At the very nose of the object the local velocity will drop to zero and the airstream dynamic pressure will be converted into an increase in static pressure at the stagnation point. In other words, there will exist a static pressure at the stagnation point which is equal to the airstream total pressure—ambient static pressure plus dynamic pressure.

Around the surface of the object the airflow will divide and the local velocity will increase from zero at the stagnation point to some maximum on the sides of the object. If friction and viscosity effects are neglected, the

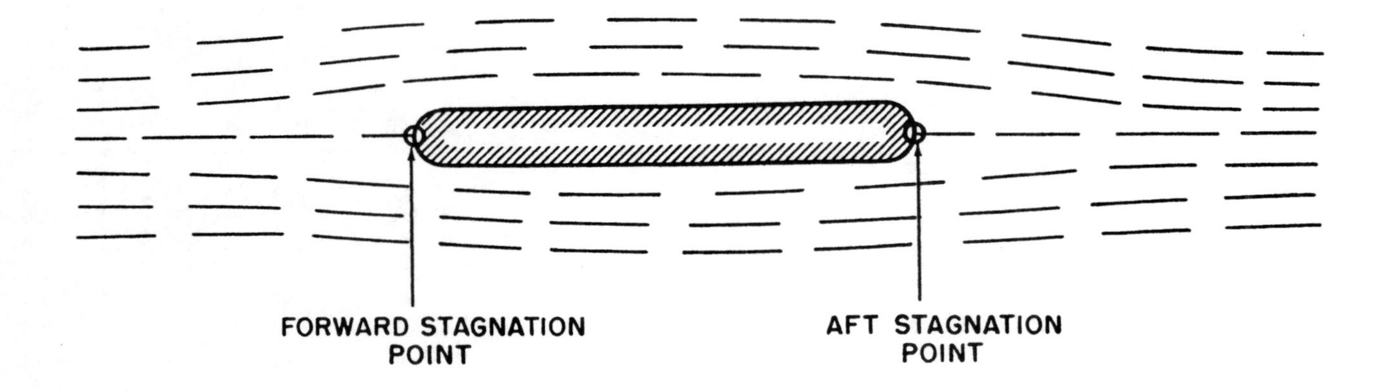

Figure 1.4. Flow Pattern on a Symmetrical Object

surface airflow continues to the aft stagnation point where the local velocity is again zero. The important point of this example of aerodynamic flow is existence of the stagnation point. The *change* in airflow static pressure which takes place at the stagnation point is equal to the free stream dynamic pressure, q.

The measurement of free stream dynamic pressure is fundamental to the indication of airspeed. In fact, airspeed indicators are simply pressure gauges which measure dynamic pressure related to various airspeeds. Typical airspeed measuring systems are illustrated in figure 1.5. The pitot head has no internal flow velocity and the pressure in the pitot tube is equal to the total pressure of the airstream. The purpose of the static ports is to sense the true static pressure of the free airstream. The total pressure and static pressure lines are attached to a differential pressure gauge and the net pressure indicated is the dynamic pressure, q. The pressure gauge is then calibrated to indicate flight speed in the standard sea level air mass. For example, a dynamic pressure of 305 psf would be realized at a sea level flight speed of 300 knots.

Actually there can be many conditions of flight where the airspeed indicator does not truly reflect the actual velocity through the air mass. The corrections that must be applied are many and listed in sequence below:

(1) The indicated airspeed (*IAS*) is the actual instrument indication for some given flight condition. Factors such as an altitude other than standard sea level, errors of the instrument and errors due to the installation, compressibility, etc. may create great variance between this instrument indication and the actual flight speed.

(2) The calibrated airspeed (*CAS*) is the result of correcting *IAS* for errors of the

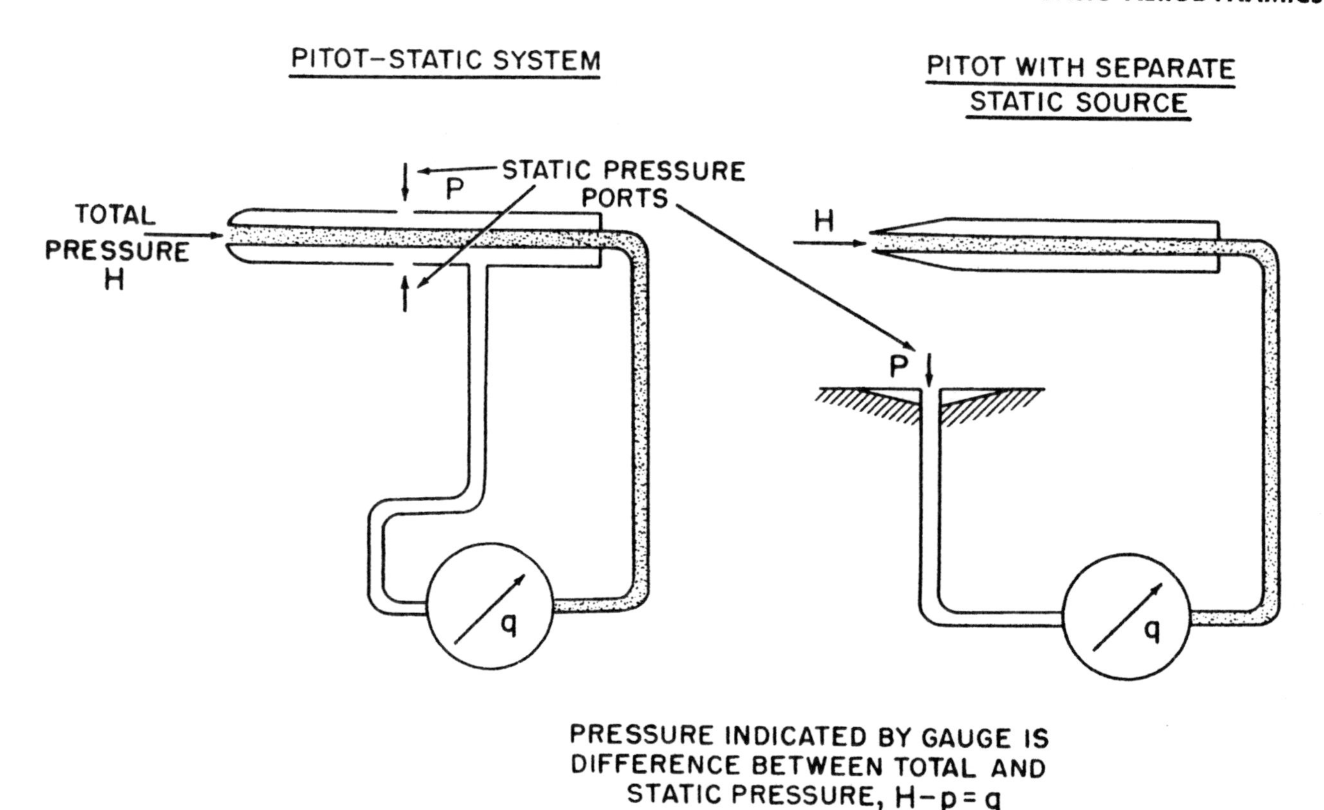

Figure 1.5. Airspeed Measurement

instrument and errors due to position or location of the installation. The instrument error must be small by design of the equipment and is usually negligible in equipment which is properly maintained and cared for. The position error of the installation must be small in the range of airspeeds involving critical performance conditions. Position errors are most usually confined to the static source in that the actual static pressure sensed at the static port may be different from the free airstream static pressure. When the aircraft is operated through a large range of angles of attack, the static pressure distribution varies quite greatly and it becomes quite difficult to minimize the static source error. In most instances a compensating group of static sources may be combined to reduce the position error. In order to appreciate the magnitude of this problem, at flight speed near 100 knots a *0.05 psi* position error is an airspeed error of 10 knots. A typical variation of airspeed system position error is illustrated in figure 1.6.

(3) The equivalent airspeed (*EAS*) is the result of correcting the (*CAS*) for compressibility effects. At high flight speeds the stagnation pressure recovered in the pitot tube is not representative of the airstream dynamic pressure due to a magnification by compressibility. Compressibility of the airflow produces a stagnation pressure in the pitot which is greater than if the flow were incompressible. As a result, the airspeed indication is given an erroneous magnification. The standard airspeed indicator is calibrated to read correct when at standard sea level conditions and thus has a compressibility correction appropriate for these conditions. However, when the aircraft is operating above standard sea level altitude,

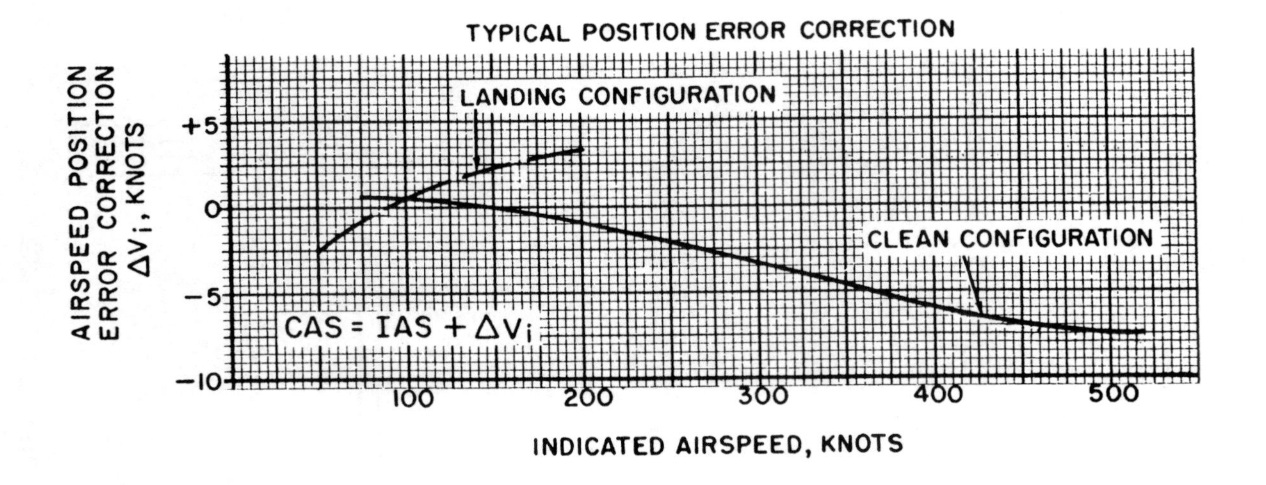

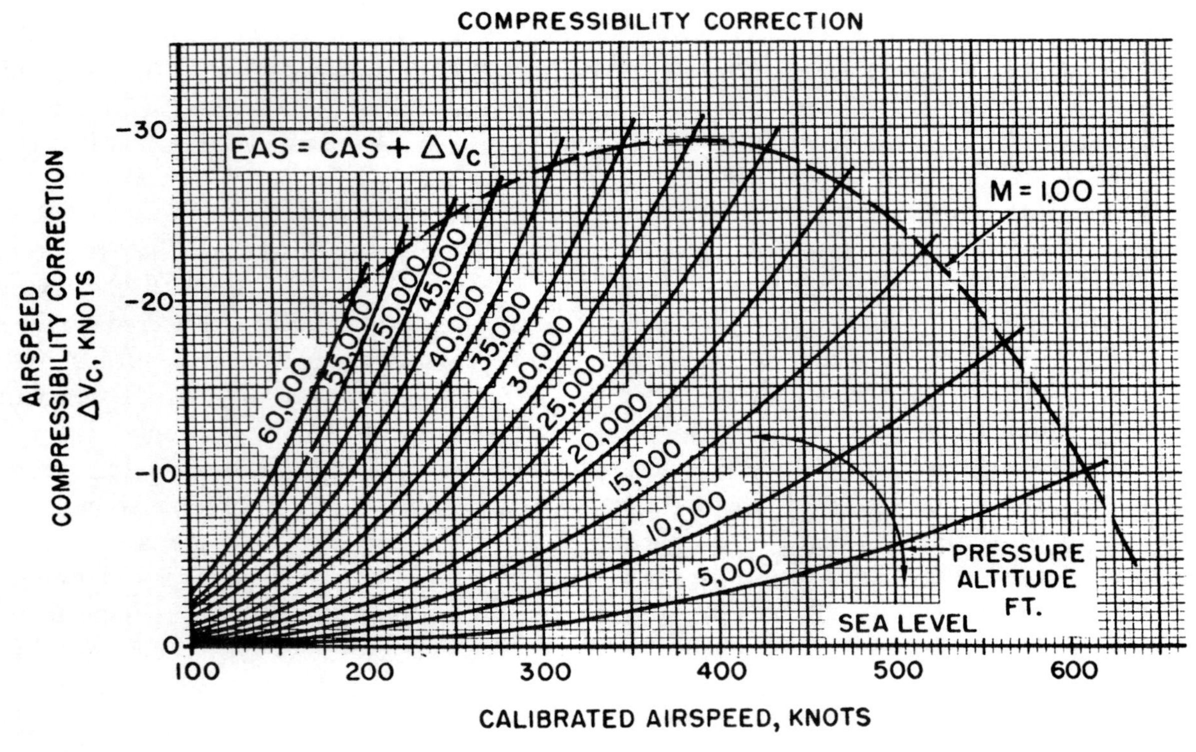

Figure 1.6. Airspeed Corrections (sheet 1 of 2)

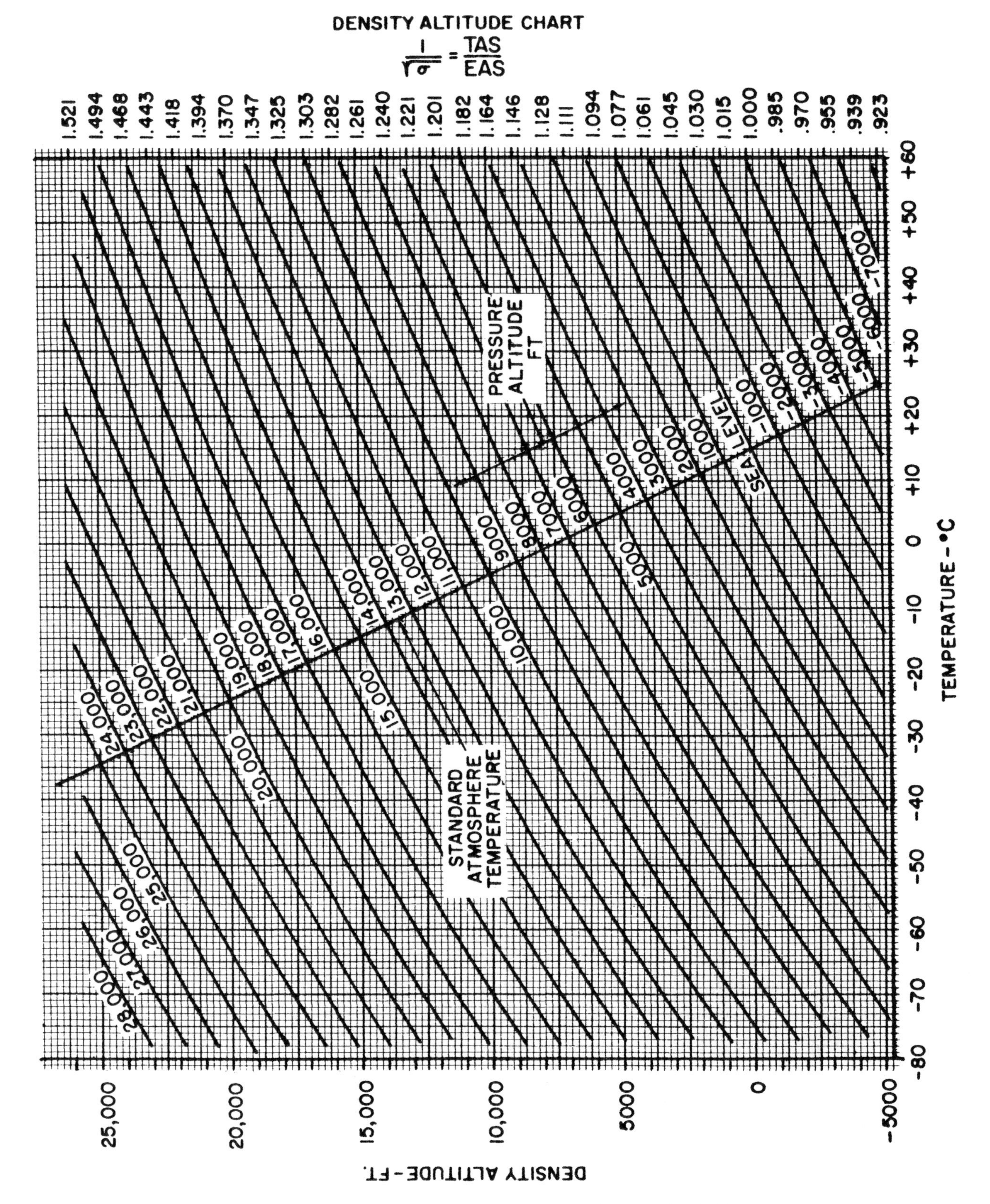

Figure 1.6. Airspeed Corrections (sheet 2 of 2)

the inherent compensation is inadequate and additional correction must be applied. The subtractive corrections that must be applied to *CAS* depend on pressure altitude and *CAS* and are shown on figure 1.6 for the subsonic flight range. The equivalent airspeed (*EAS*) is the flight speed in the standard sea level air mass which would produce the same free stream dynamic pressure as the actual flight condition.

(4) The true airspeed (*TAS*) results when the *EAS* is corrected for density altitude. Since the airspeed indicator is calibrated for the dynamic pressures corresponding to airspeeds at standard sea level conditions, variations in air density must be accounted for. To relate *EAS* and *TAS* requires consideration that the *EAS* coupled with standard sea level density produces the same dynamic pressure as the *TAS* coupled with the actual air density of the flight condition. From this reasoning, it can be shown that:

$$(TAS)^2\rho=(EAS)^2\rho_0$$

$$\text{or, } TAS=EAS\sqrt{\frac{\rho_0}{\rho}}$$

$$TAS=EAS\frac{1}{\sqrt{\sigma}}$$

where TAS = true airspeed
EAS = equivalent airspeed
ρ = actual air density
ρ_0 = standard sea level air density
σ = altitude density ratio, ρ/ρ_0

The result shows that the *TAS* is a function of *EAS* and density altitude. Figure 1.6 shows a chart of density altitude as a function of pressure altitude and temperature. Each particular density altitude fixes the proportion between *TAS* and *EAS*. The use of a navigation computer requires setting appropriate values of pressure altitude and temperature on the scales which then fixes the proportion between the scales of *TAS* and *EAS* (or *TAS* and *CAS* when compressibility corrections are applicable).

Thus, the airspeed indicator system measures dynamic pressure and will relate true flight velocity when instrument, position, compressibility, and density corrections are applied. These corrections are quite necessary for accurate determination of true airspeed and accurate navigation.

Bernoulli's principle and the concepts of static, dynamic, and total pressure are the basis of aerodynamic fundamentals. The pressure distribution caused by the variation of local static and dynamic pressures on a surface is the source of the major aerodynamic forces and moment.

DEVELOPMENT OF AERODYNAMIC FORCES

The typical airflow patterns exemplify the relationship of static pressure and velocity defined by Bernoulli. Any object placed in an airstream will have the air to impact or stagnate at some point near the leading edge. The pressure at this point of stagnation will be an absolute static pressure equal to the total pressure of the airstream. In other words, the static pressure at the stagnation point will be greater than the atmospheric pressure by the amount of the dynamic pressure of the airstream. As the flow divides and proceeds around the object, the increases in local velocity produce decreases in static pressure. This procedure of flow is best illustrated by the flow patterns and pressure distributions of figure 1.7.

STREAMLINE PATTERN AND PRESSURE DISTRIBUTION. The flow pattern of the cylinder of figure 1.7 is characterized by the streamlines which denote the local flow direction. Velocity distribution is noted by the streamline pattern since the streamlines effect a boundary of flow, and the airflow between the streamlines is similar to flow in a closed tube. When the streamlines contract and are close together, high local velocities exist; when the streamlines expand and are far apart, low local velocities exist. At the

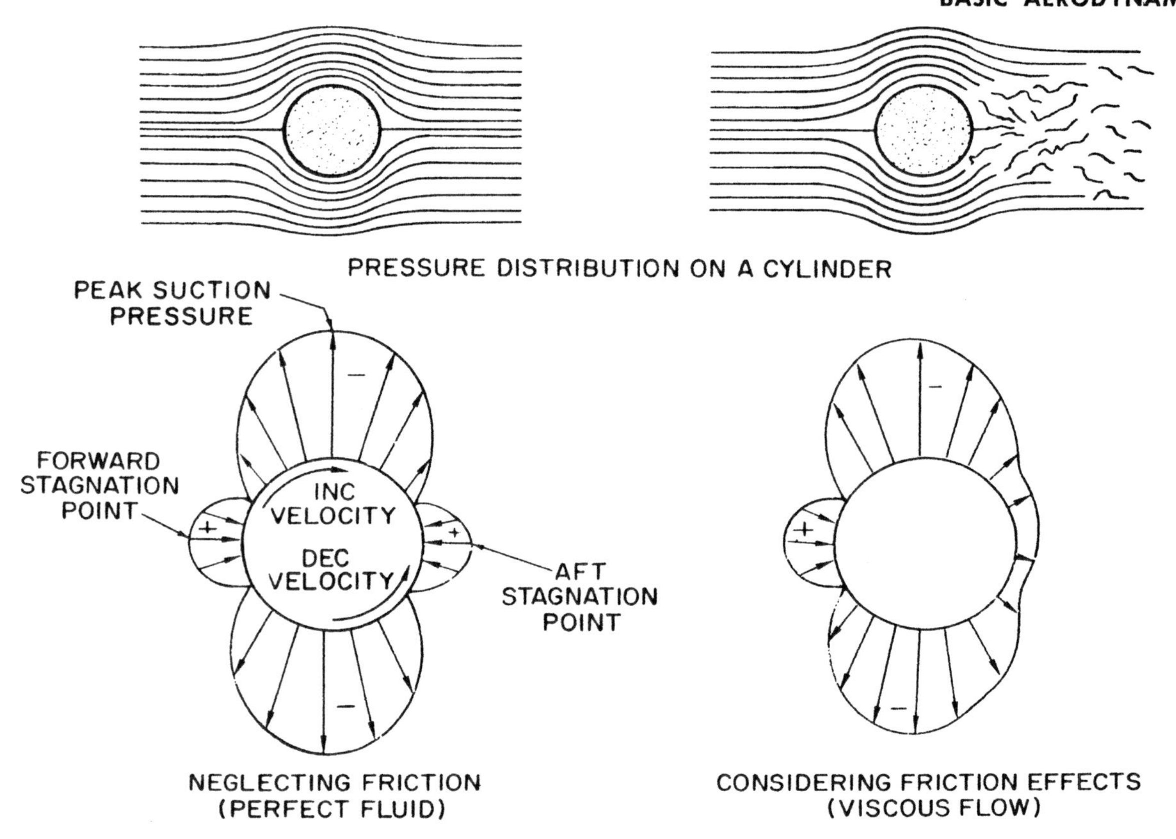

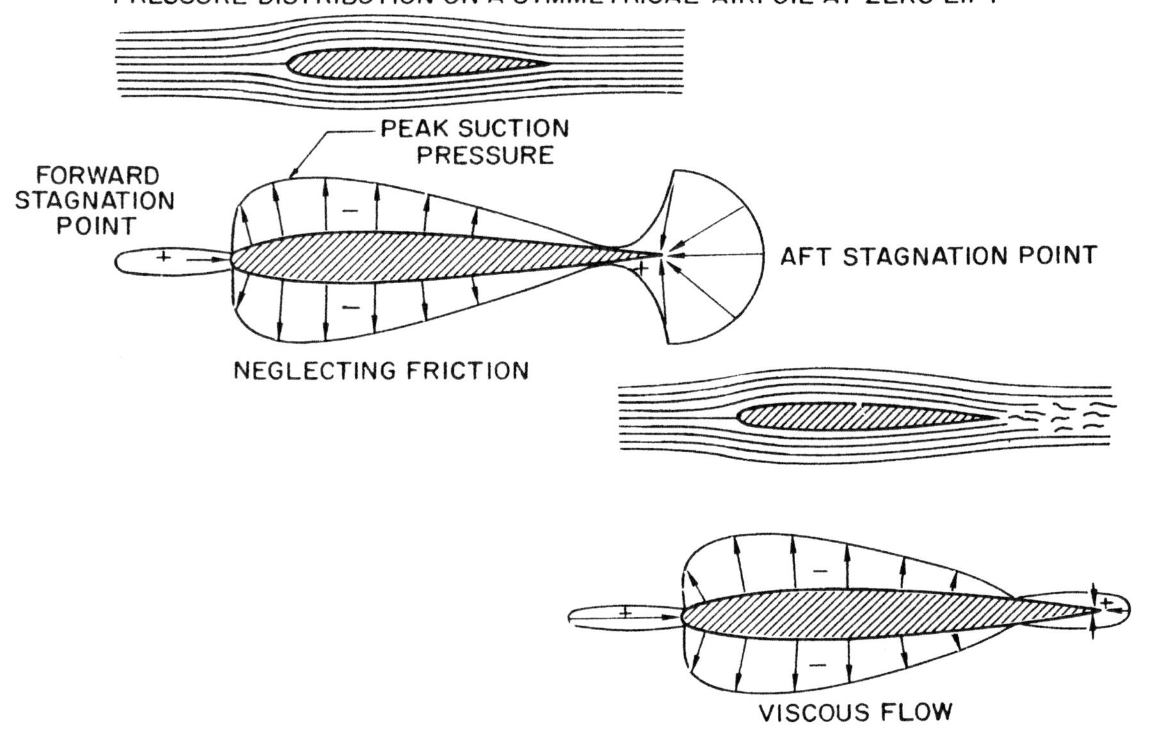

Figure 1.7. Streamline Pattern and Pressure Distribution

forward stagnation point the local velocity is zero and the maximum positive pressure results. As the flow proceeds from the forward stagnation point the velocity increases as shown by the change in streamlines. The local velocities reach a maximum at the upper and lower extremities and a peak suction pressure is produced at these points on the cylinder. (NOTE: *Positive* pressures are pressures *above* atmospheric and *negative* or *suction* pressures are *less* than atmospheric.) As the flow continues aft from the peak suction pressure, the diverging streamlines indicate decreasing local velocities and increasing local pressures. If friction and compressibility effects are not considered, the velocity would decrease to zero at the aft stagnation point and the full stagnation pressure would be recovered. The pressure distribution for the cylinder in perfect fluid flow would be symmetrical and no net force (*lift or drag*) would result. Of course, the relationship between static pressure and velocity along the surface is defined by Bernoulli's equation.

The flow pattern for the cylinder in an actual fluid demonstrates the effect of friction or viscosity. The viscosity of air produces a thin layer of retarded flow immediately adjacent to the surface. The energy expended in this "boundary layer" can alter the pressure distribution and destroy the symmetry of the pattern. The force unbalance caused by the change in pressure distribution creates a drag force which is in addition to the drag due to skin friction.

The streamline pattern for the symmetrical airfoil of figure 1.7 again provides the basis for the velocity and pressure distribution. At the leading edge the streamlines are widely diverged in the vicinity of the positive pressures. The maximum local velocities and suction (or negative) pressures exist where the streamlines are the closest together. One notable difference between the flow on the cylinder and the airfoil is that the maximum velocity and minimum pressure points on the airfoil do not necessarily occur at the point of maximum thickness. However, a similarity does exist in that the minimum pressure points correspond to the points where the streamlines are closest together and this condition exists when the streamlines are forced to the greatest curvature.

✓GENERATION OF LIFT. An important phenomenon associated with the production of lift by an airfoil is the "circulation" imparted to the airstream. The best practical illustration of this phenomenon is shown in figure 1.8 by the streamlines and pressure distributions existing on cylinders in an airstream. The cylinder without circulation has a symmetrical streamline pattern and a pressure distribution which creates no net lift. If the cylinder is given a clockwise rotation and induces a rotational or circulatory flow, a distinct change takes place in the streamline pattern and pressure distribution. The velocities due to the vortex of circulatory flow cause increased local velocity on the upper surface of the cylinder and decreased local velocity on the lower surface of the cylinder. Also, the circulatory flow produces an upwash immediately ahead and downwash immediately behind the cylinder and both fore and aft stagnation points are lowered.

The effect of the addition of circulatory flow is appreciated by the change in the pressure distribution on the cylinder. The increased local velocity on the upper surface causes an increase in upper surface suction while the decreased local velocity on the lower surface causes a decrease in lower surface suction. As a result, the cylinder with circulation will produce a net lift. This mechanically induced circulation—called Magnus effect—illustrates the relationship between circulation and lift and is important to golfers, baseball and tennis players as well as pilots and aerodynamicists. The curvature of the flight path of a golf ball or baseball requires an unbalance of force which is created by rotation of the ball. The pitcher that can accurately control a powerful

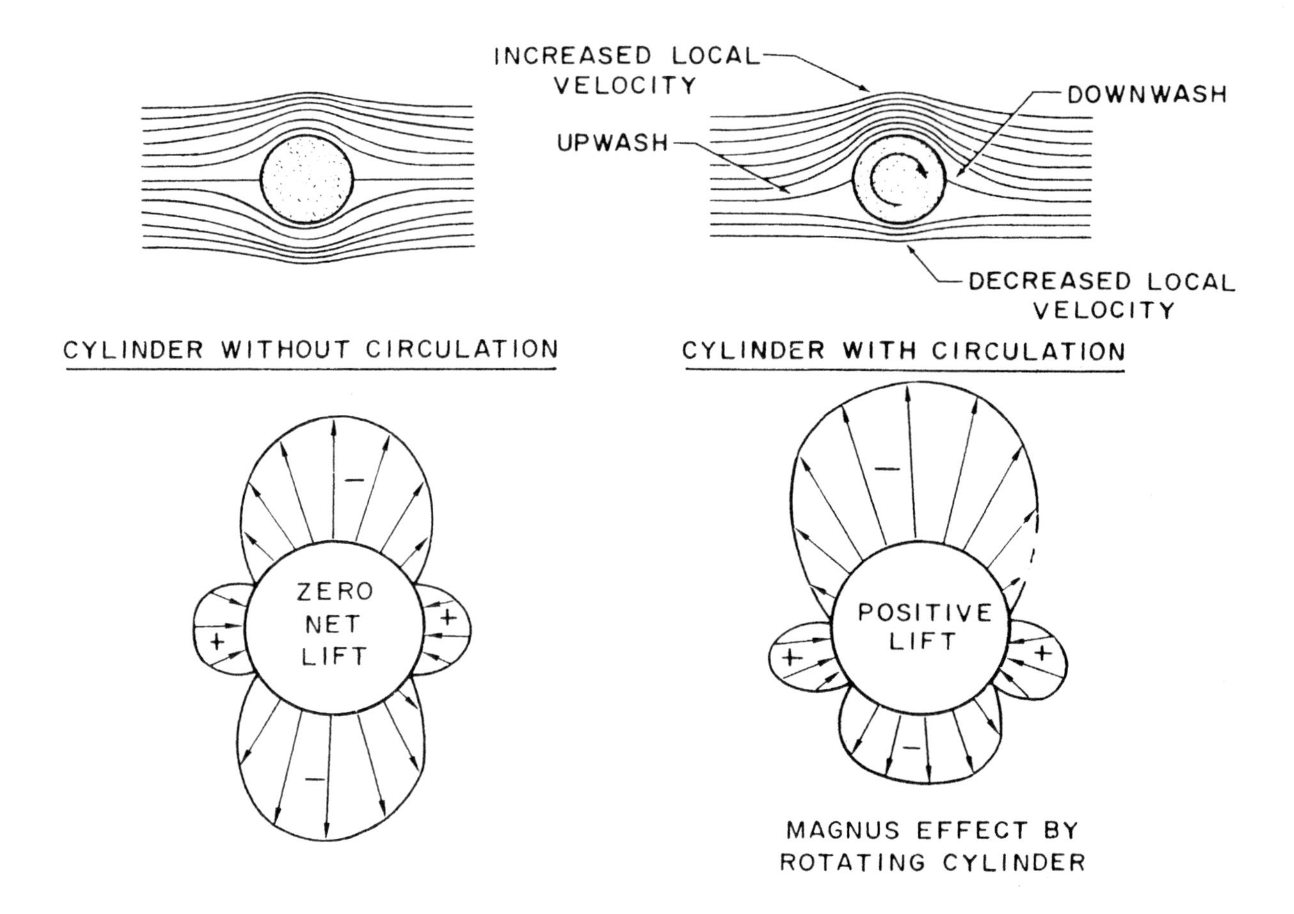

AIRFOIL LIFT

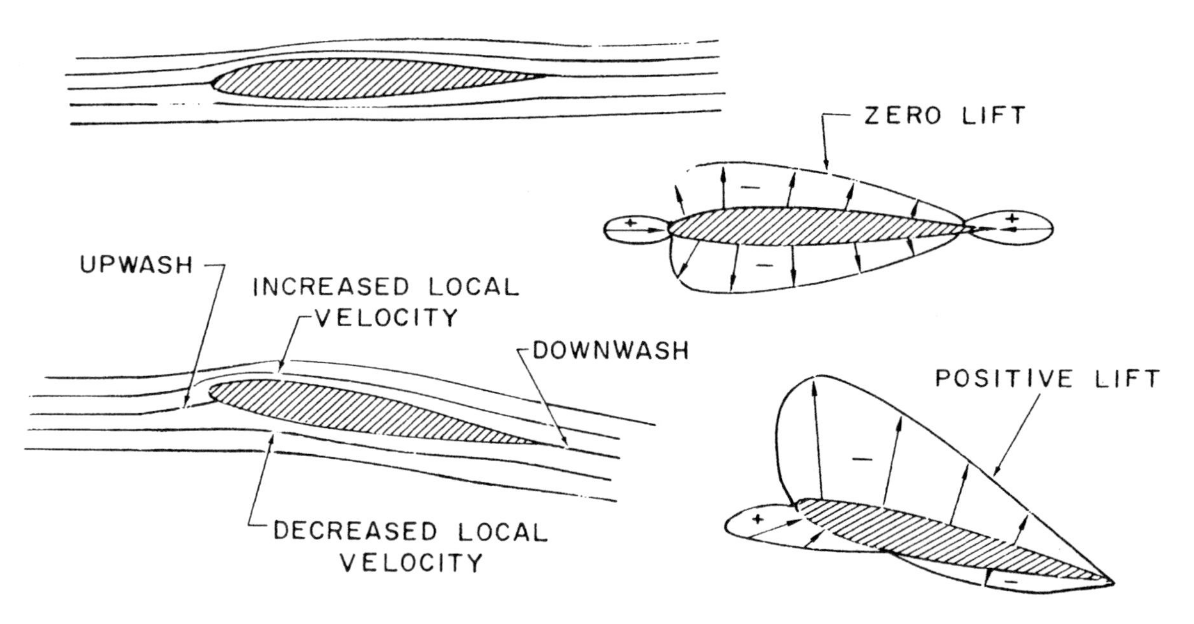

Figure 1.8. Generation of Lift (sheet 1 of 2)

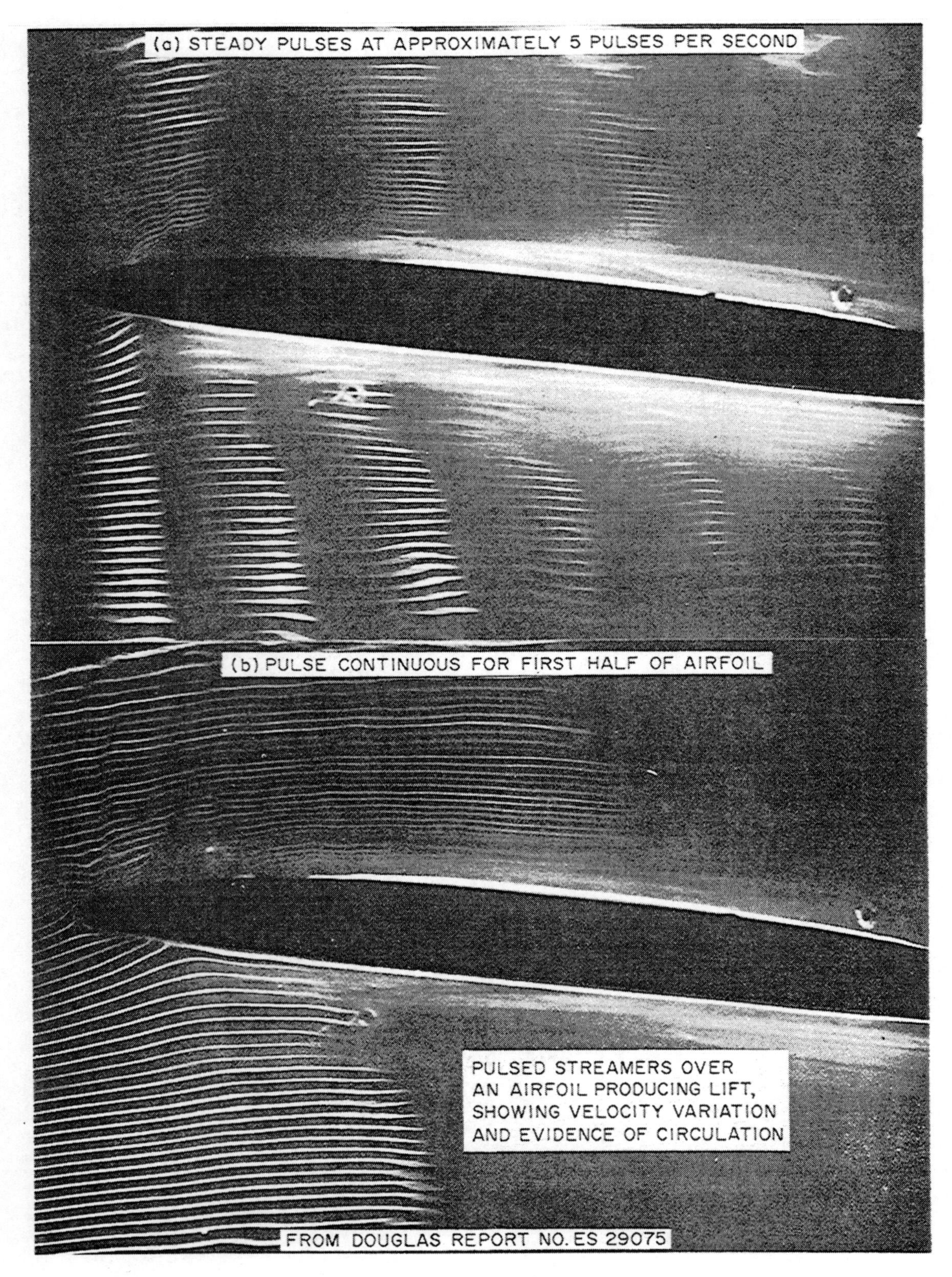

Figure 1.8. Generation of Lift (sheet 2 of 2)

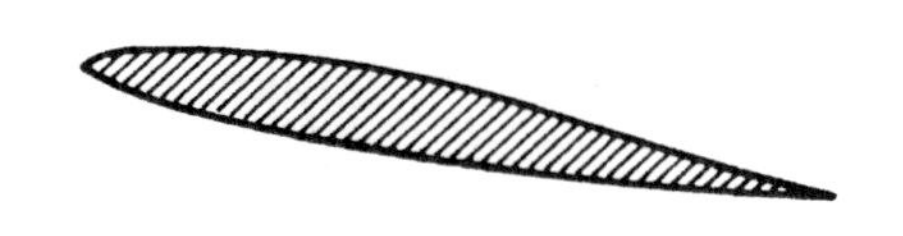

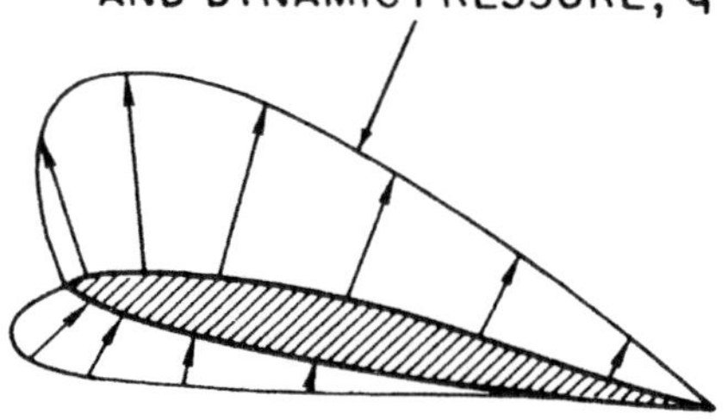

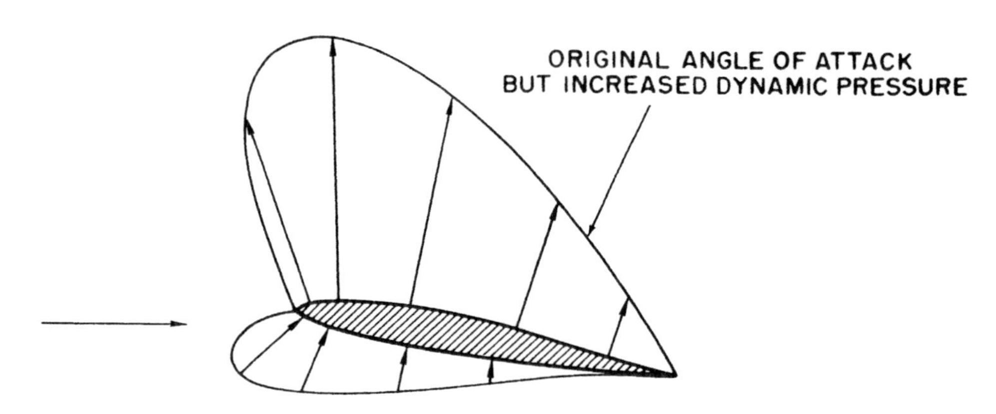

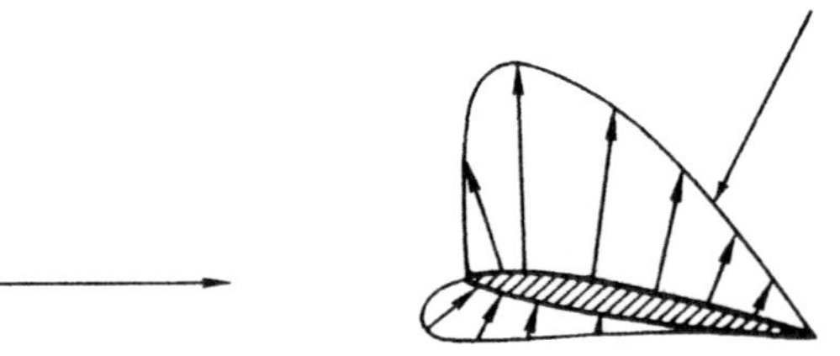

AIRFOIL SHAPE AND ANGLE OF ATTACK DEFINE
RELATIVE PRESSURE DISTRIBUTION

Figure 1.9. Airfoil Pressure Distribution

rotation will be quite a "curve ball artist"; the golfer that cannot control the lateral motion of the club face striking the golf ball will impart an uncontrollable spin and have trouble with a "hook" or "slice."

While a rotating cylinder can produce a net lift from the circulatory flow, the method is relatively inefficient and only serves to point out the relationship between lift and circulation. An airfoil is capable of producing lift with relatively high efficiency and the process is illustrated in figure 1.8. If a symmetrical airfoil is placed at zero angle of attack to the airstream, the streamline pattern and pressure distribution give evidence of zero lift. However, if the airfoil is given a positive angle of attack, changes occur in the streamline pattern and pressure distribution similar to changes caused by the addition of circulation to the cylinder. The positive angle of attack causes increased velocity on the upper surface with an increase in upper surface suction while the decreased velocity on the lower surface causes a decrease in lower surface suction. Also, upwash is generated ahead of the airfoil, the forward stagnation point moves under the leading edge, and a downwash is evident aft of the airfoil. The pressure distribution on the airfoil now provides a net force perpendicular to the airstream—lift.

The generation of lift by an airfoil is dependent upon the airfoil being able to create circulation in the airstream and develop the lifting pressure distribution on the surface. In all cases, the generated lift will be the net force caused by the distribution of pressure over the upper and lower surfaces of the airfoil. At low angles of attack, suction pressures usually will exist on both upper and lower surfaces but the upper surface suction must be greater for positive lift. At high angles of attack near that for maximum lift, a positive pressure will exist on the lower surface but this will account for approximately one-third the net lift.

The effect of free stream density and velocity is a necessary consideration when studying the development of the various aerodynamic forces. Suppose that a particular shape of airfoil is fixed at a particular angle to the airstream. The *relative* velocity and pressure distribution will be determined by the shape of the airfoil and the angle to the airstream. The effect of varying the airfoil size, air density and airspeed is shown in figure 1.9. If the same airfoil shape is placed at the same angle to an airstream with twice as great a dynamic pressure the *magnitude* of the pressure distribution will be twice as great but the *relative* shape of the pressure distribution will be the same. With twice as great a pressure existing over the surface, all aerodynamic forces and moments will double. If a half-size airfoil is placed at the same angle to the original airstream, the magnitude of the pressure distribution is the same as the original airfoil and again the *relative* shape of the pressure distribution is identical. The same pressure acting on the half-size surface would reduce all aerodynamic forces to one-half that of the original. This similarity of flow patterns means that the stagnation point occurs at the same place, the peak suction pressure occurs at the same place, and the actual *magnitude* of the aerodynamic forces and moments depends upon the airstream dynamic pressure and the surface area. This concept is extremely important when attempting to separate and analyze the most important factors affecting the development of aerodynamic forces.

AIRFOIL TERMINOLOGY. Since the shape of an airfoil and the inclination to the airstream are so important in determining the pressure distribution, it is necessary to properly define the airfoil terminology. Figure 1.10 shows a typical airfoil and illustrates the various items of airfoil terminology

(1) The *chord line* is a straight line connecting the leading and trailing edges of the airfoil.

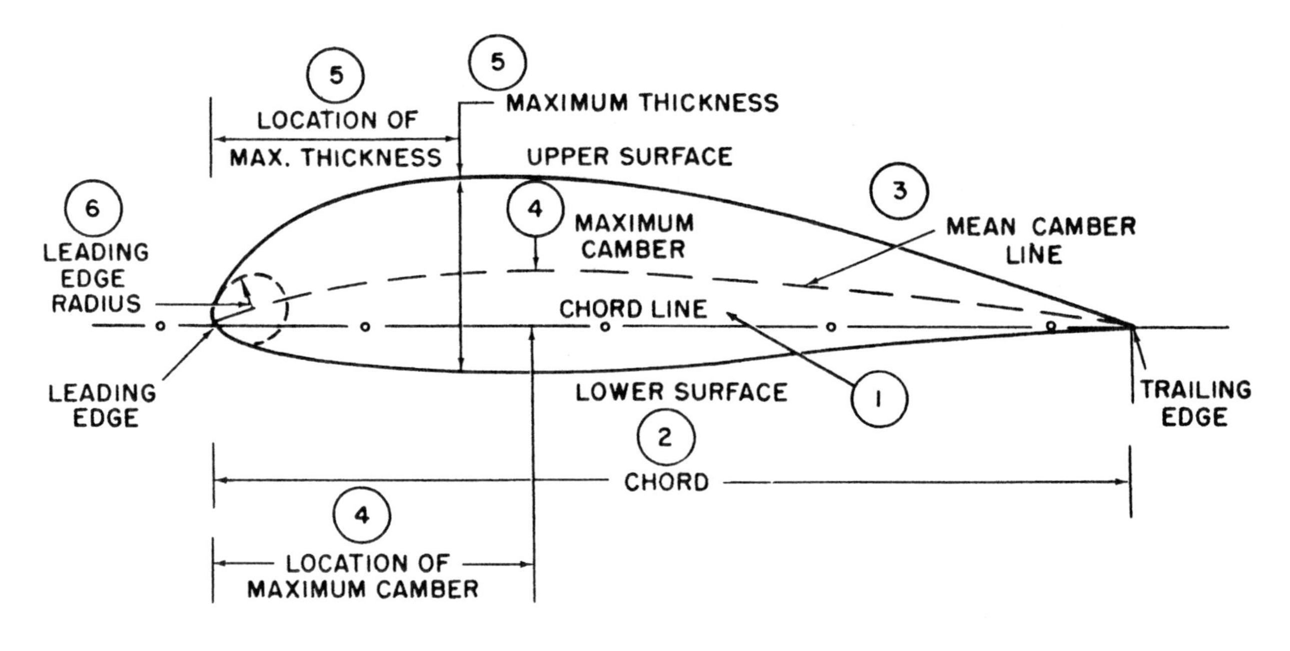

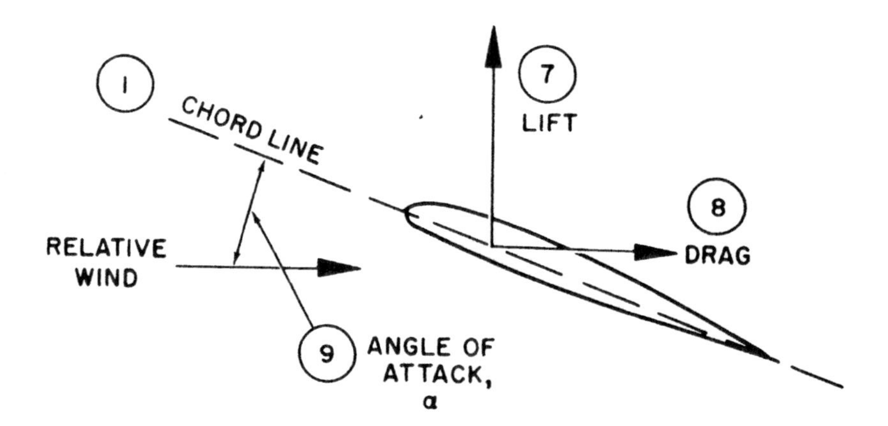

Figure 1.10. Airfoil Terminology

(2) The *chord* is the characteristic dimension of the airfoil.

(3) The *mean-camber line* is a line drawn halfway between the upper and lower surfaces. Actually, the chord line connects the ends of the mean-camber line.

(4) The shape of the mean-camber line is very important in determining the aerodynamic characteristics of an airfoil section. The *maximum camber* (displacement of the mean line from the chord line) and the *location* of the maximum camber help to define the shape of the mean-camber line. These quantities are expressed as fractions or percent of the basic chord dimension. A typical low speed airfoil may have a maximum camber of 4 percent located 40 percent aft of the leading edge.

(5) The thickness and thickness distribution of the profile are important properties of a section. The *maximum thickness* and *location* of maximum thickness define thickness and distribution of thickness and are expressed as fractions or percent of the chord. A typical low speed airfoil may have a maximum thickness of 12 percent located 30 percent aft of the leading edge.

(6) The *leading edge radius* of the airfoil is the radius of curvature given the leading edge shape. It is the radius of the circle centered on a line tangent to the leading edge camber and connecting tangency points of upper and lower surfaces with the leading edge. Typical leading edge radii are zero (knife edge) to 1 or 2 percent.

(7) The *lift* produced by an airfoil is the net force produced *perpendicular* to the *relative wind*.

(8) The *drag* incurred by an airfoil is the net force produced *parallel* to the *relative wind*.

(9) The *angle of attack* is the angle between the chord line and the relative wind. Angle of attack is given the shorthand notation α (alpha). Of course, it is important to differentiate between pitch attitude angle and angle of attack. Regardless of the condition of flight, the instantaneous flight path of the surface determines the direction of the oncoming relative wind and the angle of attack is the angle between the instantaneous relative wind and the chord line. To respect the definition of angle of attack, visualize the flight path of the aircraft during a loop and appreciate that the relative wind is defined by the flight path at any point during the maneuver.

Notice that the description of an airfoil profile is by dimensions which are fractions or percent of the basic chord dimension. Thus, when an airfoil profile is specified a *relative* shape is described. (NOTE: A numerical system of designating airfoil profiles originated by the National Advisory Committee for Aeronautics [NACA] is used to describe the main geometric features and certain aerodynamic properties. NACA Report No. 824 will provide the detail of this system.)

AERODYNAMIC FORCE COEFFICIENT. The aerodynamic forces of lift and drag depend on the combined effect of many different variables. The important single variables could be:

(1) Airstream velocity
(2) Air density
(3) Shape or profile of the surface
(4) Angle of attack
(5) Surface area
(6) Compressibility effects
(7) Viscosity effects

If the effects of viscosity and compressibility are not of immediate importance, the remaining items can be combined for consideration. Since the major aerodynamic forces are the result of various pressures distributed on a surface, the *surface area* will be a major factor. *Dynamic pressure* of the airstream is another common denominator of aerodynamic forces and is a major factor since the *magnitude* of a pressure distribution depends on the source energy of the free stream. The remaining major factor is the relative *pressure distribution* existing on the surface. Of course, the velocity distribution, and resulting pressure distribution, is determined by the shape or profile of the surface and the angle of attack. Thus, any aerodynamic force can be represented as the product of three major factors:

the surface area of the object
the dynamic pressure of the airstream
the coefficient or index of force determined by the relative pressure distribution

This relationship is expressed by the following equation:

$$F = C_F q S$$

where

F = aerodynamic force, lbs.
C_F = coefficient of aerodynamic force
q = dynamic pressure, psf
$= \frac{1}{2}\rho V^2$
S = surface area, sq. ft.

In order to fully appreciate the importance of the aerodynamic force coefficient, C_F, the above equation is rearranged to alternate forms:

$$C_F = \frac{F}{qS}$$

$$C_F = \frac{F/S}{q}$$

In this form, the aerodynamic force coefficient is appreciated as the aerodynamic force per surface area and dynamic pressure. In other words, the force coefficient is a dimensionless ratio between the average aerodynamic pressure (aerodynamic force per area) and the airstream dynamic pressure. All the aerodynamic forces of lift and drag are studied on this basis—the common denominator in each case being surface area and dynamic pressure. By such a definition, a "lift coefficient" would be the ratio between lift pressure and dynamic pressure; a "drag coefficient" would be the ratio between drag pressure and dynamic pressure. The use of the coefficient form of an aerodynamic force is necessary since the force coefficient is:

(1) An index of the aerodynamic force independent of area, density, and velocity. It is derived from the relative pressure and velocity distribution.

(2) Influenced only by the shape of the surface and angle of attack since these factors determine the pressure distribution.

(3) An index which allows evaluation of the effects of compressibility and viscosity. Since the effects of area, density, and velocity are obviated by the coefficient form, compressibility and viscosity effects can be separated for study.

THE BASIC LIFT EQUATION. Lift has been defined as the net force developed perpendicular to the relative wind. The aerodynamic force of lift on an airplane results from the generation of a pressure distribution on the wing. This lift force is described by the following equation:

$$L = C_L q S$$

where

L = lift, lbs.
C_L = lift coefficient.
q = dynamic pressure, psf
$= \frac{1}{2}\rho V^2$
S = wing surface area, sq. ft.

The lift coefficient used in this equation is the ratio of the lift pressure and dynamic pressure and is a function of the shape of the wing and angle of attack. If the lift coefficient of a conventional airplane wing planform were plotted versus angle of attack, the result would be typical of the graph of figure 1.11. Since the effects of speed, density, area, weight, altitude, etc., are eliminated by the coefficient form, an indication of the true lift capability is obtained. Each angle of attack produces a particular lift coefficient since the angle of attack is the controlling factor in the pressure distribution. Lift coefficient increases with angle of attack up to the maximum lift coefficient, $C_{L_{max}}$, and, as angle of attack is increased beyond the maximum lift angle, the airflow is unable to adhere to the upper surface. The airflow then separates from the upper surface and stall occurs.

INTERPRETATION OF THE LIFT EQUATION. Several important relationships are

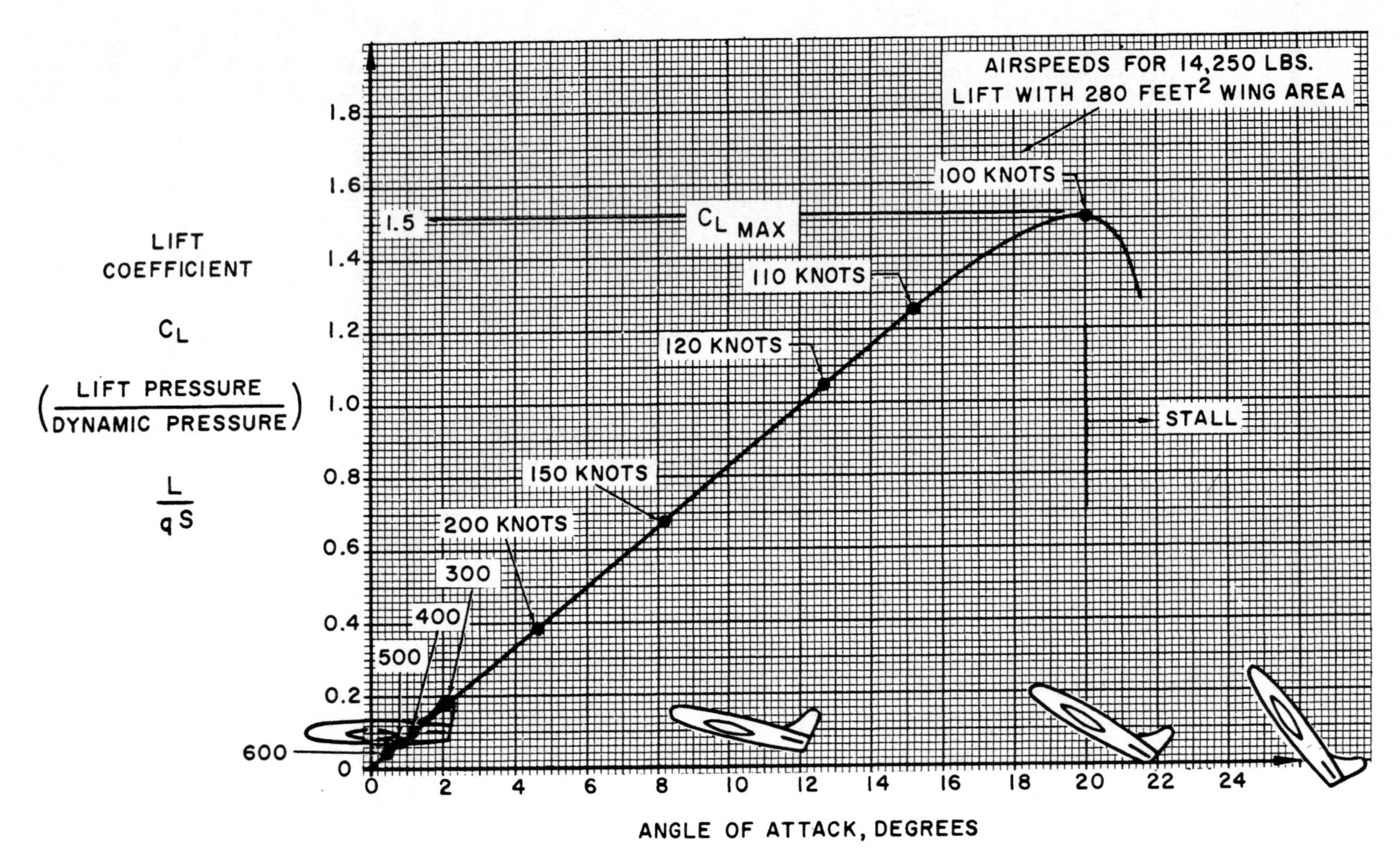

Figure 1.11. Typical Lift Characteristics

derived from study of the basic lift equation and the typical wing lift curve. One important fact to be appreciated is that the airplane shown in figure 1.11 stalls at the same angle of attack regardless of weight, dynamic pressure, bank angle, etc. Of course, the stall *speed* of the aircraft will be affected by weight, bank angle, and other factors since the product of dynamic pressure, wing area, and lift coefficient must produce the required lift. A rearrangement of the basic lift equation defines this relationship.

$$L=C_L qS$$

using $q=\frac{\sigma V^2}{295}$ (V in knots, TAS)

$$L=C_L \frac{\sigma V^2}{295} S$$

solving for V,

$$V=17.2\sqrt{\frac{L}{C_L \sigma S}}$$

Since the stall speed is the minimum flying speed necessary to sustain flight, the lift coefficient must be the maximum ($C_{L_{max}}$).

Suppose that the airplane shown in figure 1.11 has the following properties:

Weight = 14,250 lbs.

Wing area = 280 sq. ft.

$C_{L_{max}}=1.5$

If the airplane is flown in steady, level flight at sea level with lift equal to weight the stall speed would be:

$$V_s=17.2\sqrt{\frac{W}{C_{L_{max}}\sigma S}}$$

where

V_s = stall speed, knots TAS

W = weight, lbs. (lift = weight)

$$V_s=17.2\sqrt{\frac{(14{,}250)}{(1.5)(1.000)(280)}}$$

$$=100 \text{ knots}$$

Thus, a sea level airspeed (or EAS) of 100 knots would provide the dynamic pressure necessary at maximum lift to produce 14,250 lbs. of lift. If the airplane were operated at a higher weight, a higher dynamic pressure would be required to furnish the greater lift and a higher stall speed would result. If the airplane were placed in a steep turn, the greater lift required in the turn would increase the stall speed. If the airplane were flown at a higher density altitude the TAS at stall would increase. However, one factor common to each of these conditions is that the angle of attack at $C_{L_{max}}$ is the same. It is important to realize that stall warning devices must sense angle of attack (α) or pressure distribution (related to C_L).

Another important fact related by the basic lift equation and lift curve is variation of angle of attack and lift coefficient with airspeed. Suppose that the example airplane is flown in steady, wing level flight at various airspeeds with lift equal to the weight. It is obvious that an increase in airspeed above the stall speed will require a corresponding decrease in lift coefficient and angle of attack to maintain steady, lift-equal-weight flight. The exact relationship of lift coefficient and airspeed is evolved from the basic lift equation assuming constant lift (equal to weight) and equivalent airspeeds.

$$\frac{C_L}{C_{L_{max}}}=\left(\frac{V_s}{V}\right)^2$$

The example airplane was specified to have:

Weight = 14,250 lbs.

$C_{L_{max}}=1.5$

V_s = 100 knots EAS

The following table depicts the lift coefficients and angles of attack at various airspeeds in steady flight.

V, knots	$\frac{C_L}{C_{L_{max}}}=\left(\frac{V_s}{V}\right)^2$	C_L	α
100	1.000	1.50	20.0°
110	.826	1.24	15.2°
120	.694	1.04	12.7°
150	.444	.67	8.2°
200	.250	.38	4.6°
300	.111	.17	2.1°
400	.063	.09	1.1°
500	.040	.06	.7°
600	.028	.04	.5°

Note that for the conditions of steady flight, each airspeed requires a specific angle of attack and lift coefficient. This fact provides a fundamental concept of flying technique: *Angle of attack is the primary control of airspeed in steady flight.* Of course, the control stick or wheel allows the pilot to control the angle of attack and, thus, control the airspeed in steady flight. In the same sense, the throttle controls the output of the powerplant and allows the pilot to control rate of climb and descent at various airspeeds.

The real believers of these concepts are professional instrument pilots, LSO's, and glider pilots. The glider pilot (or flameout enthusiast) has no recourse but to control airspeed by angle of attack and accept whatever rate of descent is incurred at the various airspeeds. The LSO must become quite proficient at judging the flight path and angle of attack of the airplane in the pattern. The more complete visual reference field available to the LSO allows him to judge the angle of attack of the airplane more accurately than the pilot. When the airplane approaches the LSO, the precise judgment of airspeed is by the angle of attack rather than the rate of closure. If the LSO sees the airplane on the desired flight path but with too low an angle of attack, the airspeed is too high; if the angle of attack is too high, the airspeed is too low and the airplane is approaching the stall. The mirror landing system coupled with an angle of attack indicator is an obvious refinement. The mirror indicates the desired flight path and the angle of attack indicator allows precision control of the airspeed. The accomplished instrument pilot is the devotee of "attitude" flying technique—his creed being "attitude plus power equals performance." During a GCA approach, the professional instrument pilot controls airspeed with stick (angle of attack) and rate of descent with power adjustment.

Maneuvering flight and certain transient conditions of flight tend to complicate the relationship of angle of attack and airspeed. However, the majority of flight and, certainly, the most critical regime of flight (takeoff, approach, and landing), is conducted in essentially steady flight condition.

AIRFOIL LIFT CHARACTERISTICS. Airfoil section properties differ from wing or airplane properties because of the effect of the planform. Actually, the wing may have various airfoil sections from root to tip with taper, twist, sweepback and local flow components in a spanwise direction. The resulting aerodynamic properties of the wing are determined by the action of each section along the span and the three-dimensional flow. Airfoil section properties are derived from the basic shape or profile in two-dimensional flow and the force coefficients are given a notation of lower case letters. For example, a wing or airplane lift coefficient is C_L while an airfoil section lift coefficient is termed c_l. Also, wing angle of attack is α while section angle of attack is differentiated by the use of α_0. The study of section properties allows an objective consideration of the effects of camber, thickness, etc.

The lift characteristics of five illustrative airfoil sections are shown in figure 1.12. The section lift coefficient, c_l, is plotted versus section angle of attack, α_0, for five standard NACA airfoil profiles. One characteristic feature of all airfoil sections is that the slope of the various lift curves is essentially the same. At low lift coefficients, the section lift coefficient increases approximately 0.1 for each degree increase in angle of attack. For each of the airfoils shown, a 5° change in angle of

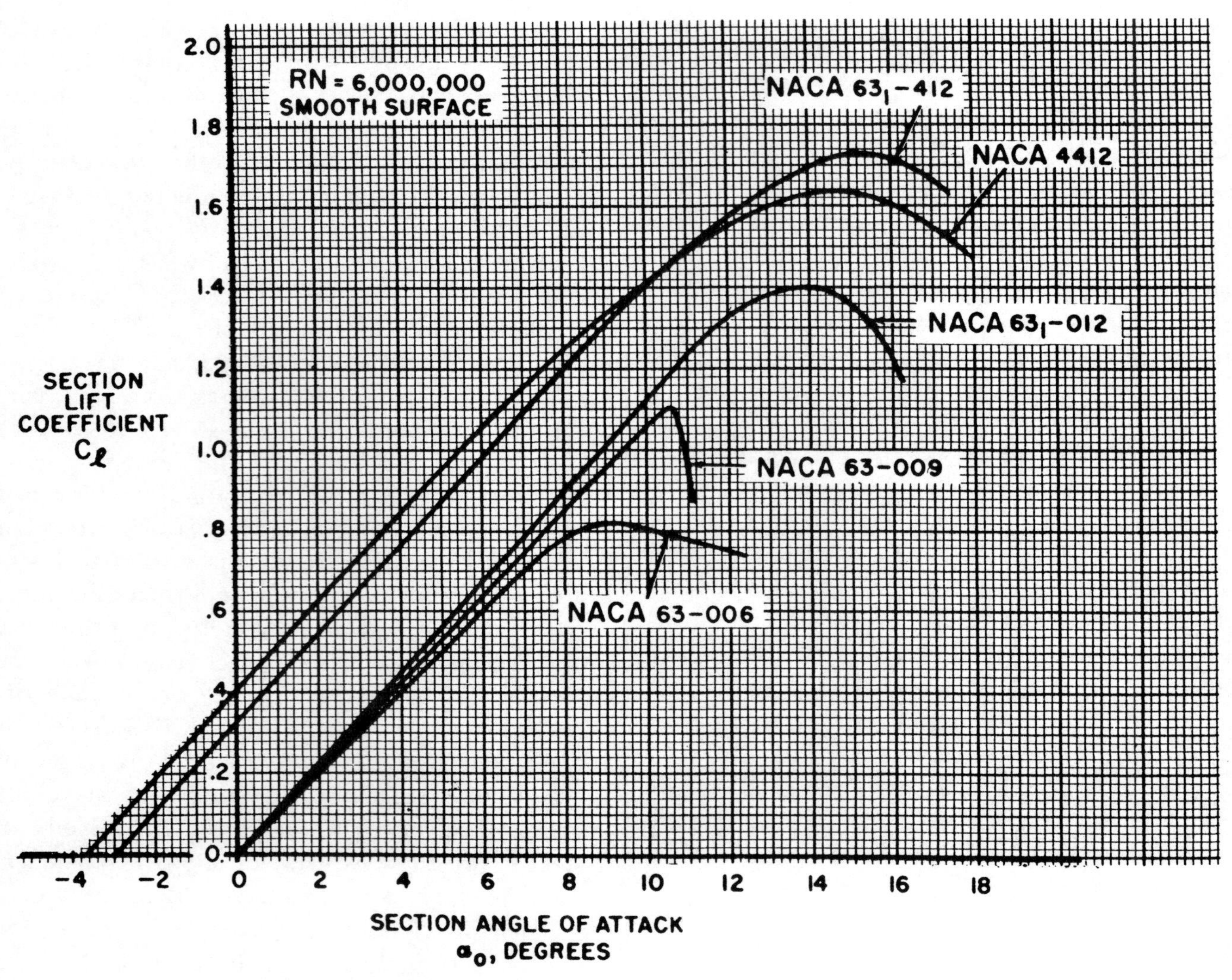

Figure 1.12. Lift Characteristics of Typical Airfoil Sections

attack would produce an approximate 0.5 change in lift coefficient. Evidently, lift curve slope is not a factor important in the selection of an airfoil.

An important lift property affected by the airfoil shape is the section maximum lift coefficient, $c_{l_{max}}$. The effect of airfoil shape on $c_{l_{max}}$ can be appreciated by comparison of the lift curves for the five airfoils of figure 1.12. The NACA airfoils 63-006, 63-009, and 63_1-012 are symmetrical sections of a basic thickness distribution but maximum thicknesses of 6, 9, and 12 percent respectively. The effect of thickness on $c_{l_{max}}$ is obvious from an inspection of these curves:

Section	$c_{l_{max}}$	α_0 for $c_{l_{max}}$
NACA 63-006	0.82	9.0°
NACA 63-009	1.10	10.5°
NACA 63_1-012	1.40	13.8°

The 12-percent section has a $c_{l_{max}}$ approximately 70 percent greater than the 6-percent thick section. In addition, the thicker airfoils have greater benefit from the use of various high lift devices.

The effect of camber is illustrated by the lift curves of the NACA 4412 and 63_1-412 sections. The NACA 4412 section is a 12 percent thick airfoil which has 4 percent maximum camber located at 40 percent of the chord. The NACA 63_1-412 airfoil has the same thickness and thickness distribution as the 63_1-012 but camber added to give a "design" lift coefficient (c_l for minimum section drag) of 0.4. The lift curves for these two airfoils show that camber has a beneficial effect on $c_{l_{max}}$.

Section	$c_{l_{max}}$	α_0 for $c_{l_{max}}$
NACA 63_1-012 (symmetrical)	1.40	13.8°
NACA 63_1-412 (cambered)	1.73	15.2°

An additional effect of camber is the change in zero lift angle. While the symmetrical sections have zero lift at zero angle of attack, the sections with positive camber have negative angles for zero lift.

The importance of maximum lift coefficient is obvious. If the maximum lift coefficient is high, the stall speed will be low. However, the high thickness and camber necessary for high section maximum lift coefficients may produce low critical Mach numbers and large twisting moments at high speed. In other words, a high maximum lift coefficient is just *one* of the many features desired of an airfoil section.

DRAG CHARACTERISTICS. Drag is the net aerodynamic force parallel to the relative wind and its source is the pressure distribution and skin friction on the surface. Large, thick bluff bodies in an airstream show a predominance of form drag due to the unbalanced pressure distribution. However, streamlined bodies with smooth contours show a predominance of drag due to skin friction. In a fashion similar to other aerodynamic forces, drag forces may be considered in the form of a coefficient which is independent of dynamic pressure and surface area. The basic drag equation is as follows:

$$D = C_D q S$$

where

D = drag, lbs.
C_D = drag coefficient
q = dynamic pressure, psf
$= \frac{\sigma V^2}{295}$ (V in knots, TAS)
S = wing surface area, sq. ft.

The force of drag is shown as the product of dynamic pressure, surface area, and drag coefficient, C_D. The drag coefficient in this equation is similar to any other aerodynamic force coefficient—it is the ratio of drag pressure to dynamic pressure. If the drag coefficient of a conventional airplane were plotted versus angle of attack, the result would be typical of the graph shown in figure 1.13. At low angles of attack the drag coefficient is low and small changes in angle of attack create only slight changes in drag coefficient. At

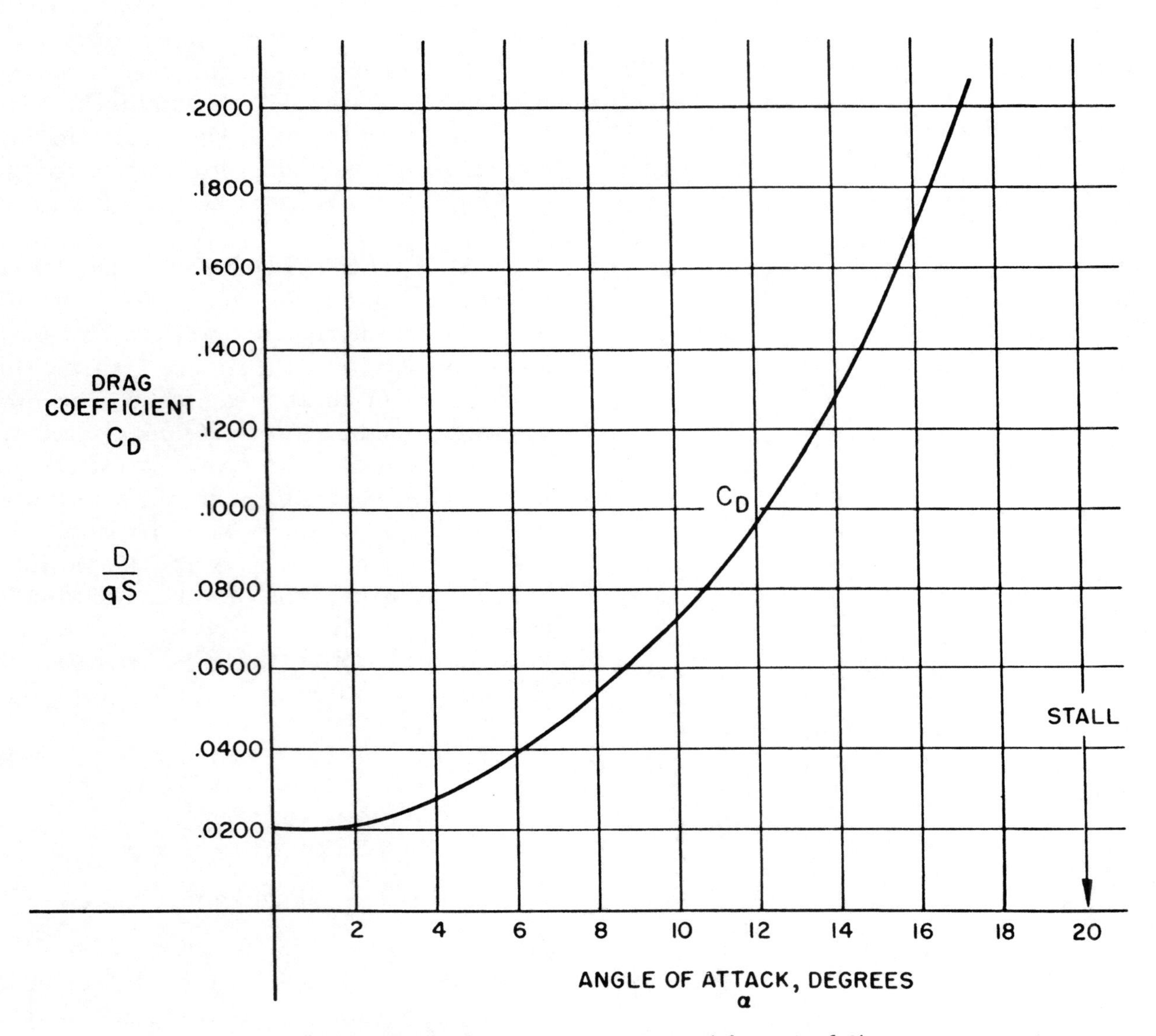

Figure 1.13. Drag Characteristics (sheet 1 of 2)

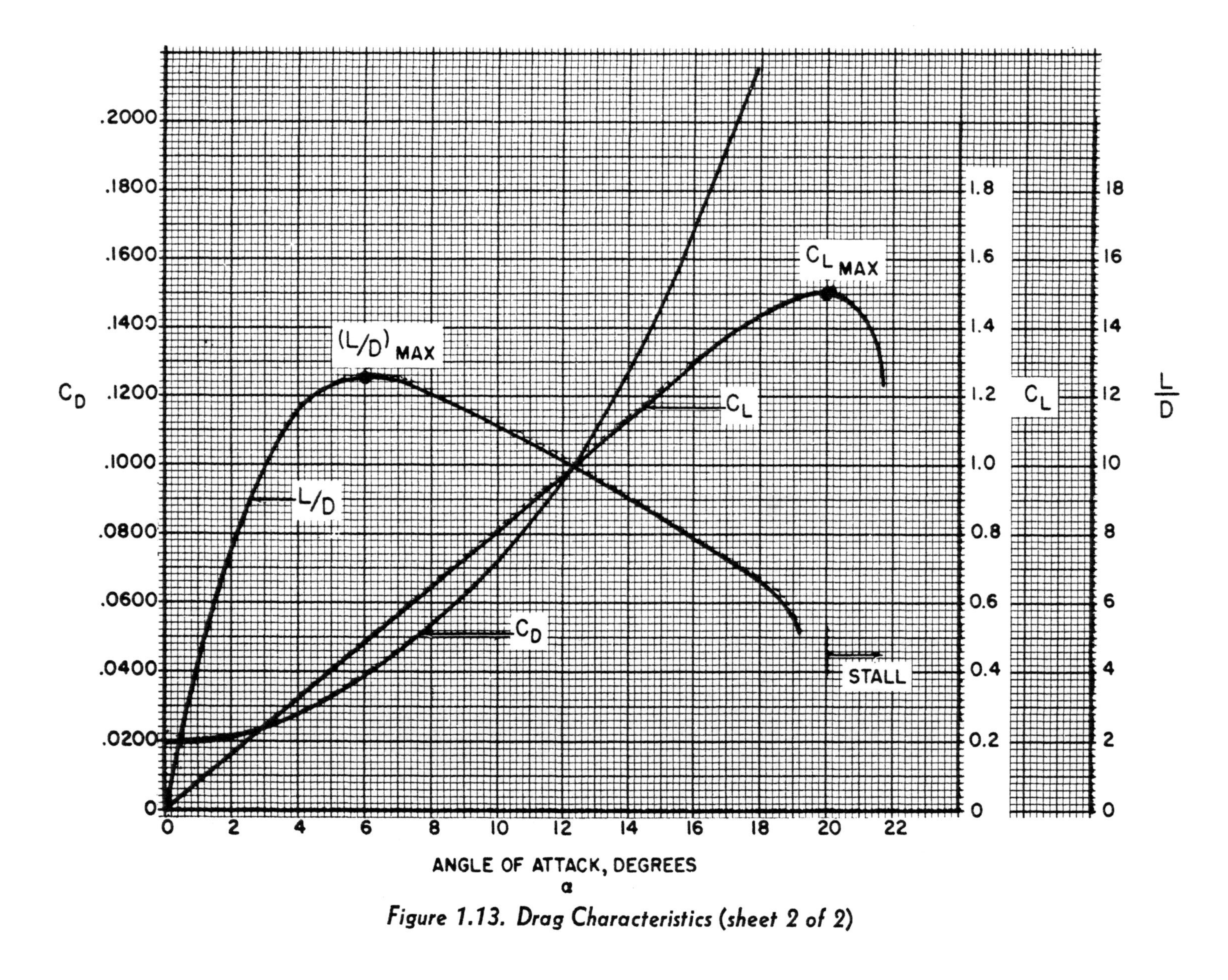

Figure 1.13. Drag Characteristics (sheet 2 of 2)

higher angles of attack the drag coefficient is much greater and small changes in angle of attack cause significant changes in drag. As stall occurs, a large increase in drag takes place.

A factor more important in airplane performance considerations is the lift-drag ratio, L/D. With the lift and drag data available for the airplane, the proportions of C_L and C_D can be calculated for each specific angle of attack. The resulting plot of lift-drag ratio with angle of attack shows that L/D increases to some maximum then decreases at the higher lift coefficients and angles of attack. Note that the maximum lift-drag ratio, $(L/D)_{max}$, occurs at one specific angle of attack and lift coefficient. If the airplane is operated in steady flight at $(L/D)_{max}$, the total drag is at a minimum. Any angle of attack lower or higher than that for $(L/D)_{max}$ reduces the lift-drag ratio and consequently increases the total drag for a given airplane lift.

The airplane depicted by the curves of Figure 1.13 has a maximum lift-drag ratio of 12.5 at an angle of attack of 6°. Suppose this airplane is operated in steady flight at a gross weight of 12,500 lbs. If flown at the airspeed and angle of attack corresponding to $(L/D)_{max}$, the drag would be 1,000 lbs. Any higher or lower airspeed would produce a drag greater than 1,000 lbs. Of course, this same airplane could be operated at higher or lower gross weights and the same maximum lift-drag ratio of 12.5 could be obtained at the same angle of attack of 6°. However, a change in gross weight would require a change in airspeed to support the new weight at the same lift coefficient and angle of attack.

Type airplane:	$(L/D)_{max}$
High performance sailplane	25-40
Typical patrol or transport	12-20
High performance bomber	20-25
Propeller powered trainer	10-15
Jet trainer	14-16
Transonic fighter or attack	10-13
Supersonic fighter or attack	4-9 (subsonic)

The configuration of an airplane has a great effect on the lift-drag ratio. Typical values of $(L/D)_{max}$ are listed for various types of airplanes. While the high performance sailplane may have extremely high lift-drag ratios, such an aircraft has no real economic or tactical purpose. The supersonic fighter may have seemingly low lift-drag ratios in subsonic flight but the airplane configurations required for supersonic flight (and high $[L/D]$'s at high Mach numbers) precipitate this situation.

Many important items of airplane performance are obtained in flight at $(L/D)_{max}$. Typical performance conditions which occur at $(L/D)_{max}$ are:

- maximum endurance of jet powered airplanes
- maximum range of propeller driven airplanes
- maximum climb *angle* for jet powered airplanes
- maximum power-off glide range, jet or prop

The most immediately interesting of these items is the power-off glide range of an airplane. By examining the forces acting on an airplane during a glide, it can be shown that the glide ratio is numerically equal to the lift-drag ratio. For example, if the airplane in a glide has an (L/D) of 15, each mile of altitude is traded for 15 miles of horizontal distance. Such a fact implies that the airplane should be flown at $(L/D)_{max}$ to obtain the greatest glide distance.

An unbelievable feature of gliding performance is the effect of airplane gross weight. Since the maximum lift-drag ratio of a given airplane is an intrinsic property of the aerodynamic configuration, gross weight will not affect the gliding performance. If a typical jet trainer has an $(L/D)_{max}$ of 15, the aircraft can obtain a maximum of 15 miles horizontal distance for each mile of altitude. This would be true of this particular airplane at any gross weight if the airplane is flown at the angle of attack for $(L/D)_{max}$. Of course, the gross weight would affect the glide airspeed necessary for this particular angle of attack but the glide ratio would be unaffected.

AIRFOIL DRAG CHARACTERISTICS. The total drag of an airplane is composed of the drags of the individual components and the forces caused by interference between these components. The drag of an airplane configuration must include the various drags due to lift, form, friction, interference, leakage, etc. To appreciate the factors which affect the drag of an airplane configuration, it is most logical to consider the factors which affect the drag of airfoil sections. In order to allow an objective consideration of the effects of thickness, camber, etc., the properties of two-dimensional sections must be studied. Airfoil section properties are derived from the basic profile in two-dimensional flow and are provided the lower case shorthand notation to distinguish them from wing or airplane properties, e.g., wing or airplane drag coefficient is C_D while airfoil section drag coefficient is c_d.

The drag characteristics of three illustrative airfoil sections are shown in figure 1.14. The section drag coefficient, c_d, is plotted versus the section lift coefficient, c_l. The drag on the airfoil section is composed of pressure drag and skin friction. When the airfoil is at low lift coefficients, the drag due to skin friction predominates. The drag curve for a conventional airfoil tends to be quite shallow in this region since there is very little variation of skin friction with angle of attack. When the airfoil is at high lift coefficients, form or pressure drag predominates and the drag coefficient varies rapidly with lift coefficient. The NACA 0006 is a thin symmetrical profile which has a maximum thickness of 6 percent located at 30 percent of the chord. This section shows a typical variation of c_d and c_l.

The NACA 4412 section is a 12 percent thick airfoil with 4 percent maximum camber at 40 percent chord. When this section is compared with the NACA 0006 section the effect of camber can be appreciated. At low lift coefficients the thin, symmetrical section has much lower drag. However, at lift coefficients above 0.5 the thicker, cambered section has the lower drag. Thus, proper camber and thickness can improve the lift-drag ratio of the section.

The NACA 63_1-412 is a cambered 12 percent thick airfoil of the "laminar flow" type. This airfoil is shaped to produce a design lift coefficient of 0.4. Notice that the drag curve of this airfoil has distinct aberrations with very low drag coefficients near the lift coefficient of 0.4. This airfoil profile has its camber and thickness distributed to produce very low uniform velocity on the forward surface (minimum pressure point well aft) at this lift coefficient. The resulting pressure and velocity distribution enhance extensive laminar flow in the boundary layer and greatly reduce the skin friction drag. The benefit of the laminar flow is appreciated by comparing the minimum drag of this airfoil with an airfoil which has one-half the maximum thickness—the NACA 0006.

The choice of an airfoil section will depend on the consideration of many different factors. While the $c_{l_{max}}$ of the section is an important quality, a more appropriate factor for consideration is the maximum lift coefficient of the section when various high lift devices are applied. Trailing edge flaps and leading edge high lift devices are applied to increase the $c_{l_{max}}$ for low speed performance. Thus, an appropriate factor for comparison is the ratio of section drag coefficient to section maximum lift coefficient with flaps—$c_d/c_{l_{m_f}}$. When this quantity is corrected for compressibility, a preliminary selection of an airfoil section is possible. The airfoil having the lowest value of $c_d/c_{l_{m_f}}$ at the design flight condition (endurance, range, high speed, etc.) will create the least section drag for a given design stall speed.

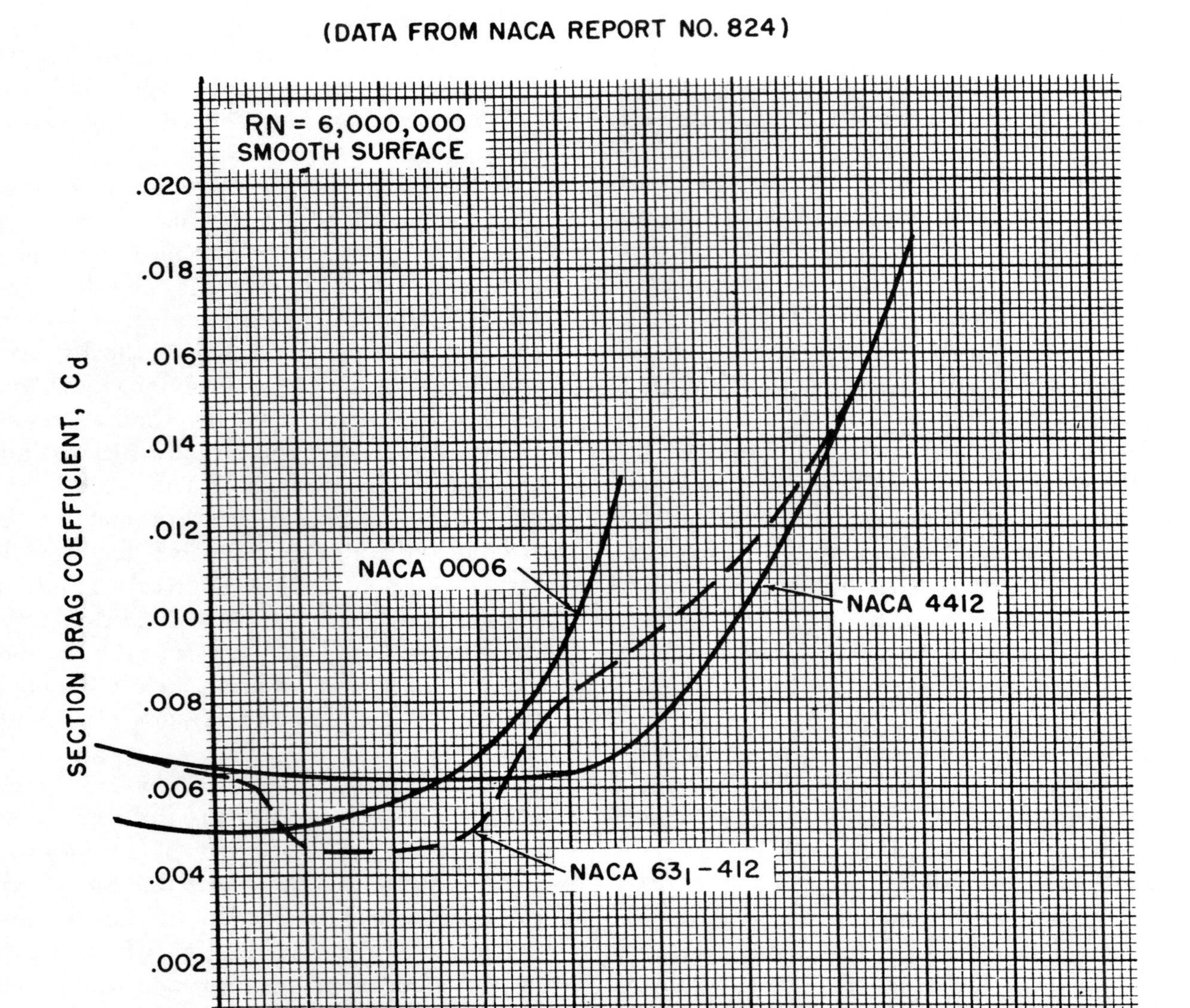

Figure 1.14. Drag Characteristics of Typical Airfoil Sections

FLIGHT AT HIGH LIFT CONDITIONS

It is frequently stated that the career Naval Aviator spends more than half his life "below a thousand feet and a hundred knots." Regardless of the implications of such a statement, the thought does connote the relationship of minimum flying speeds and carrier aviation. Only in Naval Aviation is there such importance assigned to precision control of the aircraft at high lift conditions. Safe operation in carrier aviation demands precision control of the airplane at high lift conditions.

The aerodynamic lift characteristics of an airplane are portrayed by the curve of lift coefficient versus angle of attack. Such a curve is illustrated in figure 1.15 for a specific airplane in the clean and flap down configurations. A given aerodynamic configuration experiences increases in lift coefficient with increases in angle of attack until the maximum lift coefficient is obtained. A further increase in angle of attack produces stall and the lift coefficient then decreases. Since the maximum lift coefficient corresponds to the minimum speed available in flight, it is an important point of reference. The stall speed of the aircraft in level flight is related by the equation:

$$V_s = 17.2\sqrt{\frac{W}{C_{L_{max}}\sigma S}}$$

where

V_s = stall speed, knots *TAS*
W = gross weight, lbs.
$C_{L_{max}}$ = airplane maximum lift coefficient
σ = altitude density ratio
S = wing area, sq. ft.

This equation illustrates the effect on stall speed of weight and wing area (or wing loading, W/S), maximum lift coefficient, and altitude. If the stall speed is desired in *EAS*, the density ratio will be that for sea level (σ = 1.000).

EFFECT OF WEIGHT. Modern configurations of airplanes are characterized by a large percent of the maximum gross weight being fuel. Hence, the gross weight and stall speed of the airplane can vary considerably throughout the flight. The effect of only weight on stall speed can be expressed by a modified form of the stall speed equation where density ratio, $C_{L_{max}}$, and wing area are held constant.

$$\frac{V_{s_2}}{V_{s_1}} = \sqrt{\frac{W_2}{W_1}}$$

where

V_{s_1} = stall speed corresponding to some gross weight, W_1
V_{s_2} = stall speed corresponding to a different gross weight, W_2

As an illustration of this equation, assume that a particular airplane has a stall speed of 100 knots at a gross weight of 10,000 lbs. The stall speeds of this same airplane at other gross weights would be:

Gross weight, lbs.	*Stall speed, knots EAS*
10,000	100
11,000	$100 \times \sqrt{\frac{11,000}{10,000}} = 105$
12,000	110
14,400	120
9,000	95
8,100	90

Figure 1.15 illustrates the effect of weight on stall speed on a percentage basis and will be valid for any airplane. Many specific conditions of flight are accomplished at certain fixed angles of attack and lift coefficients. The effect of weight on a percentage basis on the speeds for any specific lift coefficient and angle of attack is identical. Note that at small variations of weight, a rule of thumb may express the effect of weight on stall speed—"a 2 percent change in weight causes a 1 percent change in stall speed."

EFFECT OF MANEUVERING FLIGHT. Turning flight and maneuvers produce an effect on stall speed which is similar to the effect of weight. Inspection of the chart on figure 1.16 shows the forces acting on an airplane in a steady turn. Any steady turn requires that the *vertical* component of lift be equal to

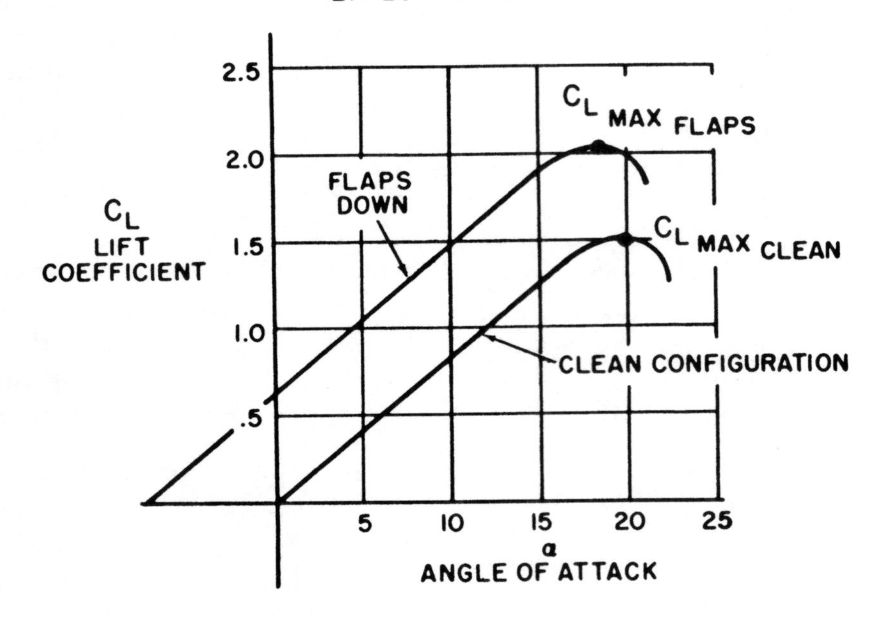

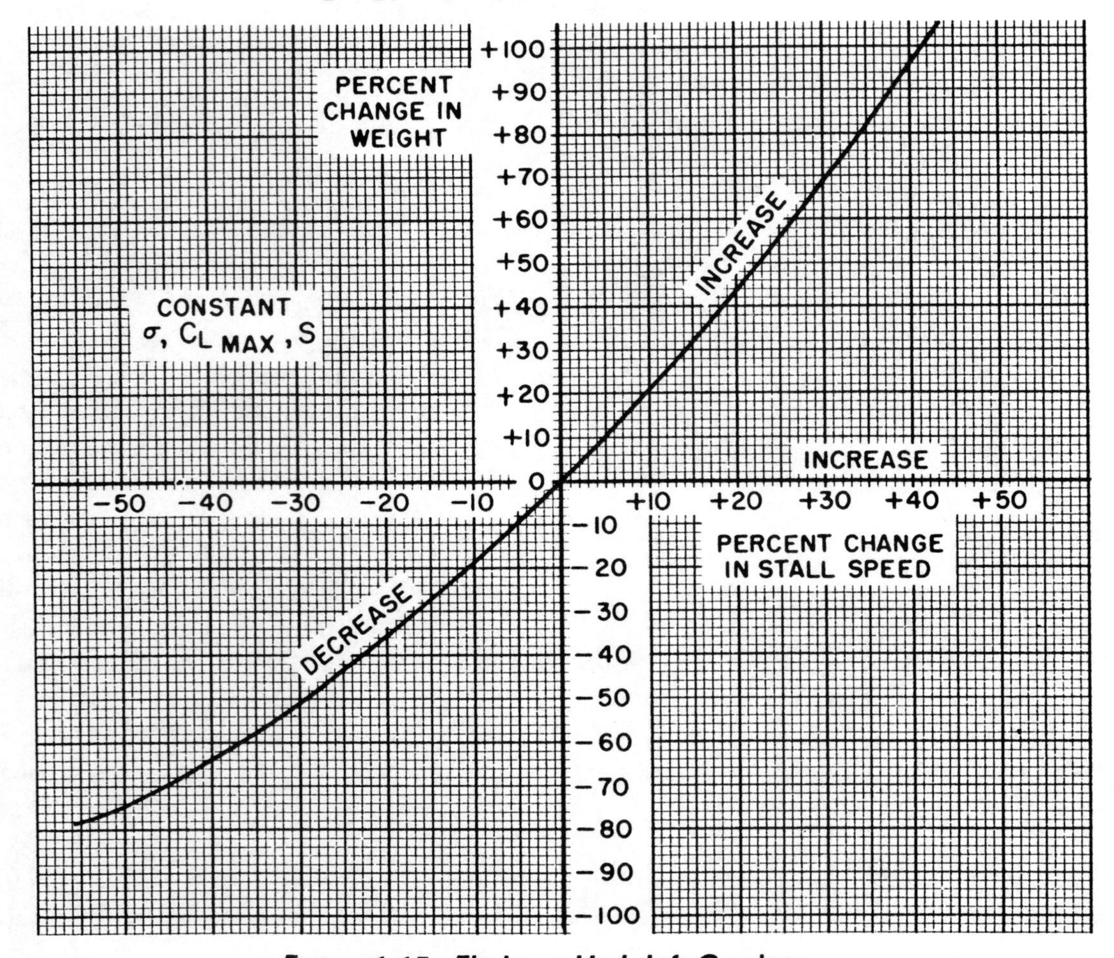

Figure 1.15. Flight at High Lift Conditions

weight of the airplane and the *horizontal* component of lift be equal to the centrifugal force. Thus, the aircraft in a steady turn develops a lift greater than weight and experiences increased stall speeds.

Trigonometric relationships allow determination of the effect of bank angle on stall speed and load factor. The load factor, n, is the proportion between lift and weight and is determined by:

$$n=\frac{L}{W}$$

$$n=\frac{1}{\cos\phi}$$

where

n = load factor (or "G")
$\cos\phi$ = cosine of the bank angle, ϕ (phi)

Typical values of load factor determined by this relationship are:

ϕ	0°	15°	30°	45°	60°	75.5°
n	1.00	1.035	1.154	1.414	2.000	4.000

The stall speed in a turn can be determined by:

$$V_{s\phi}=V_s\sqrt{n}$$

where

$V_{s\phi}$ = stall speed at some bank angle ϕ
V_s = stall speed for wing level, lift-equal-weight flight
n = load factor corresponding to the bank angle

The percent increase in stall speed in a turn is shown on figure 1.16. Since this chart is predicated on a steady turn and constant $C_{L_{max}}$, the figures are valid for any airplane. The chart shows that no appreciable change in load factor or stall speed occurs at bank angles less than 30°. Above 45° of bank the increase in load factor and stall speed is quite rapid. This fact emphasizes the need for avoiding steep turns at low airspeeds—a flight condition common to stall-spin accidents.

EFFECT OF HIGH LIFT DEVICES. The primary purpose of high lift devices (flaps, slots, slats, etc.) is to increase the $C_{L_{max}}$ of the airplane and reduce the stall speed. The take-off and landing speeds are consequently reduced. The effect of a typical high lift device is shown by the airplane lift curves of figure 1.15 and is summarized here:

Configuration	$C_{L_{max}}$	α for $C_{L_{max}}$
Clean (flaps up)	1.5	20°
Flaps down	2.0	18.5°

The principal effect of the extension of flaps is to increase the $C_{L_{max}}$ and reduce the angle of attack for any given lift coefficient. The increase in $C_{L_{max}}$ afforded by flap deflection reduces the stall speed in a certain proportion, the effect described by the equation:

$$V_{sf}=V_s\sqrt{\frac{C_{Lm}}{C_{Lmf}}}$$

where

V_{sf} = stall speed with flaps down

V_s = stall speed without flaps

C_{Lm} = maximum lift coefficient of the clean configuration

C_{Lmf} = maximum lift coefficient with flaps down

For example, assume the airplane described by the lift curves of figure 1.15 has a stall speed of 100 knots at the landing weight in the clean configuration. If the flaps are lowered the reduced stall speed is reduced to:

$$V_{sf}=100\times\sqrt{\frac{1.5}{2.0}}$$

$$=86.5 \text{ knots}$$

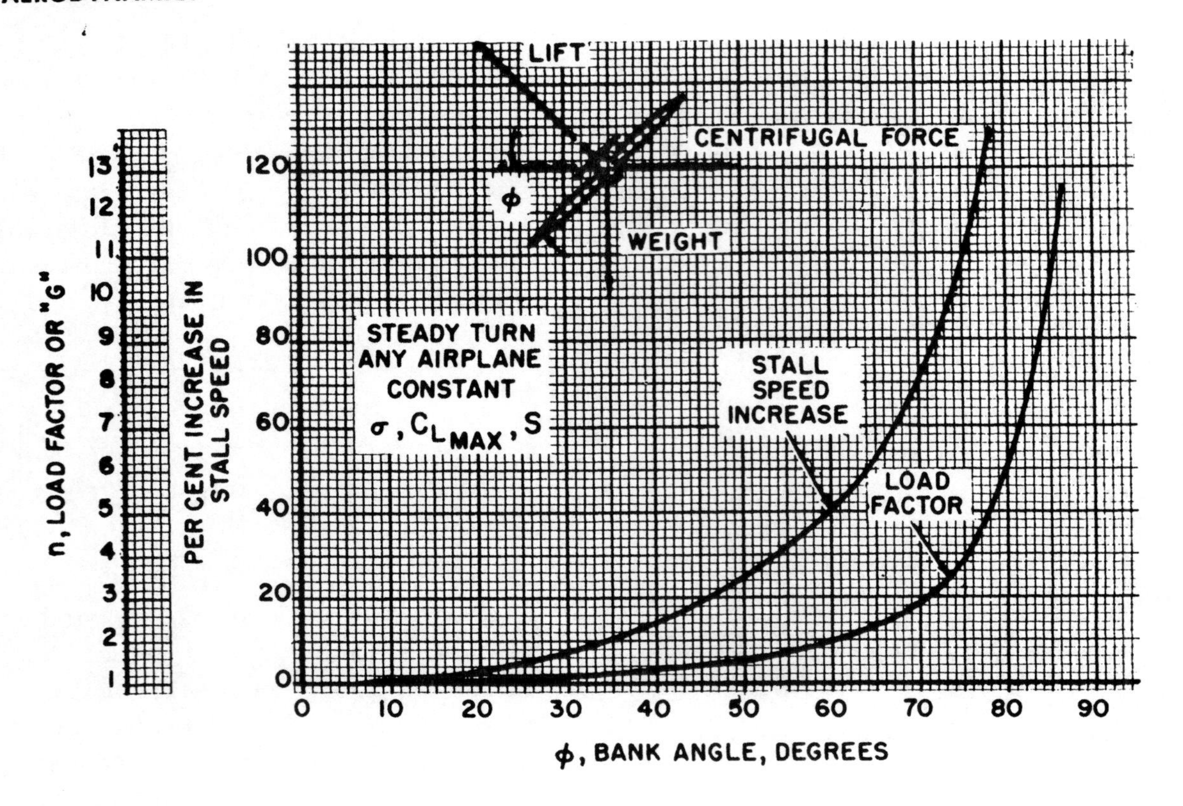

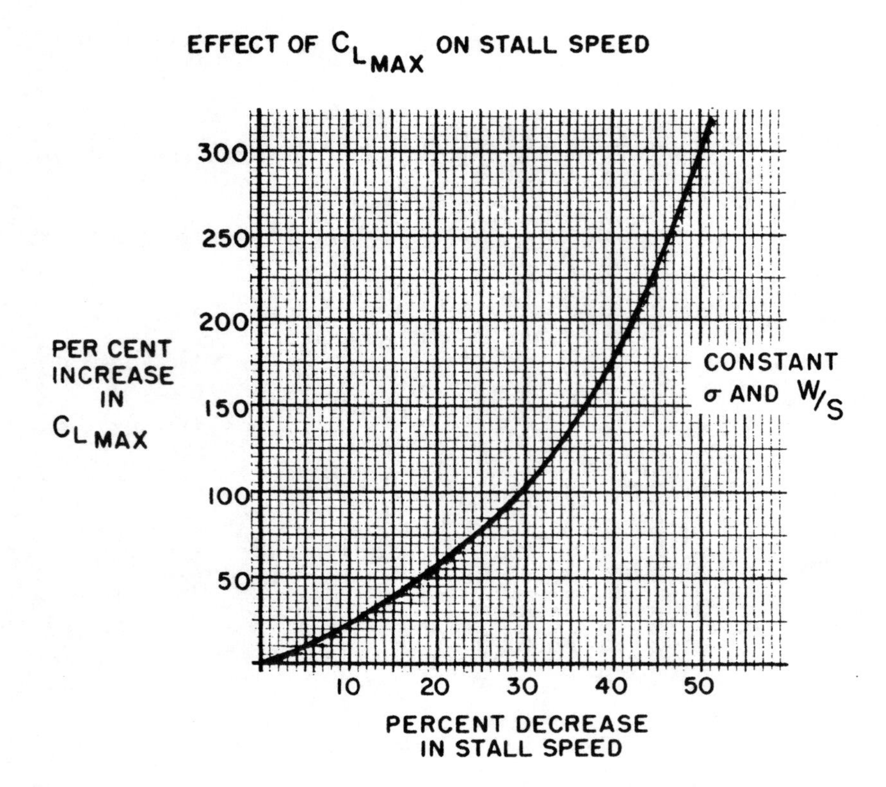

Figure 1.16. Flight at High Lift Conditions

Thus, with the higher lift coefficient available, less dynamic pressure is required to provide the necessary lift.

Because of the stated variation of stall speed with $C_{L_{max}}$, large changes in $C_{L_{max}}$ are necessary to produce significant changes in stall speed. This effect is illustrated by the graph in figure 1.16 and certain typical values are shown below:

Percent increase in $C_{L_{max}}$	2	10	50	100	300
Percent reduction in stall speed	1	5	18	29	50

The contribution of the high lift devices must be considerable to cause large reduction in stall speed. The most elaborate combination of flaps, slots, slats, and boundary layer control throughout the span of the wing would be required to increase $C_{L_{max}}$ by 300 percent. A common case is that of a typical propeller driven transport which experiences a 70 percent increase in $C_{L_{max}}$ by full flap deflection. A typical single engine jet fighter with a thin swept wing obtains a 20 percent increase in $C_{L_{max}}$ by full flap deflection. Thin airfoil sections with sweepback impose distinct limitations on the effectiveness of flaps and the 20 percent increase in $C_{L_{max}}$ by flaps is a typical—if not high—value for such a configuration.

One factor common to maximum lift condition is the angle of attack and pressure distribution. The maximum lift coefficient of a particular wing configuration is obtained at one angle of attack and one pressure distribution. Weight, bank angle, load factor, density altitude, and airspeed have no direct effect on the stall angle of attack. This fact is sufficient justification for the use of angle of attack indicators and stall warning devices which sense pressure distribution on the wing. During flight maneuvers, landing approach, takeoff, turns, etc. the airplane will stall *if the critical angle of attack is exceeded*. The airspeed at which stall occurs will be determined by weight, load factor, and altitude but the stall angle of attack is unaffected. At any particular altitude, the indicated stall speed is a function of weight and load factor. An increase in altitude will produce a decrease in density and increase the true airspeed at stall. Also, an increase in altitude will alter compressibility and viscosity effects and, generally speaking, cause the *indicated* stall speed to increase. This particular consideration is usually significant only above altitudes of 20,000 ft.

Recovery from stall involves a very simple concept. Since stall is precipitated by an excessive angle of attack, *the angle of attack must be decreased*. This is a fundamental principle which is common to any airplane.

An airplane may be designed to be "stall-proof" simply by reducing the effectiveness of the elevators. If the elevators are not powerful enough to hold the airplane to high angles of attack, the airplane cannot be stalled in any condition of flight. Such a requirement for a tactical military airplane would seriously reduce performance. High lift coefficients near the maximum are required for high maneuverability and low landing and takeoff speeds. Hence, the Naval Aviator must appreciate the effect of the many variables affecting the stall speed and regard "attitude flying," angle of attack indicators, and stall warning devices as techniques which allow more precise control of the airplane at high lift conditions.

HIGH LIFT DEVICES

There are many different types of high lift devices used to increase the maximum lift coefficient for low speed flight. The high lift devices applied to the trailing edge of a section consist of a flap which is usually 15 to 25 percent of the chord. The deflection of a flap produces the effect of a large amount of camber added well aft on the chord. The principal types of flaps are shown applied to a basic section of airfoil. The effect of a 30° deflection of a 25 percent chord flap is shown on the lift and drag curves of figure 1.17.

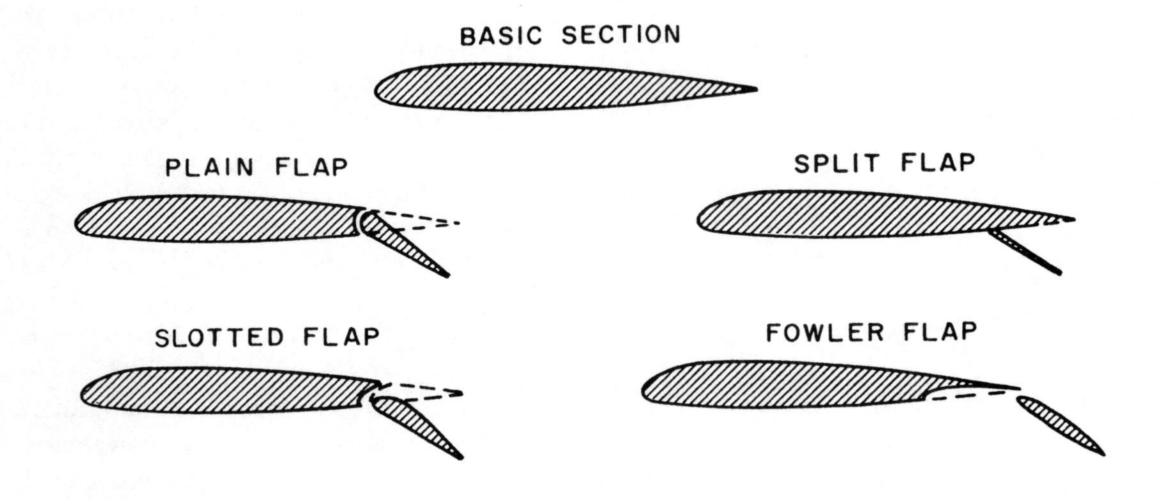

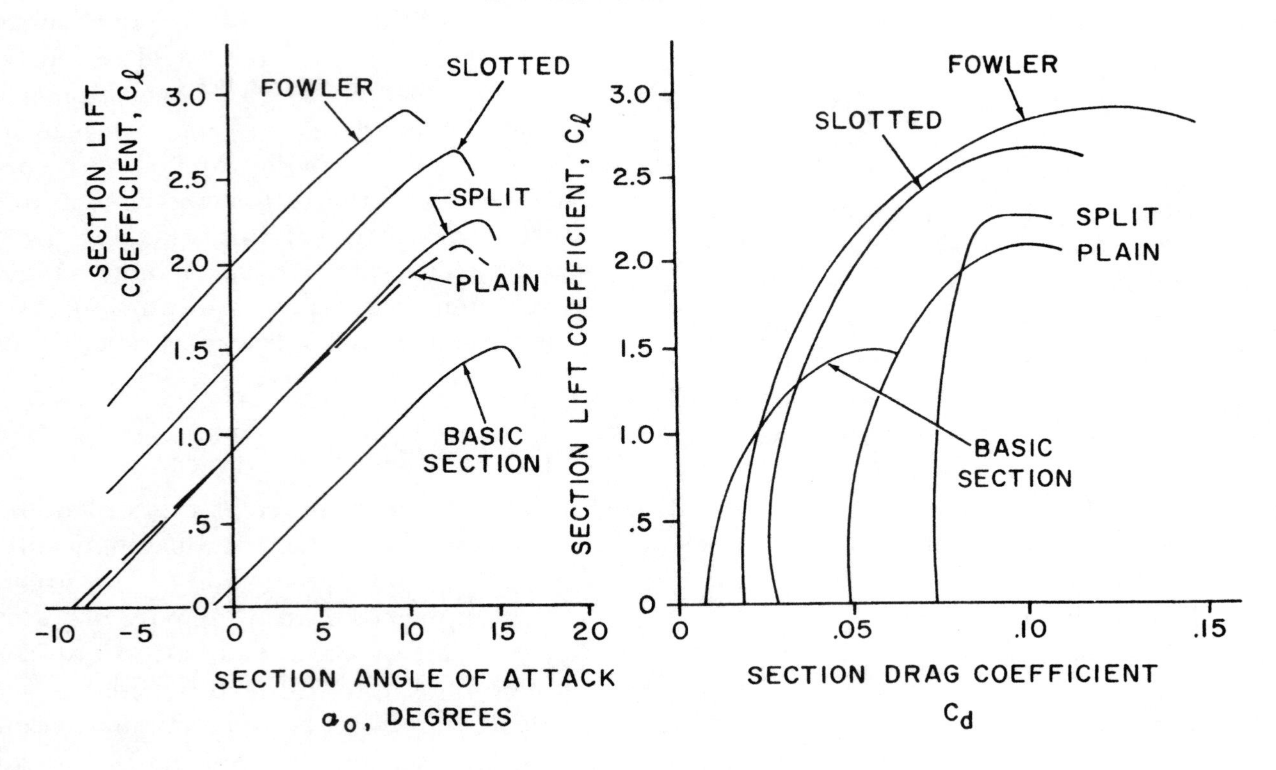

Figure 1.17. Flap Configurations

The *plain flap* shown in figure 1.17 is a simple hinged portion of the trailing edge. The effect of the camber added well aft on the chord causes a significant increase in $c_{l_{max}}$. In addition, the zero lift angle changes to a more negative value and the drag increases greatly. The *split flap* shown in figure 1.17 consist of plate deflected from the lower surface of the section and produces a slightly greater change in $c_{l_{max}}$ than the plain flap. However, a much larger change in drag results from the great turbulent wake produced by this type flap. The greater drag may not be such a disadvantage when it is realized that it may be advantageous to accomplish steeper landing approaches over obstacles or require higher power from the engine during approach (to minimize engine acceleration time for waveoff).

The *slotted flap* is similar to the plain flap but the gap between the main section and flap leading edge is given specific contours. High energy air from the lower surface is ducted to the flap upper surface. The high energy air from the slot accelerates the upper surface boundary layer and delays airflow separation to some higher lift coefficient. The slotted flap can cause much greater increases in $c_{l_{max}}$ than the plain or split flap and section drags are much lower.

The *Fowler flap* arrangement is similar to the slotted flap. The difference is that the deflected flap segment is moved aft along a set of tracks which increases the chord and effects an increase in wing area. The Fowler flap is characterized by large increases in $c_{l_{max}}$ with minimum changes in drag.

One additional factor requiring consideration in a comparison of flap types is the aerodynamic twisting moments caused by the flap. Positive camber produces a nose down twisting moment—especially great when large camber is used well aft on the chord (an obvious implication is that flaps are not practical on a flying wing or tailless airplane). The deflection of a flap causes large nose down moments which create important twisting loads on the structure and pitching moments that must be controlled with the horizontal tail. Unfortunately, the flap types producing the greatest increases in $c_{l_{max}}$ usually cause the greatest twisting moments. The Fowler flap causes the greatest change in twisting moment while the split flap causes the least. This factor-along with mechanical complexity of the installation—may complicate the choice of a flap configuration.

The effectiveness of flaps on a wing configuration depend on many different factors. One important factor is the amount of the wing area affected by the flaps. Since a certain amount of the span is reserved for ailerons, the actual wing maximum lift properties will be less than that of the flapped two-dimensional section. If the basic wing has a low thickness, any type of flap will be less effective than on a wing of greater thickness. Sweepback of the wing can cause an additional significant reduction in the effectiveness of flaps.

High lift devices applied to the leading edge of a section consist of slots, slats, and small amounts of local camber. The fixed slot in a wing conducts flow of high energy air into the boundary layer on the upper surface and delays airflow separation to some higher angle of attack and lift coefficient. Since the slot alone effects no change in camber, the higher maximum lift coefficient will be obtained at a higher angle of attack, i.e., the slot simply delays stall to a higher angle of attack. An automatic slot arrangement consists of a leading edge segment (slat) which is free to move on tracks. At low angles of attack the slat is held flush against the leading edge by the high positive local pressures. When the section is at high angles of attack, the high local suction pressures at the leading edge create a chordwise force forward to actuate the slat. The slot formed then allows the section to continue to a higher angle of attack and produce a $c_{l_{max}}$ greater than that of the

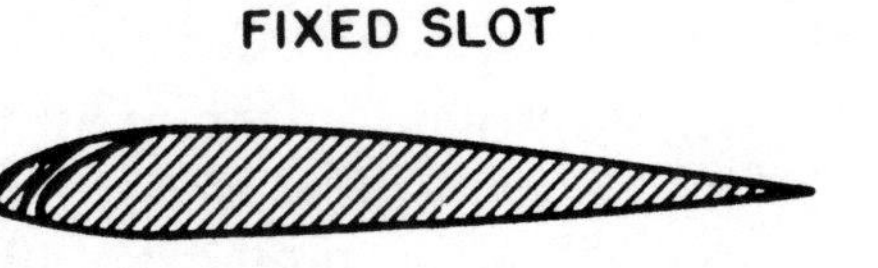

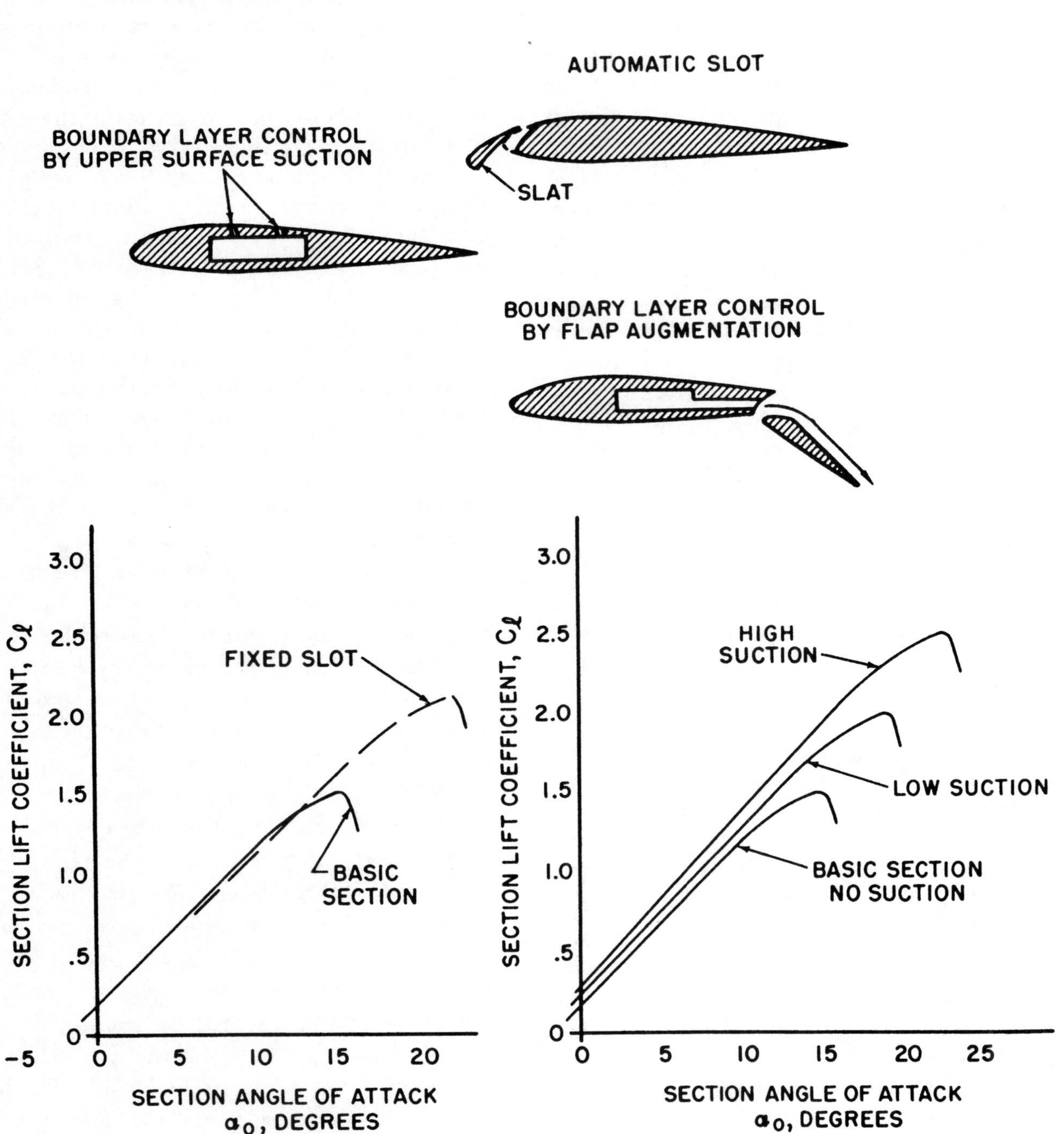

Figure 1.18. Effect of Slots and Boundary Layer Control

basic section. The effect of a fixed slot on the lift characteristics is shown in figure 1.18.

Slots and slats can produce significant increases in $c_{l_{max}}$ but the increased angle of attack for maximum lift can be a disadvantage. If slots were the only high lift device on the wing, the high take off and landing angles of attack may complicate the design of the landing gear. For this reason slots or slats are usually used in conjunction with flaps since the flaps provide reduction in the maximum lift angle of attack. The use of a slot has two important advantages: there is only a negligible change in the pitching moment due to the slot and no significant change in section drag at low angles of attack. In fact, the slotted section will have less drag than the basic section near the maximum lift angle for the basic section.

The slot-slat device finds great application in modern airplane configurations. The tailless airplane configuration can utilize only the high lift devices which have negligible effect on the pitching moments. The slot and slat are often used to increase the $c_{l_{max}}$ in high speed flight when compressibility effects are considerable. The small change in twisting moment is a favorable feature for any high lift device to be used at high speed. Leading edge high lift devices are more effective on the highly swept wing than trailing edge flaps since slats are quite powerful in controlling the flow pattern. Small amounts of local camber added to the leading edge as a high lift device is most effective on wings of very low thickness and sharp leading edges. Most usually the slope of the leading edge high lift device is used to control the spanwise lift distribution on the wing.

Boundary layer control devices are additional means of increasing the maximum lift coefficient of a section. The thin layer of airflow adjacent to the surface of an airfoil shows reduced local velocities from the effect of skin friction. When at high angles of attack this boundary layer on the upper surface tends to stagnate and come to a stop. If this happens the airflow will separate from the surface and stall occurs. Boundary layer control for high lift applications features various devices to maintain high velocity in the boundary layer to allay separation of the airflow. This control of the boundary layer kinetic energy can be accomplished in two ways. One method is the application of a suction through ports to draw off low energy boundary layer and replace it with high velocity air from outside the boundary layer. The effect of surface suction boundary layer control on lift characteristics is typified by figure 1.18. Increasing surface suction produces greater maximum lift coefficients which occur at higher angles of attack. The effect is similar to that of a slot because the slot is essentially a boundary layer control device ducting high energy air to the upper surface.

Another method of boundary layer control is accomplished by injecting a high speed jet of air into the boundary layer. This method produces essentially the same results as the suction method and is the more practical installation. The suction type *BLC* requires the installation of a separate pump while the "blown" *BLC* system can utilize the high pressure source of a jet engine compressor. The typical installation of a high pressure *BLC* system would be the augmentation of a deflected flap. Since any boundary layer control tends to increase the angle of attack for maximum lift, it is important to combine the boundary layer control with flaps since the flap deflection tends to reduce the angle of attack for maximum lift.

OPERATION OF HIGH LIFT DEVICES. The management of the high lift devices on an airplane is an important factor in flying operations. The devices which are actuated automatically—such as automatic slats and slots—are usually of little concern and cause little complication since relatively small changes in drag and pitching moments take place. However, the flaps must be properly managed by the pilot to take advantage of the capability

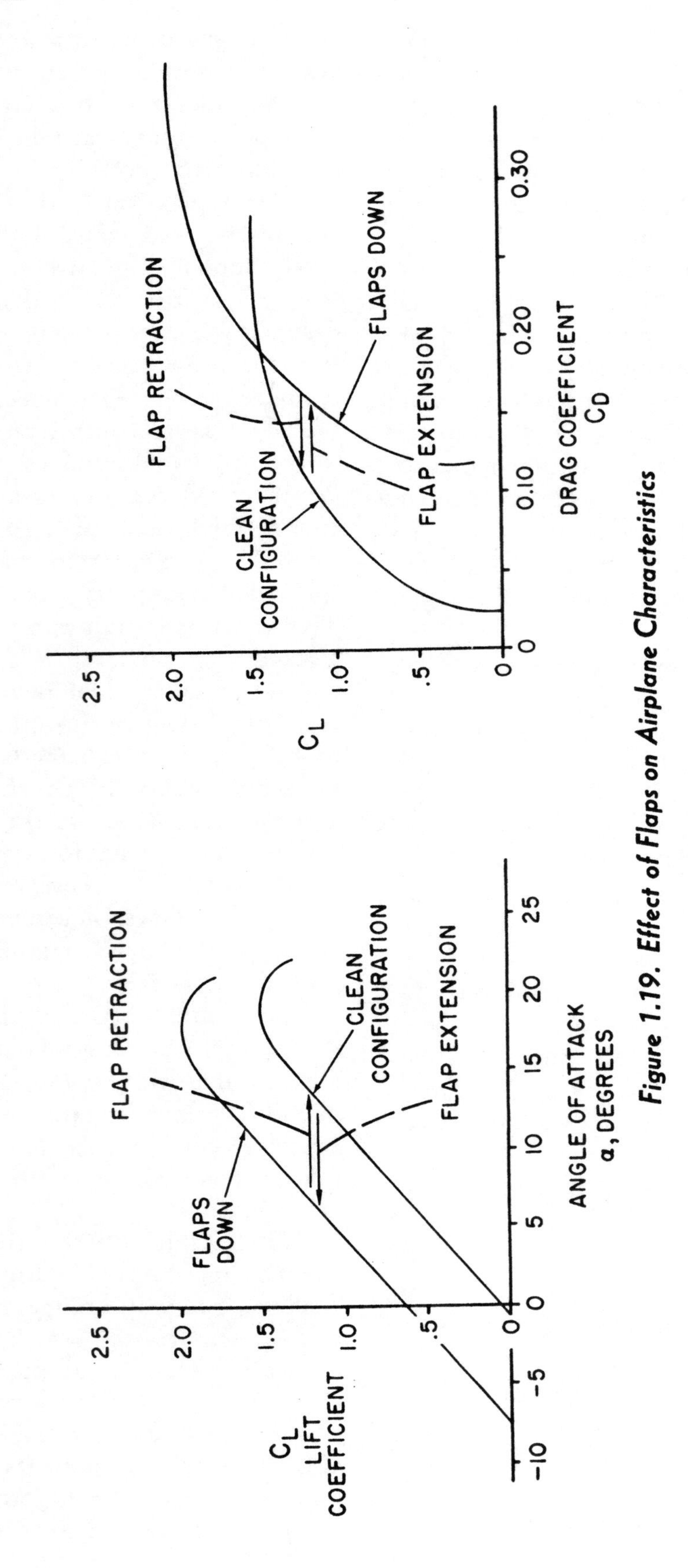

Figure 1.19. Effect of Flaps on Airplane Characteristics

of such a device. To illustrate a few principles of flap management, figure 1.19 presents the lift and drag curves of a typical airplane in the clean and flap down configurations.

In order to appreciate some of the factors involved in flap management, assume that the airplane has just taken off and the flaps are extended. The pilot should not completely retract the flaps until the airplane has sufficient speed. If the flaps are retracted prematurely at insufficient airspeed, maximum lift coefficient of the clean configuration may not be able to support the airplane and the airplane will sink or stall. Of course, this same factor must be considered for intermediate flap positions between fully retracted and fully extended. Assume that the airplane is allowed to gain speed and reduce the flight lift coefficient to the point of flap retraction indicated on figure 1.19. As the configuration is altered from the "cluttered" to the clean configuration, three important changes take place:

(1) The reduction in camber by flap retraction changes the wing pitching moment and—for the majority of airplanes—requires retrimming to balance the nose up moment change. Some airplanes feature an automatic retrimming which is programmed with flap deflection.

(2) The retraction of flaps shown on figure 1.19 causes a reduction of drag coefficient at that lift coefficient. This drag reduction improves the acceleration of the airplane.

(3) The retraction of flaps requires an increase in angle of attack to maintain the same lift coefficient. Thus, if airplane acceleration is low through the flap retraction speed range, angle of attack must be increased to prevent the airplane from sinking. This situation is typical after takeoff when gross weight, density altitude, and temperature are high. However, some aircraft have such high acceleration through the flap retraction speed that the rapid gain in airspeed requires much less noticeable attitude change.

When the flaps are lowered for landing essentially the same items must be considered. Extending the flaps will cause these changes to take place:

(1) Lowering the flaps requires retrimming to balance the nose down moment change.

(2) The increase in drag requires a higher power setting to maintain airspeed and altitude.

(3) The angle of attack required to produce the same lift coefficient is less, e.g., flap extension tends to cause the airplane to "balloon."

An additional factor which must be considered when rapidly accelerating after takeoff, or when lowering the flaps for landing, is the limit airspeed for flap extension. Excessive airspeeds in the flap down configuration may cause structural damage.

In many aircraft the effect of intermediate flap deflection is of primary importance in certain critical operating conditions. Small initial deflections of the flap cause noticeable changes in $C_{L_{max}}$ without large changes in drag coefficient. This feature is especially true of the airplane equipped with slotted or Fowler flaps (refer to fig. 1.17). Large flap deflections past 30° to 35° do not create the same rate of change of $C_{L_{max}}$ but do cause greater changes in C_D. A fact true of most airplanes is that the first 50 percent of flap deflection causes *more* than half of the total change in $C_{L_{max}}$ and the last 50 percent of flap deflection causes *more* than half of the total change in C_D.

The effect of power on the stall speed of an airplane is determined by many factors. The most important factors affecting this relationship are powerplant type (prop or jet), thrust-to-weight ratio, and inclination of the thrust vector at maximum lift. The effect of the propeller is illustrated in figure 1.20. The slipstream velocity behind the propeller is different from the free stream velocity depending on the thrust developed. Thus, when the propeller driven airplane is at low airspeeds

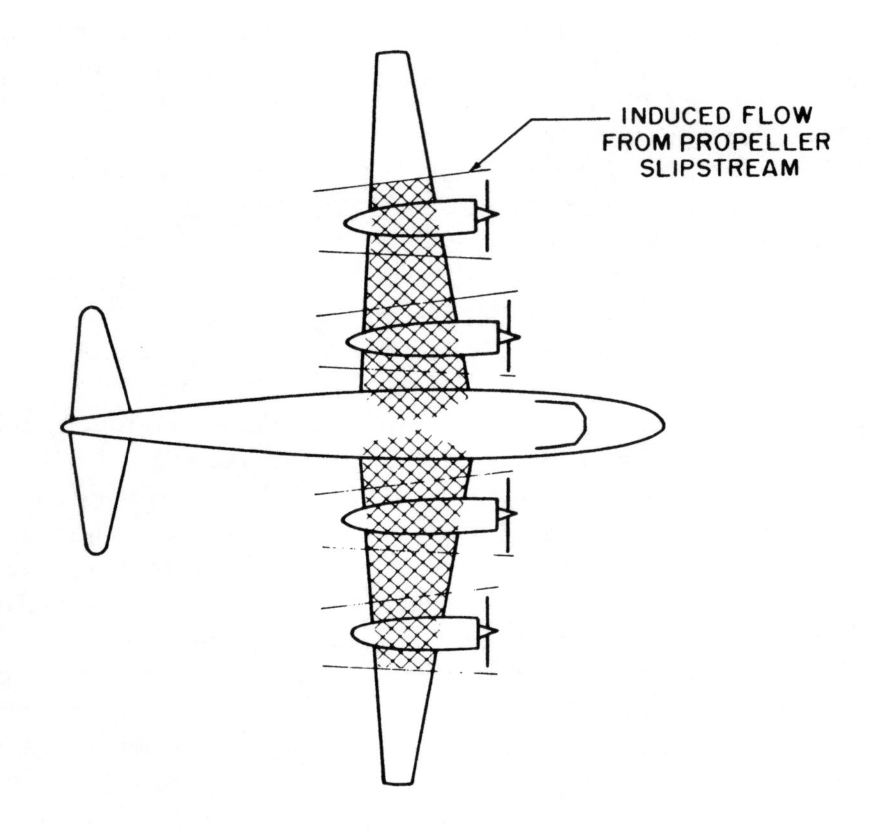

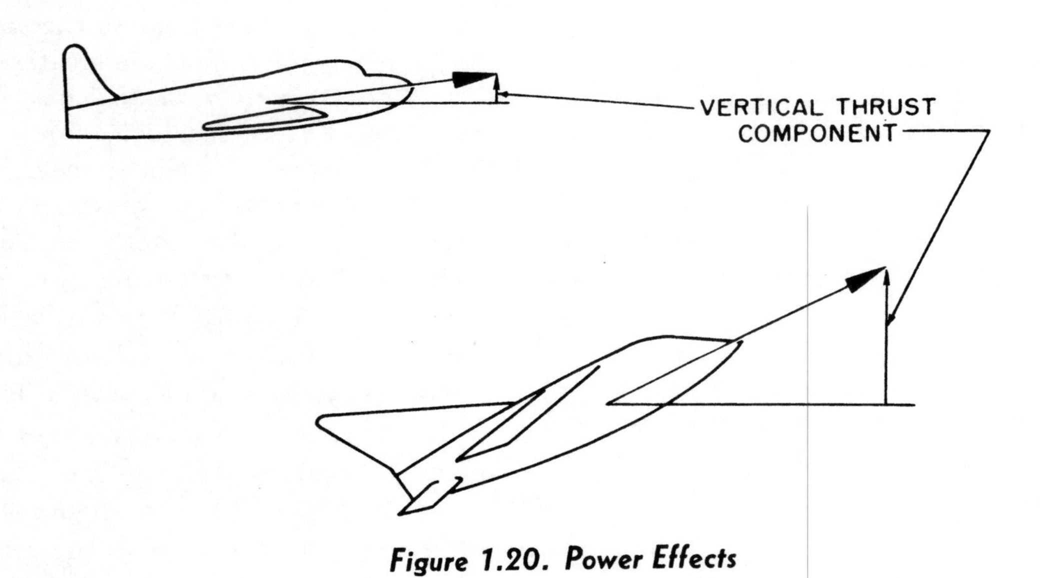

Figure 1.20. Power Effects

and high power, the dynamic pressure in the shaded area can be much greater than the free stream and this causes considerably greater lift than at zero thrust. At high power conditions the induced flow also causes an effect similar to boundary layer control and increases the maximum lift angle of attack. The typical four-engine propeller driven airplane may have 60 to 80 percent of the wing area affected by the induced flow and power effects on stall speeds may be considerable. Also, the lift of the airplane at a given angle of attack and airspeed will be greatly affected. Suppose the airplane shown is in the process of landing flare from a power-on approach. If there is a sharp, sudden reduction of power, the airplane may *drop* suddenly because of the reduced lift.

The typical jet aircraft does not experience the induced flow velocities encountered in propeller driven airplanes, thus the only significant factor is the vertical component of thrust. Since this vertical component contributes to supporting the airplane, less aerodynamic lift is required to hold the airplane in flight. If the thrust is small and the thrust inclination is slight at maximum lift angle, only negligible changes in stall speed will result. On the other hand, if the thrust is very great and is given a large inclination at maximum lift angle, the effect on stall speed can be very large. One important relationship remains—since there is very little induced flow from the jet, the angle of attack at stall is essentially the same power-on or power-off.

DEVELOPMENT OF AERODYNAMIC PITCHING MOMENTS

The distribution of pressure over a surface is the source of the aerodynamic moments as well as the aerodynamic forces. A typical example of this fact is the pressure distribution acting on the cambered airfoil of figure 1.21. The upper surface has pressures distributed which produce the upper surface lift; the lower surface has pressures distributed which produce the lower surface lift. Of course, the net lift produced by the airfoil is difference between the lifts on the upper and lower surfaces. The point along the chord where the distributed lift is effectively concentrated is termed the "center of pressure, *c.p.*" The center of pressure is essentially the "center of gravity" of the distributed lift pressure and the location of the *c.p.* is a function of camber and section lift coefficient.

Another aerodynamic reference point is the "aerodynamic center, *a.c.*" The aerodynamic center is defined as the point along the chord where all *changes* in lift effectively take place. To visualize the existence of such a point, notice the change in pressure distribution with angle of attack for the symmetrical airfoil of figure 1.21. When at zero lift, the upper and lower surface lifts are equal and located at the same point. With an increase in angle of attack, the upper surface lift increases while the lower surface lift decreases. The change of lift has taken place with no change in the center of pressure—a characteristic of symmetrical airfoils.

Next, consider the cambered airfoil of figure 1.21 at zero lift. To produce zero lift, the upper and lower surface lifts must be equal. One difference noted from the symmetrical airfoil is that the upper and lower surface lifts are not opposite one another. While no net lift exists on the airfoil, the couple produced by the upper and lower surface lifts creates a nose down moment. As the angle of attack is increased, the upper surface lift increases while the lower surface lift decreases. While a change in lift has taken place, no change in moment takes place about the point where the lift change occurs. Since the moment about the aerodynamic center is the product of a force (lift at the *c.p.*) and a lever arm (distance from *c.p.* to *a.c.*), an increase in lift moves the center of pressure toward the aerodynamic center.

It should be noted that the symmetrical airfoil at zero lift has no pitching moment about the aerodynamic center because the upper and

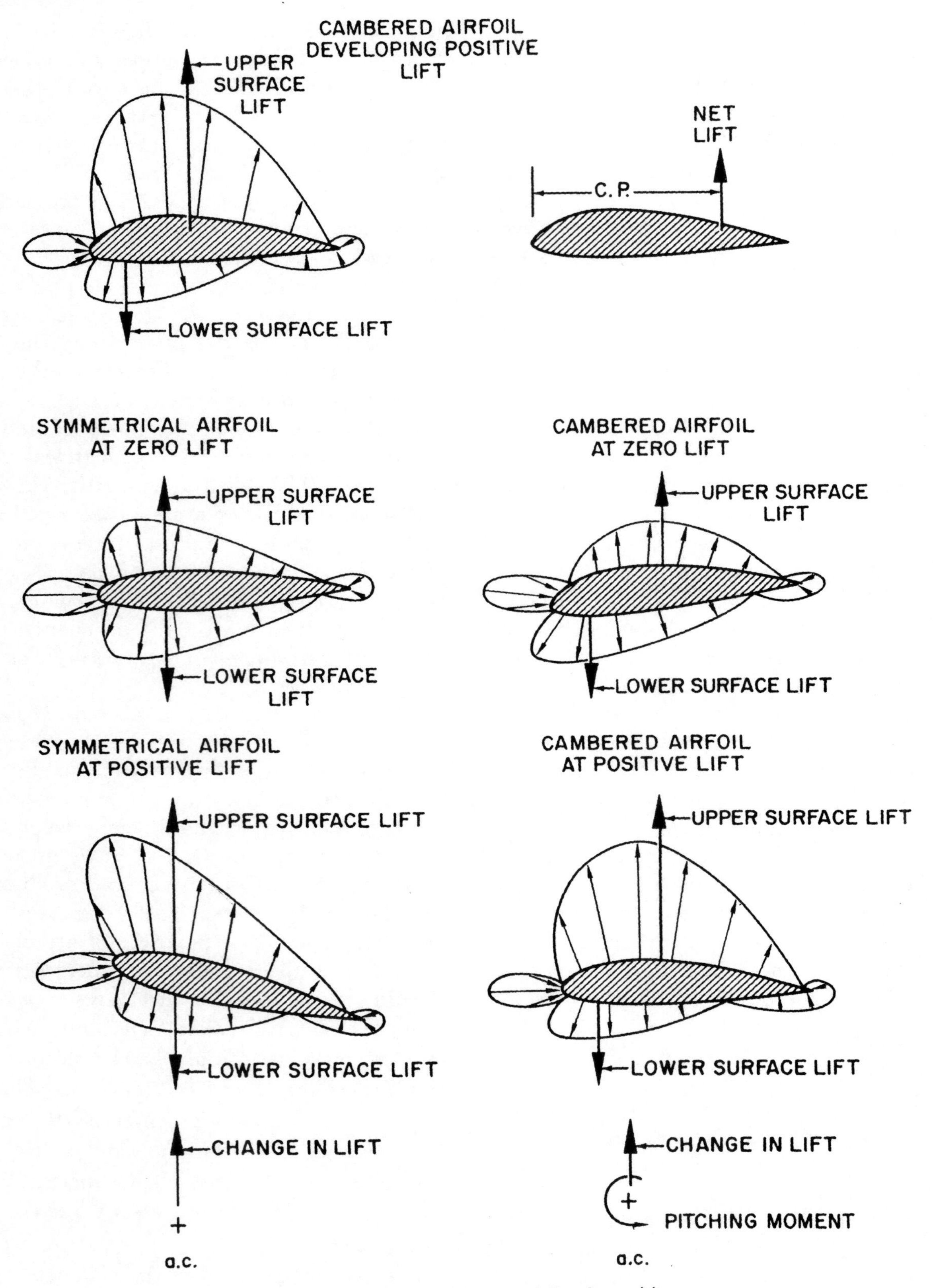

Figure 1.21. Development of Pitching Moments

lower surface lifts act along the same vertical line. An increase in lift on the symmetrical airfoil produces no change in this situation and the center of pressure remains fixed at the aerodynamic center.

The location of the aerodynamic center of an airfoil is not affected by camber, thickness, and angle of attack. In fact, two-dimensional incompressible airfoil theory will predict the aerodynamic center at the 25 percent *chord* point for any airfoil regardless of camber, thickness, and angle of attack. Actual airfoils, which are subject to real fluid flow, may not have the lift due to angle of attack concentrated at the exact 25 percent chord point. However, the actual location of the aerodynamic center for various sections is rarely forward of 23 percent or aft of 27 percent chord point.

The moment about the aerodynamic center has its source in the relative pressure distribution and requires application of the coefficient form of expression for proper evaluation. The moment about the aerodynamic center is expressed by the following equation:

$$M_{a.c.} = C_{M_{a.c.}}\, qSc$$

where

$M_{a.c.}$ = moment about the aerodynamic center, a.c., ft.-lbs.

$C_{M_{a.c.}}$ = coefficient of moment about the a.c.

q = dynamic pressure, psf

S = wing area, sq ft.

c = chord, ft.

The moment coefficient used in this equation is the dimensionless ratio of the moment pressure to dynamic pressure moment and is a function

$$C_{M_{a.c.}} = \frac{Ma.c.}{qSc}$$

of the shape of the airfoil mean camber line. Figure 1.22 shows the moment coefficient, $c_{m_{a.c.}}$ versus lift coefficient for several representative sections. The sign convention applied to moment coefficients is that the nose-up moment is positive.

The NACA 0009 airfoil is a symmetrical section of 9 percent maximum thickness. Since the mean line of this airfoil has no camber, the coefficient of moment about the aerodynamic center is zero, i.e., the c.p. is at the a.c. The departure from zero $c_{m_{a.c.}}$ occurs only as the airfoil approaches maximum lift and the stall produces a moment change in the negative (nose-down) direction. The NACA 4412 and 63_1-412 sections have noticeable positive camber which cause relatively large moments about the aerodynamic center. Notice that for each section shown in figure 1.22, the $c_{m_{a.c.}}$ is constant for all lift coefficients less than $c_{l_{max}}$.

The NACA 23012 airfoil is a very efficient conventional section which has been used on many airplanes. One of the features of the section is a relatively high $c_{l_{max}}$ with only a small $c_{m_{a.c.}}$ The pitching moment coefficients for this section are shown on figure 1.22 along with the effect of various type flaps added to the basic section. Large amounts of camber applied well aft on the chord cause large negative moment coefficients. This fact is illustrated by the large negative moment coefficients produced by the 30° deflection of a 25 percent chord flap.

The $c_{m_{a.c.}}$ is a quantity determined by the shape of the mean-camber line. Symmetrical airfoils have zero $c_{m_{a.c.}}$ and the *c.p.* remains at the *a.c.* in unstalled flight. The airfoil with positive camber will have a negative $c_{m_{a.c.}}$ which means the *c.p.* is behind the *a.c.* Since the $c_{m_{a.c.}}$ is constant in unstalled flight a certain relationship between lift coefficient and center of pressure can be evolved. An example of this relationship is shown in figure 1.22 for the NACA 63_1-412 airfoil by a plot of *c.p.* versus c_l. Note that at low lift coefficients the center of pressure is well aft—even past the trailing edge—and an increase in c_l moves the *c.p.* forward toward the *a.c.* The *c.p.* approaches the

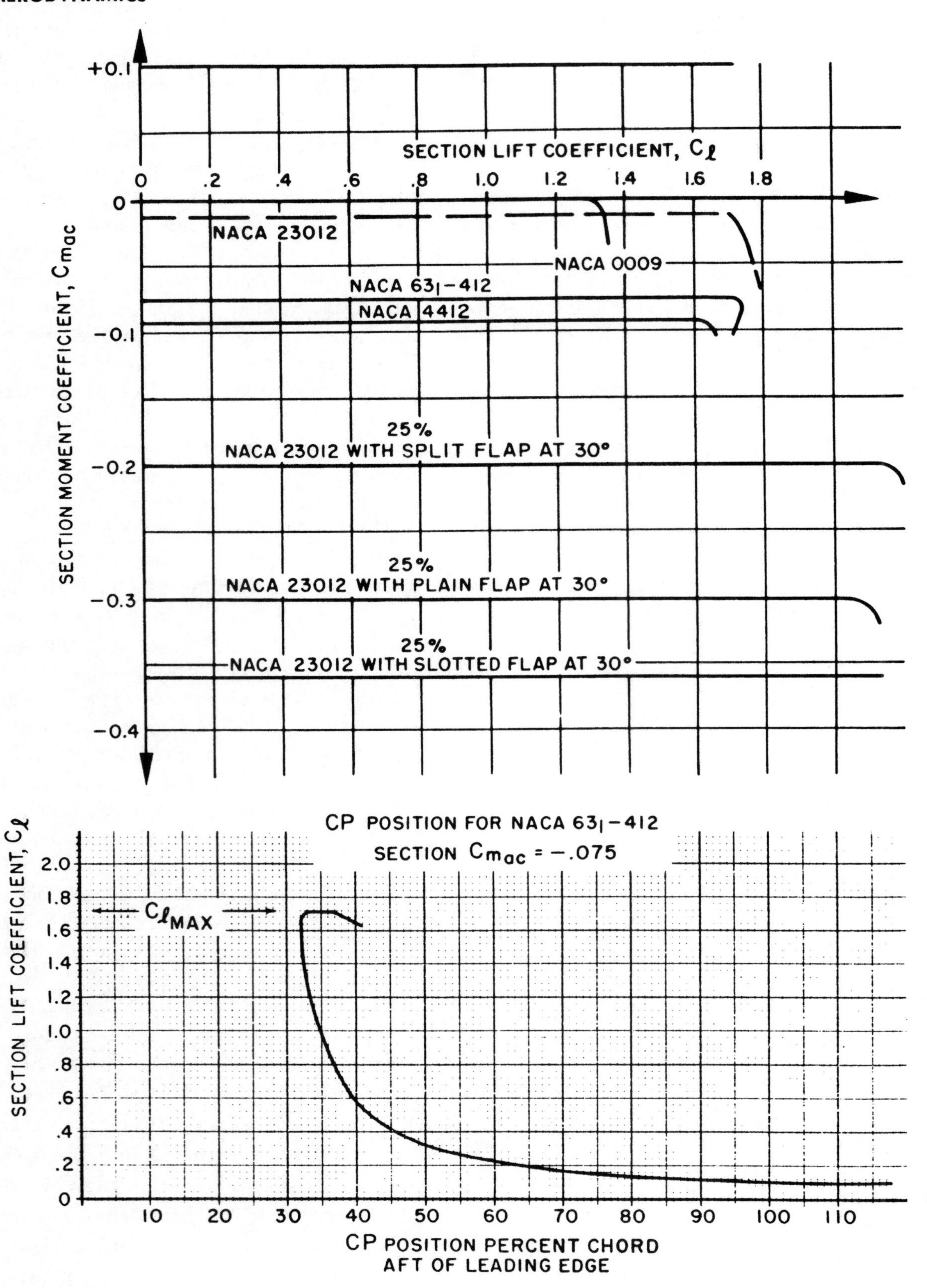

Figure 1.22. Section Moment Characteristics

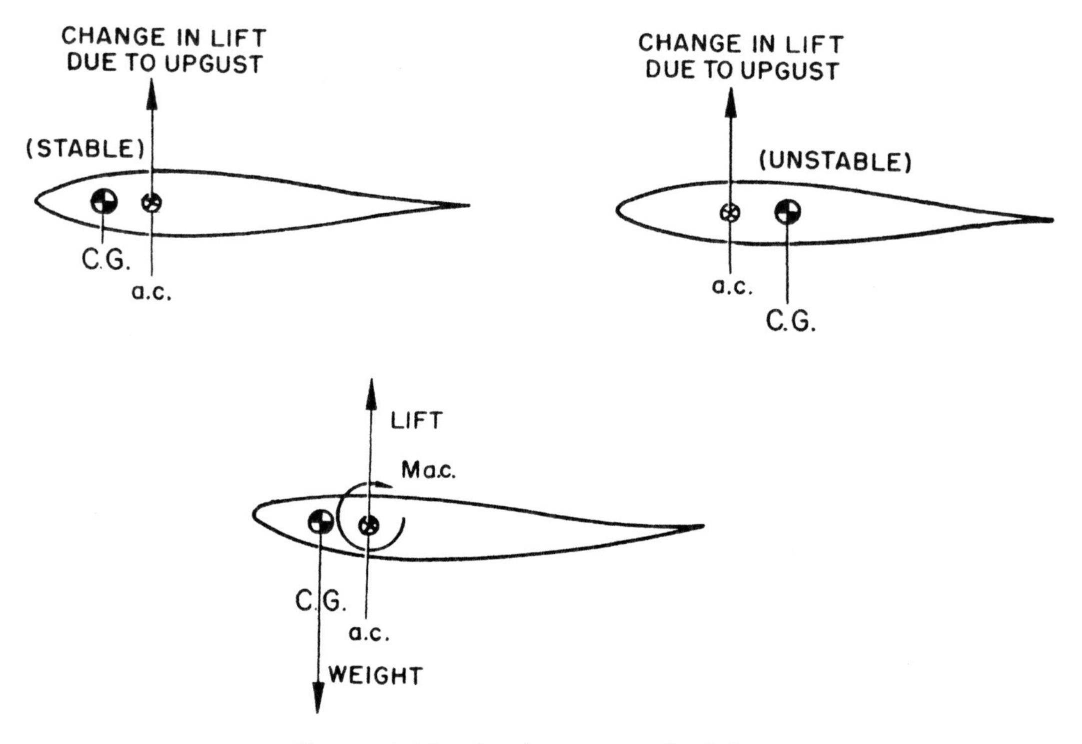

Figure 1.23. Application to Stability

a.c. as a limit but as stall occurs, the drop in suction near the leading edge cause the *c.p.* to move aft.

Of course, if the airfoil has negative camber, or a strongly reflexed trailing edge, the moment about the aerodynamic center will be positive. In this case, the location of the aerodynamic center will be unchanged and will remain at the quarter-chord position.

The aerodynamic center is the point on the chord where the coefficients of moment are constant—the point where all *changes* in lift take place. The aerodynamic center is an extremely important aerodynamic reference point and the most direct application is to the longitudinal stability of an airplane. To simplify the problem assume that the airplane is a tailless or flying wing type. In order for this type airplane to have longitudinal stability, the center of gravity must be ahead of the aerodynamic center. This very necessary feature can be visualized from the illustrations of figure 1.23.

If the two symmetrical airfoils are subject to an upgust, an increase in lift will take place at the *a.c.* If the *c.g.* is ahead of the *a.c.*, the change in lift creates a nose down moment about the *c.g.* which tends to return the airfoil to the equilibrium angle of attack. This stable, "weathercocking" tendency to return to equilibrium is a very necessary feature in any airplane. If the *c.g.* is aft of the *a.c.*, the change in lift due to the upgust takes place at the *a.c.* and creates a nose up moment about the *c.g.* This nose up moment tends to displace the airplane farther from the equilibrium and is unstable—the airplane is similar to a ball balanced on a peak. Hence, to have a stable airplane, the *c.g.* must be located ahead of the airplane *a.c.*

An additional requirement of stability is that the airplane must stabilize and be trimmed for flight at positive lift. When the *c.g.* is located ahead of *a.c.*, the weight acting at the *c.g.* is supported by the lift developed by the section. Negative camber is required to produce the positive moment about the aerodynamic center which brings about equilibrium or balance at positive lift.

Supersonic flow produces important changes in the aerodynamic characteristics of sections. The aerodynamic center of airfoils in subsonic flow is located at the 25 percent chord point. As the airfoil is subject to supersonic flow, the aerodynamic center changes to the 50 percent chord point. Thus, the airplane in transonic flight can experience large changes in longitudinal stability because of the large changes in the position of the aerodynamic center.

FRICTION EFFECTS

Because the air has viscosity, air will encounter resistance to flow over a surface. The viscous nature of airflow reduces the local velocities on a surface and accounts for the drag of skin friction. The retardation of air particles due to viscosity is greatest immediately adjacent to the surface. At the very surface of an object, the air particles are slowed to a relative velocity of near zero. Above this area other particles experience successively smaller retardation until finally, at some distance above surface, the local velocity reaches the full value of the airstream above the surface. This layer of air over the surface which shows local retardation of airflow from viscosity is termed the "boundary layer." The characteristics of this boundary layer are illustrated in figure 1.24 with the flow of air over a smooth flat plate.

The beginning flow on a smooth surface gives evidence of a very thin boundary layer with the flow occurring in smooth laminations. The boundary layer flow near the leading edge is similar to layers or laminations of air sliding smoothly over one another and the obvious term for this type of flow is the "laminar" boundary layer. This smooth laminar flow exists without the air particles moving from a given elevation.

As the flow continues back from the leading edge, friction forces in the boundary layer continue to dissipate energy of the airstream and the laminar boundary layer increases in thickness with distance from the leading edge. After some distance back from the leading edge, the laminar boundary layer begins an oscillatory disturbance which is unstable. A waviness occurs in the laminar boundary layer which ultimately grows larger and more severe and destroys the smooth laminar flow. Thus, a transition takes place in which the laminar boundary layer decays into a "turbulent" boundary layer. The same sort of transition can be noticed in the smoke from a cigarette in still air. At first, the smoke ribbon is smooth and laminar, then develops a definite waviness, and decays into a random turbulent smoke pattern.

As soon as the transition to the turbulent boundary layer takes place, the boundary layer thickens and grows at a more rapid rate. (The small scale, turbulent flow within the boundary layer should not be confused with the large scale turbulence associated with airflow separation.) The flow in the turbulent boundary layer allows the air particles to travel from one layer to another producing an energy exchange. However, some small laminar flow continues to exist in the very lower levels of the turbulent boundary layer and is referred to as the "laminar sub-layer." The turbulence which exists in the turbulent boundary layer allows determination of the point of transition by several means. Since the turbulent boundary layer transfers heat more easily than the laminar layer, frost, water, and oil films will be removed more rapidly from the area aft of the transition point. Also, a small probe may be attached to a stethoscope and positioned at various points along a surface. When the probe is in the laminar area, a low "hiss" will be heard; when the probe is in

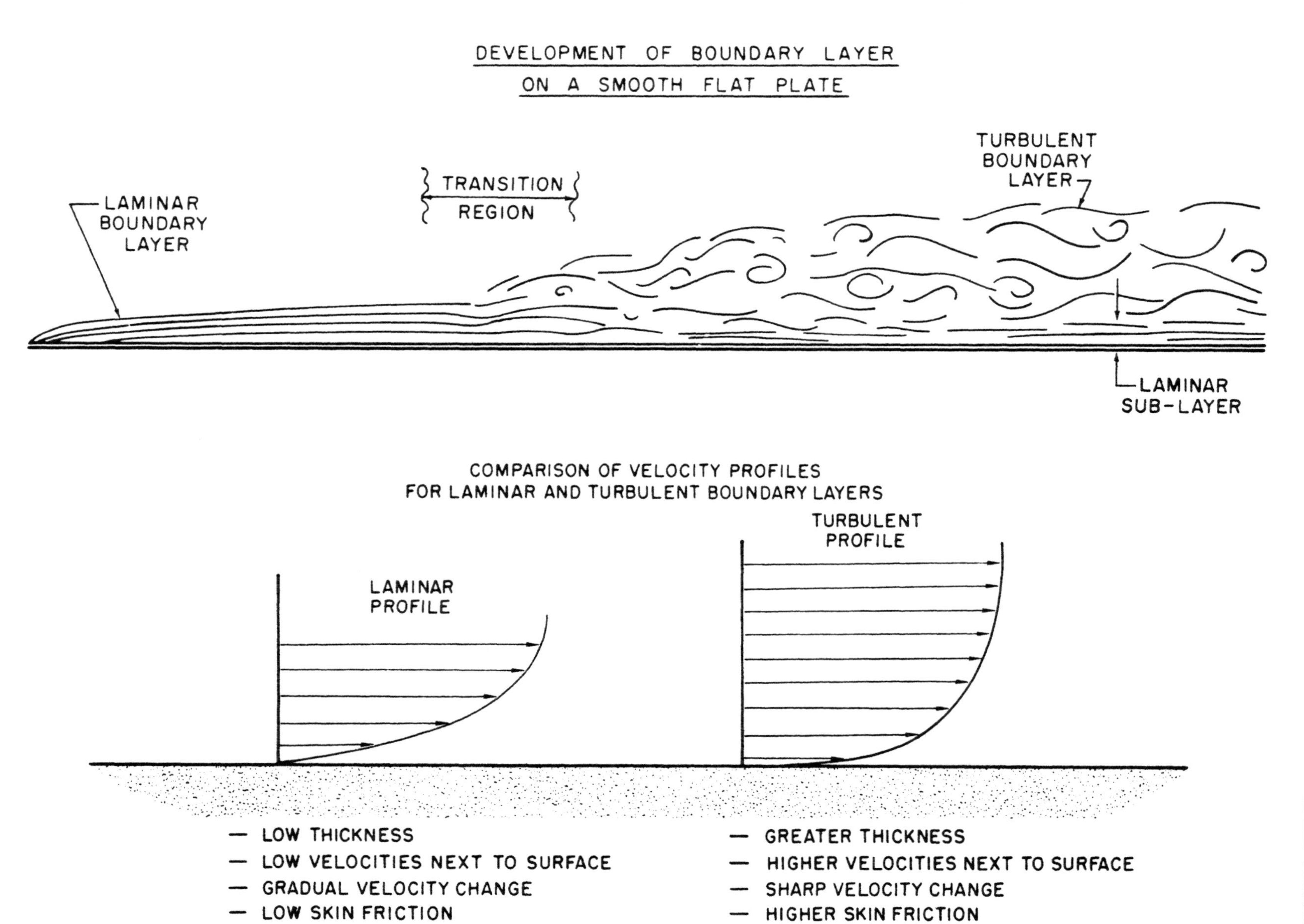

Figure 1.24. Boundary Layer Characteristics

the turbulent area, a sharp "crackling" will be audible.

In order to compare the characteristics of the laminar and turbulent boundary layers, the velocity profiles (the variation of boundary layer velocity with height above the surface) should be compared under conditions which could produce either laminar or turbulent flow. The typical laminar and turbulent profiles are shown in figure 1.24. The velocity profile of the turbulent boundary layer shows a much sharper initial change of velocity but a greater height (or boundary layer thickness) required to reach the free stream velocity. As a result of these differences, a comparison will show:

(1) The turbulent boundary layer has a fuller velocity profile and has higher local velocities immediately adjacent to the surface. The turbulent boundary layer has higher kinetic energy in the airflow next to the surface.

(2) At the surface, the laminar boundary layer has the less rapid change of velocity with distance above the plate. Since the shearing stress is proportional to the velocity gradient, the lower velocity gradient of the laminar boundary layer is evidence of a lower friction drag on the surface. If the conditions of flow were such that either a turbulent or a laminar boundary layer could exist, the laminar skin friction would be about one-third that for turbulent flow.

The low friction drag of the laminar boundary layer makes it quite desirable. However, the transition tends to take place in a natural fashion and limit the extensive development of the laminar boundary layer.

REYNOLDS NUMBER. Whether a laminar or turbulent boundary layer exists depends on the combined effects of velocity, viscosity, distance from the leading edge, density, etc. The effect of the most important factors is combined in a dimensionless parameter called "Reynolds Number, *RN*." The Reynolds Number is a dimensionless ratio which portrays the relative magnitude of dynamic and viscous forces in the flow.

$$RN=\frac{Vx}{\nu}$$

where

RN = Reynolds Number, dimensionless

V = velocity, ft. per sec.

x = distance from leading edge, ft.

ν = kinematic viscosity, sq. ft. per sec.

While the actual magnitude of the Reynolds Number has no physical significance, the quantity is used as an index to predict and correlate various phenomena of viscous fluid flow. When the *RN* is low, viscous or friction forces predominate; when the *RN* is high, dynamic or inertia forces predominate. The effect of the variables in the equation for Reynolds Number should be understood. The *RN* varies directly with velocity and distance back from the leading edge and inversely with kinematic viscosity. High *RN*'s are obtained with large chord surfaces, high velocities, and low altitude; low *RN*'s result from small chord surfaces, low velocities, and high altitudes—high altitudes producing high values for kinematic viscosity.

The most direct use of Reynolds Number is the indexing or correlating the skin friction drag of a surface. Figure 1.25 illustrates the variation of the friction drag of a smooth, flat plate with a Reynolds Number which is based on the length or chord of the plate. The graph shows separate lines of drag coefficient if the flow should be entirely laminar or entirely turbulent. The two curves for laminar and turbulent friction drag illustrate the relative magnitude of friction drag coefficient if either type of boundary layer could exist. The drag coefficients for either laminar or turbulent flow decrease with increasing *RN* since the velocity gradient decreases as the boundary layer thickens.

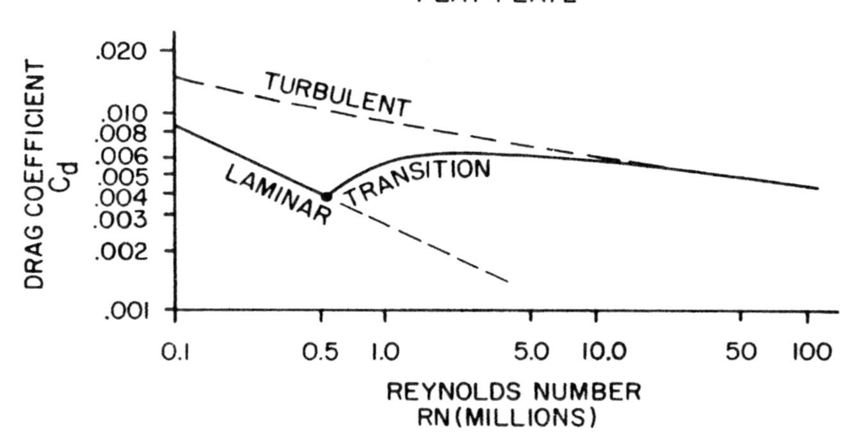

CONVENTIONAL AND LAMINAR
FLOW SECTIONS

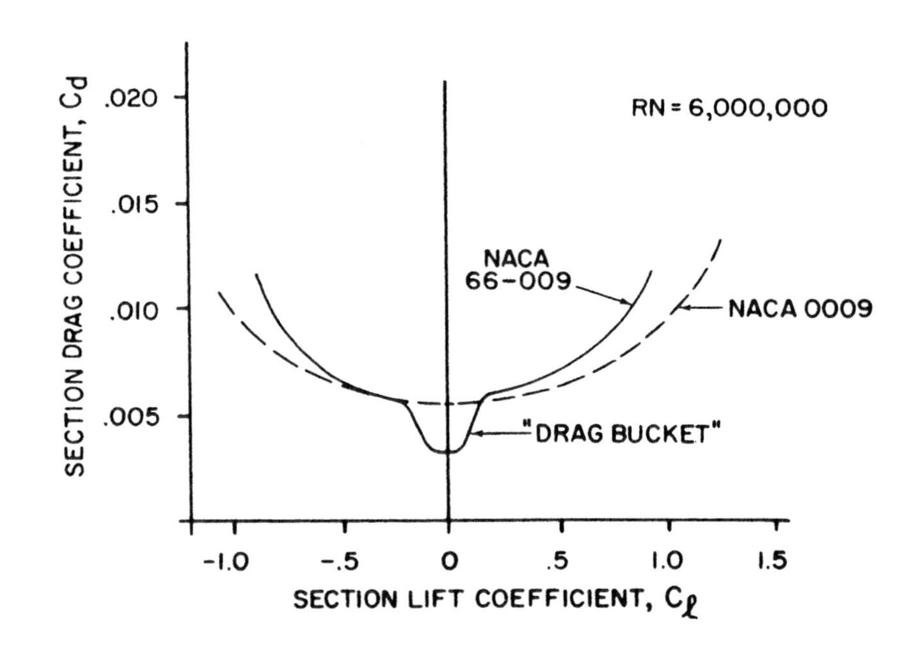

Figure 1.25. Skin Friction Drag

If the surface of the plate is smooth and the original airstream has no turbulence, the plate at low Reynolds Numbers will exist with pure laminar flow. When the *RN* is increased to approximately 530,000, transition occurs on the plate and the flow is partly turbulent. Once transition takes place, the drag coefficient of the plate increases from the laminar curve to the turbulent curve. As the *RN* approaches very high values (20 to 50 million) the drag curve of the plate approaches and nearly equals the values for the turbulent curve. At such high *RN* the boundary layer is predominantly turbulent with very little laminar flow—the transition point is very close to the leading edge. While the smooth, flat plate is not exactly representative of the typical airfoil, basic fluid friction phenomena are illustrated. At *RN* less than a half million the boundary layer will be entirely laminar unless there is extreme surface roughness or turbulence induced in the airstream. Reynolds Numbers between one and five million produce boundary layer flow which is partly laminar and partly turbulent. At *RN* above ten million the boundary layer characteristics are predominantly turbulent.

In order to obtain low drag sections, the transition from laminar to turbulent must be delayed so that a greater portion of the surface will be influenced by the laminar boundary layer. The conventional, low speed airfoil shapes are characterized by minimum pressure points very close to the leading edge. Since high local velocities enhance early transition, very little surface is covered by the laminar boundary layer. A comparison of two 9 percent thick symmetrical airfoils is presented in figure 1.25. One section is the "conventional" NACA 0009 section which has a minimum pressure point at approximately 10 percent chord at zero lift. The other section is the NACA 66-009 which has a minimum pressure point at approximately 60 percent chord at zero lift. The lower local velocities at the leading edge and the favorable pressure gradient of the NACA 66-009 delay the transition to some point farther aft on the chord. The subsequent reduction in friction drag at the low angles of attack accounts for the "drag bucket" shown on the graphs of c_d and c_l for these sections. Of course, the advantages of the laminar flow airfoil are apparent only for the smooth airfoil since surface roughness or waviness may preclude extensive development of a laminar boundary layer.

AIRFLOW SEPARATION. The character of the boundary layer on an aerodynamic surface is greatly influenced by the pressure gradient. In order to study this effect, the pressure distribution of a cylinder in a perfect fluid is repeated in figure 1.26. The airflows depict a local velocity of zero at the forward stagnation point and a maximum local velocity at the extreme surface. The airflow moves from the high positive pressure to the minimum pressure point—a favorable pressure gradient (high to low). As the air moves from the extreme surface aft, the local velocity decreases to zero at the aft stagnation point. The static pressure increases from the minimum (or maximum suction) to the high positive pressure at the aft stagnation point—an adverse pressure gradient (low to high).

The action of the pressure gradient is such that the favorable pressure gradient assists the boundary layer while the adverse pressure gradient impedes the flow of the boundary layer. The effect of an adverse pressure gradient is illustrated by the segment *X–Y* of figure 1.26. A corollary of the skin friction drag is the continual reduction of boundary layer energy as flow continues aft on a surface. The velocity profiles of the boundary layer are shown on segment *X–Y* of figure 1.26. In the area of adverse pressure gradient the boundary layer flow is impeded and tends to show a reduction in velocity next to the surface. If the boundary layer does not have sufficient kinetic energy in the presence of the adverse pressure gradient, the lower levels of the boundary layer may stagnate prematurely.

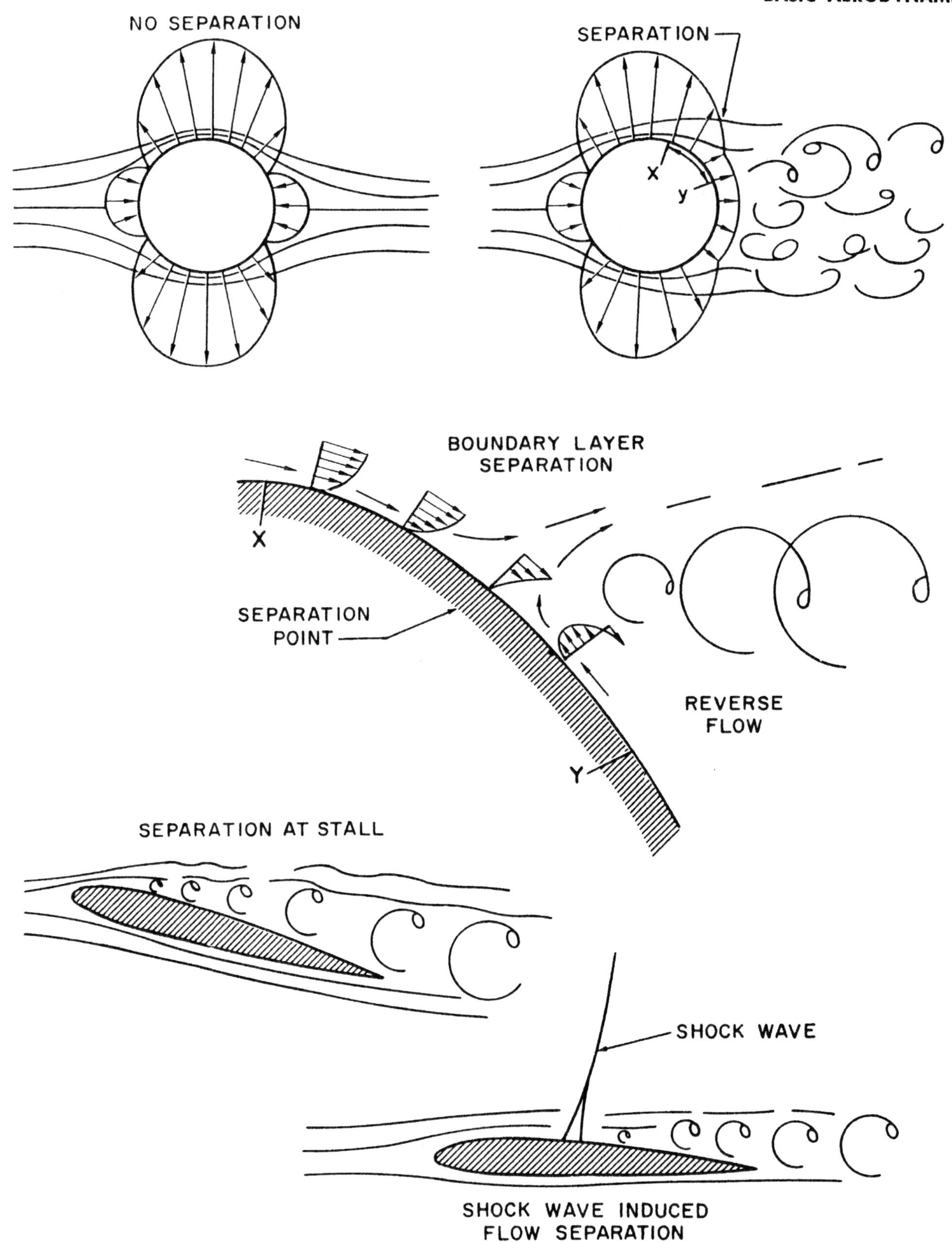

Figure 1.26. Airflow Separation (sheet 1 of 2)

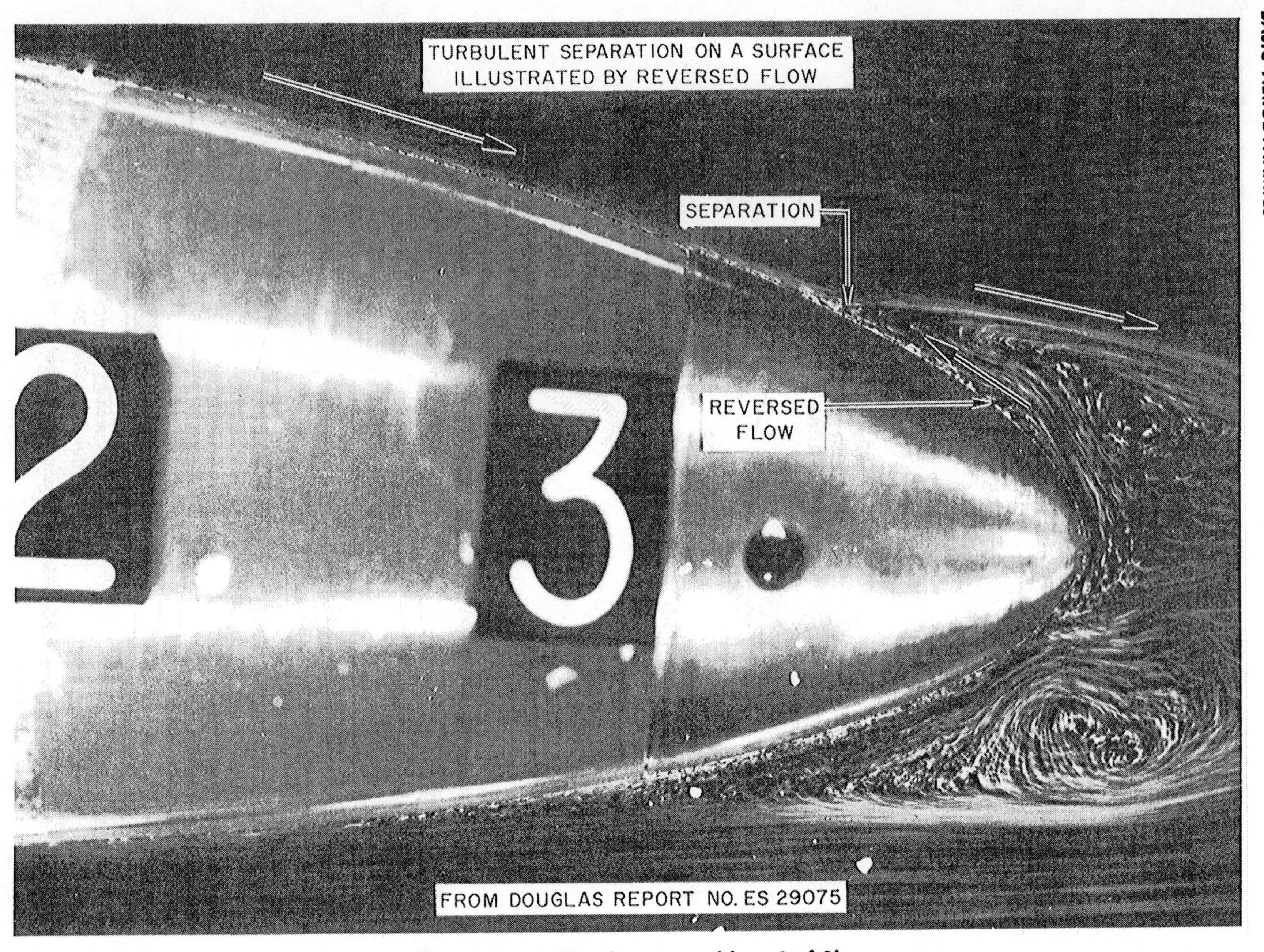

Figure 1.26. *Airflow Separation (sheet 2 of 2)*

Premature stagnation of the boundary layer means that all subsequent airflow will overrun this point and the boundary layer will separate from the surface. Surface flow which is aft of the separation point will indicate a local flow direction forward toward the separation point—a flow reversal. If separation occurs the positive pressures are not recovered and form drag results. The points of separation on any aerodynamic surface may be noted by the reverse flow area. Tufts of cloth or string tacked to the surface will lie streamlined in an area of unseparated flow but will lie forward in an area behind the separation point.

The basic feature of airflow separation is stagnation of the lower levels of the boundary layer. *Airflow separation results when the lower levels of the boundary layer do not have sufficient kinetic energy in the presence of an adverse pressure gradient.* The most outstanding cases of airflow separation are shown in figure 1.26. An airfoil at some high angle of attack creates a pressure gradient on the upper surface too severe to allow the boundary layer to adhere to the surface. When the airflow does not adhere to the surface near the leading edge the high suction pressures are lost and stall occurs. When the shock wave forms on the upper surface of a wing at high subsonic speeds, the increase of static pressure through the shock wave creates a very strong obstacle for the boundary layer. If the shock wave is sufficiently strong, separation will follow and "compressibility buffet" will result from the turbulent wake or separated flow.

In order to prevent separation of a boundary layer in the presence of an adverse pressure gradient, the boundary layer must have the highest possible kinetic energy. If a choice is available, the turbulent boundary layer would be preferable to the laminar boundary layer because the turbulent velocity profile shows higher local velocities next to the surface. The most effective high lift devices (slots, slotted flaps, *BLC*) utilize various techniques to increase the kinetic energy of the upper surface boundary layer to withstand the more severe pressure gradients common to the higher lift coefficients. Extreme surface roughness on full scale aircraft (due to surface damage, heavy frost, etc.) causes higher skin friction and greater energy loss in the boundary layer. The lower energy boundary layer may cause a noticeable change in $C_{L_{max}}$ and stall speed. In the same sense, vortex generators applied to the surfaces of a high speed airplane may allay compressibility buffet to some degree. The function of the vortex generators is to create a strong vortex which introduces high velocity, high energy air next to the surface to reduce or delay the shock induced separation. These examples serve as a reminder that separation is the result of premature stagnation of the boundary layer—insufficient kinetic energy in the presence of an adverse pressure gradient.

SCALE EFFECT. Since the boundary layer friction and kinetic energy are dependent on the characteristics of the boundary layer, Reynolds Number is important in correlating aerodynamic characteristics. The variation of the aerodynamic characteristics with Reynolds Number is termed "scale effect" and is extremely important in correlating wind tunnel test data of scale models with the actual flight characteristics of the full size aircraft. The two most important section characteristics affected by scale effects are drag and maximum lift—the effect on pitching moments usually being negligible. From the known variation of boundary layer characteristics with Reynolds Number, certain general effects may be anticipated. With increasing Reynolds Number, it may be expected that the section maximum lift coefficient will increase (from the higher energy turbulent boundary layer) and that the section drag coefficient will decrease (similar to that of the smooth plate). These effects are illustrated by the graphs of figure 1.27.

The characteristics depicted in figure 1.27 are for the NACA 4412 airfoil (4 percent

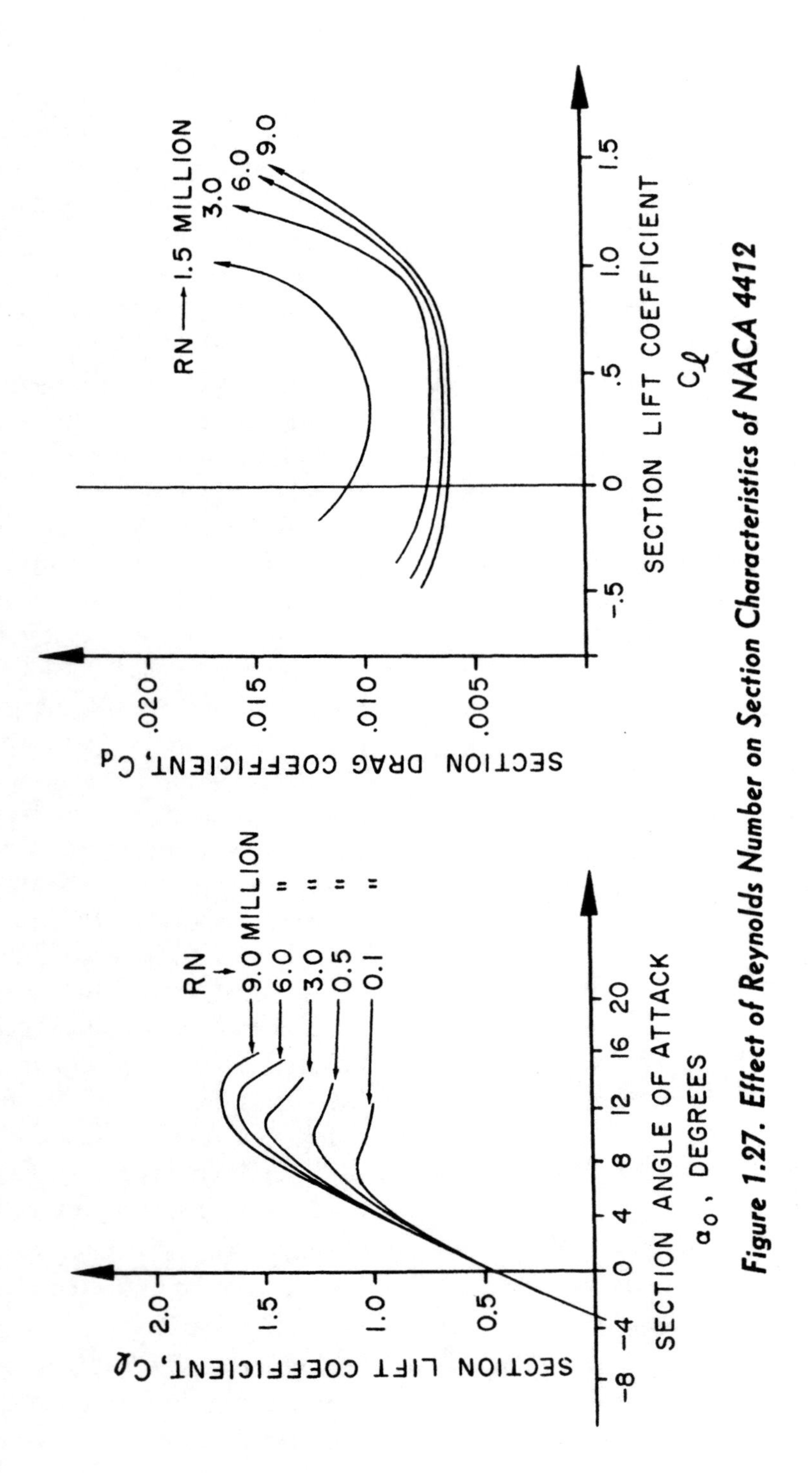

Figure 1.27. Effect of Reynolds Number on Section Characteristics of NACA 4412

camber at 40 percent chord, 12 percent thickness at 30 percent chord)—a fairly typical "conventional" airfoil section. The lift curve show a steady increase in $c_{l_{max}}$ with increasing RN. However, note that a smaller change in $c_{l_{max}}$ occurs between Reynolds Numbers of 6.0 and 9.0 million than occurs between 0.1 and 3.0 million. In other words, greater changes in $c_{l_{max}}$ occur in the range of Reynolds Numbers where the laminar (low energy) boundary layer predominates. The drag curves for the section show essentially the same feature—the greatest variations occur at very low Reynolds Numbers. Typical full scale Reynolds Numbers for aircraft in flight may be 3 to 500 million where the boundary layer is predominately turbulent. Scale model tests may involve Reynolds Numbers of 0.1 to 5 million where the boundary layer be predominately laminar. Hence, the "scale" corrections are very necessary to correlate the principal aerodynamic characteristics.

The very large changes in aerodynamic characteristics at low Reynolds Numbers are due in great part to the low energy laminar boundary layer typical of low Reynolds Numbers. Low Reynolds Numbers are the result of some combination of low velocity, small size, and high kinematic viscosity $\left(RN=\frac{Vx}{\nu}\right)$. Thus, small surfaces, low flight speeds, or very high altitudes can provide the regime of low Reynolds Numbers. One interesting phenomenon associated with low *RN* is the high form drag due to separation of the low energy boundary layer. The ordinary golf ball operates at low *RN* and would have very high form drag without dimpling. The surface roughness from dimpling disturbs the laminar boundary layer forcing a premature transition to turbulent. The forced turbulence in the boundary layer reduces the form drag by providing a higher energy boundary layer to allay separation. Essentially the same effect can be produced on a model airplane wing by roughening the leading edge—the turbulent boundary layer obtained may reduce the form drag due to separation. In each instance, the forced transition will be beneficial if the reduction in form drag is greater than the increase in skin friction. Of course, this possibility exists only at low Reynolds Numbers.

In a similar sense, "trip" wires or small surface protuberances on a wind tunnel model may be used to force transition of the boundary layer and simulate the effect of higher Reynolds Numbers.

PLANFORM EFFECTS AND AIRPLANE DRAG

EFFECT OF WING PLANFORM

The previous discussion of aerodynamic forces concerned the properties of airfoil sections in two-dimensional flow with no consideration given to the influence of the planform. When the effects of wing planform are introduced, attention must be directed to the existence of flow components in the spanwise direction. In other words, airfoil section properties deal with flow in two dimensions while planform properties consider flow in three dimensions.

In order to fully describe the planform of a wing, several terms are required. The terms having the greatest influence on the aerodynamic characteristics are illustrated in figure 1.28.

(1) The *wing area*, *S*, is simply the plan surface area of the wing. Although a portion of the area may be covered by fuselage or nacelles, the pressure carryover on these surfaces allows legitimate consideration of the entire plan area.

(2) The *wing span*, *b*, is measured tip to tip.

(3) The *average chord*, *c*, is the geometric average. The product of the span and the average chord is the wing area ($b\times c=S$).

(4) The *aspect ratio*, *AR*, is the proportion of the span and the average chord.

$$AR=b/c$$

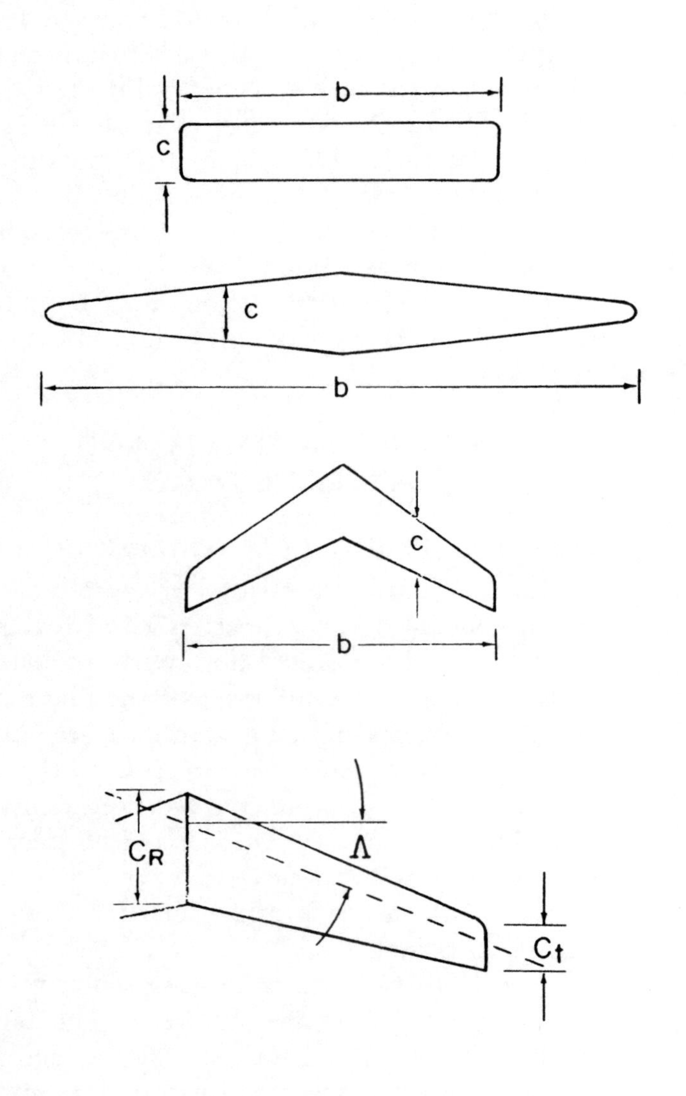

S = WING AREA, SQ. FT.

b = SPAN, FT

c = AVERAGE CHORD, FT

AR = ASPECT RATIO

$AR = b/c$

$AR = b^2/S$

C_R = ROOT CHORD, FT

C_t = TIP CHORD, FT

λ = TAPER RATIO

$\lambda = C_t/C_R$

Λ = SWEEP ANGLE, DEGREES

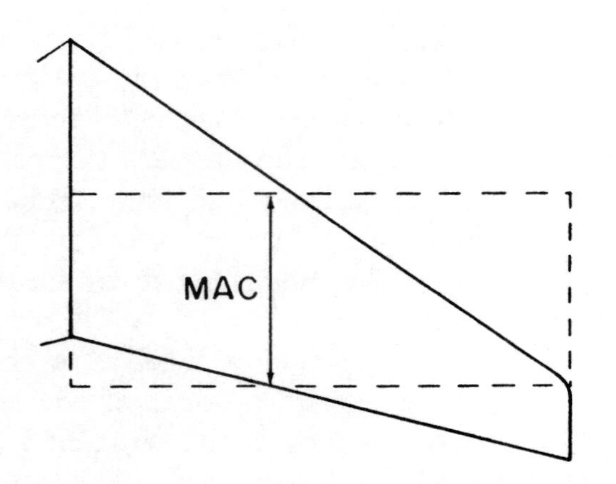

MAC = MEAN AERODYNAMIC CHORD, FT.

Figure 1.28. Description of Wing Planform

If the planform has curvature and the average chord is not easily determined, an alternate expression is:

$$AR = b^2/S$$

The aspect ratio is a fineness ratio of the wing and this quantity is very powerful in determing the aerodynamic characteristics and structural weight. Typical aspect ratios vary from 35 for a high performance sailplane to 3.5 for a jet fighter to 1.28 for a flying saucer.

(5) The *root chord*, c_r, is the chord at the wing centerline and the *tip chord*, c_t, is measured at the tip.

(6) Considering the wing planform to have straight lines for the leading and trailing edges, the *taper ratio*, λ (lambda), is the ratio of the tip chord to the root chord.

$$\lambda = c_t/c_r$$

The taper ratio affects the lift distribution and the structural weight of the wing. A rectangular wing has a taper ratio of 1.0 while the pointed tip delta wing has a taper ratio of 0.0.

(7) The *sweep angle*, Λ (cap lambda), is usually measured as the angle between the line of 25 percent chords and a perpendicular to the root chord. The sweep of a wing causes definite changes in compressibility, maximum lift, and stall characteristics.

(8) The *mean aerodynamic chord*, *MAC*, is the chord drawn through the centroid (geographical center) of plan area. A rectangular wing of this chord and the same span would have identical pitching moment characteristics. The *MAC* is located on the reference axis of the airplane and is a primary reference for longitudinal stability considerations. Note that the *MAC* is *not* the average chord but is the chord through the centroid of area. As an example, the pointed-tip delta wing with a taper ratio of zero would have an average chord equal to one-half the root chord but an *MAC* equal to two-thirds of the root chord.

The aspect ratio, taper ratio, and sweepback of a planform are the principal factors which determine the aerodynamic characteristics of a wing. These same quantities also have a definite influence on the structural weight and stiffness of a wing.

DEVELOPMENT OF LIFT BY A WING. In order to appreciate the effect of the planform on the aerodynamic characteristics, it is necessary to study the manner in which a wing produces lift. Figure 1.29 illustrates the three-dimensional flow pattern which results when the rectangular wing creates lift.

If a wing is producing lift, a pressure differential will exist between the upper and lower surfaces, i.e., for positive lift, the static pressure on the upper surface will be less than on the lower surface. At the tips of the wing, the existence of this pressure differential creates the spanwise flow components shown in figure 1.29. For the rectangular wing, the lateral flow developed at the tip is quite strong and a strong vortex is created at the tip. The lateral flow—and consequent vortex strength—reduces inboard from the tip until it is zero at the centerline.

The existence of the tip vortex is described by the drawings of figure 1.29. The rotational pressure flow combines with the local airstream flow to produce the resultant flow of the trailing vortex. Also, the downwash flow field behind a delta wing is illustrated by the photographs of figure 1.29. A tuft-grid is mounted aft of the wing to visualize the local flow direction by deflection of the tuft elements. This tuft-grid illustrates the existence of the tip vortices and the deflected airstream aft of the wing. Note that an increase in angle of attack increases lift and increases the flow deflection and strength of the tip vortices.

Figure 1.30 illustrates the principal effect of the wing vortex system. The wing producing lift can be represented by a series of

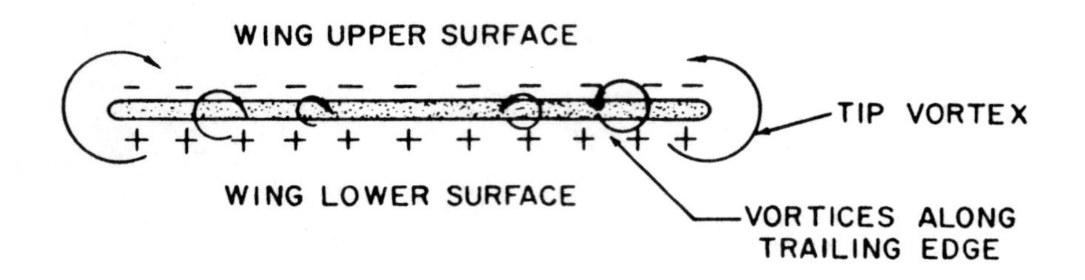

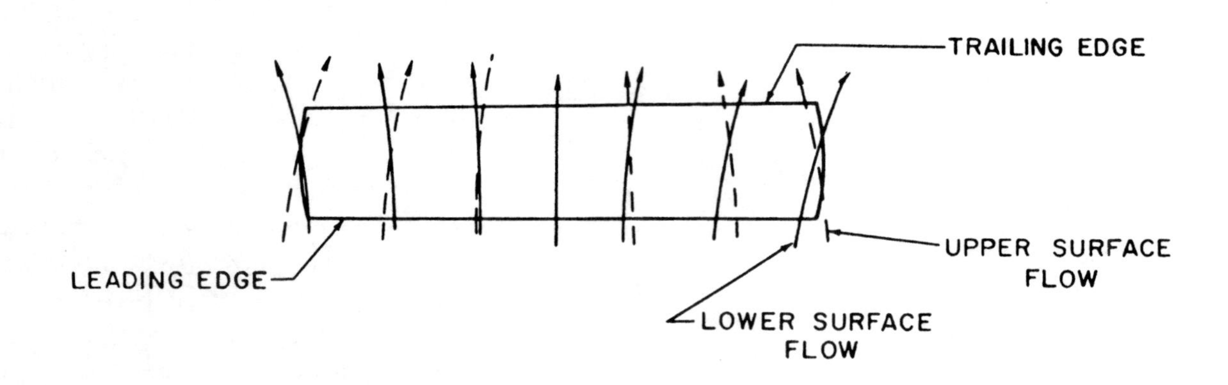

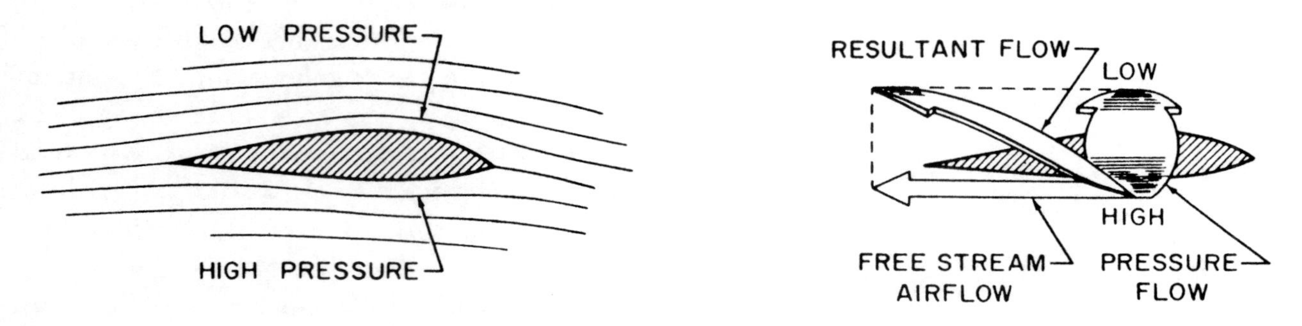

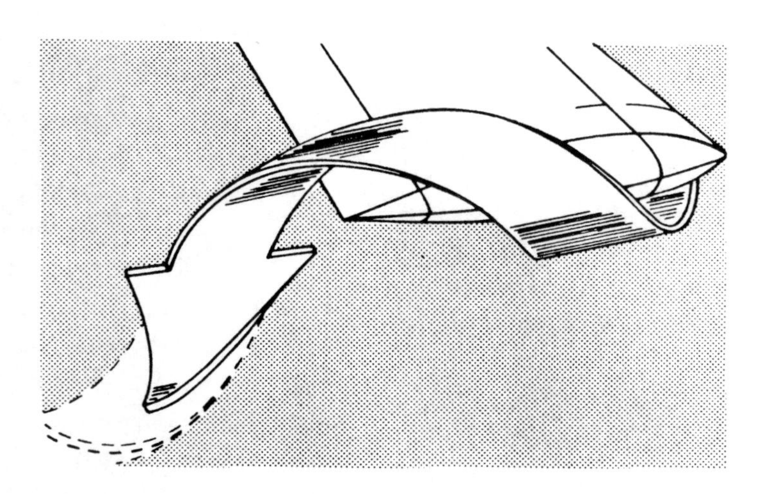

Figure 1.29. Wing Three Dimensional Flow (sheet 1 of 2)

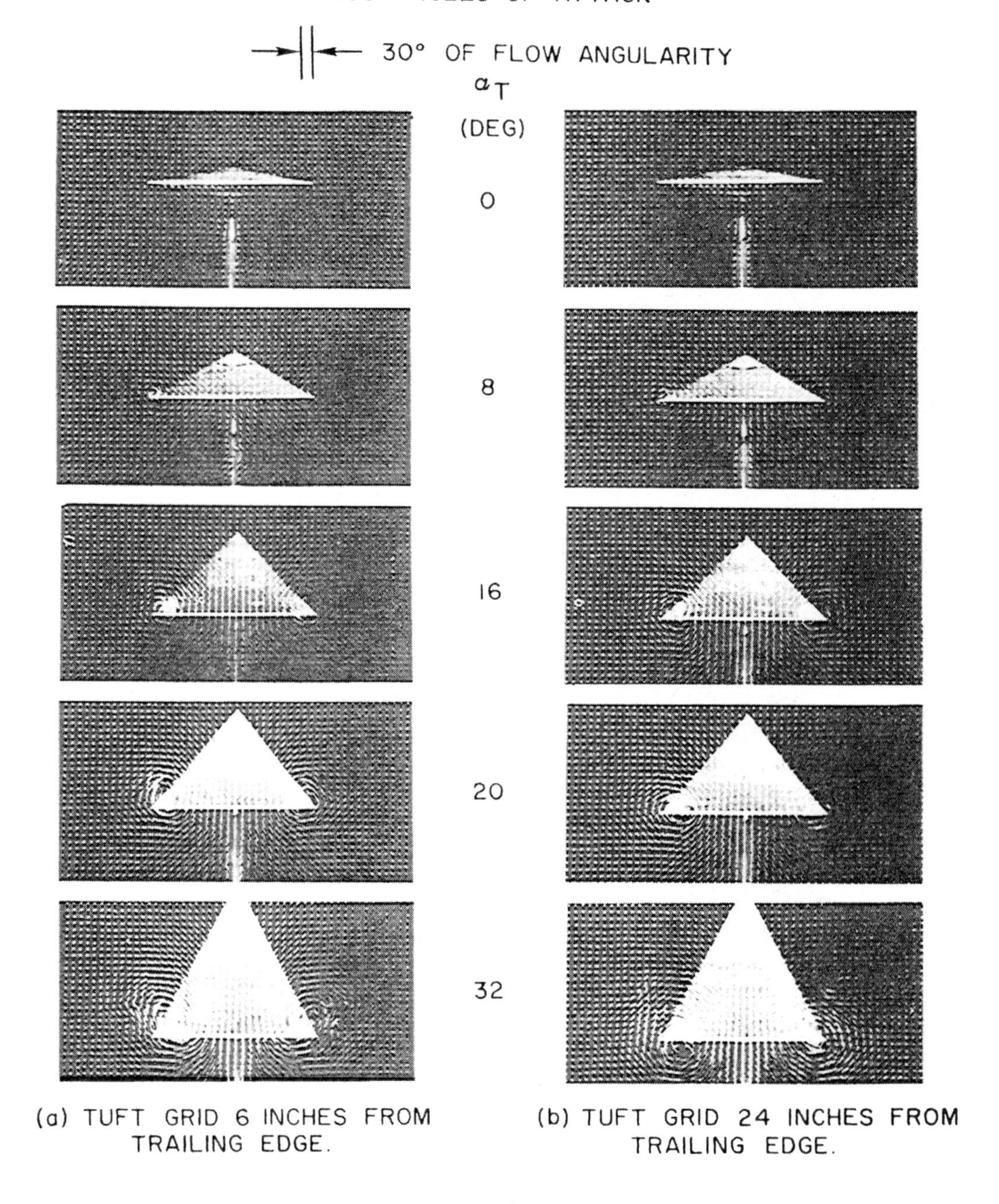

FROM NACA TN 2674

Figure 1.29. Wing Three Dimensional Flow (sheet 2 of 2)

vortex filaments which consist of the tip or trailing vortices coupled with the bound or line vortex. The tip vortices are coupled with the bound vortex when circulation is induced with lift. The effect of this vortex system is to create certain vertical velocity components in the vicinity of the wing. The illustration of these vertical velocities shows that ahead of the wing the bound vortex induces an upwash. Behind the wing, the coupled action of the bound vortex and the tip vortices induces a downwash. With the action of tip and bound vortices coupled, a final vertical velocity ($2w$) is imparted to the airstream by the wing producing lift. This result is an inevitable consequence of a finite wing producing lift. The wing producing lift applies the equal and opposite force to the airstream and deflects it downward. One of the important factors in this system is that a downward velocity is created at the aerodynamic center (w) which is one half the final downward velocity imparted to the airstream ($2w$).

The effect of the vertical velocities in the vicinity of the wing is best appreciated when they are added vectorially to the airstream velocity. The remote free stream well ahead of the wing is unaffected and its direction is opposite the flight path of the airplane. Aft of the wing, the vertical velocity ($2w$) adds to the airstream velocity to produce the downwash angle ϵ (epsilon). At the aerodynamic center of the wing, the vertical velocity (w) adds to the airstream velocity to produce a downward deflection of the airstream one-half that of the downwash angle. In other words, the wing producing lift by the deflection of an airstream incurs a downward slant to the wind in the immediate vicinity of the wing. *Hence, the sections of the wing operate in an* average *relative wind which is inclined downward one-half the final downwash angle*. This is one important feature which distinguishes the aerodynamic properties of a *wing* from the aerodynamic properties of an airfoil *section*.

The induced velocities existing at the aerodynamic center of a finite wing create an average relative wind which is different from the remote free stream wind. Since the aerodynamic forces created by the airfoil sections of a wing depend upon the immediate airstream in which they operate, consideration must be given to the effect of the inclined average relative wind.

To create a certain lift coefficient with the airfoil section, a certain angle must exist between the airfoil chord line and the *average* relative wind. This angle of attack is α_0, the section angle of attack. However, as this lift is developed on the wing, downwash is incurred and the average relative wind is inclined. Thus, the wing must be given some angle attack greater than the required section angle of attack to account for the inclination of the average relative wind. Since the wing must be given this additional angle of attack because of the induced flow, the angle between the average relative wind and the remote free stream is termed the induced angle of attack, α_i. From this influence, the wing angle of attack is the sum of the section and induced angles of attack.

$$\alpha = \alpha_0 + \alpha_i$$

where α = wing angle of attack
α_0 = section angle of attack
α_i = induced angle of attack

INDUCED DRAG

Another important influence of the induced flow is the orientation of the actual lift on a wing. Figure 1.30 illustrates the fact that the lift produced by the wing sections is perpendicular to the average relative wind. Since the average relative wind is inclined downward, the section lift is inclined aft by the same amount—the induced angle of attack, α_i. The lift and drag of a wing must continue to be referred perpendicular and parallel to the remote free stream ahead of the wing. In this respect, the lift on the wing has a component of force parallel to the remote free stream. This component of lift in the drag direction is the undesirable—but unavoidable—conse-

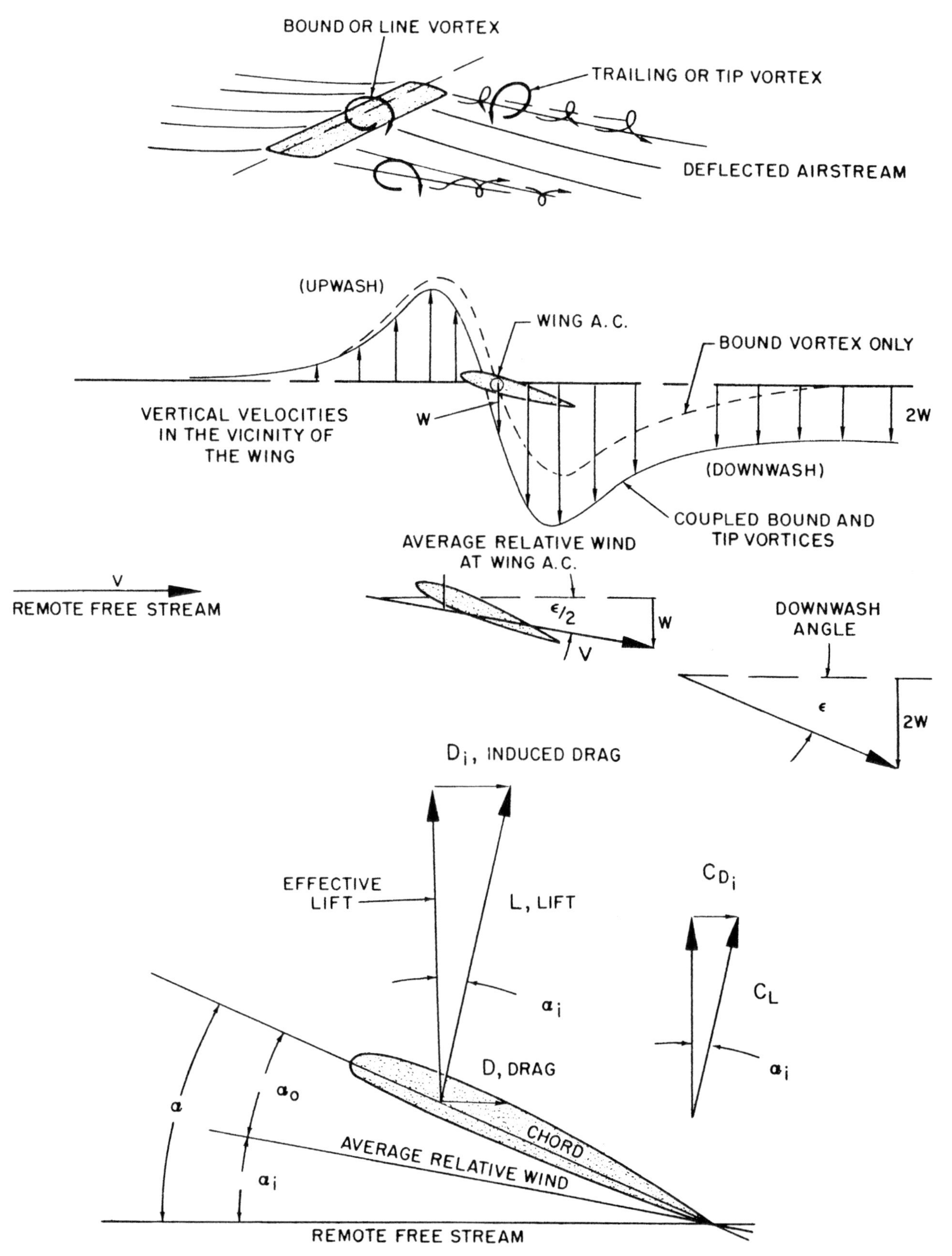

Figure 1.30. Wing Vortex System and Induced Flow

quence of developing lift with a finite wing and is termed INDUCED DRAG, D_i. Induced drag is separate from the drag due to form and friction and is due simply to the development of lift.

By inspection of the force diagram of figure 1.30, a relationship between induced drag, lift, and induced angle of attack is apparent. The induced drag coefficient, C_{D_i}, will vary directly with the wing lift coefficient, C_L, and the induced angle of attack, α_i. The effective lift is the vertical component of the actual lift and, if the induced angle of attack is small, will be essentially the same as the actual lift. The horizontal and vertical component of drag is insignificant under the same conditions. By a detailed study of the factors involved, the following relationships can be derived for a wing with an elliptical lift distribution:

(1) The induced drag equation follows the same form as applied to any other aerodynamic force.

$$D_i = C_{D_i} q S$$

where

D_i = induced drag, lbs.
q = dynamic pressures, psf
$= \frac{\sigma V^2}{295}$
C_{D_i} = induced drag coefficient
S = wing area, sq. ft.

(2) The induced drag coefficient can be derived as:

$$C_{D_i} = C_L \sin \alpha_i$$

or

$$C_{D_i} = \frac{C_L^2}{\pi AR}$$

$$= 0.318 \left(\frac{C_L^2}{AR}\right)$$

where

C_L = lift coefficient
$\sin \alpha_i$ = natural sine of the induced angle of attack, α_i, degrees
π = 3.1416, constant
AR = wing aspect ratio

(3) The induced angle of attack can be derived as:

$$\alpha_i = 18.24 \left(\frac{C_L}{AR}\right) \text{ (degrees)}$$

(NOTE: the derivation of these relationships may be found in any of the standard engineering aerodynamics textbooks.)

These relationships facilitate an understanding and appreciation of induced drag.

The induced angle of attack $\left(\alpha_i = 18.24 \frac{C_L}{AR}\right)$ depends on the lift coefficient and aspect ratio. Flight at high lift conditions such as low speed or maneuvering flight will create high induced angles of attack while high speed, low lift flight will create very small induced angles of attack. The inference is that high lift coefficients require large downwash and result in large induced angles of attack. The effect of aspect ratio is significant since a very high aspect ratio would produce a negligible induced angle of attack. If the aspect ratio were infinite, the induced angle of attack would be zero and the aerodynamic characteristics of the wing would be identical with the airfoil section properties. On the other hand, if the wing aspect ratio is low, the induced angle of attack will be large and the low aspect ratio airplane must operate at high angles of attack at maximum lift. Essentially, the low aspect ratio wing affects a relatively small mass of air and consequently must provide a large deflection (downwash) to produce lift.

EFFECT OF LIFT. The induced drag coefficient $\left(C_{D_i} = 0.318 \frac{C_L^2}{AR}\right)$ shows somewhat similar effects of lift coefficient and aspect ratio. Because of the power of variation of induced drag coefficient with lift coefficient, high lift coefficients provide very high induced drag and low lift coefficients very low induced drag. The direct effect of C_L can be best appreciated by assuming an airplane is flying at a given weight, altitude, and airspeed. If the airplane is maneuvered from steady level flight to a load factor of two, the lift coefficient is doubled and the induced drag *is four times as great*. If the flight load factor is changed from one to five, the induced drag is twenty-five times as great. If all other factors are held constant to single out this effect, it could be stated that "induced drag varies as the square of the lift".

$$\frac{D_{i_2}}{D_{i_1}}=\left(\frac{L_2}{L_1}\right)^2$$

where

D_{i_1}= induced drag corresponding to some original lift, L_1

D_{i_2}= induced drag corresponding to some new lift, L_2

(and q (or EAS), S, AR are constant)

This expression defines the effect of gross weight, maneuvers, and steep turns on the induced drag, e.g., 10 percent higher gross weight increases induced drag 21 percent, 4G maneuvers cause 16 times as much induced drag, a turn with 45° bank requires a load factor of 1.41 and this doubles the induced drag.

EFFECT OF ALTITUDE. The effect of altitude on induced drag can be appreciated by holding all other factors constant. The general effect of altitude is expressed by:

$$\frac{Di_2}{Di_1}=\left(\frac{\sigma_1}{\sigma_2}\right)$$

where

Di_1= induced drag corresponding to some original altitude density ratio, σ_1

Di_2= induced drag corresponding to some new altitude density ratio, σ_2

(and L, S, AR, V are constant)

This relationship implies that induced drag would increase with altitude, e.g., a given airplane flying in level flight at a given TAS at 40,000 ft. (σ=0.25) would have four times as much induced drag than when at sea level (σ=1.00). This effect results when the lower air density requires a greater deflection of the airstream to produce the same lift. However, if the airplane is flown at the same EAS, the dynamic pressure will be the same and induced drag will not vary. In this case, the TAS would be higher at altitude to provide the same EAS.

EFFECT OF SPEED. The general effect of speed on induced drag is unusual since low airspeeds are associated with high lift coefficients and high lift coefficients create high induced drag coefficients. The immediate implication is that *induced drag increases with decreasing airspeed*. If all other factors are held constant to single out the effect of airspeed, a rearrangement of the previous equations would predict that "induced drag varies inversely as the square of the airspeed."

$$\frac{Di_2}{Di_1}=\left(\frac{V_1}{V_2}\right)^2$$

where

Di_1= induced drag corresponding to some original speed, V_1

Di_2= induced drag corresponding to some new speed, V_2

(and L, S, AR, σ are constant)

Such an effect would imply that a given airplane in steady flight would incur one-fourth as great an induced drag at twice as great a speed or four times as great an induced drag at half the original speed. This variation may be illustrated by assuming that an airplane in steady level flight is slowed from 300 to 150 knots. The dynamic pressure at 150 knots is one-fourth the dynamic pressure at 300 knots and the wing must deflect the airstream four times as greatly to create the same lift. The same lift force is then slanted aft four times as greatly and the induced drag is four times as great.

The expressed variation of induced drag with speed points out that induced drag will be of

NAVY
NM

greatest importance at low speeds and practically insignificant in flight at high dynamic pressures. For example, a typical single engine jet airplane at low altitude and maximum level flight airspeed has an induced drag which is less than *1 percent* of the total drag. However, this same airplane in steady flight just above the stall speed could have an induced drag which is approximately *75 percent* of the total drag.

EFFECT OF ASPECT RATIO. The effect of aspect ratio on the induced drag

$$\left(C_{D_i}=0.318\frac{C_L^2}{AR}\right)$$

is the principal effect of the wing planform. The relationship for induced drag coefficient emphasizes the need of a high aspect ratio for the airplane which is continually operated at high lift coefficients. In other words, airplane configurations designed to operate at high lift coefficients during the major portion of their flight (sailplanes, cargo, transport, patrol, and antisubmarine types) demand a high aspect ratio wing to minimize the induced drag. While the high aspect ratio wing will minimize induced drag, long, thin wings increase structural weight and have relatively poor stiffness characteristics. This fact will temper the preference for a very high aspect ratio. Airplane configurations which are developed for very high speed flight (especially supersonic flight) operate at relatively low lift coefficients and demand great aerodynamic cleanness. These configurations of airplanes do not have the same preference for high aspect ratio as the airplanes which operate continually at high lift coefficients. This usually results in the development of low aspect ratio planforms for these airplane configurations.

The effect of aspect ratio on the lift and drag characteristics is shown in figure 1.31 for wings of a basic 9 percent symmetrical section. The basic airfoil section properties are shown on these curves and these properties would be typical only of a wing planform of extremely high (infinite) aspect ratio. When a wing of some finite aspect ratio is constructed of this basic section, the principal differences will be in the lift and drag characteristics—the moment characteristics remain essentially the same. The effect of decreasing aspect ratio on the lift curve is to increase the wing angle of attack necessary to produce a given lift coefficient. The difference between the wing angle of attack and the section angle of attack is the induced angle of attack, $\alpha_i=18.24\frac{C_L}{AR}$, which increases with decreasing aspect ratio. The wing with the lower aspect ratio is less sensitive to changes in angle of attack and requires higher angles of attack for maximum lift. When the aspect ratio is very low (below 5 or 6) the induced angles of attack are not accurately predicted by the elementary equation for α_i and the graph of C_L versus α develops distinct curvature. This effect is especially true at high lift coefficients where the lift curve for the very low aspect ratio wing is very shallow and $C_{L_{max}}$ and stall angle of attack are less sharply defined.

The effect of aspect ratio on wing drag characteristics may be appreciated from inspection of figure 1.31. The basic section properties are shown as the drag characteristics of an infinite aspect ratio wing. When a planform of some finite aspect ratio is constructed, the wing drag coefficient is the *sum* of the induced drag coefficient, $C_{D_i}=0.318\frac{C_L^2}{AR}$, and the section drag coefficient. Decreasing aspect ratio increases the wing drag coefficient at any lift coefficient since the induced drag coefficient varies inversely with aspect ratio. When the aspect ratio is very low, the induced drag varies greatly with lift and at high lift coefficients, the induced drag is very high and increases very rapidly with lift coefficient.

While the effect of aspect ratio on lift curve slope and drag due to lift is an important relationship, it must be realized that design for

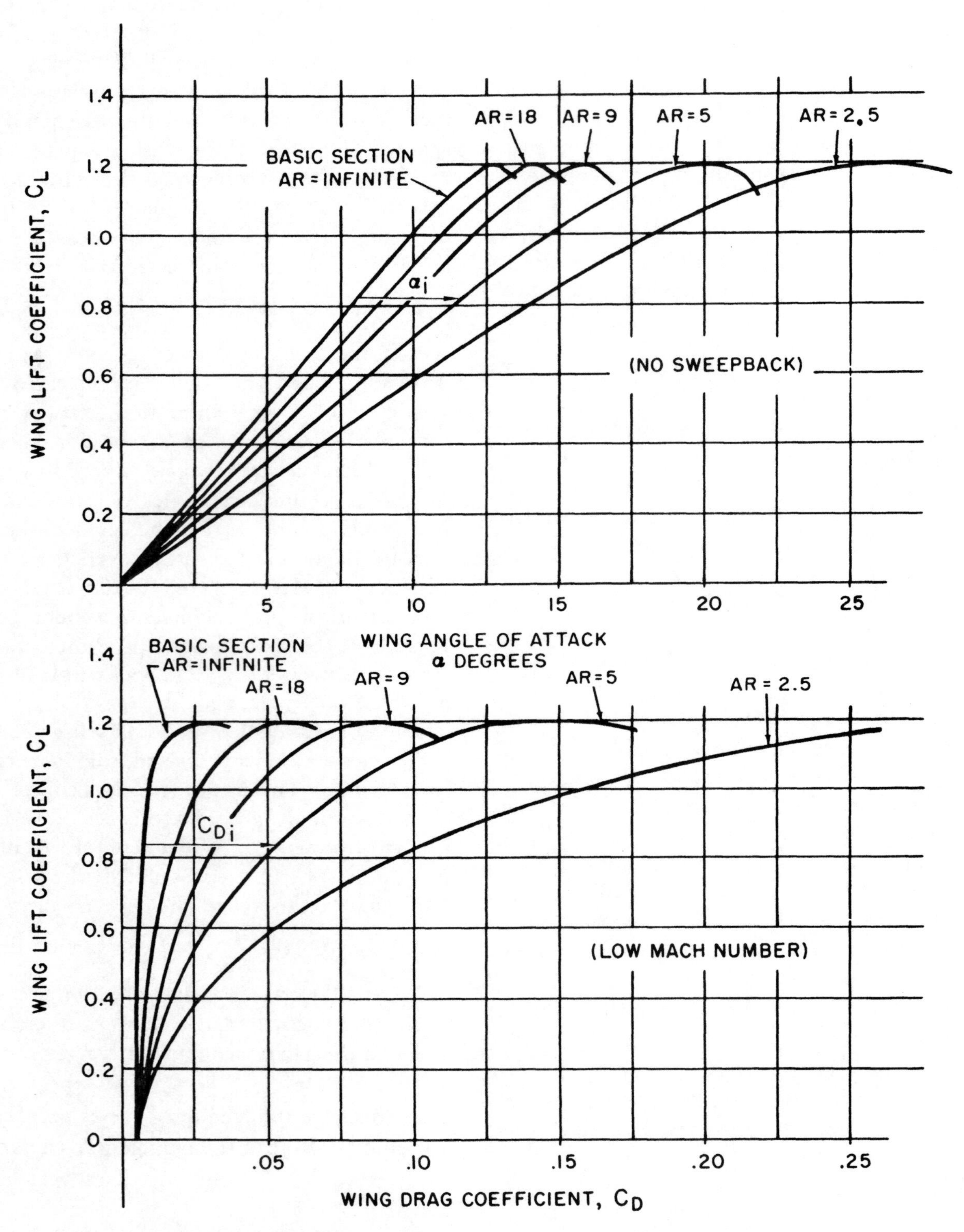

Figure 1.31. Effect of Aspect Ratio on Wing Characteristics

very high speed flight does not favor the use of high aspect ratio planforms. Low aspect ratio planforms have structural advantages and allow the use of thin, low drag sections for high speed flight. The aerodynamics of transonic and supersonic flight also favor short span, low aspect ratio surfaces. Thus, the modern configuration of airplane designed for high speed flight will have a low aspect ratio planform with characteristic aspect ratios of two to four. The most important impression that should result is that the typical modern configuration will have high angles of attack for maximum lift and very prodigious drag due to lift at low flight speeds. This fact is of importance to the Naval Aviator because the majority of pilot-caused accidents occur during this regime of flight—during takeoff, approach, and landing. Induced drag predominates in these regimes of flight.

The modern configuration of high speed airplane usually has a low aspect ratio planform with high wing loading. When wing sweepback is coupled with low aspect ratio, the wing lift curve has distinct curvature and is very flat at high angles of attack, i.e., at high C_L, C_L increases very slowly with an increase in α. In addition, the drag curve shows extremely rapid rise at high lift coefficients since the drag due to lift is so very large. These effects produce flying qualities which are distinctly different from a more "conventional" high aspect ratio airplane configuration.

Some of the most important ramifications of the modern high speed configuration are:

(1) During *takeoff* where the airplane must not be over-rotated to an excessive angle of attack. Any given airplane will have some fixed angle of attack (and C_L) which produces the best takeoff performance and this angle of attack will not vary with weight, density altitude, or temperature. An excessive angle of attack produces additional induced drag and may have an undesirable effect on takeoff performance. Takeoff acceleration may be seriously reduced and a large increase in takeoff distance may occur. Also, the initial climb performance may be marginal at an excessively low airspeed. There are modern configurations of airplanes of very low aspect ratio (plus sweepback) which—if over-rotated during a high altitude, high gross weight takeoff—cannot fly out of ground effect. With the more conventional airplane configuration, an excess angle of attack produces a well defined stall. However, the modern airplane configuration at an excessive angle of attack has no sharply defined stall but developes an excessive amount of induced drag. To be sure that it will not go unsaid, an excessively low angle of attack on takeoff creates its own problems—excess takeoff speed and distance and critical tire loads.

(2) During *approach* where the pilot must exercise proper technique to control the flight path. "Attitude plus power equals performance." The modern high speed configuration at low speeds will have low lift-drag ratios due to the high induced drag and can require relatively high power settings during the power approach. If the pilot interprets that his airplane is below the desired glide path, his first reaction *must not* be to just ease the nose up. An increase in angle of attack without an increase in power will lower the airspeed and greatly increase the induced drag. Such a reaction could create a high rate of descent and lead to very undesirable consequences. The angle of attack indicator coupled with the mirror landing system provides reference to the pilot and emphasizes that during the steady approach "angle of attack is the primary control of airspeed and power is the primary control of rate of climb or descent." Steep turns during approach at low airspeed are always undesirable in any type of airplane because of the increased stall speed and induced drag. Steep turns at low airspeeds in a low aspect ratio airplane can create extremely high induced drag and can incur dangerous sink rates.

(3) During the *landing* phase where an excessive angle of attack (or excessively low airspeed) would create high induced drag and a high power setting to control rate of descent. A common error in the technique of landing modern configurations is a steep, low power approach to landing. The steep flight path requires considerable maneuver to flare the airplane for touchdown and necessitates a definite increase in angle of attack. Since the maneuver of the flare is a transient condition, the variation of both lift and drag with angle of attack must be considered. The lift and drag curves for a high aspect ratio wing (fig. 1.31) show continued strong increase in C_L with α up to stall and large changes in C_D only at the point of stall. These characteristics imply that the high aspect ratio airplane is usually capable of flare without unusual results. The increase in angle of attack at flare provides the increase in lift to change the flight path direction without large changes in drag to decelerate the airplane.

The lift and drag curves for a low aspect ratio wing (fig. 1.31) show that at high angles of attack the lift curve is shallow, i.e., small changes in C_L with increased α. This implies a large rotation needed to provide the lift to flare the airplane from a steep approach. The drag curve for the low aspect ratio wing shows large, powerful increases in C_D with C_L well below the stall. These lift and drag characteristics of the low aspect ratio wing create a distinct change in the flare characteristics. If a flare is attempted from a steep approach at low airspeed, the increased angle of attack may provide such increased induced drag and rapid loss of airspeed that the airplane does not actually flare. A possible result is that an even higher sink rate may be incurred. This is one factor favoring the use of the "no-flare" or "minimum flare" type landing technique for certain modern configurations. These same aerodynamic properties set the best glide speeds of low aspect ratio airplanes above the speed for $(L/D)_{max}$. The additional speed provides a more favorable margin of flare capability for flameout landing from a steep glide path (low aspect ratio, low $(L/D)_{max}$, low glide ratio).

The landing technique must emphasize proper control of angle of attack and rate of descent to prevent high sink rates and hard landings. As before, to be sure that it will not go unsaid, excessive airspeed at landing creates its own problems—excessive wear and tear on tires and brakes, excessive landing distance, etc.

The effect of the low aspect ratio planform of modern airplanes emphasizes the need for proper flying techniques at low airspeeds. Excessive angles of attack create enormous induced drag which can hinder takeoff performance and incur high sink rates at landing. Since such aircraft have intrinsic high minimum flying speeds, an excessively low angle of attack at takeoff or landing creates its own problems. These facts underscore the importance of a "thread-the-needle," professional flying technique.

EFFECT OF TAPER AND SWEEPBACK

The aspect ratio of a wing is the primary factor in determining the three-dimensional characteristics of the ordinary wing and its drag due to lift. However, certain local effects take place throughout the span of the wing and these effects are due to the distribution of area throughout the span. The distribution of lift along the span of a wing cannot have sharp discontinuities. (Nature just doesn't arrange natural forces with sharp discontinuities.) The typical lift distribution is arranged in some elliptical fashion. A representative distribution of the lift per foot of span along the span of a wing is shown in figure 1.32.

The natural distribution of lift along the span of a wing provides a basis for appreciating the effect of area distribution and taper along the span. If the elliptical lift distribution is

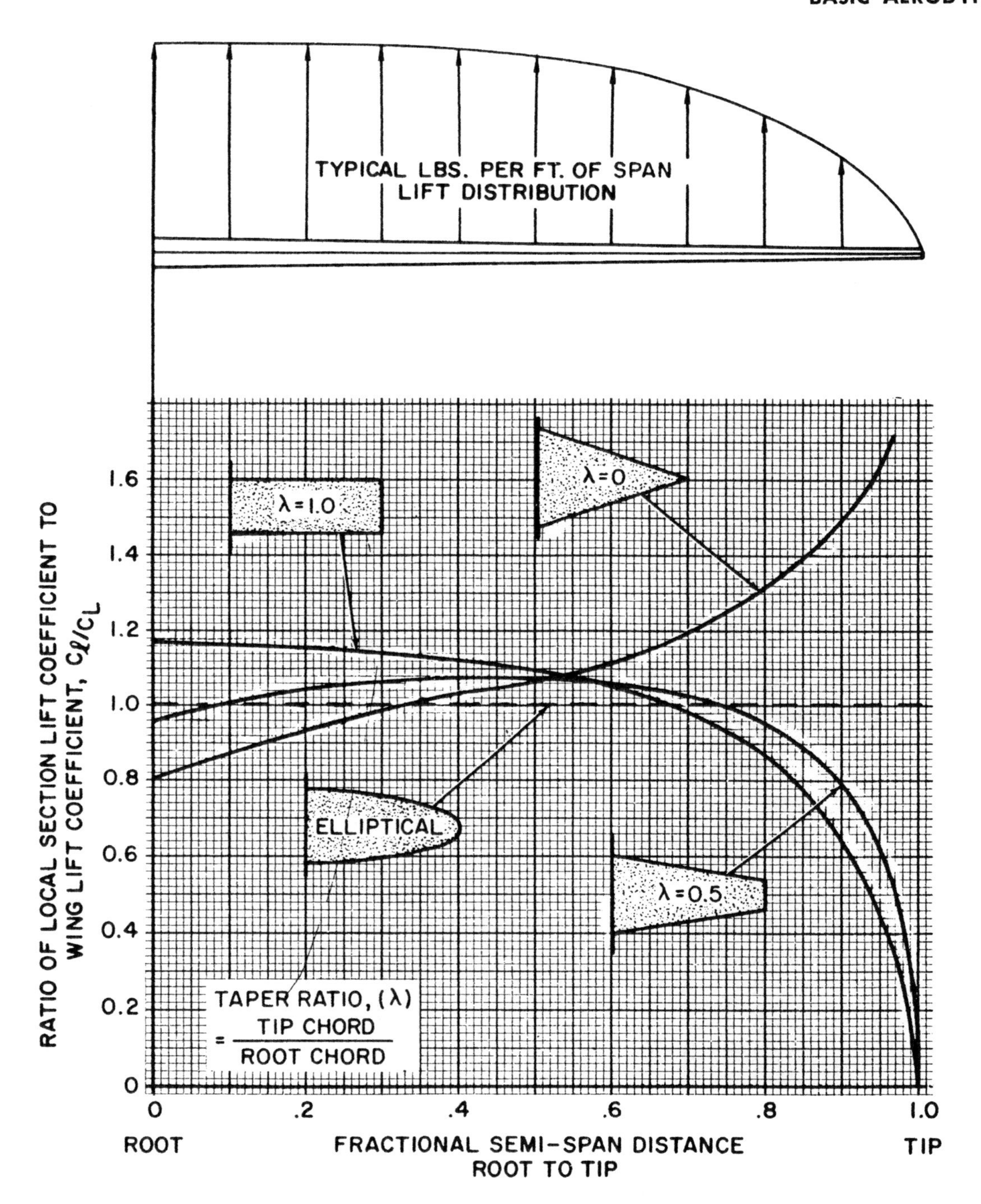

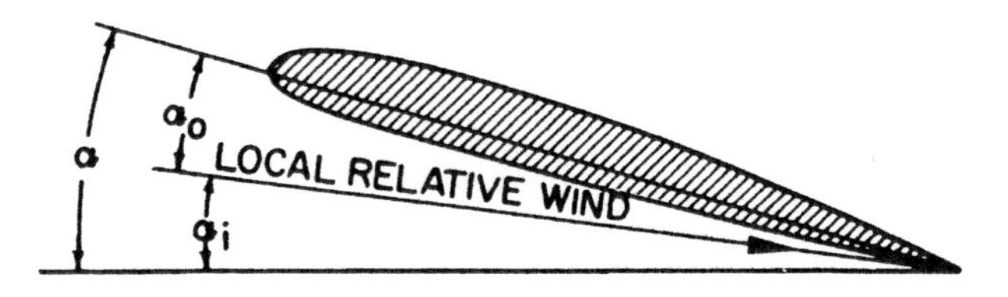

Figure 1.32. Spanwise Lift Distribution

matched with a planform whose chord is distributed in an elliptical fashion (the elliptical wing), each square foot of area along the span produces exactly the same lift pressure. The elliptical wing planform then has each section of the wing working at exactly the same local lift coefficient and the induced downflow at the wing is uniform throughout the span. In the aerodynamic sense, the elliptical wing is the most efficient planform because the uniformity of lift coefficient and downwash incurs *the least induced drag for a given aspect ratio*. The merit of any wing planform is then measured by the closeness with which the distribution of lift coefficient and downwash approach that of the elliptical planform.

The effect of the elliptical planform is illustrated in figure 1.32 by the plot of local lift coefficient to wing lift coefficient, $\frac{c_l}{C_L}$, versus semispan distance. The elliptical wing produces a constant value of $\frac{c_l}{C_L}=1.0$ throughout the span from root to tip. Thus, the local section angle of attack, α_0, and local induced angle of attack, α_i, are constant throughout the span. If the planform area distribution is anything other than elliptical, it may be expected that the local section and induced angles of attack will not be constant along the span.

A planform previously considered is the simple rectangular wing which has a taper ratio of 1.0. A characteristic of the rectangular wing is a strong vortex at the tip with local downwash behind the wing which is high at the tip and low at the root. This large nonuniformity in downwash causes similar variation in the local induced angles of attack along the span. At the tip, where high downwash exists, the local induced angle of attack is greater than the average for the wing. Since the wing angle of attack is composed of the sum of α_i and α_0, a large local α_i reduces the local α_0 creating low local lift coefficients at the tip. The reverse is true at the root of the rectangular wing where low local downwash exists. This situation creates an induced angle of attack at the root which is less than the average for the wing and a local section angle of attack higher than the average for the wing. The result is shown by the graph of figure 1.32 which depicts a local lift coefficient at the root almost 20 percent greater than the wing lift coefficient.

The effect of the rectangular planform may be appreciated by matching a near elliptical lift distribution with a planform with a constant chord. The chords near the tip develop less lift pressure than the root and consequently have lower section lift coefficients. The great nonuniformity of local lift coefficient along the span implies that some sections carry more than their share of the load while others carry less than their share of the load. Hence, for a given aspect ratio, the rectangular planform will be less efficient than the elliptical wing. For example, a rectangular wing of $AR=6$ would have 16 percent higher induced angle of attack for the wing and 5 percent higher induced drag than an elliptical wing of the same aspect ratio.

At the other extreme of taper is the pointed wing which has a taper ratio of zero. The extremely small parcel of area at the pointed tip is not capable of holding the main tip vortex at the tip and a drastic change in downwash distribution results. The pointed wing has greatest downwash at the root and this downwash decreases toward the tip. In the immediate vicinity of the pointed tip, an *upwash* is encountered which indicates that negative induced angles of attack exist in this area. The resulting variation of local lift coefficient shows low c_l at the root and very high c_l at the tip. This effect may be appreciated by realizing that the wide chords at the root produce low lift pressures while the very narrow chords toward the tip are subject to very high lift pressures. The variation of $\frac{c_l}{C_L}$ throughout the span of the wing of taper ratio$=0$ is shown on the graph of figure

1.32. As with the rectangular wing, the nonuniformity of downwash and lift distribution result in inefficiency of this planform. For example, a pointed wing of $AR=6$ would have 17 percent higher induced angle of attack for the wing and 13 percent higher induced drag than an elliptical wing of the same aspect ratio.

Between the two extremes of taper will exist planforms of more tolerable efficiency. The variations of $\frac{c_l}{C_L}$ for a wing of taper ratio $=0.5$ closely approximates the lift distribution of the elliptical wing and the drag due to lift characteristics are nearly identical. A wing of $AR=6$ and taper ratio$=0.5$ has only 3 percent higher α_i and 1 percent greater C_{D_i} than an elliptical wing of the same aspect ratio.

A separate effect on the spanwise lift distribution is contributed by wing sweepback. Sweepback of the planform tends to alter the lift distribution similar to decreasing the taper ratio. Also, large sweepback tends to increase induced drag.

The elliptical wing is the ideal of the subsonic aerodynamic planform since it provides a minimum of induced drag for a given aspect ratio. However, the major objection to the elliptical planform is the extreme difficulty of mechanical layout and construction. A highly tapered planform is desirable from the standpoint of structural weight and stiffness and the usual wing planform may have a taper ratio from 0.45 to 0.20. Since structural considerations are quite important in the development of an airplane configuration, the tapered planform is a necessity for an efficient configuration. In order to preserve the aerodynamic efficiency, the resulting planform is tailored by wing twist and section variation to obtain as near as possible the elliptic lift distribution.

STALL PATTERNS

An additional effect of the planform area distribution is on stall pattern of wing. The desirable stall pattern of any wing is a stall which begins on the root sections first. The advantages of root stall first are that ailerons remain effective at high angles of attack, favorable stall warning results from the buffet on the empennage and aft portion of the fuselage, and the loss of downwash behind the root usually provides a stable nose down moment to the airplane. Such a stall pattern is favored but may be difficult to obtain with certain wing configurations. The types of stall patterns inherent with various planforms are illustrated in figure 1.33. The various planform effects are separated as follows:

(A) The elliptical planform has constant local lift coefficients throughout the span from root to tip. Such a lift distribution means that all sections will reach stall at essentially the same wing angle of attack and stall will begin and progress uniformly throughout the span. While the elliptical wing would reach high lift coefficients before incipient stall, there would be little advance warning of complete stall. Also, the ailerons may lack effectiveness when the wing operates near the stall and lateral control may be difficult.

(B) The lift distribution of the rectangular wing exhibits low local lift coefficients at the tip and high local lift coefficients at the root. Since the wing will initiate stall in the area of highest local lift coefficients, the rectangular wing is characterized by a strong root stall tendency. Of course, this stall pattern is favorable since there is adequate stall warning buffet, adequate aileron effectiveness, and usually strong stable moment changes on the airplane. Because of the great aerodynamic and structural inefficiency of this planform, the rectangular wing finds limited application only to low cost, low speed light planes. The simplicity of construction and favorable stall characteristics are predominating requirements of such an airplane. The stall sequence for a rectangular wing is shown by the tuft-grid pictures. The progressive flow separation illustrates the strong root stall tendency.

(C) The wing of moderate taper (taper ratio$=0.5$) has a lift distribution which closely

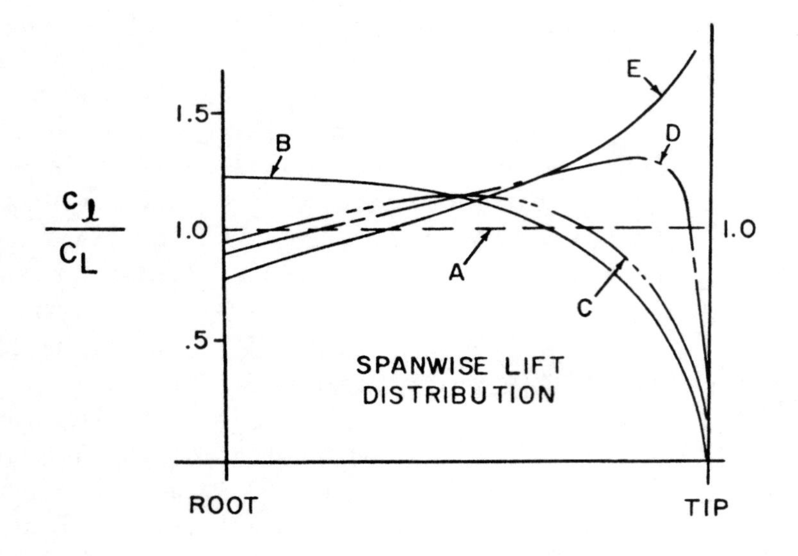

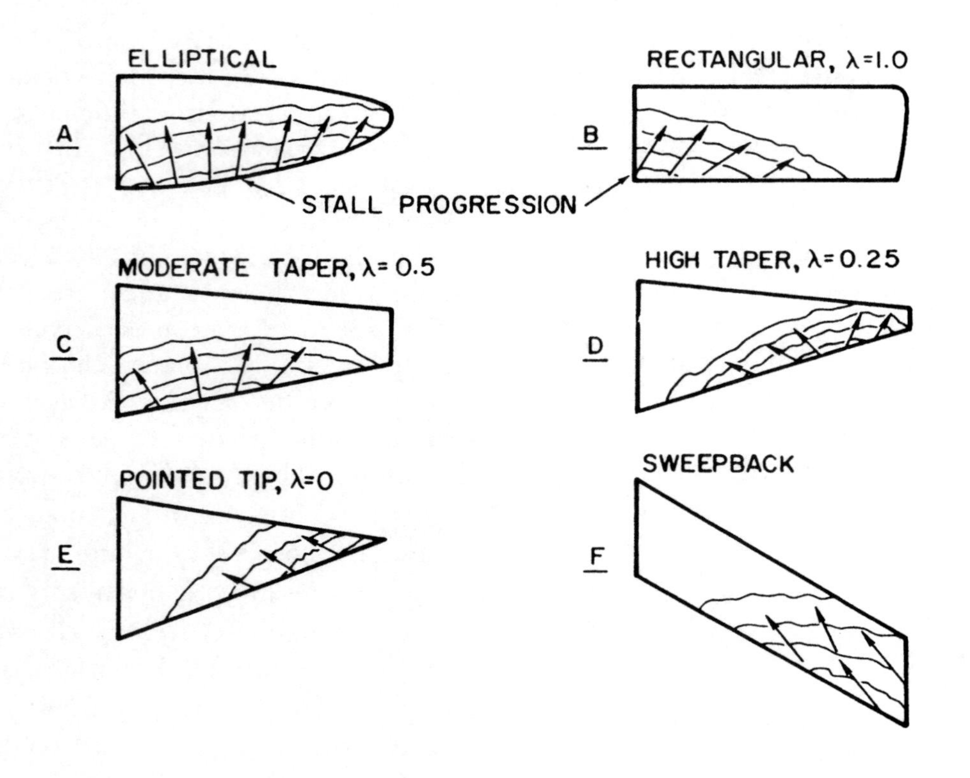

Figure 1.33. Stall Patterns (sheet 1 of 8)

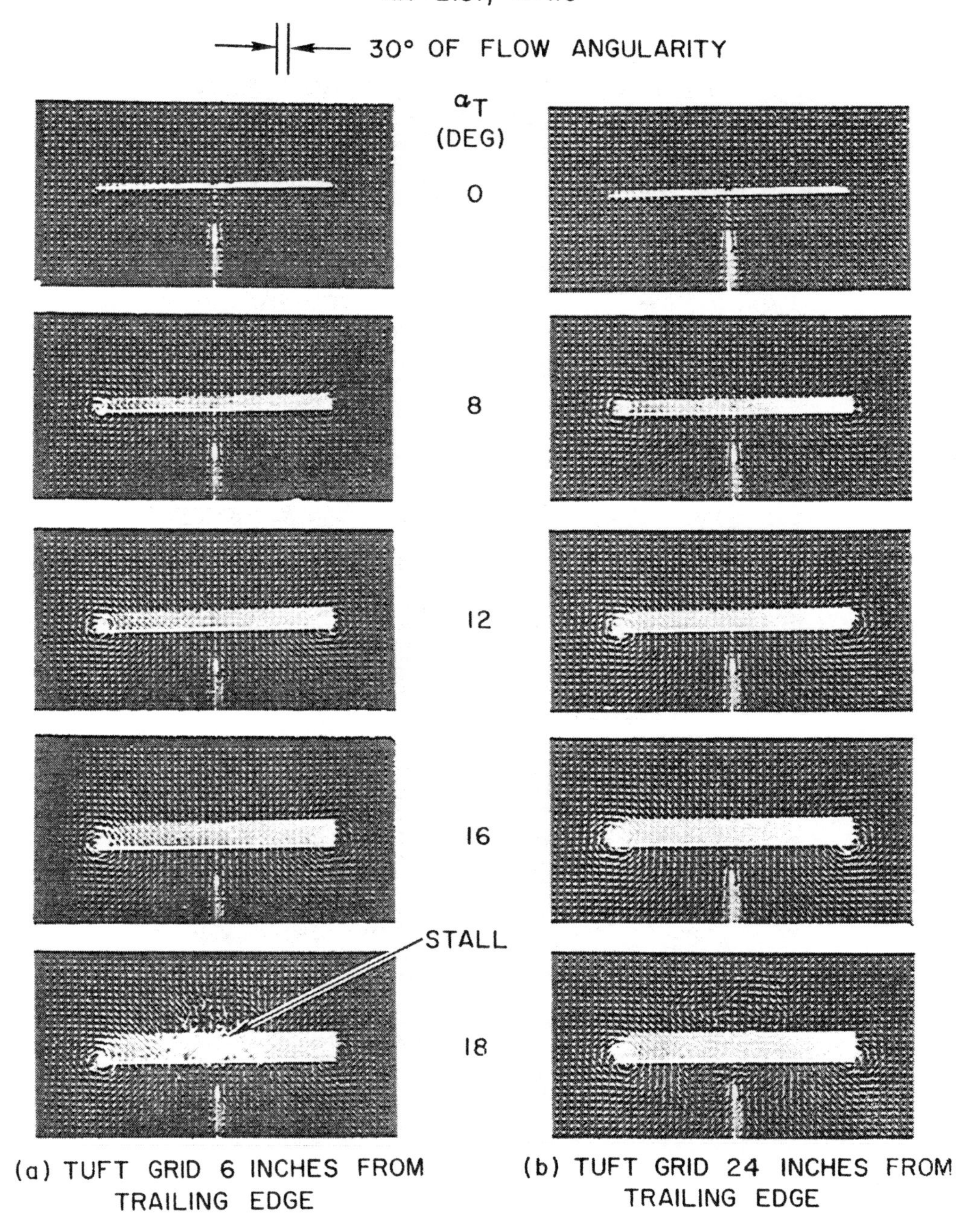

Figure 1.33. Stall Patterns (sheet 2 of 8)

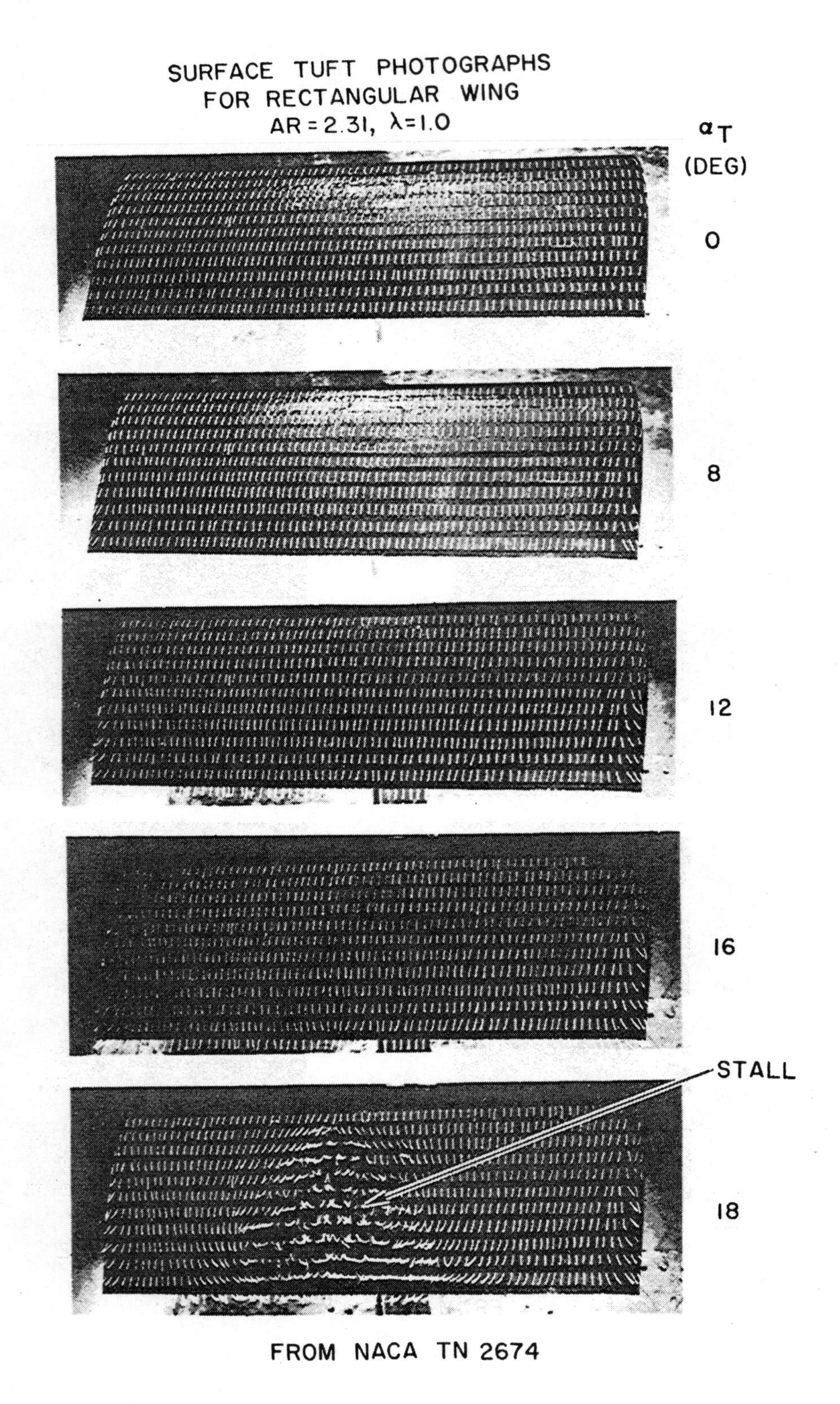

Figuse 1.33. Stall Patterns (sheet 3 of 8)

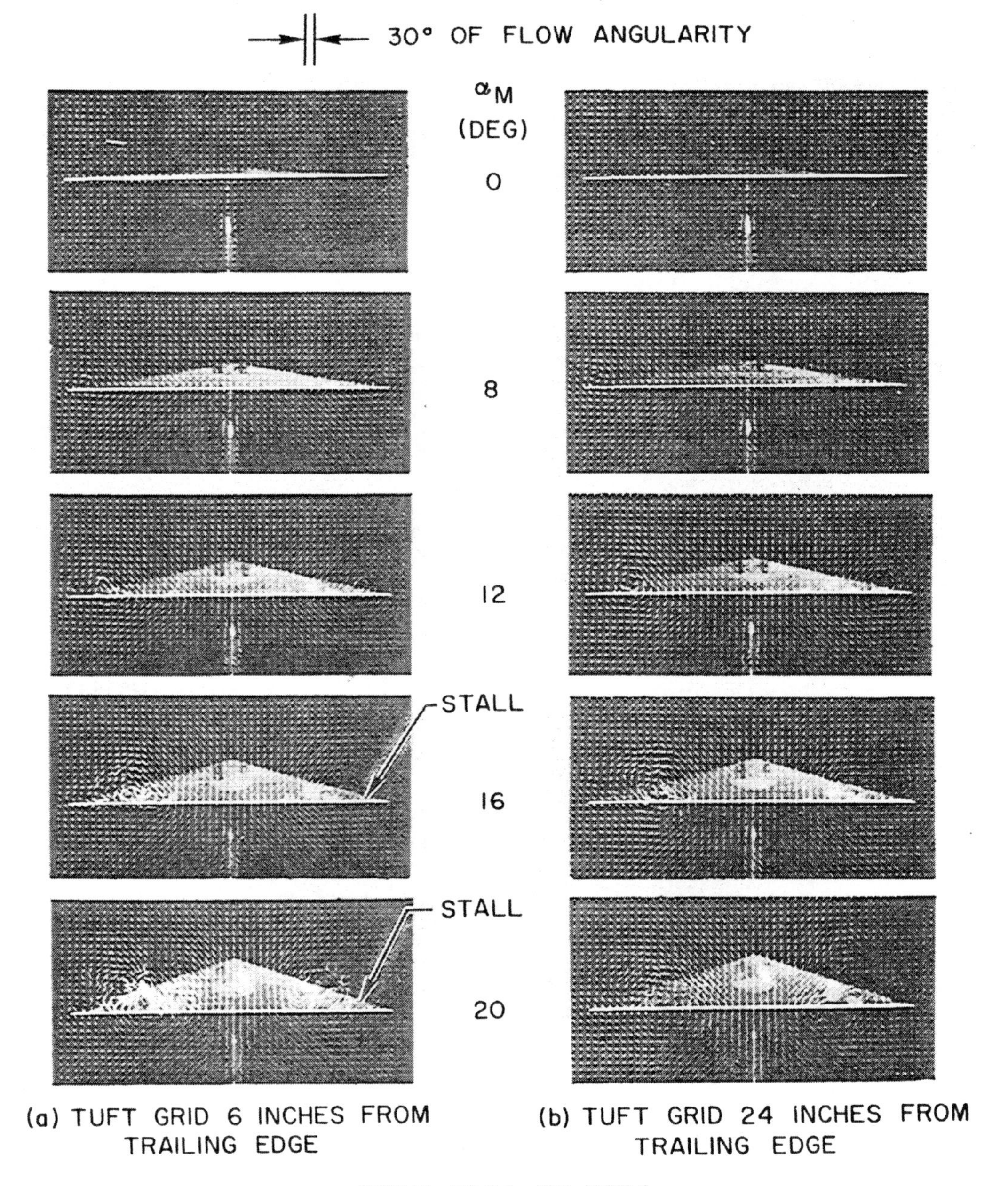

Figure 1.33. Stall Patterns (sheet 4 of 8)

SURFACE TUFT PHOTOGRAPHS
FOR A SWEPT, TAPERED WING
45° DELTA, AR=4.0, λ=0

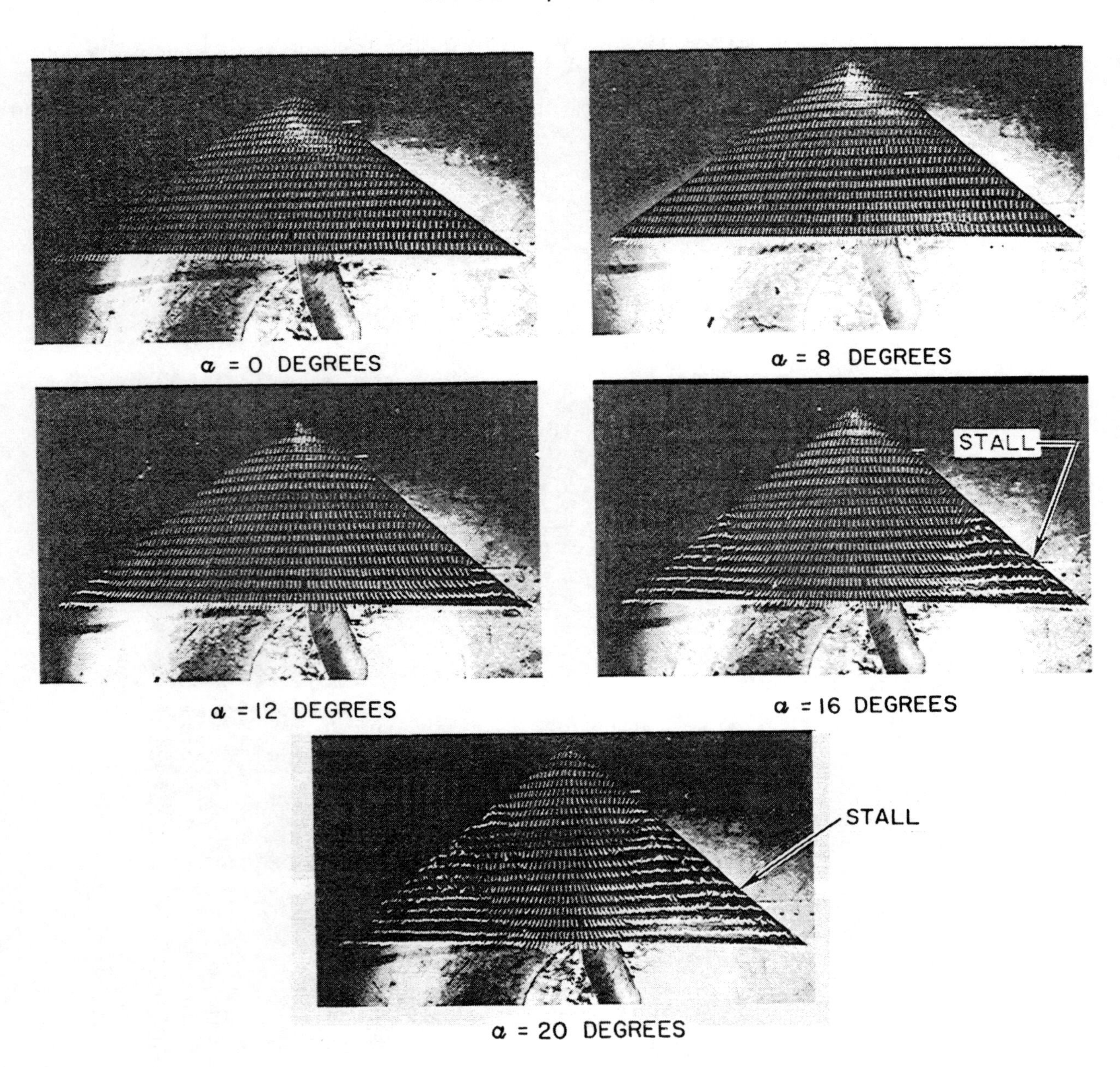

FROM NACA TN 2674

Figure 1.33. Stall Patterns (sheet 5 of 8)

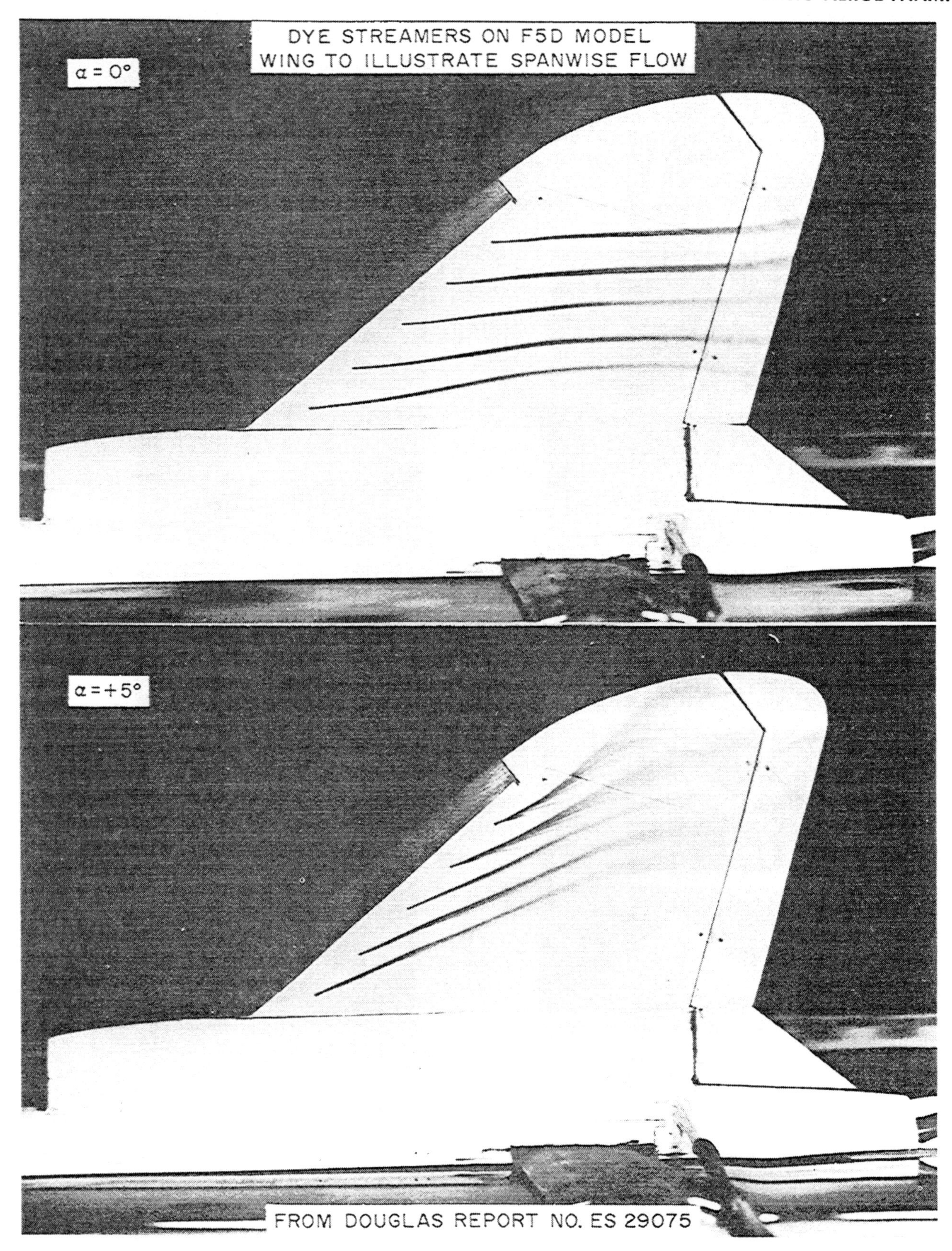

Figure 1.33. Stall Patterns (sheet 6 of 8)

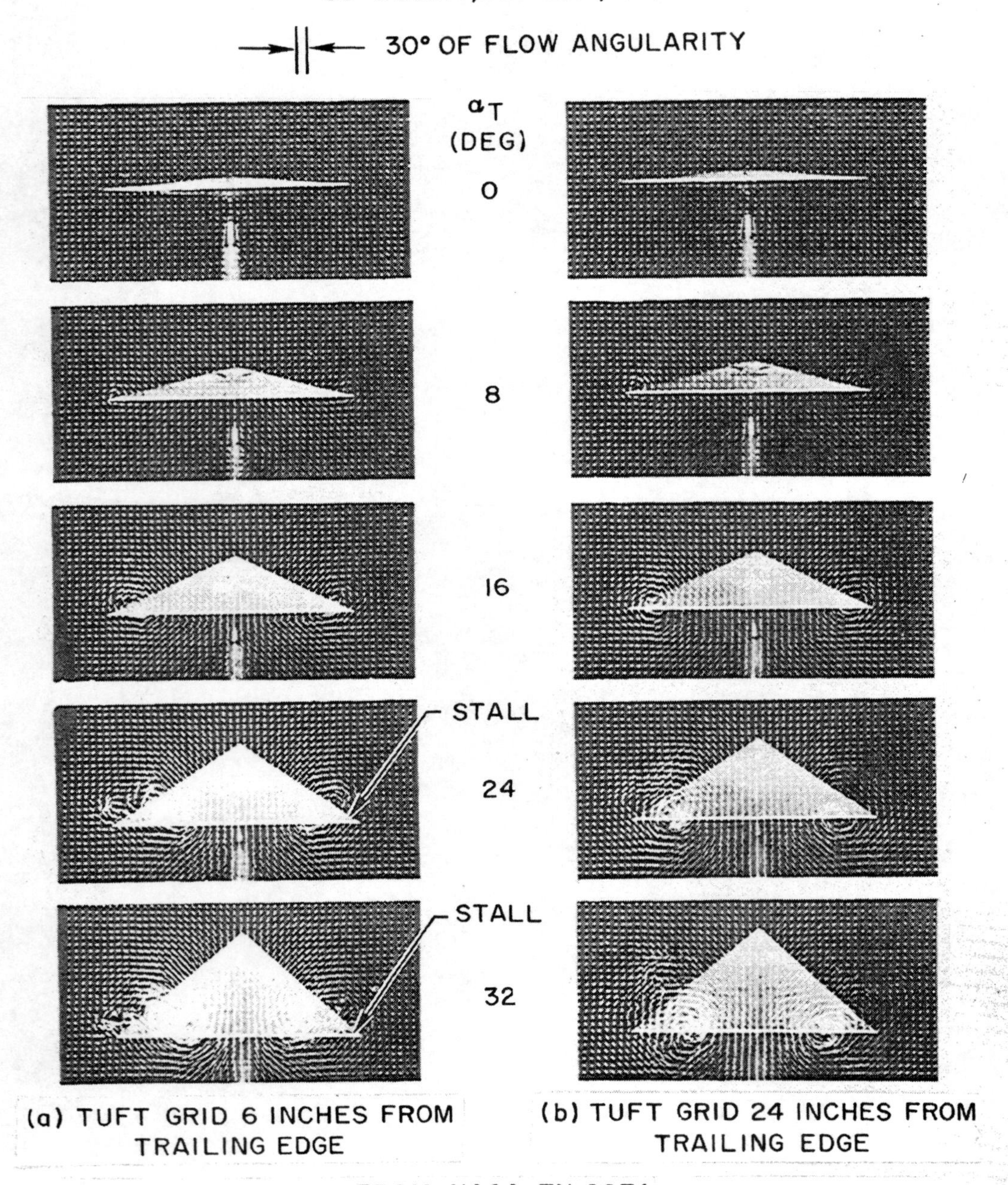

Figure 1.33. Stall Patterns (sheet 7 of 8)

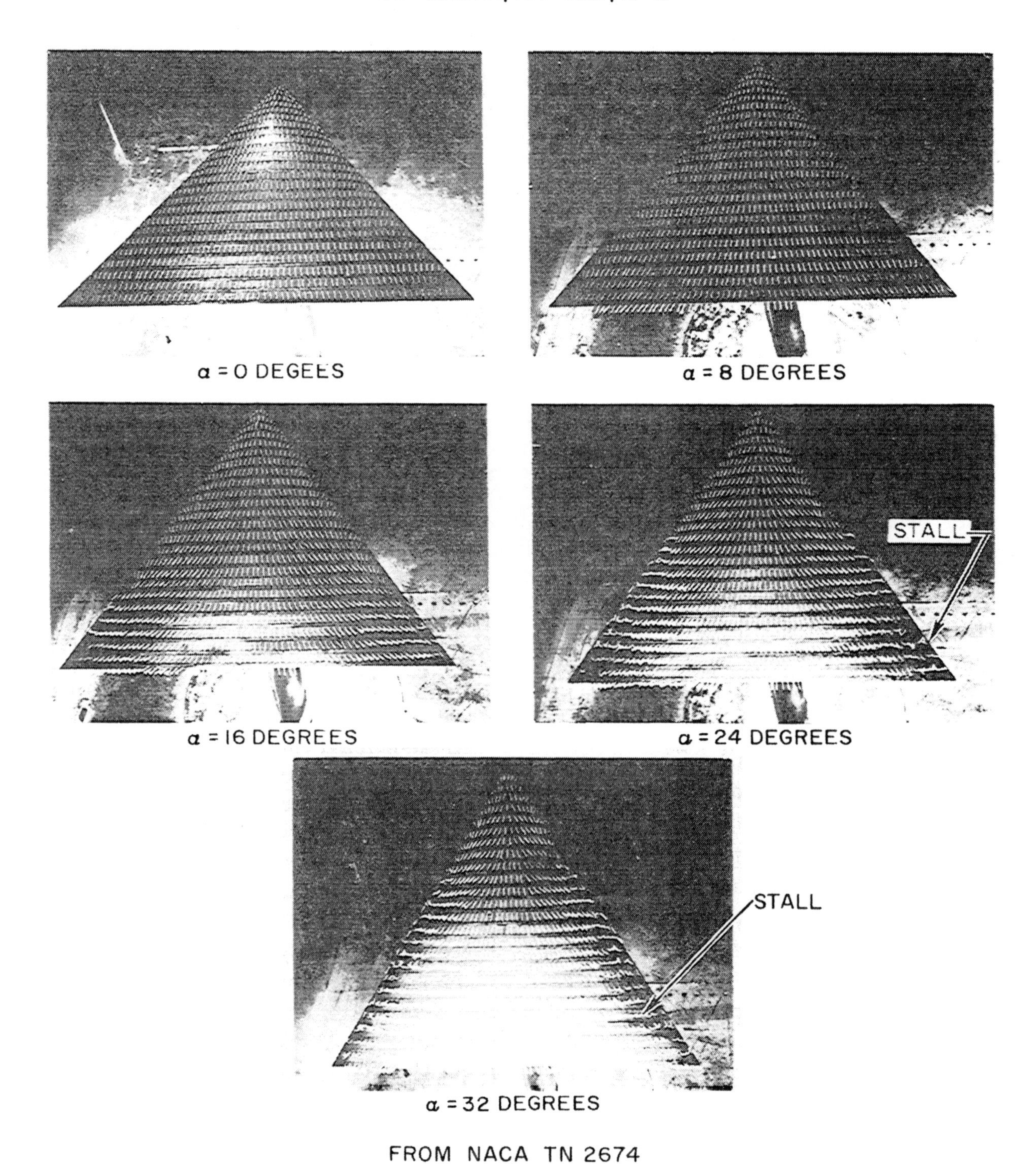

Figure 1.33. Stall Patterns (sheet 8 of 8)

approximates that of the elliptical wing. Hence, the stall pattern is much the same as the elliptical wing.

(D) The highly tapered wing of taper ratio=0.25 shows the stall tendency inherent with high taper. The lift distribution of such a wing has distinct peaks just inboard from the tip. Since the wing stall is started in the vicinity of the highest local lift coefficient, this planform has a strong "tip stall" tendency. The initial stall is not started at the exact tip but at the station inboard from the tip where highest local lift coefficients prevail. If an actual wing were allowed to stall in this fashion the occurrence of stall would be typified by aileron buffet and wing drop. There would be no buffet of the empennage or aft fuselage, no strong nose down moment, and very little—if any—aileron effectiveness. In order to prevent such undesirable happenings, the wing must be tailored to favor the stall pattern. The wing may be given a geometric twist or "washout" to decrease the local angles of attack at the tip. In addition, the airfoil section may be varied throughout the span such that sections with greater thickness and camber are located in the areas of highest local lift coefficients. The higher $c_{l_{max}}$ of such sections can then develop the higher local c_l's and be less likely to stall. The addition of leading edge slots or slats toward the tip increase the local $c_{l_{max}}$ and stall angle of attack and are useful in allaying tip stall and loss of aileron effectiveness. Another device for improving the stall pattern would be the *forcing* of stall in the desired location by *decreasing* the section $c_{l_{max}}$ in this vicinity. The use of sharp leading edges or "stall strips" is a powerful device to control the stall pattern.

(E) The pointed tip wing of taper ratio equal to zero develops extremely high local lift coefficients at the tip. For all practical purposes, the pointed tip will be stalled at any condition of lift unless extensive tailoring is applied to the wing. Such a planform has no practical application to an airplane which is definitely subsonic in performance.

(F) Sweepback applied to a wing planform alters the lift distribution similar to decreasing taper ratio. Also, a predominating influence of the swept planform is the tendency for a strong crossflow of the boundary layer at high lift coefficients. Since the outboard sections of the wing trail the inboard sections, the outboard suction pressures tend to draw the boundary layer toward the tip. The result is a thickened low energy boundary layer at the tips which is easily separated. The development of the spanwise flow in the boundary layer is illustrated by the photographs of figure 1.33. Note that the dye streamers on the upper surface of the swept wing develop a strong spanwise crossflow at high angles of attack. Slots, slats, and flow fences help to allay the strong tendency for spanwise flow.

When sweepback and taper are combined in a planform, the inherent tip stall tendency is considerable. If tip stall of any significance is allowed to occur on the swept wing, an additional complication results: the forward shift in the wing center of pressure creates an unstable nose up pitching moment. The stall sequence of a swept, tapered wing is indicated by the tuft-grid photographs of figure 1.33.

An additional effect on sweepback is the reduction in the slope of the lift curve and maximum lift coefficient. When the sweepback is large and combined with low aspect ratio the lift curve is very shallow and maximum lift coefficient can occur at tremendous angles of attack. The lift curve of one typical low aspect ratio, highly tapered, swept wing airplane depicts a maximum lift coefficient at approximately 45° angle of attack. Such drastic angles of attack are impractical in many respects. If the airplane is operated at such high angles of attack an extreme landing gear configuration is required, induced drag is extremely high, and the stability of the airplane may seriously deteriorate. Thus, the modern configuration of airplane may have "minimum control speeds" set by these factors rather than simple stall speeds based on $C_{L_{max}}$.

When a wing of a given planform has various high lift devices added, the lift distribution and stall pattern can be greatly affected. Deflection of trailing edge flaps increases the local lift coefficients in the flapped areas and since the stall angle of the flapped section is decreased, initial stall usually begins in the flapped area. The extension of slats simply allows the slatted areas to go to higher lift coefficients and angles of attack and generally delays stall in that vicinity. Also, power effects may adversely affect the stall pattern of the propeller powered airplane. When the propeller powered airplane is at high power and low speed, the flow induced at the wing root by the slipstream may cause considerable delay in the stall of the root sections. Hence, the propeller powered airplane may have its most undesirable stall characteristics during the power-on stall rather than the power-off stall.

PARASITE DRAG

In addition to the drag caused by the development of lift (induced drag) there is the obvious drag which is *not* due to the development of lift. A wing surface even at zero lift will have "profile" drag due to skin friction and form. The other components of the airplane such as the fuselage, tail, nacelles, etc., contribute to drag because of their own form and skin friction. Any loss of momentum of the airstream due to powerplant cooling, air conditioning, or leakage through construction or access gaps is, in effect, an additional drag. When the various components of the airplane are put together the total drag will be greater than the sum of the individual components because of "interference" of one surface on the other.

The most usual interference of importance occurs at the wing-body intersection where the growth of boundary layer on the fuselage reduces the boundary layer velocities on the wing root surface. This reduction in energy allows the wing root boundary layer to be more easily separated in the presence of an adverse pressure gradient. Since the upper wing surface has the more critical pressure gradients, a low wing position on a circular fuselage would create greater interference drag than a high wing position. Adequate filleting and control of local pressure gradients is necessary to minimize such additional drag due to interference.

The sum of all the drags due to form, friction, leakage and momentum losses, and interference drag is termed "parasite" drag since it is not directly associated with the development of lift. While this parasite drag is not *directly* associated with the production of lift it is a variable with lift. The variation of parasite drag coefficient, C_{D_P}, with lift coefficient, C_L, is shown for a typical airplane in figure 1.34. The minimum parasite drag coefficient, $C_{D_{P_{min}}}$, usually occurs at or near zero lift and parasite drag coefficient increases above this point in a smooth curve. The induced drag coefficient is shown on the same graph for purposes of comparison since the total drag of the airplane is a sum of the parasite and induced drag.

In many parts of airplane performance it is necessary to completely distinguish between drag due to lift and drag not due to lift. The total drag of an airplane is the sum of the parasite and induced drags.

$$C_D = C_{D_P} + C_{D_i}$$

where

C_D = airplane drag coefficient

C_{D_P} = parasite drag coefficient

C_{D_i} = induced drag coefficient

$$= 0.318 \frac{C_L^2}{AR}$$

From inspection of figure 1.34 it is seen that both C_{D_P} and C_{D_i} vary with lift coefficient. However, the usual variation of parasite drag allows a simple correlation with the induced drag term. In effect, the part of parasite drag above the minimum at zero lift can be "lumped"

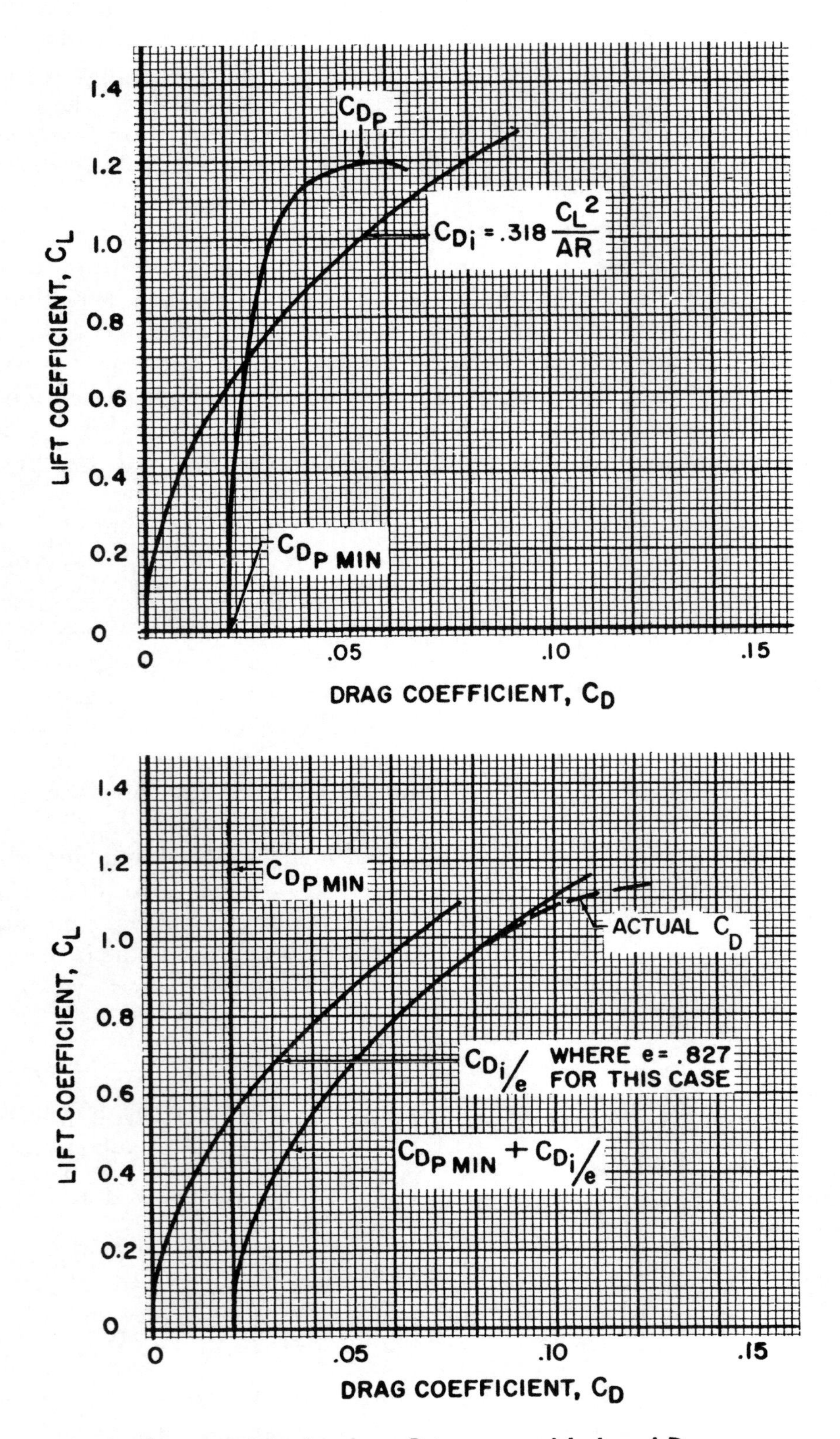

Figure 1.34. Airplane Parasite and Induced Drag

in with the induced drag coefficient by a constant factor which is defined as the "airplane efficiency factor", e. By this method of accounting the airplane drag coefficient is expressed as:

$$C_D = C_{D_{P_{min}}} + \frac{C_{D_i}}{e}$$

$$C_D = C_{D_{P_{min}}} + 0.318 \left(\frac{C_L^2}{ARe}\right)$$

where

$C_{D_{P_{min}}}$ = minimum parasite drag coefficient

C_{D_i} = induced drag coefficient

e = airplane efficiency factor

In this form, the airplane drag coefficient is expressed as the sum of drag not due to lift $(C_{D_{P_{min}}})$ and drag due to lift $(\frac{C_{D_i}}{e})$. The airplane efficiency factor is some constant (usually less than unity) which includes parasite drag due to lift with the drag induced by lift. $C_{D_{P_{min}}}$ is invariant with lift and represents the parasite drag at zero lift. A typical value of $C_{D_{P_{min}}}$ would be 0.020, of which the wing may account for 50 percent, the fuselage and nacelles 40 percent, and the tail 10 percent. The term of $\left(0.318 \frac{C_L^2}{ARe}\right)$ accounts for all drag due to lift—the drag induced by lift and the extra parasite drag due to lift. Typical values of the airplane efficiency factor range from 0.6 to 0.9 depending on the airplane configuration and its characteristics. While the term of drag due to lift does include some parasite drag, it is still generally referred to as induced drag.

The second graph of figure 1.34 shows that the sum of $C_{D_{P_{min}}}$ and $\frac{C_{D_i}}{e}$ can approximate the actual airplane C_D through a large range of lift coefficients. For airplanes of moderate aspect ratio, this representation of the airplane total drag is quite accurate in the ordinary range of lift coefficients up to near 70 percent of $C_{L_{max}}$. At high lift coefficients near $C_{L_{max}}$, the procedure is not too accurate because of the sharper variation of parasite drag at high angles of attack. In a sense, the airplane efficiency factor would change from the constant value and decrease. The deviation of the actual airplane drag from the approximating curve is quite noticeable for airplanes with low aspect ratio and sweepback. Another factor to consider is the effect of compressibility. Since compressibility effects would destroy this relationship, the greatest application is for subsonic performance analysis.

The total airplane drag is the sum of the parasite and induced drags.

$$D = D_p + D_i$$

where

D_i = induced drag

$$= \left(0.318 \frac{C_L^2}{ARe}\right) qS$$

and

D_p = parasite drag

$$= C_{D_{P_{min}}} qS$$

When expressed in this form the induced drag, D_i, includes all drags due to lift and is solely a function of lift. The parasite drag, D_p, is the parasite drag and is completely independent of lift—it could be called the "barn door" drag of the airplane.

An alternate expression for the parasite drag is:

$$D_p = fq$$

where

f = equivalent parasite area, sq. ft.

$f = C_{D_{P_{min}}} S$

q = dynamic pressure, psf

$$= \frac{\sigma V^2}{295}$$

or

$$D_p = \frac{f \sigma V^2}{295}$$

In this form, the equivalent parasite area, f, is the product of $C_{D_{P_{min}}}$ and S and relates an

NAVY

impression of the "barn door" size. Hence, parasite drag can be appreciated as the result of the dynamic pressure, q, acting on the equivalent parasite area, f. The "equivalent" parasite area is defined by this relationship as a hypothetical surface with a $C_D=1.0$ which produces the same parasite drag as the airplane. An analogy would be a barn door in the airstream which is equivalent to the airplane. Typical values for the equivalent parasite area range from 4 sq. ft. for a clean fighter type airplane to 40 sq. ft. for a large transport type airplane. Of course, when any airplane is changed from the clean configuration to the landing configuration, the equivalent parasite area increases.

EFFECT OF CONFIGURATION. The parasite drag, D_p, is unaffected by lift, but is variable with dynamic pressure and equivalent parasite area. This principle furnishes the basis for illustrating the variation of parasite drag with the various conditions of flight. If all other factors are held constant, the parasite drag varies directly with the equivalent parasite area.

$$\frac{D_{p_2}}{D_{p_1}}=\left(\frac{f_2}{f_1}\right)$$

where

D_{p_1} = parasite drag corresponding to some original parasite area, f_1

D_{p_2} = parasite drag corresponding to some new parasite area, f_1

(V and σ are constant)

As an example, the lowering of the landing gear and flaps may increase the parasite area 80 percent. At any given speed and altitude this airplane would experience an 80 percent increase in parasite drag.

EFFECT OF ALTITUDE. In a similar manner the effect of altitude on parasite drag may be appreciated. The general effect of altitude is expressed by:

$$\frac{D_{p_2}}{D_{p_1}}=\frac{\sigma_2}{\sigma_1}$$

where

D_{p_1} = parasite drag corresponding to some original altitude density ratio, σ_1

D_{p_2} = parasite drag corresponding to some new altitude density ratio, σ_2

(and f, V are constant)

This relationship implies that parasite drag would decrease at altitude, e.g., a given airplane in flight at a given *TAS* at 40,000 ft. ($\sigma=0.25$) would have one-fourth the parasite drag when at sea level ($\sigma=1.00$). This effect results when the lower air density produces less dynamic pressure. However, if the airplane is flown at a constant *EAS*, the dynamic pressure and, thus, parasite drag do not vary. In this case, the *TAS* would be higher at altitude to provide the same *EAS*.

EFFECT OF SPEED. The effect of speed alone on parasite drag is the most important. If all other factors are held constant, the effect of velocity on parasite drag is expressed as:

$$\frac{D_{p_2}}{D_{p_1}}=\left(\frac{V_2}{V_1}\right)^2$$

where

D_{p_1} = parasite drag corresponding to some original speed, V_1

D_{p_2} = parasite drag corresponding to some new speed, V_2

(f and σ are constant)

This relationship expresses a powerful effect of speed on parasite drag. As an example, a given airplane in flight at some altitude would have four times as much parasite drag at twice as great a speed or one-fourth as much parasite drag at half the original speed. This fact may be appreciated by the relationship of dynamic pressure with speed—twice as much V, four times as much q, and four times as much D_p. This expressed variation of parasite drag with speed points out that parasite drag will be of greatest importance at high speeds and practically insignificant in flight at low dynamic pressures. To illustrate this fact, an airplane in flight just above the stall speed could have a parasite drag which is only 25 percent of the total drag. However, this same airplane at maximum level flight speed at low altitude would have a parasite drag which is very nearly 100 percent of the total drag. The predominance of parasite drag at high flight speeds emphasizes the necessity for great aerodynamic cleanness (low f) to obtain high speed performance.

In the subsonic regime of flight, the ordinary configuration of airplane has a very large portion of the equivalent parasite area determined by skin friction drag. As the wing contributes nearly half of the total parasite drag, the profile drag of the wing can be minimized by the use of the airfoil sections which produce extensive laminar flow. A subtle effect on parasite drag occurs from the influence of the wing area. Since the wing area (S) appears directly in the parasite drag equation, a reduction in wing area would reduce the parasite drag if all other factors were unchanged. While the exact relationship involves consideration of many factors, most optimum airplane configurations have a strong preference for the highest practical wing loading and minimum wing surface area.

As the flight speeds of aircraft approach the speed of sound, great care must be taken to delay and alleviate compressibility effects. In order to delay and reduce the drag rise associated with compressibility effects, the components of the airplanes must be arranged to reduce the early formation of shock waves on the airplane. This will generally require fuselage and nacelles of high fineness ratio, well faired canopies, and thin wing sections which have very smooth uniform pressure distributions. Low aspect ratios and sweepback are favorable in delaying and reducing the compressibility drag rise. In addition, interference effects are quite important in transonic and supersonic flight and the airplane cross section area distribution must be controlled to minimize local velocity peaks which could create premature strong shock wave formation.

The modern configuration of airplane will illustrate the features required to effect very high speed performance—low aspect ratio, sweepback, thin low drag sections, etc. These same features produce flight characteristics at low airspeeds which necessitate proper flying technique.

AIRPLANE TOTAL DRAG

The total drag of an airplane in flight is the sum of the induced and parasite drag. Figure 1.35 illustrates the variation of total drag with speed for a given airplane in level flight at a particular weight, configuration, and altitude. The parasite drag increases with speed varying as the square of the velocity while the induced drag decreases with speed varying inversely as the square of the velocity. The total drag of the airplane shows the predominance of induced drag at low speed and parasite drag at high speed. Specific points of interest on the drag curve are as follows:

(A) Stall of this particular airplane occurs at 100 knots and is indicated by a sharp rise in the actual drag. Since the generalized equations for induced and parasite do not account for conditions at stall, the actual drag of the airplane is depicted by the "hook" of the dotted line.

(B) At a speed of 124 knots, the airplane would incur a minimum rate of descent in power-off flight. Note that at this speed the induced drag comprises 75 percent of the total drag. If this airplane were powered with a reciprocating-propeller type powerplant, maximum endurance would occur at this airspeed.

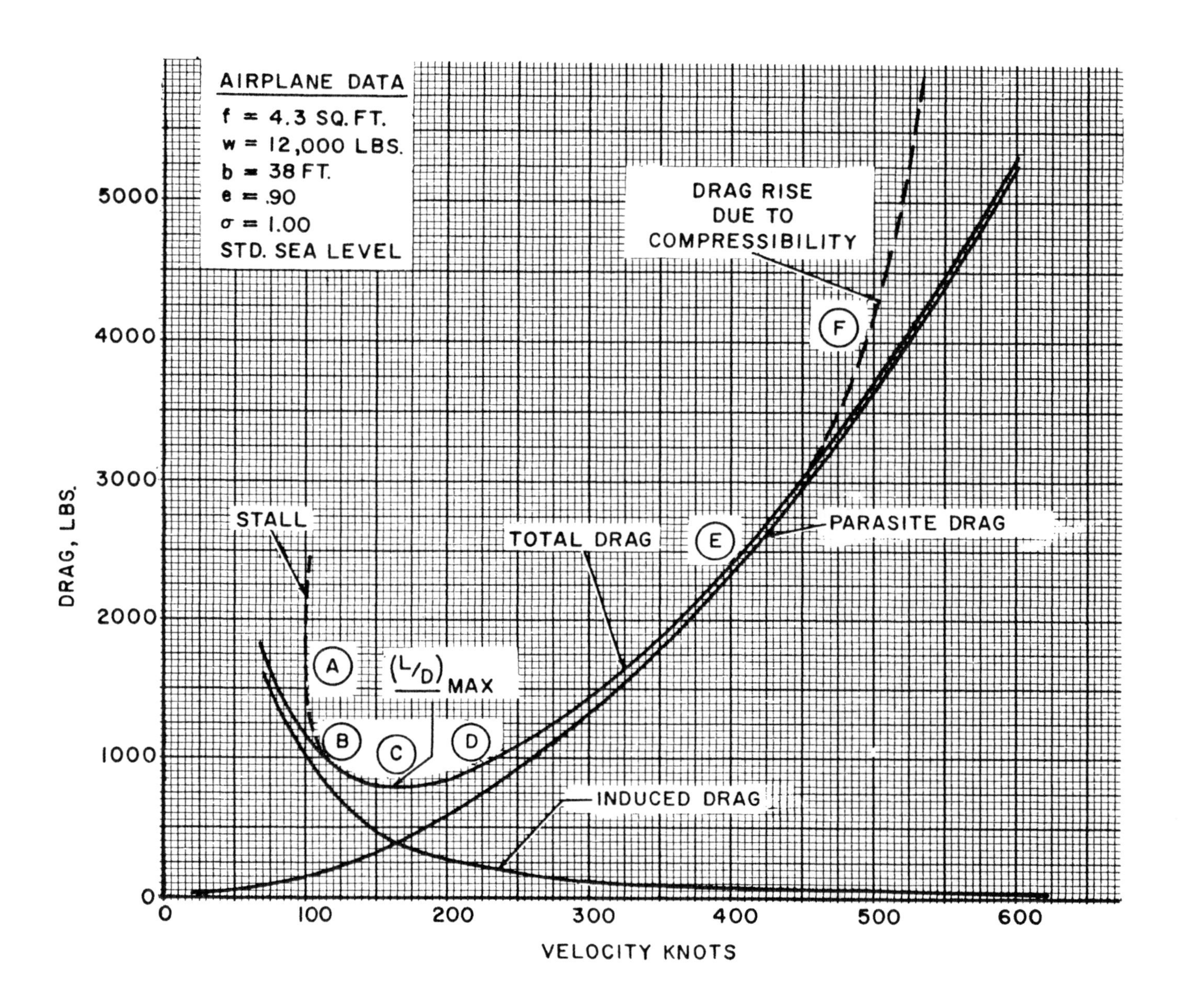

Figure 1.35. Typical Airplane Drag Curves

(C) The point of minimum total drag occurs at a speed of 163 knots. Since this speed incurs the least total drag for lift-equal-weight flight, the airplane is operating at $(L/D)_{max}$. Because of the particular manner in which parasite and induced drags vary with speed (parasite drag directly as the speed squared; induced drag inversely as the speed squared) the minimum total drag occurs when the induced and parasite drags are equal. The speed for minimum drag is an important reference for many items of airplane performance. One item previously presented related glide performance and lift-drag ratio. At the speed of 163 knots this airplane incurs a total drag of 778 lbs. while producing 12,000 lbs. of lift. These figures indicate a maximum lift-drag ratio of 15.4 and relate a glide ratio of 15.4. In addition, if this airplane were jet powered, the airplane would achieve maximum endurance at this airspeed for the specified altitude. If this airplane were propeller powered, the airplane would achieve maximum range at this airspeed for the specified altitude.

(D) Point (*D*) is at an airspeed approximately 32 percent greater than the speed for $(L/D)_{max}$. Note that the parasite drag comprises 75 percent of the total drag at a speed of 215 knots. This point on the drag curve produces the highest proportion between velocity and drag and would be the point for maximum range if the airplane were jet powered. Because of the high proportion of parasite drag at this point the long range jet airplane has great preference for great aerodynamic cleanness and less demand for a high aspect ratio than the long range propeller powered airplane.

(E) At a speed of 400 knots, the induced drag is an extremely small part of the total drag and parasite drag predominates.

(F) As the airplane reaches very high flight speeds, the drag rises in a very rapid fashion due to compressibility. Since the generalized equation for parasite drag does not account for compressibility effects, the actual drag rise is typified by the dashed line.

The airplane drag curve shown in figure 1.34 is particular to one weight, configuration, and altitude in level flight. Any change in one of these variables will affect the specific drags at specific velocities.

The airplane drag curve is a major factor in many items of airplane performance. Range, endurance, climb, maneuver, landing, takeoff, etc., performance are based on some relationship involving the airplane drag curve.

Chapter 2
AIRPLANE PERFORMANCE

The performance of an aircraft is the most important feature which defines its suitability for specific missions. The principal items of airplane performance deserve detailed consideration in order to better understand and appreciate the capabilities of each airplane. Knowledge of the various items of airplane performance will provide the Naval Aviator with a more complete appreciation of the operating limitations and insight to obtain the design performance of his aircraft. The performance section of the flight handbook provides the specific information regarding the capabilities and limitations of each airplane. Every Naval Aviator must rely upon these handbook data as the guide to safe and effective operation of his aircraft.

REQUIRED THRUST AND POWER

DEFINITIONS

All of the principal items of flight performance involve steady state flight conditions and equilibrium of the airplane. For the airplane to remain in steady level flight, equilibrium must be obtained by a lift equal to the airplane weight and a powerplant thrust equal to the airplane drag. Thus, the airplane drag defines the thrust required to maintain steady level flight.

The total drag of the airplane is the sum of the parasite and induced drags. Parasite drag is the sum of pressure and friction drag which is due to the basic configuration and, as defined, is independent of lift. Induced drag is the undesirable but unavoidable consequence of the development of lift. In the process of creating lift by the deflection of an airstream, the actual lift is inclined and a component of lift is incurred parallel to the flight path direction. This component of lift combines with any change in pressure and friction drag due to change in lift to form the induced drag. While the parasite drag predominates at high speed, induced drag predominates at low speed. Figure 2.1 illustrates the variation with speed of the induced, parasite, and total drag for a specific airplane configuration in steady level flight.

The power required for flight depends on the thrust required and the flight velocity. By definition, the propulsive horsepower required is related to thrust required and flight velocity by the following equation:

$$Pr=\frac{TrV}{325}$$

where

Pr=power required, h.p.
Tr=thrust required (total drag), lbs.
V=true airspeed, knots

By inspection of this relationship, it is apparent that each pound of drag incurred at 325 knots requires one horsepower of propulsive power. However, each pound of drag at 650 knots requires two horsepower while each pound of drag at 162.5 knots requires one-half horsepower. The term "power" implies work rate and, as such, will be a function of the speed at which a particular force is developed.

Distinction between *thrust* required and *power* required is necessary for several reasons. For the items of performance such as range and endurance, it is necessary to relate powerplant fuel flow with the propulsive requirement for steady level flight. Some powerplants incur fuel flow rate according to output thrust while other powerplants incur fuel flow rate depending on output power. For example, the turbojet engine is principally a thrust producing machine and fuel flow is most directly related to thrust output. The reciprocating engine is principally a power producing machine and fuel flow is most directly related to power output. For these reasons the variation of thrust required will be of greatest interest in the performance of the turbojet powered airplane while the variation of power required will be of greatest interest in the performance of the propeller powered airplane. Also, distinction between power and thrust required is necessary in the study of climb performance. During a steady climb, the rate of climb will depend on excess power while the angle of climb is a function of excess thrust.

The total power required for flight can be considered as the sum of induced and parasite effects similar to the total drag of the airplane. The induced power required is a function of the induced drag and velocity.

$$Pr_i=\frac{D_iV}{325}$$

where

Pr_i=induced power required, h.p.
D_i=induced drag, lbs.
V=true airspeed, knots

Thus, induced power required will vary with lift, aspect ratio, altitude, etc., in the same manner as the induced drag. The only difference will be the variation with speed. If all other factors remain constant, the induced power required varies inversely with velocity while induced drag varies inversely with the square of the velocity.

$$\frac{Pr_{i_2}}{Pr_{i_1}}=\frac{V_1}{V_2}$$

where

Pr_{i_1} = induced power required corresponding to some original speed, V_1

Pr_{i_2} = induced power required corresponding to some different speed, V_2

For example, if an airplane in steady level flight is operated at twice as great a speed, the induced drag is one-fourth the original value but the induced power required is one-half the original value.

The parasite power required is a function of the parasite drag and velocity.

$$Pr_p=\frac{D_pV}{325}$$

where

Pr_p = parasite power required, h.p.
D_p = parasite drag, lbs.
V = true airspeed, knots

Thus, parasite power required will vary with altitude and equivalent parasite area (f) in the same manner as the parasite drag. However, the variation with speed will be different. If all other factors are constant, the parasite drag varies as the square of velocity but parasite power varies as the cube of velocity.

$$\frac{Pr_{P2}}{Pr_{P1}}=\left(\frac{V_2}{V_1}\right)^3$$

where

Pr_{P1} = parasite power required corresponding to some original speed, V_1

Pr_{P2} = parasite power required corresponding to some different speed, V_2

For example, if an airplane in steady flight is operated at twice as great a speed, the parasite drag is four times as great but the parasite power required is eight times the original value.

Figure 2.1 presents the thrust required and power required for a specific airplane configuration and altitude. The curves of figure 2.1 are applicable for the following airplane data:

gross weight, W = 15,000 lbs.
span, b = 40 ft.
equivalent parasite area, f = 7.2 sq. ft.
airplane efficiency factor, e = .827
sea level altitude, σ = 1.000
compressibility corrections neglected

The curve of drag or thrust required versus velocity shows the variation of induced, parasite, and total drag. Induced drag predominates at low speeds. When the airplane is operated at maximum lift-drag ratio, $(L/D)_{max}$, the total drag is at a minimum and the induced and parasite drags are equal. For the specific airplane of figure 2.1, $(L/D)_{max}$ and minimum total drag are obtained at a speed of 160 knots.

The curve of power required versus velocity shows the variation of induced, parasite, and total power required. As before, induced power required predominates at low speeds and parasite power required predominates at high speeds and the induced and parasite power are equal at $(L/D)_{max}$. However, the condition of $(L/D)_{max}$ defines only the point of minimum drag and does not define the point of minimum *power* required. Ordinarily, the point of minimum power required will occur at a speed which is 76 percent of the speed for minimum drag and, in the case of the airplane configuration of figure 2.1, the speed for minimum power required would be 122 knots. The total drag at the speed for minimum power required is 15 percent higher than the drag at $(L/D)_{max}$ but the minimum power required is 12 percent lower than the power required at $(L/D)_{max}$.

530454 O—60——8

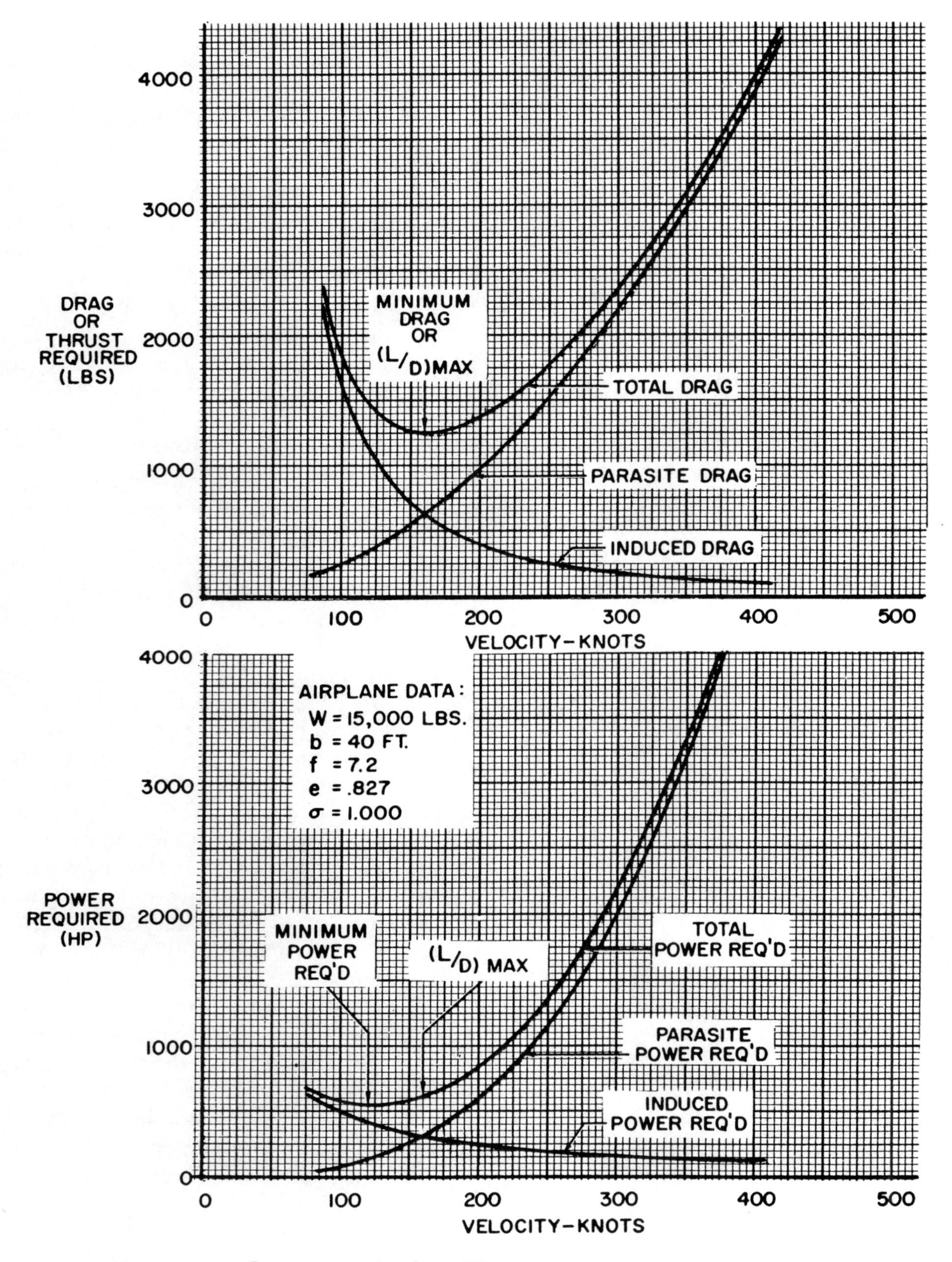

Figure 2.1. Airplane Thrust and Power Required

Induced drag predominates at speeds below the point of minimum total drag. When the airplane is operated at the condition of minimum power required, the total drag is 75 percent induced drag and 25 percent parasite drag. Thus, the induced drag is three times as great as the parasite drag when at minimum power required.

VARIATIONS OF THRUST REQUIRED AND POWER REQUIRED

The curves of thrust required and power required versus velocity provide the basis for comprehensive analysis of all the major items of airplane performance. The changes in the drag and power curves with variations of air-plane gross weight, configuration, and altitude furnish insight for the variation of range, endurance, climb performance, etc., with these same items.

The effect of a change in weight on the thrust and power required is illustrated by figure 2.2. The primary effect of a weight change is a change in the induced drag and induced power required at any given speed. Thus, the greatest changes in the curves of thrust and power required will take place in the range of low speed flight where the induced effects predominate. The changes in thrust and power required in the range of high speed flight are relatively slight because parasite effects predominate at high speed. The induced effects at high speed are relatively small and changes in these items produce a small effect on the total thrust or power required.

In addition to the general effect on the induced drag and power required at particular speeds, a change in weight will require that the airplane operate at different airspeeds to maintain conditions of a specific lift coefficient and angle of attack. If the airplane is in steady flight at a particular C_L, the airpseed required for this C_L will vary with weight in the following manner:

$$\frac{V_2}{V_1}=\sqrt{\frac{W_2}{W_1}}$$

where

V_1 = speed corresponding to a specific C_L and weight, W_1

V_2 = speed corresponding to the same C_L but a different weight, W_2

For the example airplane of figure 2.2, a change of gross weight from 15,000 to 22,500 lbs. requires that the airplane operate at speeds which are 22.5 percent greater to maintain a specific lift coefficient. For example, if the 15,000-lb. airplane operates at 160 knots for $(L/D)_{max}$, the speed for $(L/D)_{max}$ at 22,500 lbs. is:

$$V_2=V_1\sqrt{\frac{W_2}{W_1}}$$

$$=160\sqrt{\frac{22,500}{15,000}}$$

$$=(160)\,(1.225)$$

$$=196 \text{ knots}$$

The same situation exists with respect to the curves of power required where a change in weight requires a change of speed to maintain flight at a particular C_L. For example, if the 15,000-lb. airplane achieves minimum power required at 122 knots, an increase in weight to 22,500 lbs. increases the speed for minimum power required to 149 knots.

Of course, the thrust and power required at specific lift coefficients are altered by changes in weight. At a specific C_L, any change in weight causes a like change in thrust required, e.g., a 50-percent increase in weight causes a 50-percent increase in thrust required at the same C_L. The effect of a weight change on the power required at a specific C_L is a bit more complex because a change in speed accompanies the change

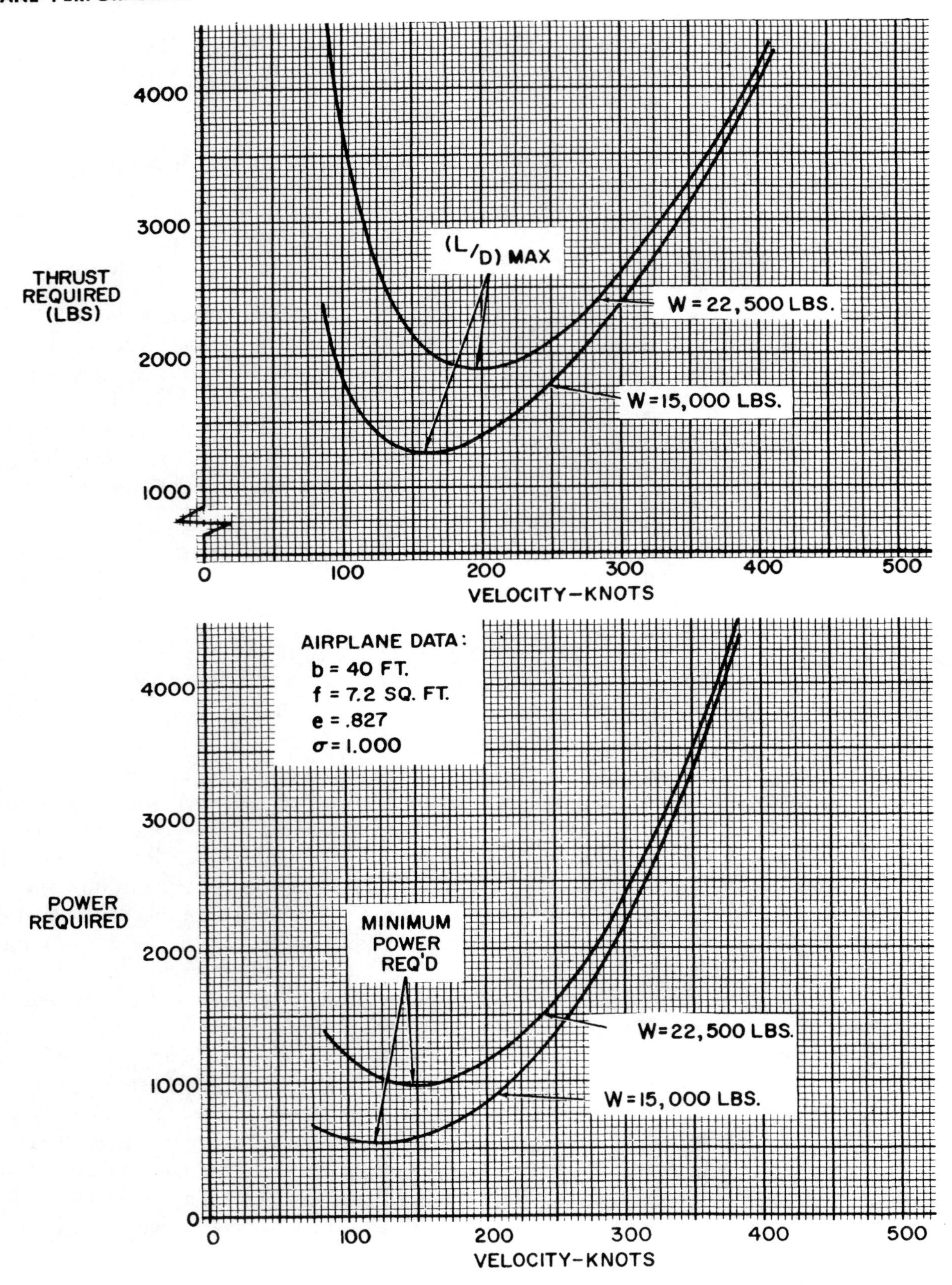

Figure 2.2. Effect of Weight on Thrust and Power Required

in drag and there is a two-fold effect. A 50-percent increase in weight produces an increase of 83.8 percent in the power required to maintain a specific C_L. This is the result of a 50-percent increase in thrust required coupled with a 22.5-percent increase in speed. The effect of a weight change on thrust required, power required, and airspeed at specific angles of attack and lift coefficients provides an important basis for various techniques of cruise and endurance conditions of flight.

Figure 2.3 illustrates the effect on the curves of thrust and power required of a change in the equivalent parasite area, f, of the configuration. Since parasite drag predominates in the region of high flight speed, a change in f will produce the greatest change in thrust and power required at high speed. Since parasite drag is relatively small in the region of low speed flight, a change in f will produce relatively small changes in thrust and power required at low speeds. The principal effect of a change in equivalent parasite area of the configuration is to change the parasite drag at any given airspeed.

The curves of figure 2.3 depict the changes in the curves of thrust and power required due to a 50 percent increase in equivalent parasite area of the configuration. The minimum total drag is increased by an increase in f and the $(L/D)_{max}$ is reduced. Also, the increase in f will increase the C_L for $(L/D)_{max}$ and require a reduction in speed at the new, but decreased, $(L/D)_{max}$. The point of minimum power required occurs at a lower airspeed and the value of the minimum power required is increased slightly. Generally, the effect on the minimum power required is slight because the parasite drag is only 25 percent of the total at this specific condition of flight.

An increase in the equivalent parasite area of an airplane may be brought about by the deflection of flaps, extension of landing gear, extension of speed brakes, addition of external stores, etc. In such instances a decrease in the airplane efficiency factor, e, may accompany an increase in f to account for the additional changes in parasite drag which may vary with C_L.

A change in altitude can produce significant changes in the curves of thrust and power required. The effects of altitude on these curves provide a great part of the explanation of the effect of altitude on range and endurance. Figure 2.4 illustrates the effect of a change in altitude on the curves of thrust and power required for a specific airplane configuration and gross weight. As long as compressibility effects are negligible, the principal effect of increased altitude on the curve of thrust required is that specific aerodynamic conditions occur at higher true airspeeds. For example, the subject airplane at sea level has a minimum drag of 1,250 lbs. at 160 knots. The same airplane would incur the same drag at altitude if operated at the same *equivalent airspeed* of 160 knots. However, the equivalent airspeed of 160 knots at 22,000 ft. altitude would produce a true airspeed of 227 knots. Thus, an increase in altitude will cause the curve of thrust required to flatten out and move to the direction of higher velocity. Note that altitude alone will not alter the value of minimum drag.

The effect of altitude on the curve of power required can best be considered from the effect on true airspeed to achieve a specific aerodynamic condition. The sea level power required curve of figure 2.4 indicates that $(L/D)_{max}$ occurs at 160 knots and requires 615 h.p. If this same airplane is operated at $(L/D)_{max}$ at an altitude of 22,000 ft., the same drag is incurred at a higher velocity and requires a higher power. The increase in velocity to 227 knots accounts for the increase in power required to 872 h.p. Actually, the various points on the curve of power required can be considered affected in this same fashion. At specific lift coefficients and angles of attack, a change in altitude will alter the true airspeed particular to these points and cause a change in power required because of the change in true airspeed. An increase in altitude will

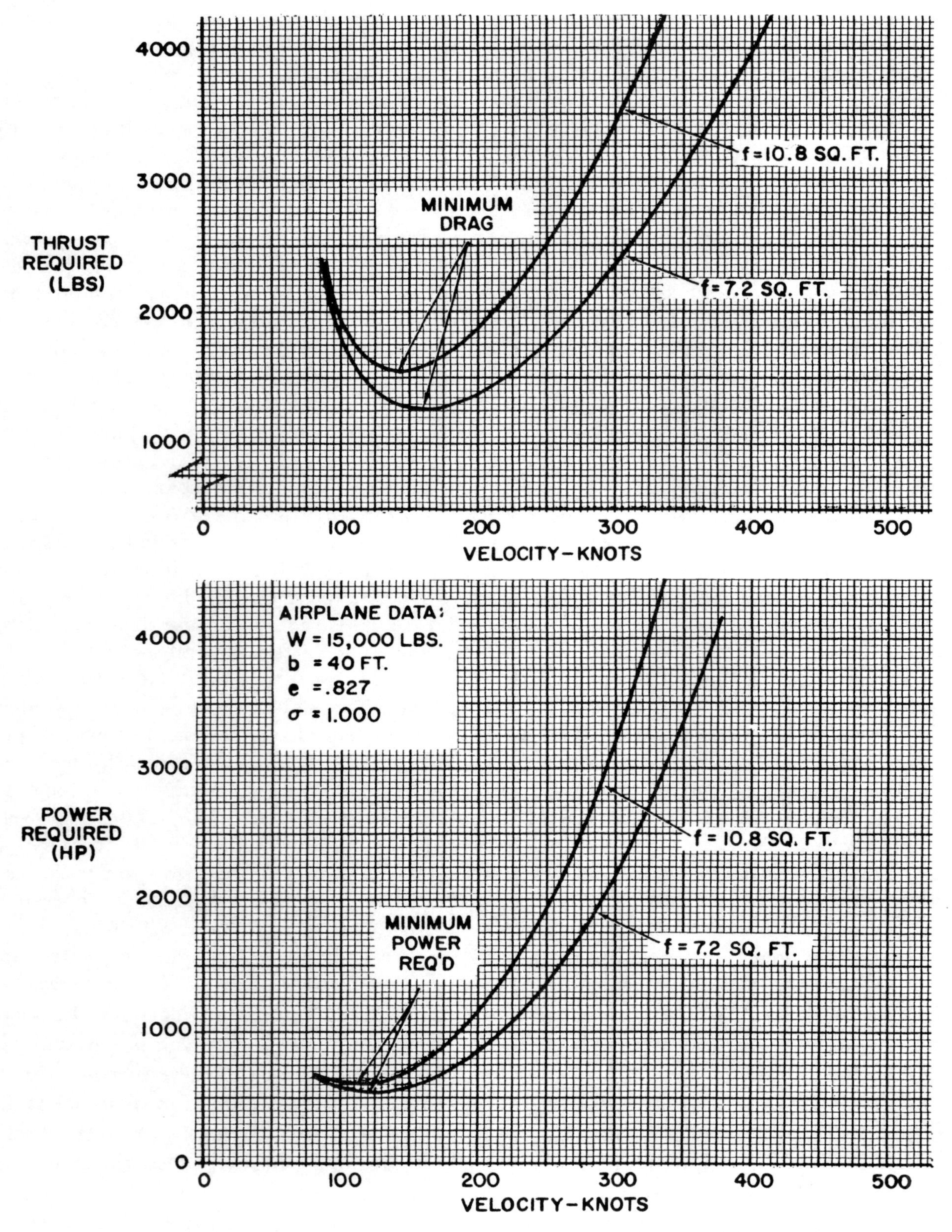

Figure 2.3. Effect of Equivalent Parasite Area, f, on Thrust and Power Required

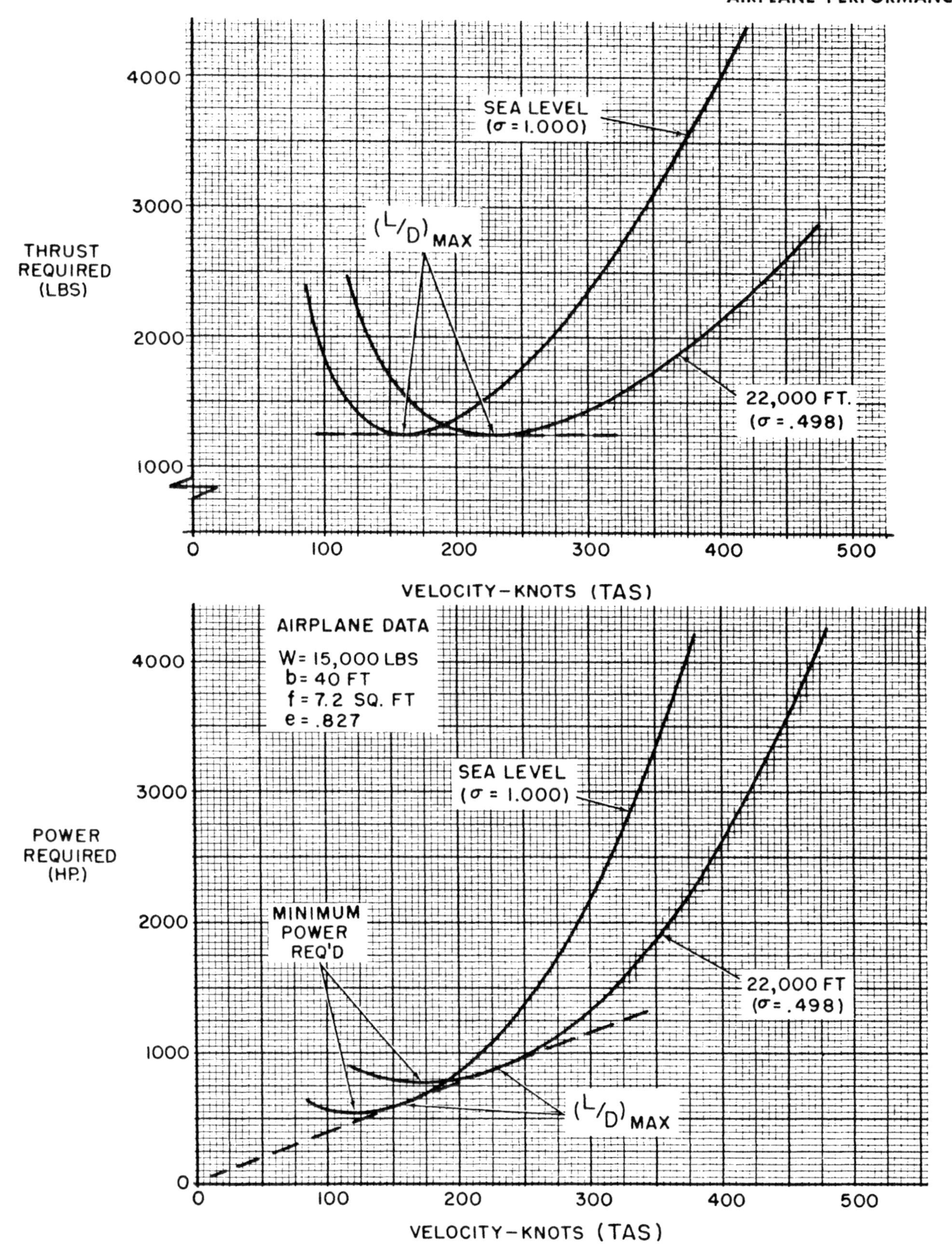

Figure 2.4. Effect of Altitude on Thrust and Power Required

cause the power required curve to flatten out and move to higher velocities and powers required.

The curves of thrust and power required and their variation with weight, altitude, and configuration are the basis of all phases of airplane performance. These curves define the *requirements* of the airplane and must be considered with the power and thrust available from the powerplants to provide detailed study of the various items of airplane performance.

AVAILABLE THRUST AND POWER

PRINCIPLES OF PROPULSION

All powerplants have in common certain general principles. Regardless of the type of propulsion device, the development of thrust is related by Newton's laws of motion.

$$F=ma$$

or

$$F=\frac{d(mV)}{dt}$$

where

F=force or thrust, lbs.

m=mass, slugs

a=acceleration, ft. per sec.^{2}

$\frac{d}{dt}$=derivative with respect to time, e.g., rate of change with time

mV=momentum, lb.-sec., product of mass and velocity

The force of thrust results from the acceleration provided the mass of working fluid. The magnitude of thrust is accounted for by the rate of change of momentum produced by the powerplant. A rocket powerplant creates thrust by creating a very large change in velocity of a relatively small mass of propellants. A propeller produces thrust by creating a comparatively small change in velocity of a relatively large mass of air.

The development of thrust by a turbojet or ramjet powerplant is illustrated by figure 2.5. Air approaches at a velocity, V_1, depending on the flight speed and the powerplant operates on a certain mass flow of air, Q, which passes through the engine. Within the powerplant the air is compressed, energy is added by the burning of fuel, and the mass flow is expelled from the nozzle finally reaching a velocity, V_2. The momentum change accomplished by this action produces the thrust,

$$Ta=Q\,(V_2-V_1)$$

where

Ta=thrust, lbs.

Q=mass flow, slugs per sec.

V_1=inlet (or flight) velocity, ft. per sec.

V_2=jet velocity, ft. per sec.

The typical ramjet or turbojet powerplant derives its thrust by working with a mass flow relatively smaller than that of a propeller but a relatively greater change of velocity. From the previous equation it should be appreciated that the jet thrust varies directly with the mass flow Q, and velocity change, V_2-V_1. This fact is useful in accounting for many of the performance characteristics of the jet powerplant.

In the process of creating thrust by momentum change of the airstream, a relative velocity, V_2-V_1, is imparted to the airstream. Thus, some of the available energy is essentially wasted by this addition of kinetic energy to the airstream. The change of kinetic energy per time can account for the power wasted in the airstream.

$$Pw=KE/t$$

$$=\frac{Q}{2}\,(V_2-V_1)^2$$

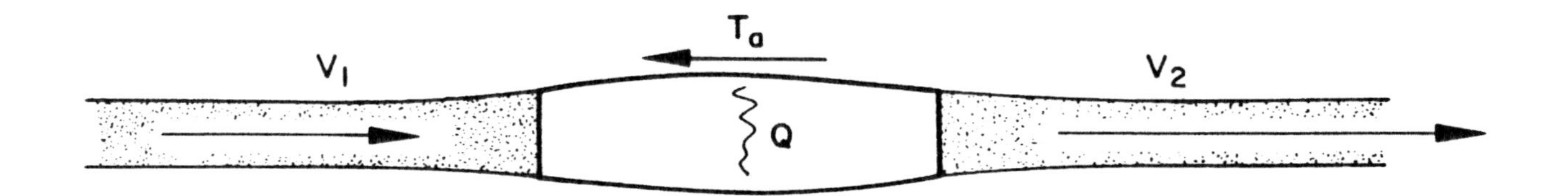

$$F = ma$$

$$F = \frac{d}{dt}(mV)$$

$$T_a = Q\,(V_2 - V_1)$$

$$P_a = T_a V_1$$

$$P_w = Q/2\,(V_2 - V_1)^2$$

$$\eta_p = \frac{2V_1}{V_2 + V_1}$$

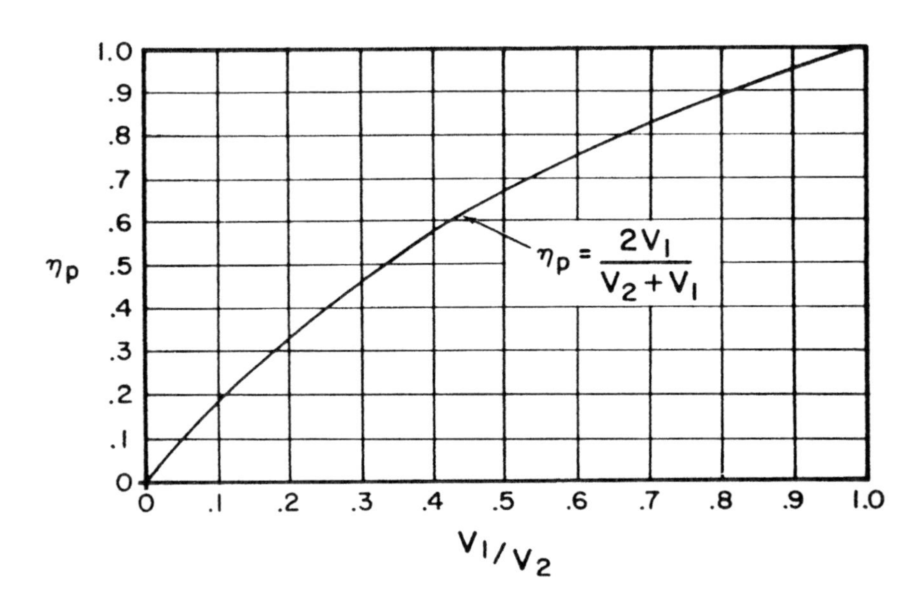

Figure 2.5. Principles of Propulsion

Of course, the development of thrust with some finite mass flow will require some finite velocity change and there will be the inevitable waste of power in the airstream. In order to achieve high efficiency of propulsion, the thrust should be developed with a minimum of wasted power.

The propulsion efficiency of the jet powerplant can be evaluated by comparing the propulsive output power with the input power. Since the input power is the sum of the output power and wasted power, an expression for propulsion efficiency can be derived.

$$\eta_p = \frac{Pa}{Pa+Pw}$$

$$\eta_p = \frac{2V_1}{V_2+V_1}$$

where

η_p = propulsion efficiency

η = "eta"

Pa = propulsive power available

$= TaV_1$

Pw = power wasted

The resulting expression for propulsion efficiency, η_p, shows a dependency on the flight velocity, V_1, and the jet velocity, V_2. When the flight velocity is zero, the propulsion efficiency is zero since all power generated is wasted in the slipstream and the propulsive power is zero. The propulsion efficiency would be 1.00 (or 100 percent) only when the flight velocity, V_1, equals the jet velocity, V_2. Actually, it would not be possible to produce thrust under such conditions with a finite mass flow. While 100 percent efficiency of propulsion can not be attained practically, some insight is furnished to the means of creating high values of propulsion efficiency. To obtain high propulsion efficiency it is necessary to produce the required thrust with the highest possible mass flow and lowest possible velocity change.

The graph of figure 2.5 shows the variation of propulsion efficiency, η_p, with the ratio of flight speed to jet velocity, V_1/V_2. To achieve a propulsion efficiency of 0.85 requires that the flight velocity be approximately 75 percent of the slipstream speed relative to the airplane. Such a propulsive efficiency could be typical of a propeller powered airplane which derives its thrust by the propeller handling a large mass flow of air. The typical turbojet powerplant cannot achieve such high propulsive efficiency because the thrust is derived with a relatively smaller mass flow and larger velocity change. For example, if the jet velocity is 1,200 ft. per sec. at a flight velocity of 600 ft. per sec., the propulsion efficiency is 0.67. The ducted fan, bypass jet, and turboprop are variations which improve the propulsive efficiency of a type of powerplant which has very high power capability.

When the conditions of range, endurance, or economy of operation are predominant, high propulsion efficiency is necessary. Thus, the propeller powered airplane with its inherent high propulsive efficiency will always find application. The requirements of very high speed and high altitude demand very high propulsive power from relatively small powerplants. When there are practical limits to the increase of mass flow, high output is obtained by large velocity changes and low propulsive efficiency is an inevitable consequence.

TURBOJET ENGINES

The turbojet engine has found widespread use in aircraft propulsion because of the relatively high power output per powerplant weight and size. Very few aircraft powerplants can compare with the high output, flexibility, simplicity, and small size of the aircraft gas turbine. The coupling of the propeller and reciprocating engine is one of the most efficient means known for converting fuel energy into propulsive energy. However, the intermittent action of the reciprocating engine places practical limits to the airflow that can be processed and restricts the development of power. The continuous, steady flow feature of the gas turbine allows such a powerplant to process considerably greater airflow and, thus, utilize a greater expenditure of fuel energy. While the propulsive efficiency of the turbojet engine is considerably below that of the reciprocating engine-propeller combination, the specific power output of the turbojet at high speeds is quite superior.

The operation of the turbojet engine involves a relatively large change in velocity being imparted to the mass flow through the engine. Figure 2.6 illustrates the operation of a typical turbojet engine by considering the processing given a unit weight of inlet airflow. Consider a unit weight of ambient air approaching the inlet to the engine then experiencing the changes in pressure and volume as it is processed by the turbojet. The chart of pressure versus volume of figure 2.6 shows that the unit weight of airflow at atmospheric condition *A* is delivered to the inlet entrance at condition *B*. The purpose of the inlet or diffuser is to reduce the velocity and increase the pressure of the flow entering the compressor section. Thus, the aerodynamic compression produces an increase in pressure and decrease in volume of the unit weight of air and delivers air to the compressor at condition *C*. The work done by the aerodynamic compression of the inlet or diffuser is represented by the area *ABCX*. Generally, most conventional turbojet engines require that the compressor inlet flow be subsonic and supersonic flight will involve considerable aerodynamic compression in the inlet.

Air delivered to the compressor inlet at condition *C* is then subject to further compression through the compressor section. As a result of the function of the compressor, the unit weight of air is subject to a decrease in volume and increase in pressure to condition *D*. The compressor pressure ratio should be high to produce a high thermal efficiency in the engine The area *XCDZ* represents the work done by the compressor during the compression of the unit weight of air. Of course, certain losses and inefficiencies are incurred during the compression and the power required to operate the compressor will be greater than that indicated by the work done on the engine airflow.

Compressed air is discharged from the compressor to the combustion chamber at condition *D*. Fuel is added in the combustion chamber and the combustion of fuel liberates considerable heat energy. The combustion process in the gas turbine differs from that of the reciprocating engine in that the process is essentially a constant pressure addition of heat energy. As a result, the combustion of fuel causes a large change in temperature and large change of volume of the unit weight of airflow. The process in the combustion chamber is represented by the change from point *D* to point *E* of the pressure-volume diagram of figure 2.6.

The combustion products are delivered to the turbine section where sufficient work must be extracted to power the compressor section. The combustion chamber discharges high temperature, high pressure gas to the turbine where a partial expansion is accomplished with a drop in pressure and increase in volume to point *F* on the pressure-volume diagram. The work extracted from the unit weight of air by the turbine section is represented by the area *ZEFY*. As with the compressor, the actual shaft work extracted by the turbine will differ from that indicated by the pressure-volume diagram because of certain losses incurred through the turbine section. For steady, stabilized operation of the turbojet engine the power extracted by the turbine will equal the power required to operate the compressor. If the turbine power exceeds the compressor power required, the engine will accelerate; if the turbine power is less than the compressor power required, the engine will decelerate.

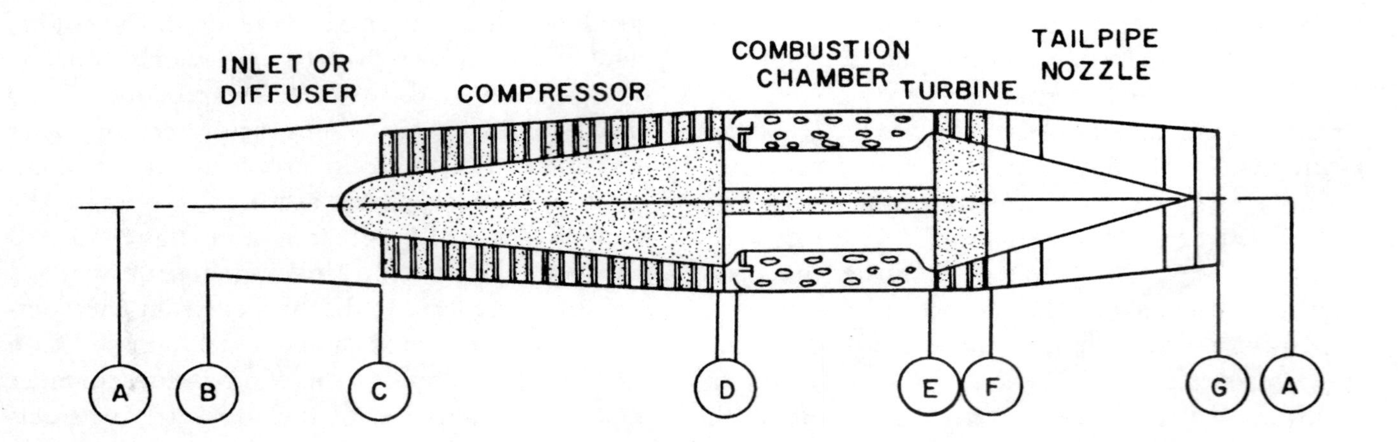

TURBOJET ENGINE CYCLE

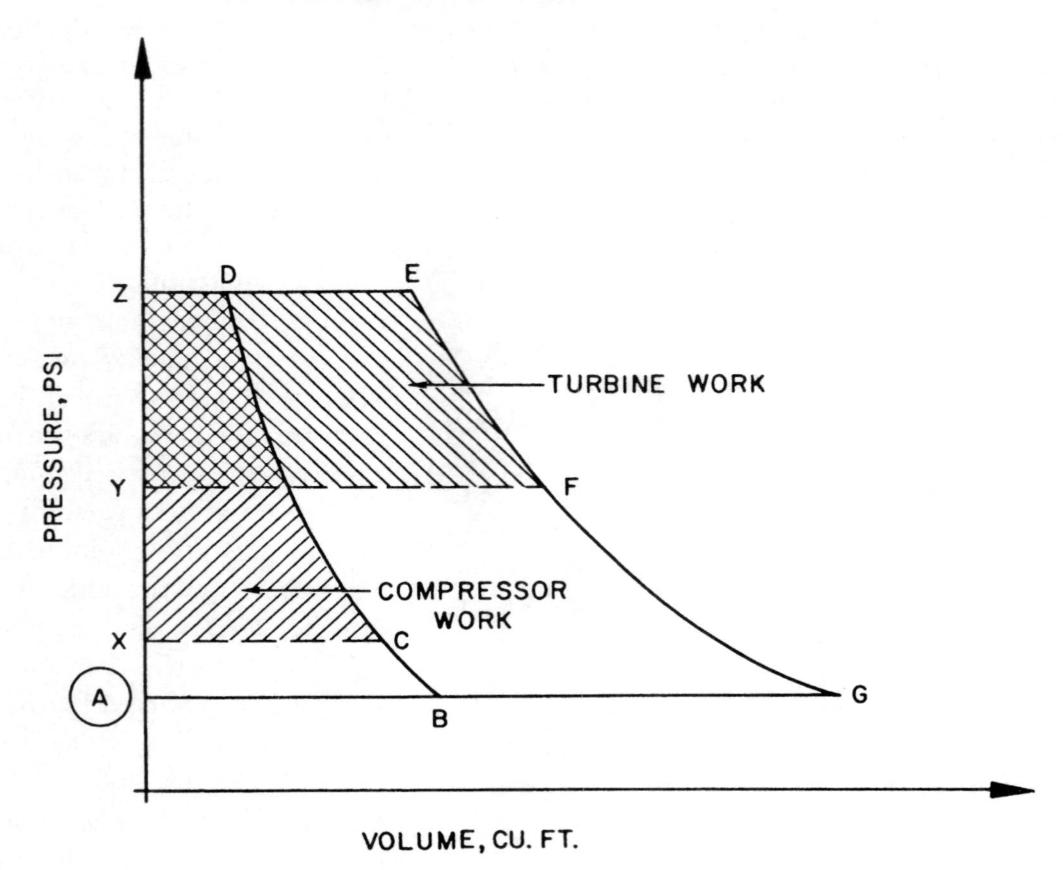

Figure 2.6. Turbojet Engines

The partial expansion of the gases through the turbine will provide the power to operate the engine. As the gases are discharged from the turbine at point F, expansion will continue through the tailpipe nozzle until atmospheric pressure is achieved in the exhaust. Thus, continued expansion in the jet nozzle will reduce the pressure and increase the volume of the unit weight of air to point G on the pressure volume diagram. As a result, the final jet velocity is greater than the inlet velocity and the momentum change necessary for the development of thrust has been created. The area $YFGA$ represents the work remaining to provide the expansion to jet velocity after the turbine has extracted the work required to operate the compressor.

Of course, the combustion chamber discharge could be more completely expanded through a larger turbine section and the net power could be used to operate a propeller rather than provide high exhaust gas velocity. For certain applications, the gas turbine-propeller combination could utilize the high power capability of the gas turbine with greater propulsive efficiency.

FUNCTION OF THE COMPONENTS. Each of the engine components previously described will contribute some function affecting the efficiency and output of the turbojet engine. For this reason, each of these components should be analyzed to determine the requirements for satisfactory operating characteristics.

The *inlet or diffuser* must be matched to the powerplant to provide the compressor entry with the required airflow. Generally, the compressor inlet must receive the required airflow at subsonic velocity with uniform distribution of velocity and direction at the compressor face. The diffuser must capture high energy air and deliver it at low Mach number uniformly to the compressor. When the inlet is along the sides of the fuselage, the edges of the inlet must be located such that the inlet receives only high energy air and provision must be made to dispose of the boundary layer along the fuselage surface. At supersonic flight speeds, the diffuser must slow the air to subsonic with the least waste of energy in the inlet air and accomplish the process with a minimum of aerodynamic drag. In addition, the inlet must be efficient and stable in operation throughout the range of angles of attack and Mach numbers of which the airplane is capable.

The operation of the compressor can be affected greatly by the uniformity of flow at the compressor face. When large variations in flow velocity and direction exist at the face of the axial compressor, the efficiency and stall-surge limits are lowered. Thus, the flight conditions which involve high angle of attack and high sideslip can cause deterioration of inlet performance.

The *compressor section* is one of the most important components of the turbojet engine. The compressor must furnish the combustion chamber with large quantities of high pressure air in a most efficient manner. Since the compressor of a jet engine has no direct cooling, the compression process takes place with a minimum of heat loss of the compressed air. Any friction loss or inefficiency of the compression process is manifested as an undesirable additional increase in the temperature of the compressor discharge air. Hence, compressor efficiency will determine the compressor power necessary to create the pressure rise of a given airflow and will affect the temperature change which can take place in the combustion chamber.

The compressor section of a jet engine may be an axial flow or centrifugal flow compressor. The centrifugal flow compressor has great utility, simplicity, and flexibility of operation. The operation of the centrifugal compressor requires relatively low inlet velocities and a plenum chamber or expansion space must be provided for the inlet. The impeller rotating at high speed receives the inlet air and provides high acceleration by virtue of centrifugal force. As a result, the air leaves the impeller

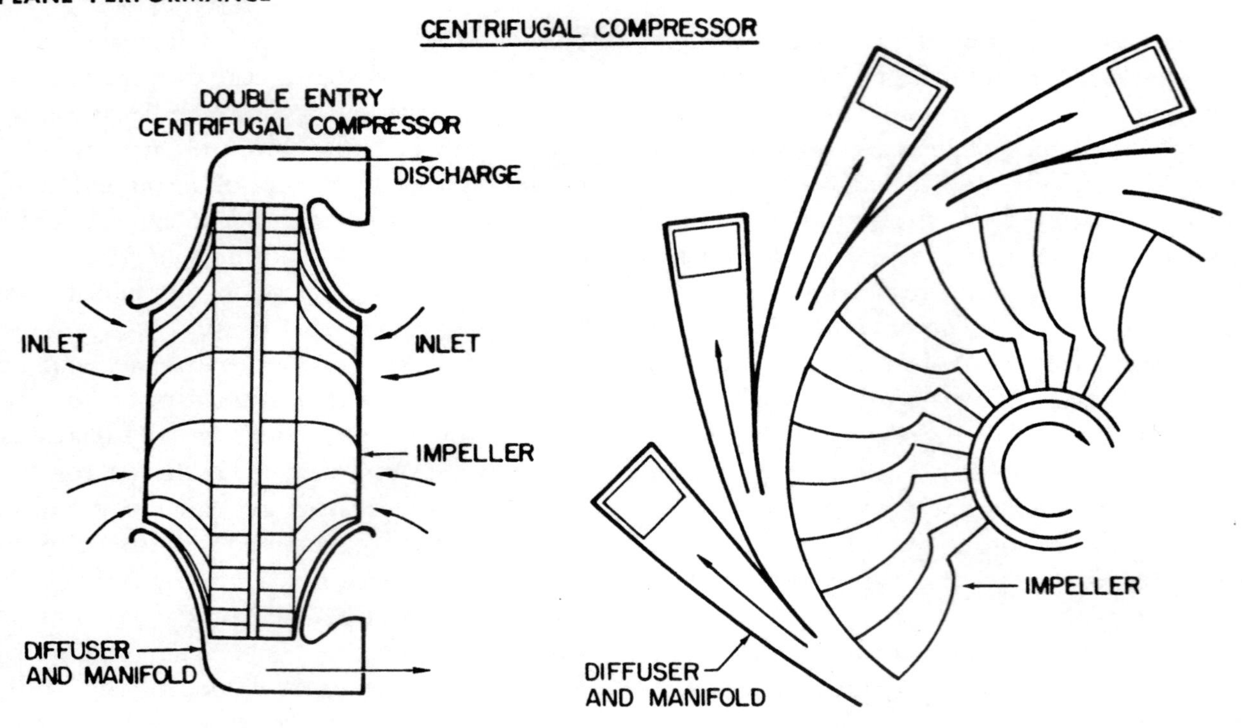

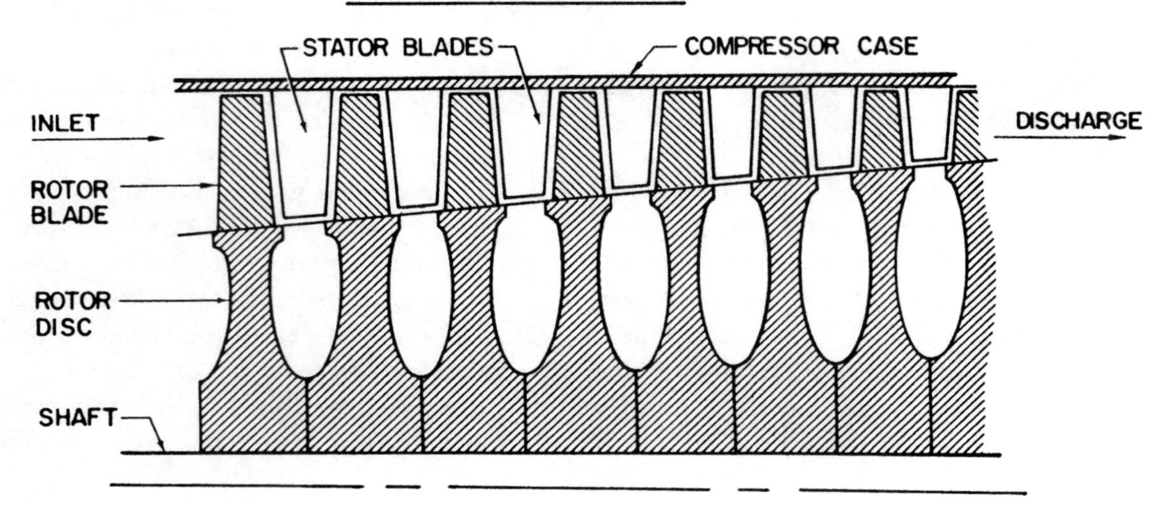

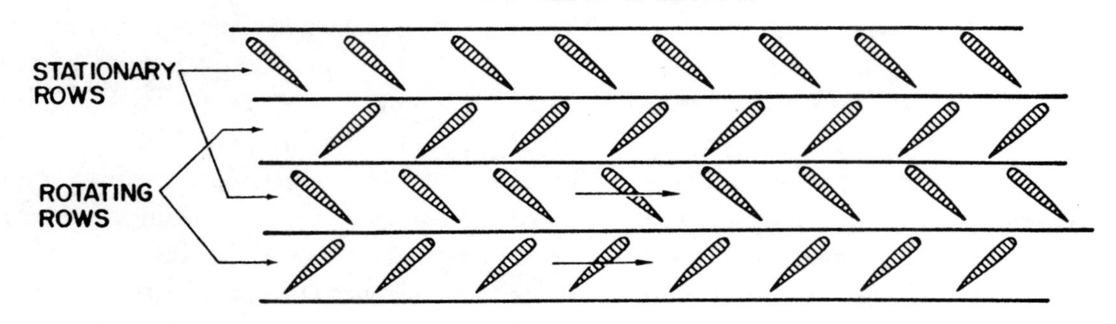

Figure 2.7. Compressor Types

at very high velocity and high kinetic energy. A pressure rise is produced by subsequent expansion in the diffuser manifold by converting the kinetic energy into static pressure energy. The manifold then distributes the high pressure discharge to the combustion chambers. A double entry impeller allows a given diameter compressor to process a greater airflow. The major components of the centrifugal compressor are illustrated in figure 2.7.

The centrifugal compressor can provide a relatively high pressure ratio per stage but the provision of more than one or two stages is rarely feasible for aircraft turbine engines. The single stage centrifugal compressor is capable of producing pressure ratios of about three or four with reasonable efficiency. Pressure ratios greater than four require such high impeller tip speed that compressor efficiency decreases very rapidly. Since high pressure ratios are necessary to achieve low fuel consumption, the centrifugal compressor finds greatest application to the smaller engines where simplicity and flexibility of operation are the principal requirements rather than high efficiency.

The axial flow compressor consists of alternate rows of rotating and stationary airfoils. The major components of the axial flow compressor are illustrated in figure 2.7. A pressure rise occurs through the row of rotating blades since the airfoils cause a decrease in velocity relative to the blades. Additional pressure rise takes place through the row of stationary blades since these airfoils cause a decrease in the absolute velocity of flow. The decrease in velocity, relative or absolute, effects a compression of the flow and causes the increase in static pressure. While the pressure rise per stage of the axial compressor is relatively low, the efficiency is very high and high pressure ratios can be obtained efficiently by successive axial stages. Of course, the efficient pressure rise in each stage is limited by excessive gas velocities. The multistage axial flow compressor is capable of providing pressure ratios from five to ten (or greater) with efficiencies which cannot be approached with a multistage centrifugal compressor.

The axial flow compressor can provide efficiently the high pressure ratios necessary for low fuel consumption. Also, the axial compressor is capable of providing high airflow with a minimum of compressor diameter. When compared with the centrifugal compressor, the design and construction of the axial compressor is relatively complex and costly and the high efficiency is sustained over a much narrower range of operating conditions. For these reasons, the axial compressor finds greatest application where the demands of efficiency and output predominate over considerations of cost, simplicity, flexibility of operation, etc. Multispool compressors and variable stator blades serve to improve the operating characteristics of the axial compressor and increase the flexibility of operation.

The *combustion chamber* must convert the fuel chemical energy into heat energy and cause a large increase in the total energy of the engine airflow. The combustion chamber will operate with one principal limitation: the discharge from the combustion chamber must be at temperatures which can be tolerated by the turbine section. The combustion of liquid hydrocarbon fuels can produce gas temperatures which are in excess of 1,700 to 1,800° C. However, the maximum continuous turbine blade operating temperatures rarely exceed 800° to 1,000° C and considerable excess air must be used in the combustion chamber to prevent exceeding these temperature limits.

While the combustion chamber design may take various forms and configurations, the main features of a typical combustion chamber are illustrated by figure 2.8. The combustion chamber receives the high pressure discharge from the compressor and introduces approximately one half of this air into the immediate area of the fuel spray. This primary combustion air must be introduced with relatively high turbulence and quite low velocities to

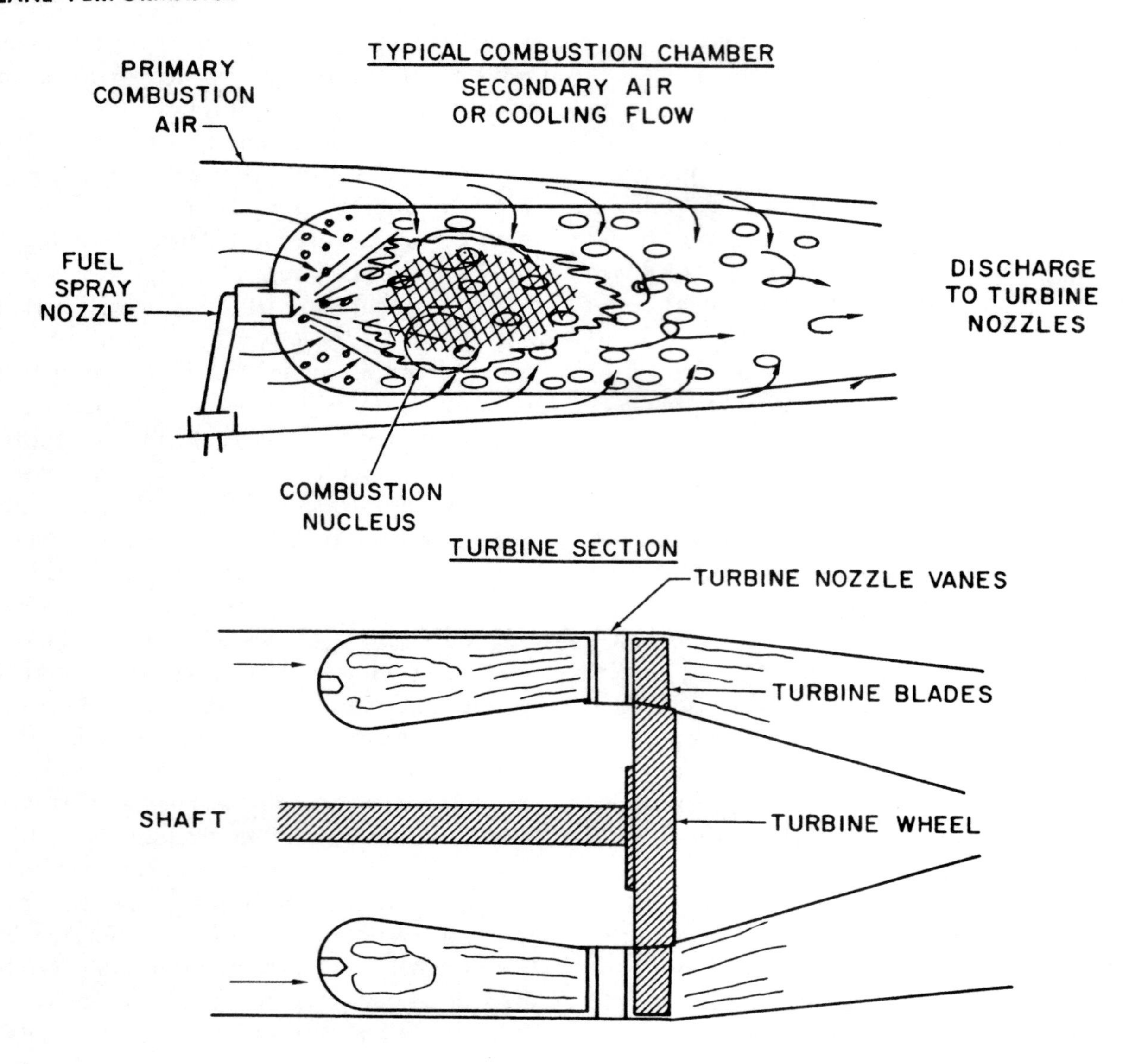

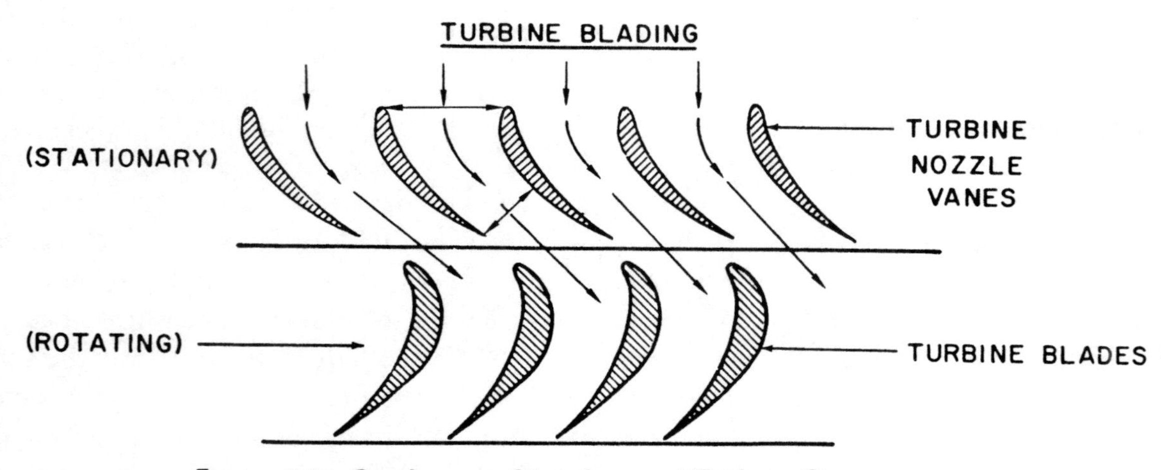

Figure 2.8. Combustion Chamber and Turbine Components

maintain a nucleus of combustion in the combustion chamber. In the normal combustion process, the speed of flame propagation is quite low and, if the local velocities are too high at the forward end of the combustion chamber, poor combustion will result and it is likely that the flame will blow out. The secondary air—or cooling flow—is introduced downstream from the combustion nucleus to dilute the combustion products and lower the discharge gas temperature.

The fuel nozzle must provide a finely atomized, evenly distributed spray of fuel through a wide range of flow rates. Very specialized design is necessary to provide a nozzle with suitable characteristics. The spray pattern and circulation in the combustion chamber must make efficient use of the fuel by complete combustion. The temperatures in the combustion nucleus can exceed 1,700° to 1,800° C but the secondary air will dilute the gas and reduce the temperature to some value which can be tolerated in the turbine section. A pressure drop will occur through the combustion chamber to accelerate the combustion gas rearward. In addition, turbulence and fluid friction will cause a pressure drop but this loss must be held to the minimum incurred by providing complete combustion. Heat transferred through the walls of the combustion chamber constitutes a loss of thermal energy and should be held to a minimum. Thus, the combustion chamber should enclose the combustion space with a minimum of surface area to minimize heat and friction losses. Hence, the "annular" type combustion chamber offers certain advantages over the multiple "can" type combustion chamber.

The *turbine section* is the most critical element of the turbojet engine. The function of the turbine is to extract energy from the combustion gases and furnish power to drive the compressor and accessories. In the case of the turboprop engine, the turbine section must extract a very large portion of the exhaust gas energy to drive the propeller in addition to the compressor and accessories.

The combustion chamber delivers high energy combustion gases to the turbine section at high pressure and tolerable temperature. The turbine nozzle vanes are a row of stationary blades immediately ahead of the rotating turbine. These blades form the nozzles which discharge the combustion gases as high velocity jets onto the rotating turbine. In this manner, the high pressure energy of the combustion gases is converted into kinetic energy and a pressure and temperature drop takes place. The function of the turbine blades operating in these jets is to develop a tangential force along the turbine wheel thus extracting mechanical energy from the combustion gases. This is illustrated in figure 2.8.

The form of the turbine blades may be a combination of two distinct types. The *impulse* type turbine relies upon the nozzle vanes to accomplish the conversion of combustion gas static pressure to high velocity jets. The impulse turbine blades are shaped to produce a large deflection of the gas and develop the tangential force by the flow direction change. In such a design, negligible velocity and pressure drop occurs with the flow across the turbine rotor blades. The *reaction* type turbine differs in that large velocity and pressure changes occur across the turbine rotor blades. In the reaction turbine, the stationary nozzle vanes serve only to guide the combustion gas onto the turbine rotor with negligible changes in velocity and pressure. The reaction turbine rotor blades are shaped to provide a pressure drop and velocity increase across the blades and the reaction from this velocity increase provides the tangential force on the wheel. Generally, the turbine design is a form utilizing some feature of each of the two types.

The turbine blade is subjected to high centrifugal stresses which vary as the square of the rotative speed. In addition, the blade is subjected to the bending and torsion of the tangential impulse-reaction forces. The blade must withstand these stresses which are generally of a vibratory and cyclic nature while at high temperatures. The elevated temperatures at which the turbine must function produce extreme conditions for structural creep and fatigue considerations. Consequently, the engine speed and temperature operating limits demand very careful consideration. Excessive engine temperatures or speeds may produce damage which is immediately apparent. However, creep and fatigue damage is cumulative and even though damage may not be immediately apparent by visual inspection, proper inspection methods (other than visual) must be utilized and proper records kept regarding the occurrence.

Actually, the development of high temperature alloys for turbines is a critical factor in the development of high efficiency, high output aircraft gas turbines. The higher the temperature of gases entering the turbine, the higher can be the temperature and pressure of the gases at discharge from the turbine with greater exhaust jet velocity and thrust.

The function of the *tailpipe or exhaust nozzle* is to discharge the exhaust gases to the atmosphere at the highest possible velocity to produce the greatest momentum change and thrust. If a majority of the expansion occurs through the turbine section, there remains only to conduct the exhaust gases rearward with a minimum energy loss. However, if the turbine operates against a noticeable back pressure, the nozzle must convert the remaining pressure energy into exhaust gas velocity. Under ideal conditions, the nozzle would expand the flow to the ambient static pressure at the exhaust and the area distribution in the nozzle must provide these conditions. When the ratio of exhaust gas pressure to ambient pressure is relatively low and incapable of producing sonic flow, a converging nozzle provides the expansion. The exit area must be of proper size to bring about proper exit conditions. If the exit area is too large, incomplete expansion will take place; if the exit area is too small, an over expansion tendency results. The exit area can affect the upstream conditions and must be properly proportioned for overall performance.

When the ratio of exhaust gas pressure to ambient pressure is greater than some critical value, sonic flow can exist and the nozzle will be choked or limited to some maximum flow. When supersonic exhaust gas velocities are required to produce the necessary momentum change, the expansion process will require the convergent-divergent nozzle illustrated in figure 2.9. With sufficient pressure available the initial expansion in the converging portion is subsonic increasing to sonic velocity at the throat. Subsequent expansion in the divergent portion of the nozzle is supersonic and the result is the highest exit velocity for a given pressure ratio and mass flow. When the pressure ratio is very high the final exit diameter required to expand to ambient pressure may be very large but is practically limited to the fuselage or nacelle afterbody diameter. If the exhaust gases exceed sonic velocity, as is possible in a ramjet combustion chamber or afterburner section, only the divergent portion of the nozzle may be necessary.

Figure 2.9 provides illustration of the function of the various engine components and the changes in static pressure, temperature, and velocity through the engine. The conditions at the inlet provide the initial properties of the engine airflow. The compressor section furnishes the compression pressure rise with a certain unavoidable but undesirable increase in temperature. High pressure air delivered to combustion chamber receives heat from the combustion of fuel and experiences a rise in temperature. The fuel flow is limited so that the turbine inlet temperature is within limits which can be tolerated by the turbine structure. The combustion takes place at relatively constant pressure and initially low velocity. Heat addition then causes large increases in gas volume and flow velocity.

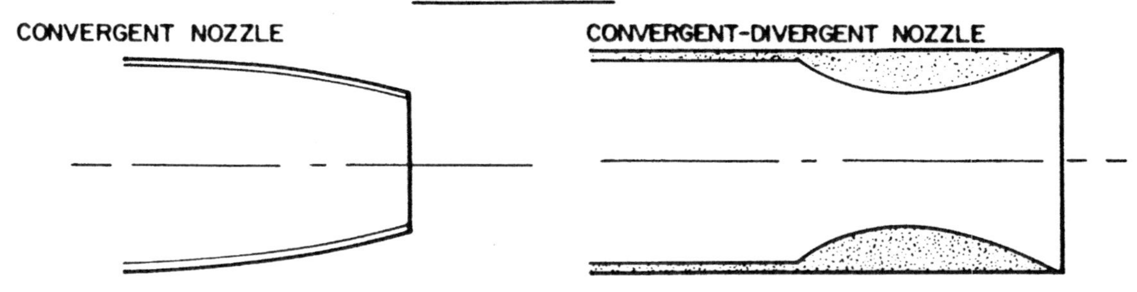

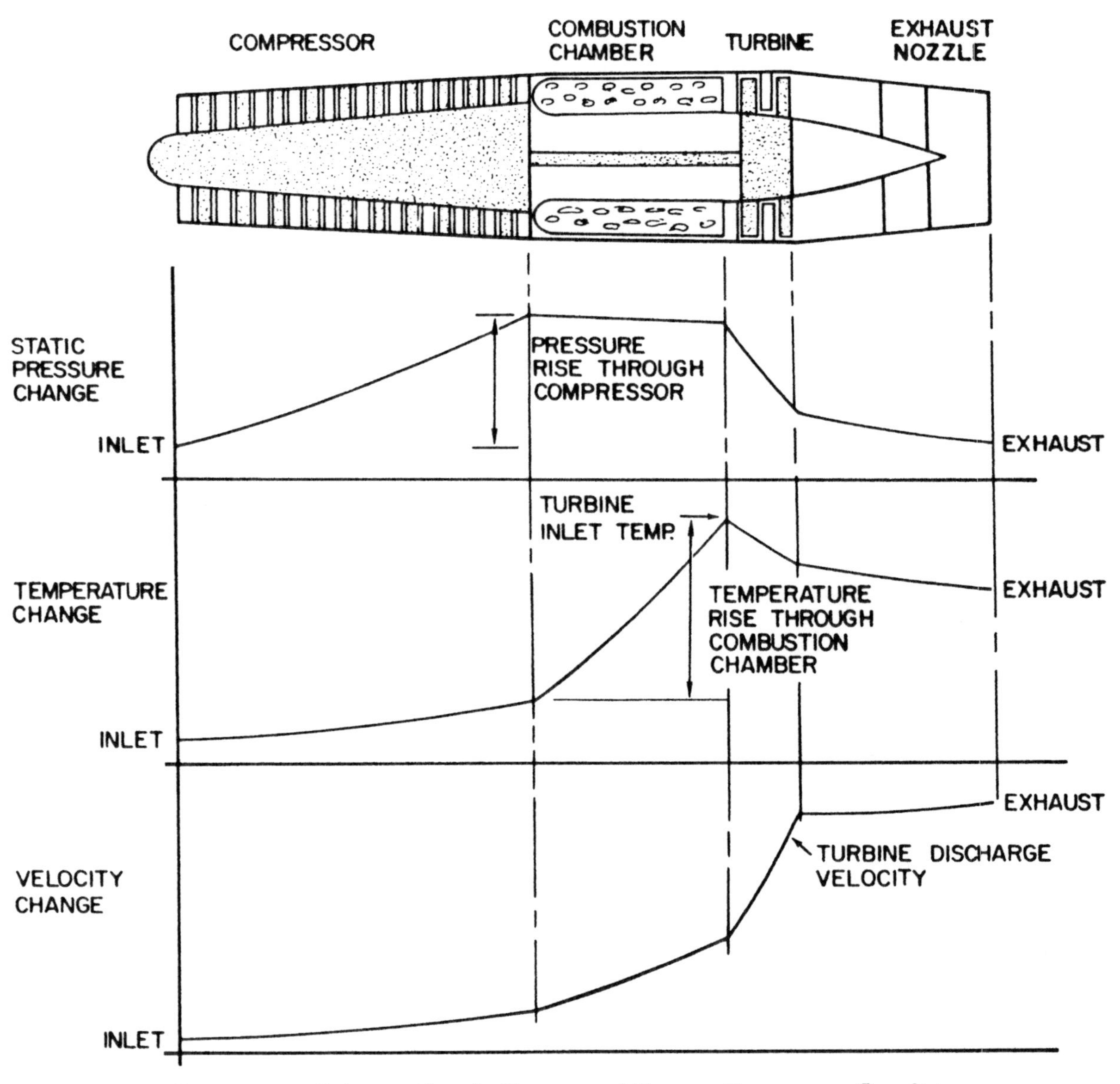

Figure 2.9. Exhaust Nozzle Types and Engine Operating Conditions

Generally, the overall fuel-air ratio of the turbojet is quite low because of the limiting turbine inlet temperature. The overall air-fuel ratio is usually some value between 80 to 40 during ordinary operating conditions because of the large amount of secondary air or cooling flow.

High temperature, high energy combustion gas is delivered to the turbine section where power is extracted to operate the compressor section. Partial or near-complete expansion can take place through the turbine section with the accompanying pressure and temperature drop. The exhaust nozzle completes the expansion by producing the final jet velocity and momentum change necessary in the development of thrust.

TURBOJET OPERATING CHARACTERISTICS. The turbojet engine has many operating characteristics which are of great importance to the various items of jet airplane performance. Certain of these operating characteristics will provide a strong influence on the range, endurance, etc., of the jet-powered airplane. Other operating characteristics will require operating techniques which differ greatly from more conventional powerplants.

The turbojet engine is essentially a thrust-producing powerplant and the propulsive power produced is a result of the flight speed. The variation of available thrust with speed is relatively small and the engine output is very nearly constant with flight speed. The momentum change given the engine airflow develops thrust by the following relationship:

$$Ta=Q(V_2-V_1)$$

where

Ta= thrust available, lbs.

Q= mass flow, slugs per sec.

V_1= inlet or flight velocity, ft. per sec.

V_2= jet velocity, ft. per see.

Since an increase in flight speed will increase the magnitude of V_1, a constant thrust will be obtained only if there is an increase in mass flow, Q, or jet velocity, V_2, When at low velocity, an increase in velocity will reduce the velocity change through the engine without a corresponding increase in mass flow and the available thrust will decrease. At higher velocity, the beneficial ram helps to overcome this effect and the available thrust no longer decreases, but increases with speed.

The propulsive power available from the turbojet engine is the product of available thrust and velocity. The propulsive horsepower available from the turbojet engine is related by the following expression:

$$Pa=\frac{TaV}{325}$$

where

Pa= propulsive power available, h.p.

Ta= thrust available, lbs.

V= flight velocity, knots

The factor of 325 evolves from the use of the nautical unit of velocity and implies that each pound of thrust developed at 325 knots is the equivalent of one horsepower of propulsive power. Since the thrust of the turbojet engine is essentially constant with speed, the power available increases almost linearly with speed. In this sense, a turbojet with 5000 lbs. of thrust available could produce a propulsive power of 5,000 h.p. at 325 knots or 10,000 h.p. at 650 knots. The tremendous propulsive power at high velocities is one of the principal features of the turbojet engine. When the engine RPM and operating altitude are fixed, the variation with speed of turbojet thrust and power available is typified by the first graph of figure 2.10.

The variation of thrust output with engine speed is a factor of great importance in the operation of the turbojet engine. By reasoning that static pressure changes depend on the square of the flow velocity, the changes of pressure throughout the turbojet engine would be expected to vary as the square of the rotative speed, N. However, since a variation in rotative speed will alter airflow, fuel flow, compressor and turbine efficiency, etc., the thrust variation will be much greater than just the second power of rotative speed. Instead of thrust being proportional to N^2, the typical fixed geometry engine develops thrust approximately proportional to $N^{3.5}$. Of course, such a variation is particular to constant altitude and speed.

Figure 2.10 illustrates the variation of percent maximum thrust with percent maximum RPM for a typical fixed geometry engine. Typical values from this graph are as follows:

Percent max. RPM	*Percent max. thrust*
100	100 (of course)
99	96.5
95	83.6
90	69.2
80	45.8
70	28.7

Note that in the top end of power output, each 1 percent RPM change causes a 3.5-percent change in thrust output. This illustrates the power of variation of thrust with rotative speed which, in this example, is $N^{3.5}$. Also note that the top 20 percent of RPM controls more than half of the output thrust.

While the fixed geometry engine develops thrust approximately proportional to $N^{3.5}$, the engine with variable geometry will demonstrate a much more powerful effect of rotative speed. When the jet engine is equipped with a variable nozzle, multispool compressor, variable stator blades, etc., the engine is more likely to develop thrust proportional to rotative speed from values of $N^{4.5}$ to $N^{6.0}$. For example, if a variable geometry engine develops thrust proportional to $N^{5.0}$, each one per cent RPM change causes a 5.0-percent thrust change at the top end of power output. Also, the top 13 percent of RPM would control the top 50 percent of thrust output.

The powerful variation of thrust with engine speed has certain ramifications which should be appreciated. If the turbojet powerplant operates at less than the "trimmed" or adjusted speed for maximum thrust, the deficiency of thrust for takeoff may cause a considerable increase in takeoff distance. During approach, an excessively low RPM may cause very low thrust and produce a very steep glide path. In addition, the low RPM range involves the much greater engine acceleration time to produce thrust for a waveoff. Another complication exists when the thrust is proportional to some large power of rotative speed, e.g., $N^{5.0}$. The small changes in RPM produce such large variations in thrust that instruments other than the tachometer must be furnished for accurate indication of thrust output.

The "specific fuel consumption, c_t" is an important factor for evaluating the performance and efficiency of operation of a turbojet engine. The specific fuel consumption is the proportion between the fuel flow (in lbs. per hr.) and the thrust (in lbs.). For example, an engine which has a fuel flow of 14,000 lbs. per hr. and a thrust of 12,500 lbs. has a specific fuel consumption of:

$$c_t = \frac{\text{Fuel flow}}{\text{Thrust}}$$

$$c_t = \frac{14{,}000 \text{ lbs./hr.}}{12{,}500 \text{ lbs.}}$$

$$c_t = 1.12 \text{ lbs./hr./lb.}$$

Thus, each unit pound of thrust requires 1.12 lbs. per hr. fuel flow. Obviously, high engine efficiency would be indicated by a low value of c_t. Typical values for turbojet engines with relatively high pressure ratios range from 0.8 to 1.2 at design operating conditions in subsonic flight. High energy fuels and greater pressure ratios tend to produce the lower values of c_t. Supersonic flight with the attendant inlet losses and high compressor inlet air temperatures tend to increase the specific fuel consumption to values of 1.2 to 2.0. Of course, the use of an afterburner is quite inefficient

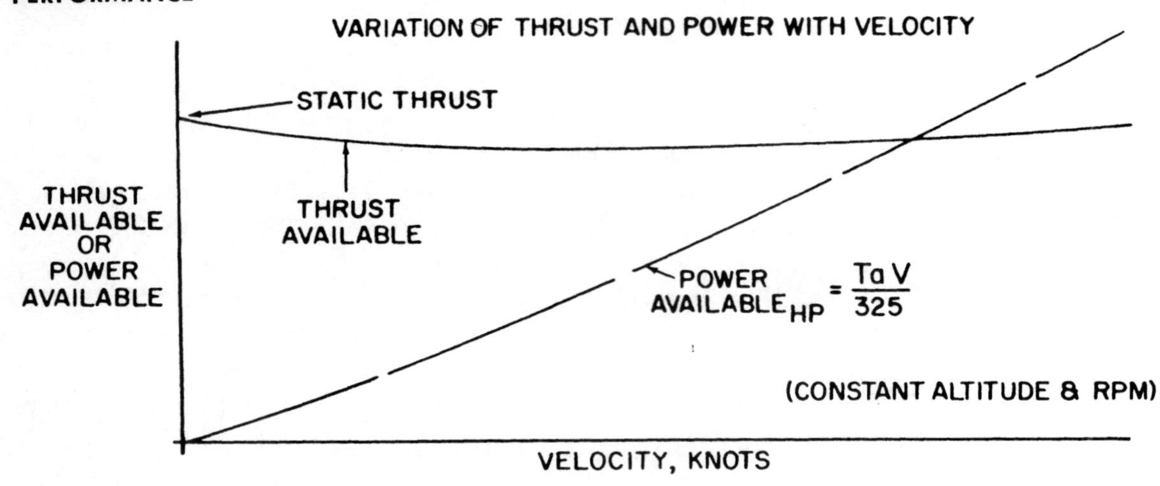

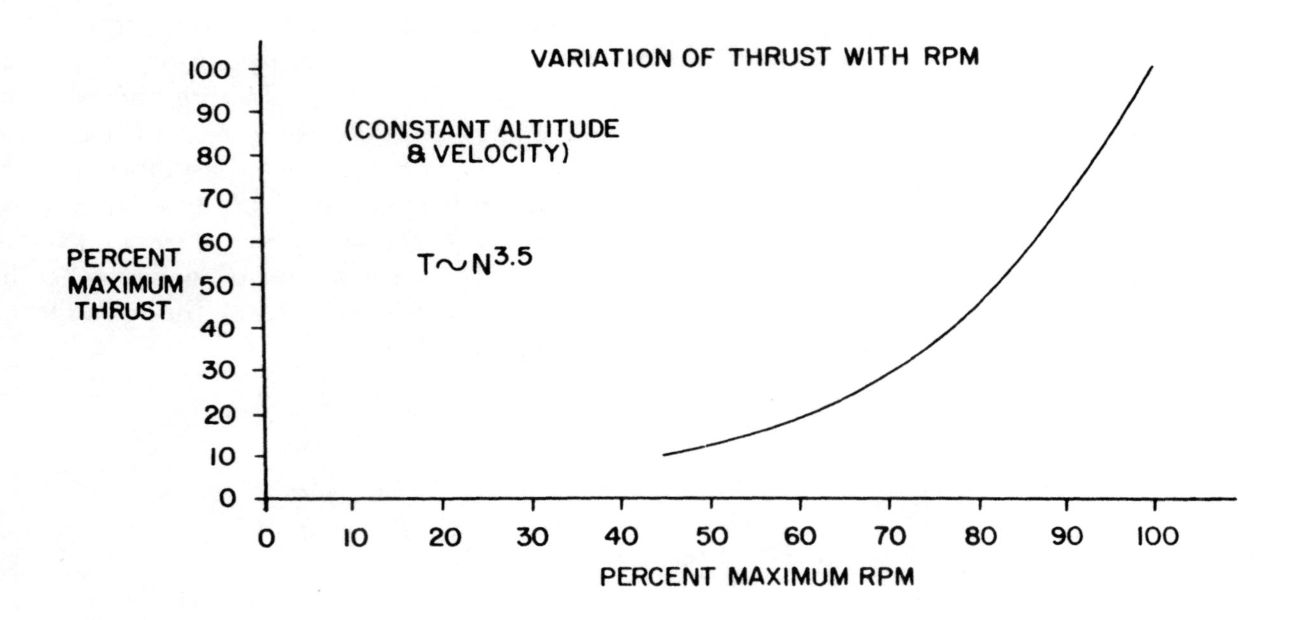

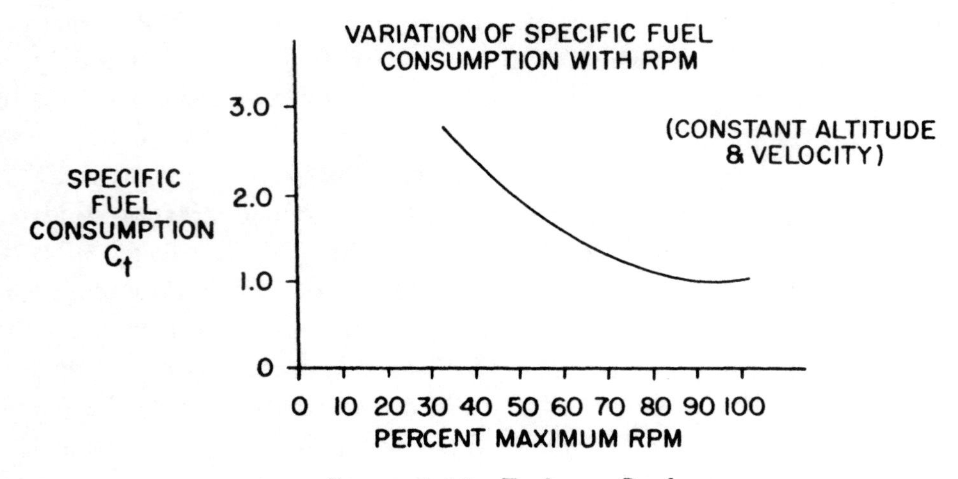

Figure 2.10. Turbojet Performance

due to the low combustion pressure and values of c_t from 2.0 to 4.0 are typical with afterburner operation.

The turbojet engine usually has a strong preference for high RPM to produce low specific fuel consumption. Since the normal rated thrust condition is a particular design point for the engine, the minimum value of c_t will occur at or near this range of RPM. The illustration of figure 2.10 shows a typical variation of c_t with percent maximum RPM where values of RPM less than 80 to 85 percent produce a specific fuel consumption much greater than the minimum obtainable. This preference for high RPM to obtain low values of c_t is very pronounced in the fixed geometry engine. Turbojet engines with multispool compressors tend to be less sensitive in this respect and are more flexible in their operating characteristics. Whenever low values of c_t are necessary to obtain range or endurance, the preference of the turbojet engine for the design operating RPM can be a factor of great influence.

Altitude is one factor which strongly affects the performance of the turbojet engine. An increase in altitude produces a decrease in density and pressure and, if below the tropopause, a decrease in temperature. If a typical nonafterburning turbojet engine is operated at a constant RPM and true airspeed, the variation of thrust and specific fuel consumption with altitude can be approximated from figure 2.11. The variation of density in the standard atmosphere is shown by the values of density ratio at various altitudes. Typical values of the density ratio at specific altitudes are as follows:

Altitude, ft.:	*Density ratio*
Sea level	1.000
5,000	.8617
10,000	.7385
22,000	.4976
35,000	.3099
40,000	.2462
50,000	.1532

If the fixed geometry engine is operated at a constant V (TAS) in subsonic flight and constant N (RPM) the inlet velocity, inlet ram, and compressor pressure ratio are essentially constant with altitude. An increase in altitude then causes the engine air mass flow to decrease in a manner very nearly identical to the altitude density ratio. Of course, this decrease in mass flow will produce a significant effect on the output thrust of the engine. Actually, the variation of thrust with altitude is not quite as severe as the density variation because favorable decreases in temperature occur. The decrease in inlet air temperature will provide a relatively greater combustion gas energy and allow a greater jet velocity. The increase in jet velocity somewhat offsets the decrease in mass flow. Of course, an increase in altitude provides lower temperatures below the tropopause. Above the tropopause, no further favorable decrease in temperature takes place so a more rapid variation of thrust will take place. The approximate variation of thrust with altitude is represented by figure 2.11 and some typical values at specific altitudes are as follows:

Altitude, ft.:	Ratio of $\left(\frac{\textit{Thrust at altitude}}{\textit{Thrust at sea level}}\right)$
Sea level	1.000
5,000	.888
10,000	.785
20,000	.604
35,000	.392
40,000	.315
50,000	.180

Since the change in density with altitude is quite rapid at low altitude turbojet takeoff performance will be greatly affected at high altitude. Also note that the thrust at 35,000 ft. is approximately 39 percent of the sea level value.

The thrust added by the afterburner of a turbojet engine is not affected so greatly by altitude as the basic engine thrust. The use of afterburner may provide a thrust increase of 50 percent at low altitude or as much as 100 percent at high altitude.

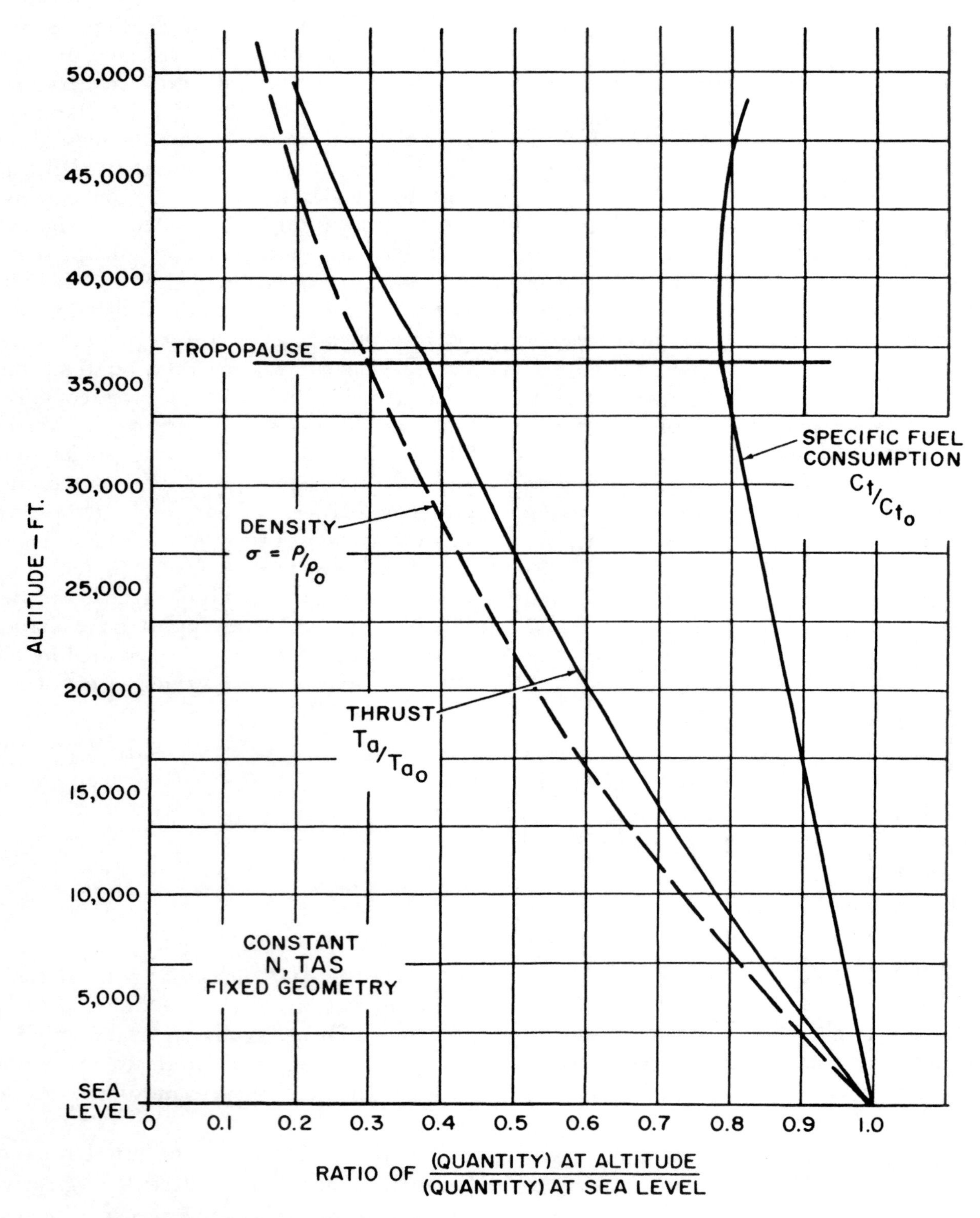

Figure 2.11. Approximate Effect of Altitude on Engine Performance

When the inlet ram and compressor pressure ratio is fixed, the principal factor affecting the specific fuel consumption is the inlet air temperature. When the inlet air temperature is lowered, a given heat addition can provide relatively greater changes in pressure or volume. As a result, a given thrust output requires less fuel flow and the specific fuel consumption, c_t, is reduced. While the effect of altitude on specific fuel consumption does not compare with the effect on thrust output, the variation is large enough to strongly influence range and endurance conditions. Figure 2.11 illustrates a typical variation of specific fuel consumption with altitude. Generally, the specific fuel consumption decreases steadily with altitude until the tropopause is reached and the specific fuel consumption at this point is approximately 80 percent of the sea level value.

Above the tropopause the temperature is constant and altitudes slightly above the tropopause cause no further decrease in specific fuel consumption. Actually, altitudes much above the tropopause bring about a general deterioration of overall engine efficiency and the specific fuel consumption begins an increase with altitude. The extreme altitudes above the tropopause produce low combustion chamber pressures, low compressor Reynolds Numbers, low fuel flow, etc. which are not conducive to high engine efficiency.

Because of the variation of c_t with altitude, the majority of turbojet engines achieve maximum efficiency at or above 35,000 ft. For this reason, the turbojet airplane will find optimum range and endurance conditions at or above 35,000 ft. provided the aircraft is not thrust or compressibility limited at these altitudes.

The *governing apparatus* of the turbojet engine consists primarily of the items which control the flow of fuel to the engine. In addition, there may be included certain functions which operate variable nozzles, variable stator vanes, variable inlets, etc. Generally, the fuel control and associated items should regulate fuel flow, nozzle area, etc. to provide engine performance scheduled by the throttle or power lever. These regulatory functions provided must account for variations in altitude, temperature, and flight velocity.

One principal governing factor which must be available is that a selected power setting (RPM) must be maintained throughout a wide range of flight conditions. Figure 2.12 illustrates the variation of fuel flow with RPM for a turbojet operating at a particular set of flight conditions. Curve *1* depicts the variation with RPM of the fuel flow required for stabilized, steady state operation of the engine. Each point along this curve *1* defines the fuel flow which is necessary to achieve equilibrium at a given RPM. The steady state fuel flow produces a turbine power to equal the compressor power requirement at a particular RPM. The throttle position primarily commands a given engine speed and, as changes occur in the ambient pressure, temperature, and flight speed, the steady state fuel flow will vary. The governing apparatus must account for these variations in flight conditions and maintain the power setting scheduled by throttle position.

In addition to the maintenance of steady state operation, the fuel control and associated engine control items must provide for the transient conditions of engine acceleration and deceleration. In order to accelerate the engine, the fuel control must supply a fuel flow greater than that required for steady state operation to produce a turbine power greater than the compressor power requirement. However, the additional fuel flow to accelerate the engine must be controlled and regulated to prevent any one or combination of the following items:

(1) compressor stall or surge
(2) excessive turbine inlet temperature
(3) excessively rich fuel-air ratio which may not sustain combustion

Generally, the stall-surge and turbine temperature limits predominate to form an acceleration fuel flow boundary typified by curve

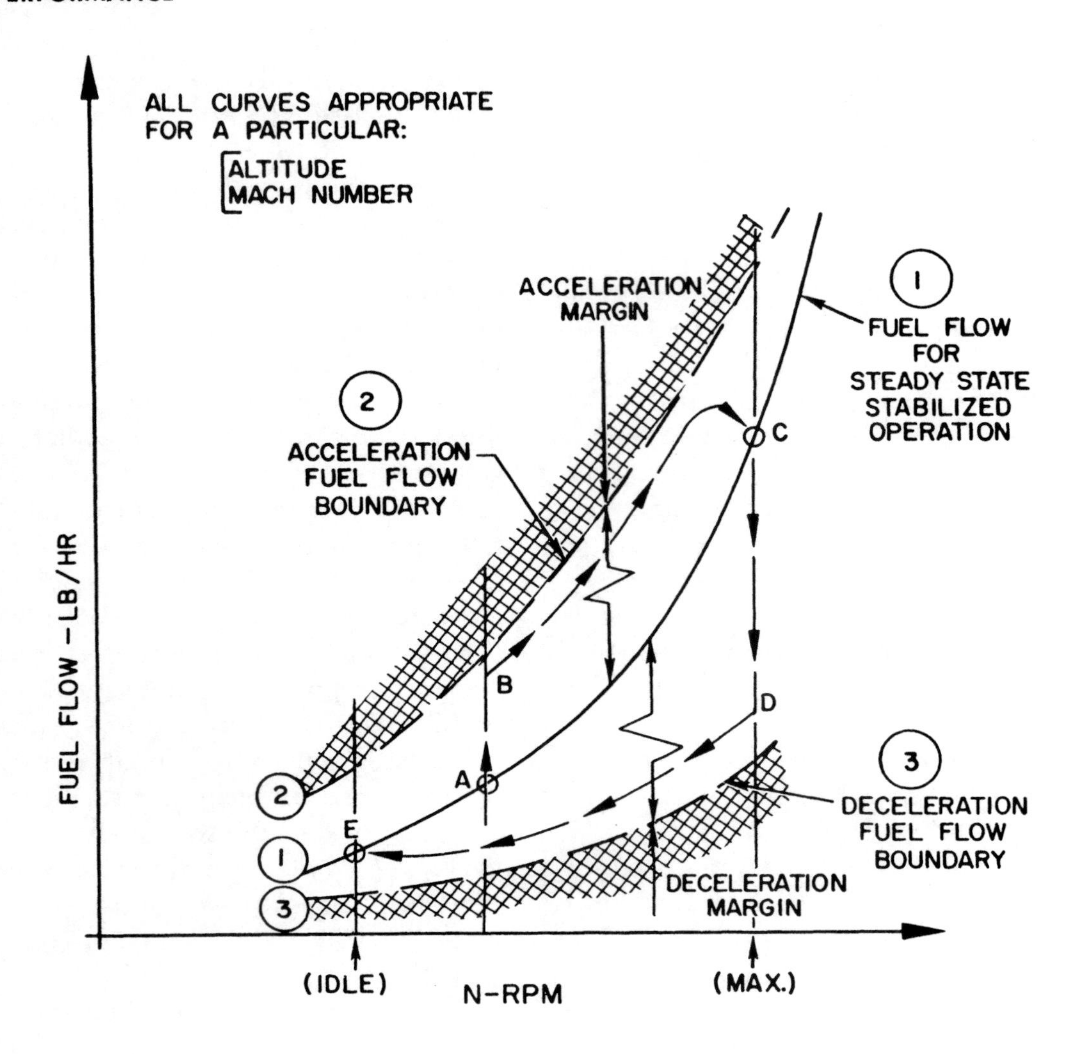

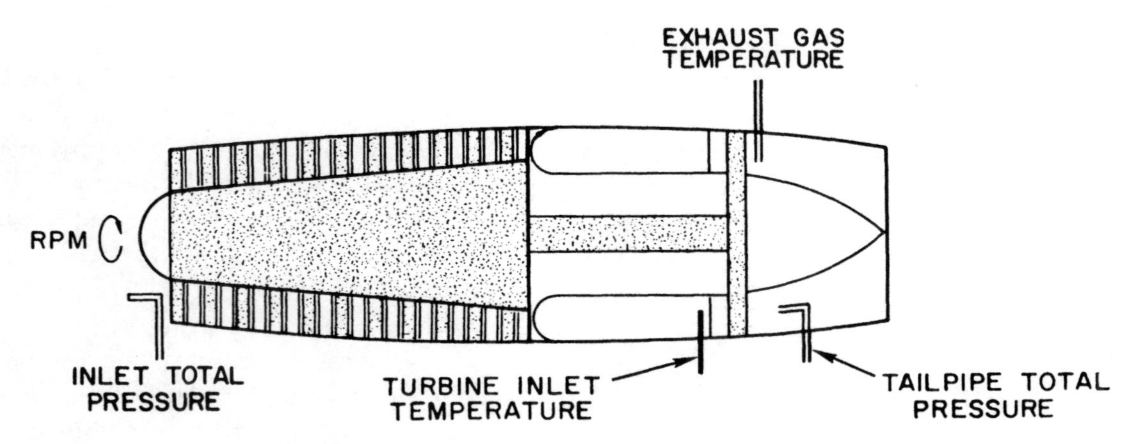

Figure 2.12. Engine Governing and Instrumentation

2 of figure 2.12. Curve 2 of this illustration defines an upper limit of fuel flow which can be tolerated within stall-surge and temperature limits. The governing apparatus of the engine must limit the acceleration fuel flow within this boundary.

To appreciate the governing requirements during the acceleration process, assume the engine described in figure 2.12 is in steady state stabilized operation at point *A* and it is desired to accelerate the engine to maximum RPM and stabilize at point *C*. As the throttle is placed at the position for maximum RPM, the fuel control will increase the fuel flow to point *B* to provide acceleration fuel flow. As the engine accelerates and increases RPM, the fuel control will continue to increase the fuel flow within the acceleration boundary until the engine speed approaches the controlled maximum RPM at point *C*. As the engine speed nears the maximum at point *C*, the fuel control will reduce fuel flow to produce stabilized operation at this point and prevent the engine overspeeding the commanded RPM. Of course, if the throttle is opened very gradually, the acceleration fuel flow is barely above the steady state condition and the engine does not approach the acceleration fuel flow boundary. While this technique is recommended for ordinary conditions to achieve trouble free operation and good service life, the engine must be capable of good acceleration to produce rapid thrust changes for satisfactory flight control.

In order for the powerplant to achieve minimum acceleration times, the fuel control must provide acceleration fuel flow as close as practical to the acceleration boundary. Thus, a maximum controlled acceleration may produce limiting turbine inlet temperatures or slight incipient stall-surge of the compressor. Proper maintenance and adjustment of the engine governing apparatus is essential to produce minimum acceleration times without incurring excessive temperatures or heavy stall-surge conditions.

During deceleration conditions, the minimum allowable fuel flow is defined by the lean limit to support combustion. If the fuel flow is reduced below some critical value at each RPM, lean blowout or flameout will occur. This condition is illustrated by curve 3 of figure 2.12 which forms the deceleration fuel flow boundary. The governing apparatus must regulate the deceleration fuel flow within this boundary.

To appreciate the governing requirements during the deceleration process, assume the engine described in figure 2.12 is in stabilized, steady state operation at point *C* and it is desired to decelerate to idle conditions and stabilize at point *E*. As the throttle is placed at the position for idle RPM, the fuel control will decrease the fuel flow to point *D* to provide the deceleration fuel flow. As the engine decelerates and decreases RPM, the fuel governing will continue to decrease the fuel flow within the deceleration boundary until the idle fuel flow is reached and RPM is established at point *E*. Of course, if the throttle is closed very slowly, the deceleration fuel flow is barely below the steady state condition and the engine does not approach the deceleration fuel flow boundary. The fuel control must provide a deceleration flow close to the boundary to provide rapid decrease in thrust and satisfactory flight control.

In most cases, the deceleration fuel flow boundary is considerably below the steady state fuel flow and no great problem exists in obtaining satisfactory deceleration characteristics. In fact, the greater problem is concerned with obtaining proper acceleration characteristics. For the majority of centrifugal flow engines, the acceleration boundary is set usually by temperature limiting conditions rather than compressor surge conditions. Peak operating efficiency of the centrifugal compressor is obtained at flow conditions which are below the surge limit, hence acceleration fuel flow boundary is determined by turbine temperature limits. The usual result is that the centrifugal flow engine has relatively large acceleration margins and good acceleration characteristics result with the low rotational inertia. The axial flow compressor must operate relatively close to the stall-surge limit to obtain peak efficiency. Thus, the acceleration fuel flow boundary for the axial flow engine is set by these stall-surge limits which are more immediate to steady state conditions than turbine temperature limits. The fixed geometry axial flow engine encounters relatively small acceleration margins and, when compared to the centrifugal flow engine with larger acceleration margins and lower rotational inertia, has inferior acceleration characteristics. Certain variation of the axial flow engine such as variable nozzles, variable stator blades, multiple-spool compressors, etc., greatly improve the acceleration characteristics.

A note of caution is appropriate at this point. If the main fuel control and governing apparatus should malfunction or become inoperative and an unmodulated secondary or emergency system be substitued, extreme care must be taken to avoid abrupt changes in throttle position. In such a case, very gradual movement of the throttle is necessary to accomplish changes in power setting without excessive turbine temperatures, compressor stall or surge, or flameout.

There are various instruments to relate important items of turbojet engine performance. Certain combinations of these instruments are capable of immediately relating the thrust output of the powerplant in a *qualitative* manner. It is difficult to provide an instrument or combination of instruments which immediately relate the thrust output in a *quantitative* manner. As a result, the pilot must rely on a combination of instrument readings and judge the output performance according to standard values particular to the powerplant. Some of the usual engine indicating instruments are as follows:

(1) The tachometer provides indication of engine speed, N, by percent of the maximum RPM. Since the variation of thrust with RPM is quite powerful, the tachometer indication is a powerful reference.

(2) The exhaust gas temperature gauge provides an important reference for engine operating limitations. While the temperature probe may be located downstream from the turbine (tailpipe or turbine discharge temperature) the instrument should provide an accurate reflection of temperatures upstream in the turbine section. The exhaust gas temperature relates the energy change accomplished by fuel addition.

(3) The fuel flowmeter can provide a fair reflection of thrust output and operating efficiency. Operation at high density altitude or high inlet air temperatures reduces the output thrust and this effect is related by a reduction of fuel flow.

(4) The tailpipe total pressure ($p+q$ in the tailpipe) can be correlated with the jet thrust for a given engine geometry and set of operating conditions. The output thrust can be related accurately with various combinations of compressor inlet total pressure, tailpipe total pressure, ambient pressure and temperature. Hence, pressure differential (Δp), pressure ratio, and tailpipe total pressure instruments can provide more accurate immediate indications of output thrust than combined indications of RPM and EGT. This is especially true with variable geometry or multiple spool engines.

Many other specialized instruments furnish additional information for more detailed items of engine performance. Various additional engine information is realized from fuel pressure, nozzle positions, compressor inlet air temperature, etc.

TURBOJET OPERATING LIMITATIONS.

The operating characteristics of the turbojet engine provide various operating limitations which must be given due respect. Operation of the powerplant within the specified limitations is absolutely necessary in order to obtain the design service life with trouble-free operation. The following items describe the critical areas encountered during the operational use of the turbojet engine:

(1) The limiting *exhaust gas temperatures* provide the most important restrictions to the operation of the turbojet engine. The turbine components are subject to centrifugal loads of rotation, impulse and reaction loads on the blades, and various vibratory loads which may be inherent with the design. When the turbine components are subject to this variety of stress in the presence of high temperature, two types of structural phenomena must be considered. When a part is subject to a certain stress at some high temperature, *creep* failure will take place after a period of time. Of course, an increase in temperature or stress will increase the rate at which creep damage is accumulated and reduce the time required to cause failure. Another problem results when a part is subjected to a repeated or cyclic stress. *Fatigue* failure will occur after a number of cycles of a varying stress. An increase in temperature or magnitude of cyclic stress will increase the rate of fatigue damage and reduce the number of cycles necessary to produce failure. It is important to note that both fatigue and creep damage are cumulative.

A gross overstress or overtemperature of the turbine section will produce damage that is immediately apparent. However, the creep and fatigue damage accumulated through periods of less extreme overstress or overtemperature is more subtle. If the turbine is subject to repeated excessive temperatures, the greatly increased rate of creep and fatigue damage will produce failure early within the anticipated service life.

Generally, the operations which produce the highest exhaust gas temperatures are starting, acceleration, and maximum thrust at high altitude. The time spent at these temperatures must be limited arbitrarily to prevent excessive accumulation of creep and fatigue. Any time spent at temperatures in excess of the operational limits for these conditions will increase the possibility of early failure of the turbine components.

While the turbine components are the most critically stressed high temperature elements they are not the only items. The combustion chamber components may be critical at low altitude where high combustion chamber pressures exist. Also, the airframe structure and equipment adjacent to the engine may be subject to quite high temperatures and require provision to prevent damage by excess time at high temperature.

(2) The *compressor stall or surge* has the possibility of producing damaging temperatures in the turbine and combustion chamber or unusual transient loads in the compressor. While the stall-surge phenomenon is possible with the centrifugal compressor, the more common occurrence is with the axial flow compressor. Figure 2.13 depicts the pressure distribution that may exist for steady state operation of the engine. In order to accelerate the engine to a greater speed, more fuel must be added to increase the turbine power above that required to operate the compressor.

Suppose that the fuel flow is increased beyond the steady state requirement without a change in rotative speed. The increased combustion chamber pressure due to the greater fuel flow requires that the compressor discharge pressure be higher. For the instant before an engine speed change occurs, an increase in compressor discharge pressure will be accompanied by a decrease in compressor flow velocity. The equivalent effect is illustrated by the flow components onto the rotating compressor blade of figure 2.13. One component of velocity is due to rotation and this component remains unchanged for a given rotative velocity of the single blade. The axial flow velocity for steady state operation combines with rotational component to define a resultant velocity and direction. If the axial flow component is reduced, the resultant velocity and direction provide an increase in angle of

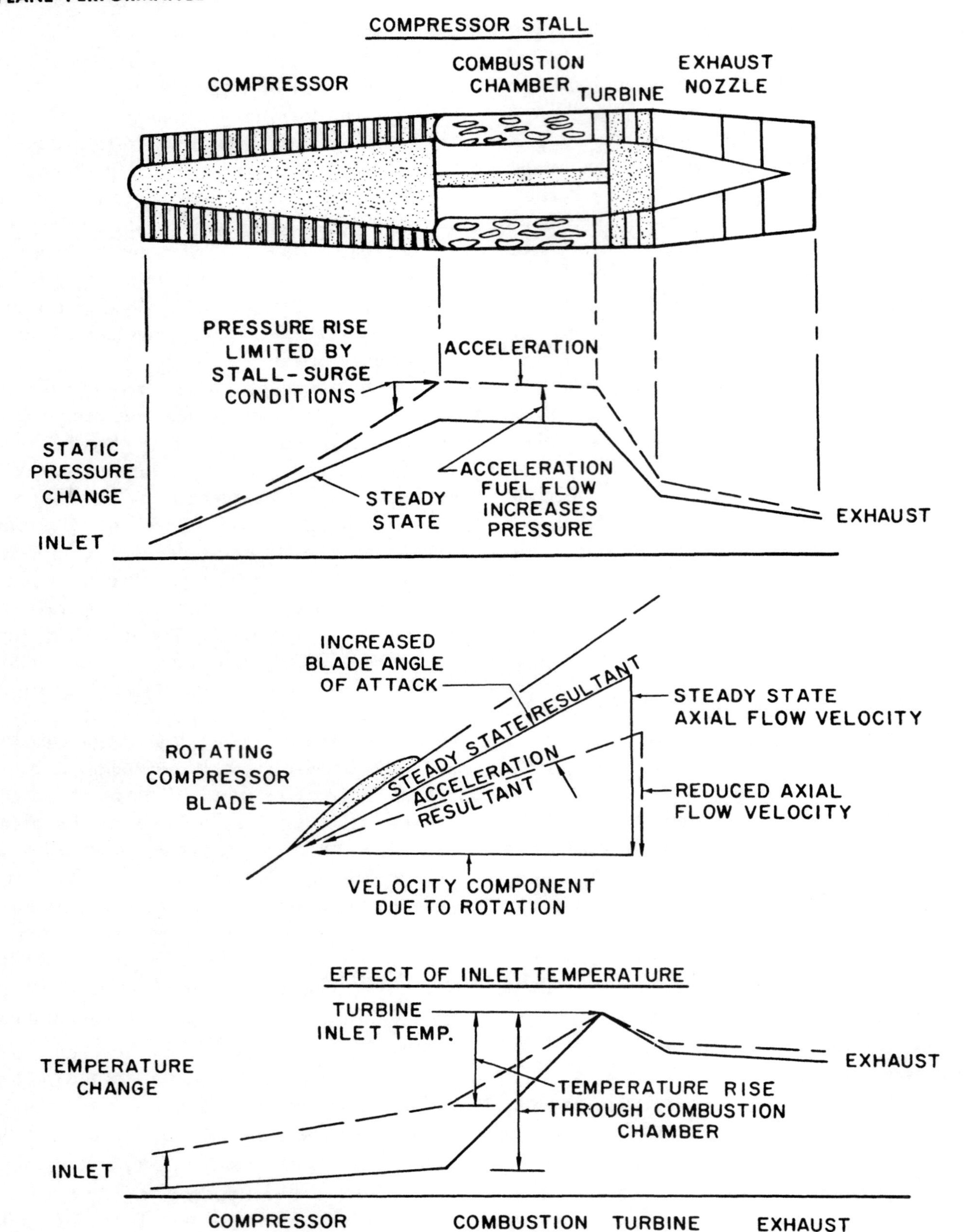

Figure 2.13. Effect of Compressor Stall and Inlet Temperature on Engine Operation

attack for the rotating blade with a subsequent increase in pressure rise. Of course, if the change in angle of attack or pressure rise is beyond some critical value, stall will occur. While the stall phenomenon of a series of rotating compressor blades differs from that of a single airfoil section in a free airstream, the cause and effect are essentially the same.

If an excessive pressure rise is required through the compressor, stall may occur with the attendant breakdown of stable, steady flow through the compressor. As stall occurs, the pressure rise drops and the compressor does not furnish discharge at a pressure equal to the combustion chamber pressure. As a result, a flow reversal or backfire takes place. If the stall is transient and intermittent, the indication will be the intermittent "bang" as backfire and flow reversal take place. If the stall develops and becomes steady, strong vibration and a loud (and possibly expensive) roar develops from the continuous flow reversal. The increase in compressor power required tends to reduce RPM and the reduced airflow and increased fuel flow cause rapid, immediate rise in exhaust gas temperature. The possibility of damage is immediate with the steady stall and recovery must be accomplished quickly by reducing throttle setting, lowering the airplane angle of attack, and increasing airspeed. Generally, the compressor stall is caused by one or a combination of the following items:

(*a*) A malfunctioning fuel control or governing apparatus is a common cause. Proper maintenance and adjustment is a necessity for stall-free operation. The malfunctioning is most usually apparent during engine acceleration.

(*b*) Poor inlet conditions are typical at high angles of attack and sideslip. These conditions reduce inlet airflow and create nonuniform flow conditions at the compressor face. Of course, these conditions are at the immediate control of the pilot.

(*c*) Very high altitude flight produces low compressor Reynolds numbers and an effect similar to that of airfoil sections. As a decrease to low Reynolds numbers reduces the section $c_{l_{max}}$, very high altitudes reduce the maximum pressure ratio of the compressor. The reduced stall margins increase the likelihood of compressor stall.

Thus, the recovery from a compressor stall must entail reduction of throttle setting to reduce fuel flow, lowering angle of attack and sideslip and increasing airspeed to improve inlet condition, and reducing altitude if high altitude is a contributing factor.

(3) While the *flameout* is a rare occurrence with modern engines, various malfunctions and operating conditions allow the flameout to remain a possibility. A uniform mixture of fuel and air will sustain combustion within a relatively wide range of fuel-air ratios. Combustion can be sustained with a fuel-air ratio as rich as one to five or as lean as one to twenty-five. Fuel air ratios outside these limits will not support combustion due to the deficiency of air or deficiency of fuel. The characteristics of the fuel nozzle and spray pattern as well as the governing apparatus must insure that the nucleus of combustion is maintained throughout the range of engine operation.

If the rich limit of fuel-air ratio is exceeded in the combustion chamber, the flame will blow out. While this condition is a possibility the more usual cause of a flameout is exceeding the lean blowout limit. Any condition which produces some fuel-air ratio leaner than the lean limit of combustion will produce a flameout. Any interruption of the fuel supply could bring on this condition. Fuel system failure, fuel system icing, or prolonged unusual attitudes could starve the flow of fuel to the engine. It should be noted the majority of aviation fuels are capable of holding in solution a certain small amount of water. If the aircraft is refueled with relatively warm fuel then flown to high altitude, the lower temperatures can precipitate this water out of solution in liquid or ice crystal form.

High altitude flight produces relatively small air mass flow through the engine and the relatively low fuel flow rate. At these conditions a malfunction of the fuel control and governing apparatus could cause flameout. If the fuel control allows excessively low fuel flow during controlled deceleration, the lean blow out limit may be exceeded. Also, if the governed idle condition allows any deceleration below the idle condition the engine will usually continue to lose speed and flameout.

Restarting the engine in flight requires sufficient RPM and airflow to allow stabilized operation. Generally, the extremes of altitude are most critical for attempted airstart.

(4) An increased *compressor inlet air temperature* can have a profound effect on the output thrust of a turbojet engine. As shown in figure 2.13, an increase in compressor inlet temperature produces an even greater increase in the compressor discharge temperature. Since the turbine inlet temperature is limited to some maximum value, any increase in compressor discharge temperature will reduce the temperature change which can take place in the combustion chamber. Hence, the fuel flow will be limited and a reduction in thrust is incurred.

The effect of inlet air temperature on thrust output has two special ramifications. At *takeoff*, a high ambient air temperature at a given pressure altitude relates a high density altitude. Thus, the takeoff thrust is reduced because of low density and low mass flow. In addition to the loss of thrust due to reduced mass flow, thrust and fuel flow are reduced further because of the high compressor inlet temperature. In flight at *high Mach number*, the aerodynamic heating will provide an increase in compressor inlet temperature. Since the compressor inlet temperature will reflect the compressor discharge temperature and the allowable fuel flow, the compressor inlet air temperature may provide a convenient limit to sustained high speed flight.

(5) The effect of *engine overspeed or critical vibration speed ranges* is important in the service life of an engine. One of the principal sources of turbine loads is the centrifugal loads due to rotation. Since the centrifugal loads vary as the square of the rotative speed, a 5 percent overspeed would produce 10.25 percent overstress ($1.05^2=1.1025$). The large increase in stress with rotative speed could produce very rapid accumulation of creep and fatigue damage at high temperature. Repeated overspeed and, hence, overstress can cause failure early in the anticipated service life.

Since the turbojet engine is composed of many different distributed masses and elastic structure, there are certain vibratory modes and frequencies for the shaft, blades, etc. While it is necessary to prevent any resonant conditions from existing within the normal operating range, there may be certain vibratory modes encountered in the low power range common to ground operation, low altitude endurance, acceleration or deceleration. If certain operating RPM range restrictions are specified due to vibratory conditions, operations must be conducted with a minimum of time in this area. The greatly increased stresses common to vibratory conditions are quite likely to cause fatigue failures of the offending components.

The operating limitations of the engine are usually specified by various combinations of RPM, exhaust gas temperature, and allowable time. The conditions of high power output and acceleration have relatively short times allowable to prevent abuse of the powerplant and obtain good service life. While the allowable times at various high power and acceleration condition appear arbitrary, the purpose is to reduce the spectrum of loading which contributes the most rapid accumulation of creep and fatigue damage. In fact, in some instances, the arbitrary time standards can be set to suit the particular requirements of a certain type of operation. Of course, the effect on service life of any particular load spectrum must be anticipated.

One exception to the arbitrary time standard for operation at high temperatures or sustained high powers is the case of the afterburner operation. When the cooling flow is only that necessary to prevent excessive temperatures for adjacent structure and equipment, sustained operation past a time limit may cause damage to these items.

THRUST AUGMENTATION. Many operating performance conditions may require that additional thrust be provided for short periods of time. Any means of augmenting the thrust of the turbojet engine must be accomplished without an increase in engine speed or maximum turbine section temperature. The various forms of afterburning or water injection allow the use of additional fuel to provide thrust augmentation without increase in engine speed or turbine temperature.

The *afterburner* is a relatively simple means of thrust augmentation and the principal features are light weight and large thrust increase. A typical afterburner installation may add only 10 to 20 percent of the basic engine weight but can provide a 40- to 60-percent increase in the static sea level thrust. The afterburner consists of an additional combustion area aft of the turbine section with an arrangement of fuel nozzles and flameholders. Because the local flow velocities in the afterburner are quite high, the flameholders are necessary to provide the turbulence to maintain combustion within the afterburner section. The turbojet engine operates with airflows greatly in excess of that chemically required to support combustion of engine fuel. This is necessary because of cooling requirements and turbine temperature limitations. Since only 15 to 30 percent of the engine airflow is used in the combustion chamber, the large excess air in the turbine discharge can support combustion of large amounts of additional fuel. Also, there are no highly stressed, rotating members in the afterburner and very high temperatures can be tolerated. The combustion of fuel in the afterburner brings additional increase in temperature and volume and adds considerable energy to the exhaust gases producing increased jet velocity. The major components of the afterburner are illustrated in figure 2.14.

One necessary feature of the turbojet engine equipped with afterburner is a variable nozzle area. As the afterburner begins functioning, the exit nozzle area must increase to accommodate the increased combustion products. If the afterburner were to begin functioning without an increase in exit area, the mass flow through the engine would drop and the temperatures would increase rapidly. The nozzle area must be controlled to increase as afterburner combustion begins. As a result, the engine mass flow is given a large increase in jet velocity with the corresponding increase in thrust.

The combustion of fuel in the afterburner takes place at low pressures and is relatively inefficient. This basic inefficiency of the low pressure combustion is given evidence by the large increase in specific fuel combustion. Generally, the use of afterburner at least will double the specific fuel consumption. As an example, consider a turbojet engine capable of producing 10,000 lbs. of thrust which can develop 15,000 lbs. of thrust with the use of afterburner. Typical values for specific fuel consumption would be $c_t = 1.05$ for the basic engine or $c_t = 2.1$ when the afterburner is in use. The fuel flow during operation would be as follows:

fuel flow = (thrust) (specific fuel consumption)

without afterburner,

fuel flow = (10,000) (1.05)
= 10,500 lbs./hr.

with afterburner,

fuel flow = (15,000) (2.1)
= 31,500 lbs./hr.

The low efficiency of the afterburner is illustrated by the additional 21,000 lbs./hr. of fuel flow to create the additional 5,000 lbs. of

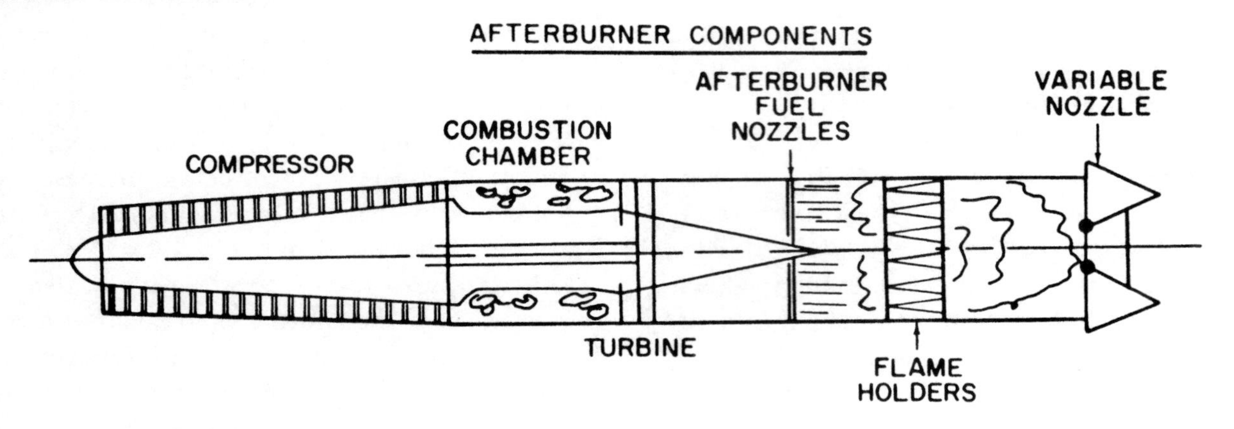

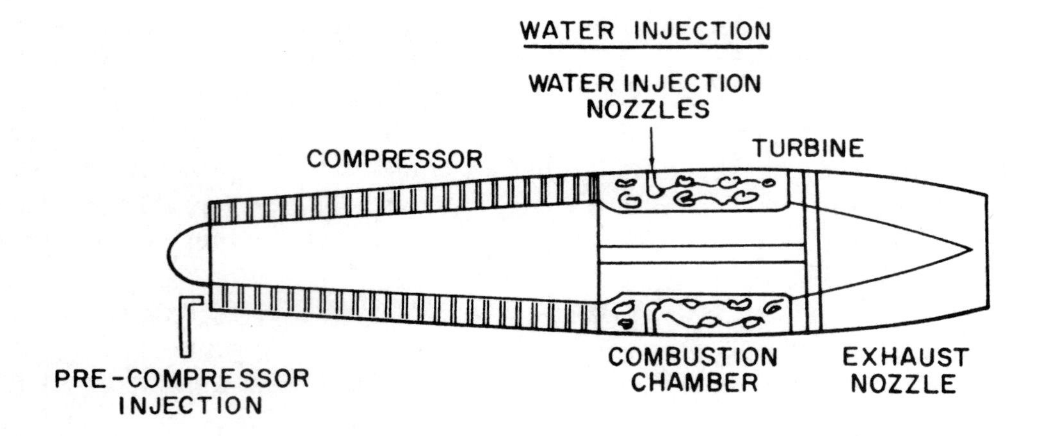

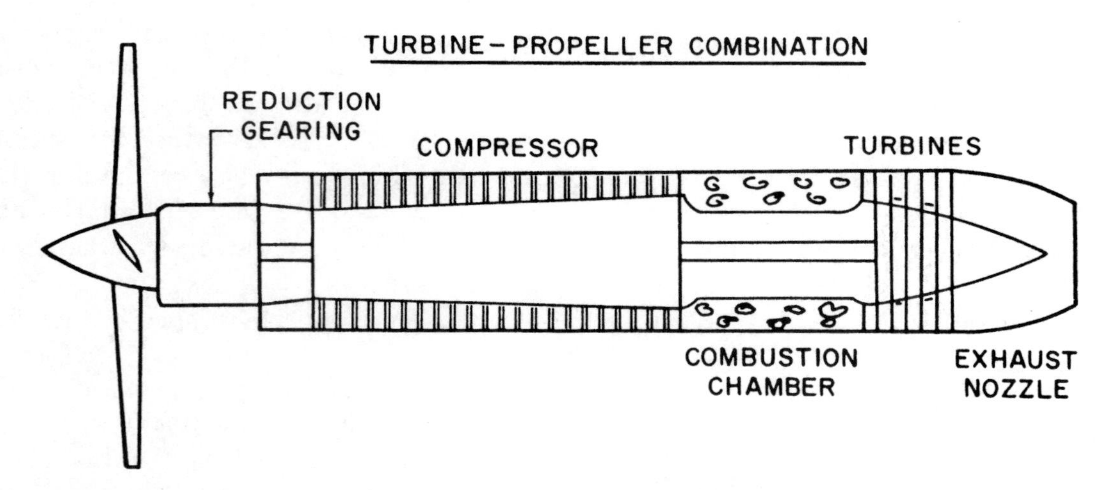

Figure 2.14. Thrust Augmentation and the Gas Turbine-Propeller Combination

thrust. Because of the high fuel consumption during afterburner operation and the adverse effect on endurance, the use of the afterburner should be limited to short periods of time. In addition, there may be limited time for the use of the afterburner due to critical heating of supporting or adjacent structure in the vicinity of the afterburner.

The specific fuel consumption of the basic engine will increase with the addition of the afterburner apparatus. The losses incurred by the greater fluid friction, nozzle and flameholder pressure drop, etc. increase the specific fuel consumption of the basic engine approximately 5 to 10 percent.

The principal advantage of afterburner is the ability to add large amounts of thrust with relatively small weight penalty. The application of the afterburner is most common to the interceptor, fighter, and high speed type aircraft.

The use of *water injection* in the turbojet engine is another means of thrust augmentation which allows the combustion of additional fuel within engine speed and temperature limits. The most usual addition of water injection devices is to supplement takeoff and climbout performance, especially at high ambient temperatures and high altitudes. The typical water injection device can produce a 25 to 35 percent increase in thrust.

The most usual means of water injection is direct flow of the fluid into the combustion chamber. This is illustrated in figure 2.14. The addition of the fluid directly into the combustion chamber increases the mass flow and reduces the turbine inlet temperature. The drop in temperature reduces the turbine power and a greater fuel flow is required to maintain engine speed. Thus, the mass flow is increased, more fuel flow is allowed within turbine limits, and greater energy is imparted to the exhaust gases.

The fluid injected into the combustion chambers is generally a mixture of water and alcohol. The water-alcohol solution has one immediate advantage in that it prevents fouling of the plumbing from the freezing of residual fluid at low temperatures. In addition, a large concentration of alcohol in the mixture can provide part of the additional chemical energy required to maintain engine speed. In fact, the large concentration of alcohol in the injection mixture is a preferred means of adding additional fuel energy. If the added chemical energy is included with the water flow, no abrupt changes in governed fuel flow are necessary and there is less chance of underspeed with fluid injection and overspeed or overtemperature when fluid flow is exhausted. Of course, strict proportions of the mixture are necessary. Since most water injection devices are essentially an unmodulated flow, the use of this device is limited to high engine speed and low altitude to prevent the water flow from quenching combustion.

THE GAS TURBINE-PROPELLER COMBINATION. The turbojet engine utilizes the turbine to extract sufficient power to operate the compressor. The remaining exhaust gas energy is utilized to provide the high exhaust gas velocity and jet thrust. The propulsive efficiency of the turbojet engine is relatively low because thrust is produced by creating a large velocity change with a relatively small mass flow. The gas turbine-propeller combination is capable of producing higher propulsive efficiency in subsonic flight by having the propeller operate on a much greater mass flow.

The turboprop or propjet powerplant requires additional turbine stages to continue expansion in the turbine section and extract a very large percent of the exhaust gas energy as shaft power. In this sense, the turboprop is primarily a power producing machine and the jet thrust is a small amount of the output propulsive power. Ordinarily, the jet thrust of the turboprop accounts for 15 to 25 percent of the total thrust output. Since the turboprop is primarily a power producing machine,

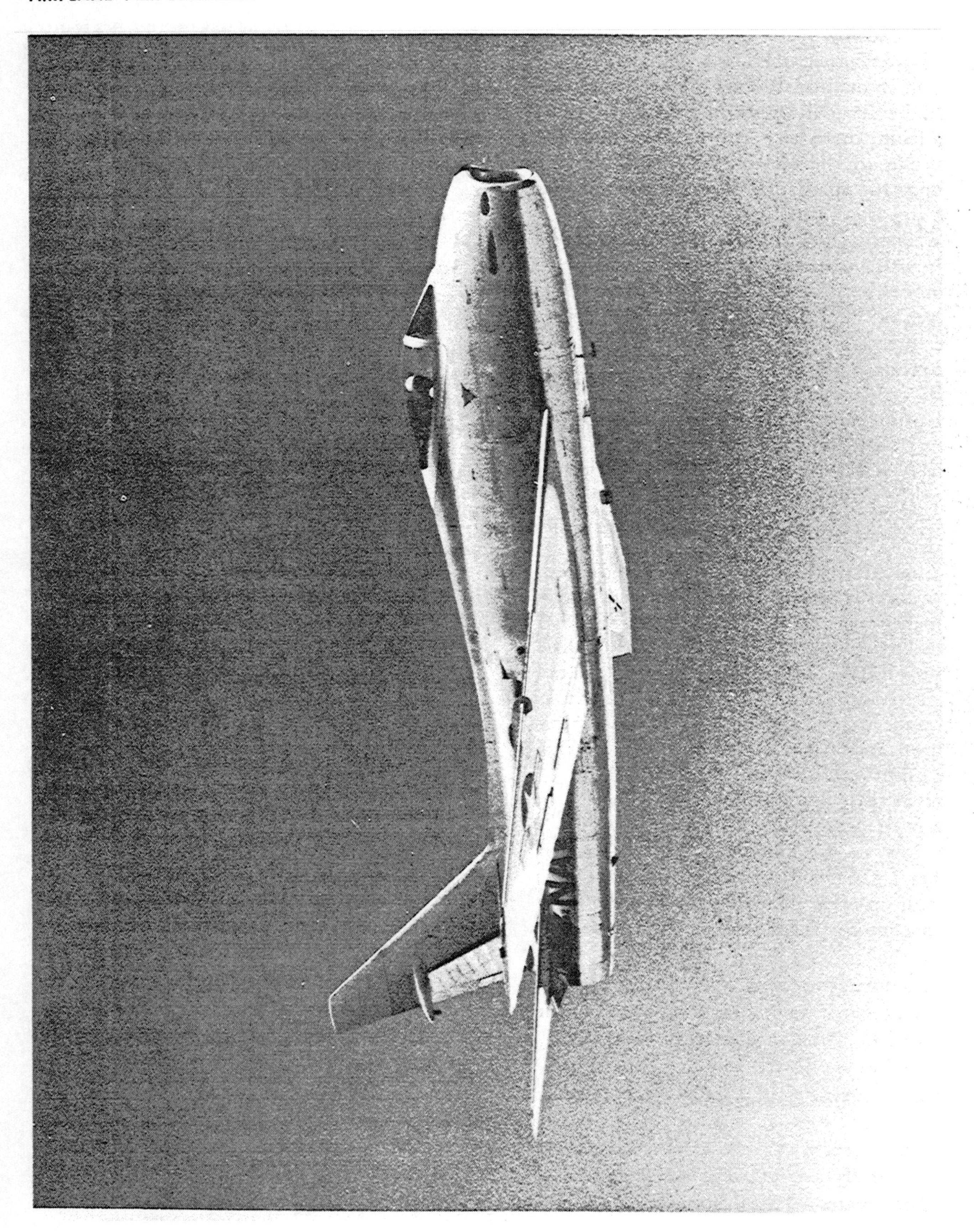

the turboprop powerplant is rated by an "equivalent shaft horsepower."

$$ESHP = BHP + \frac{T_j V}{325 \eta_p}$$

where

$ESHP$ = equivalent shaft horsepower
BHP = brake horsepower, or shaft horsepower applied to the propeller
T_j = jet thrust, lbs.
V = flight velocity, knots, *TAS*
η_p = propeller efficiency

The gas turbine engine is capable of processing large quantities of air and can produce high output power for a given engine size. Thus, the principal advantage of the turboprop powerplant is the high specific power output, high power per engine weight and high power per engine size.

The gas turbine engine must operate at quite high rotative speed to process large airflows and produce high power. However, high rotative speeds are not conducive to high propeller efficiency because of compressibility effects. A large reduction of shaft speed must be provided in order to match the powerplant and the propeller. The reduction gearing must provide a propeller shaft speed which can be utilized effectively by the propeller and, because of the high rotative speeds of the turbine, gearing ratios of 6 to 15 may be typical. The transmission of large shaft horsepower with such high gearing involves considerable design problems to provide good service life. The problems of such gearing were one of the greatest difficulties in the development of turboprop powerplants.

The governing apparatus for the turboprop powerplant must account for one additional variable, the propeller blade angle. If the propeller is governed separately from the turbine, an interaction can exist between the engine and propeller governers and various "hunting," overspeed, and overtemperature conditions are possible. For this reason, the engine-propeller combination is operated at a constant RPM throughout the major range of output power and the principal variables of control are fuel flow and propeller blade angle. In the major range of power output, the throttle commands a certain fuel flow and the propeller blade angle adjusts to increase the propeller load and remain at the governed speed.

The operating limitations of the turboprop powerplant are quite similar in nature to the operating limitations of the turbojet engine. Generally, the turbine temperature limitations are the most critical items. In addition, overspeed conditions can produce overstress of the gearing and propeller as well as overstress of the turbine section.

The performance of the turboprop illustrates the typical advantages of the propeller-engine combination. Higher propulsive efficiency and high thrust and low speeds provide the characteristic of range, endurance, and takeoff performance superior to the turbojet. As is typical of all propeller equipped powerplants, the power available is nearly constant with speed. Because the power from the jet thrust depends on velocity, the power available increases slightly with speed. However, the thrust available decreases with speed. The equivalent shaft horsepower, *ESHP*, of the turboprop is affected by mass flow and inlet temperature in fashion similar to that of the turbojet. Thus, the *ESHP* will vary with altitude much like the thrust output of the turbojet because the higher altitude produces much lower density and engine mass flow. The gas turbine-propeller combination utilizes a number of turbine stages to extract shaft power from the exhaust gases and, as high compressor inlet temperatures reduce the fuel flow allowable within turbine temperature limits, hot days will cause a noticeable loss of output power. Generally, the turboprop is just as sensitive, if not more sensitive, to compressor inlet air temperature as the turbojet engine.

The specific fuel consumption of the turboprop powerplant is defined as follows:

$$\text{specific fuel consumption} = \frac{\text{engine fuel flow}}{\text{equivalent shaft horsepower}}$$

$$c = \frac{\text{lbs. per hr.}}{ESHP}$$

Typical values for specific fuel consumption, c, range from 0.5 to 0.8 lbs. per hr. per *ESHP*. The variation of specific fuel consumption with operating conditions is similar to that of the turbojet engine. The minimum specific fuel consumption is obtained at relatively high power setting and high altitudes. The low inlet air temperature reduces the specific fuel consumption and the lowest values of c are obtained near altitudes of 25,000 to 35,000 ft. Thus, the turboprop as well as the turbojet has a preference for high altitude operation.

THE RECRIPROCATING ENGINE

The reciprocating engine is one of the most efficient powerplants used for aircraft power. The combination of the reciprocating engine and propeller is one of the most efficient means of converting the chemical energy of fuel into flying time or distance. Because of the inherent high efficiency, the reciprocating engine is an important type of aircraft powerplant.

OPERATING CHARACTERISTICS. The function of the typical reciprocating engine involves four strokes of the piston to complete one operating cycle. This principal operating cycle is illustrated in figure 2.15 by the variation of pressure and volume within the cylinder. The first stroke of the operating cycle is the downstroke of the piston with the intake valve open. This stroke draws in a charge of fuel-air mixture along *AB* of the pressure-volume diagram. The second stroke accomplishes compression of the fuel-air mixture along line *BC*. Combustion is initiated by a spark ignition apparatus and combustion takes place in essentially a constant volume. The combustion of the fuel-air mixture liberates heat and causes the rise of pressure along line *CD*. The power stroke utilizes the increased pressure through the expansion along line *DE*. Then the exhaust begins by the initial rejection along line *EB* and is completed by the upstroke along line *BA*.

The net work produced by the cycle of operation is idealized by the area *BCDE* on the pressure-volume diagram of figure 2.15. During the actual rather than ideal cycle of operation, the intake pressure is lower than the exhaust pressure and the negative work represents a pumping loss. The incomplete expansion during the power stroke represents a basic loss in the operating cycle because of the rejection of combustion products along line *EB*. The area *EFB* represents a basic loss in the operating cycle because of the rejection of combustion products along line *EB*. The area *EFB* represents a certain amount of energy of the exhaust gases, a part of which can be extracted by exhaust turbines as additional shaft power to be coupled to the crankshaft (turbo-compound engine) or to be used in operating a supercharger (turbosupercharger). In addition, the exhaust gas energy may be utilized to augment engine cooling flow (ejector exhaust) and reduce cowl drag.

Since the net work produced during the operating cycle is represented by the enclosed area of pressure-volume diagram, the output of the engine is affected by any factor which influences this area. The weight of fuel-air mixture will determine the energy released by combustion and the weight of charge can be altered by altitude, supercharging, etc. Mixture strength, preignition, spark timing, etc., can affect the energy release of a given airflow and alter the work produced during the operating cycle.

The mechanical work accomplished during the power stroke is the result of the gas pressure sustained on the piston. The linkage of the piston to a crankshaft by the connecting rod applies torque to the output shaft. During this conversion of pressure energy to mechanical energy, certain losses are inevitable because

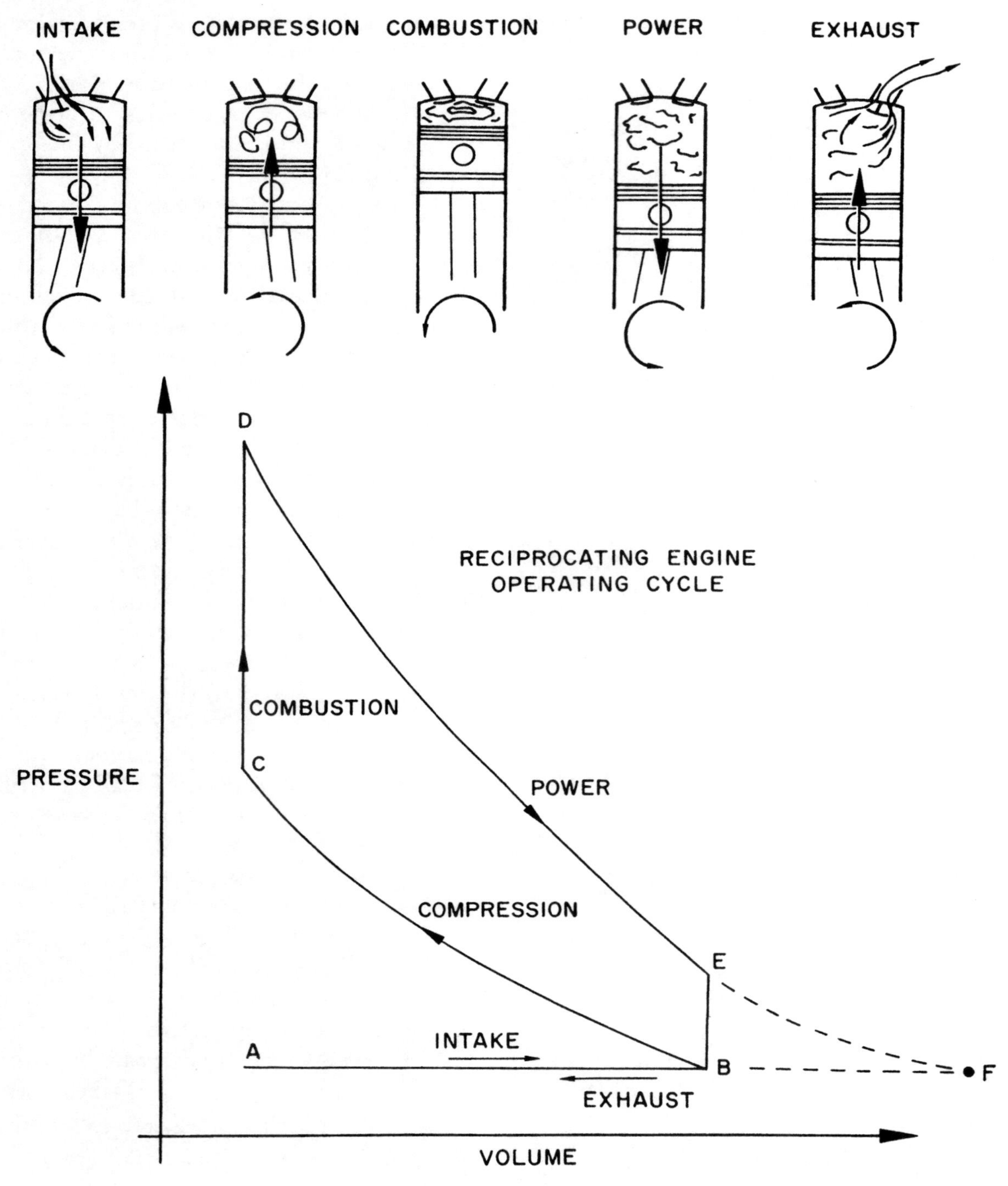

Figure 2.15. Reciprocating Engines

of friction and the mechanical output is less than the available pressure energy. The power output from the engine will be determined by the magnitude and rate of the power impulses. In order to determine the power output of the reciprocating engine, a brake or load device is attached to the output shaft and the operating characteristics are determined. Hence, the term "brake" horsepower, *BHP*, is used to denote the output power of the powerplant.

From the physical definition of "power" and the particular unit of "horsepower" (1 h.p.= 33,000 ft.-lbs. per min.), the brake horsepower can be expressed in the following form.

$$BHP=\frac{2\pi TN}{33,000}$$

or

$$BHP=\frac{TN}{5255}$$

where

BHP= brake horsepower

T= output torque, ft.-lbs.

N= output shaft speed, RPM

In this relationship, the output power is appreciated as some direct variable of torque, T, and RPM. Of course, the output torque is some function of the combustion gas pressure during the power stroke. Thus, it is helpful to consider the mean effective gas pressure during the power stroke, the "brake mean effective pressure" or *BMEP*. With use of this term, the *BHP* can be expressed in the following form.

$$BHP=\frac{(BMEP)(D)(N)}{792,000}$$

where

BHP= brake horsepower

$BMEP$= brake mean effective pressure, psi

D= engine displacement, cu. in.

N= engine speed, RPM

The *BMEP* is not actual pressure within the cylinder, but an effective pressure representing the mean gas load acting on the piston during the power stroke. As such, *BMEP* is a convenient index for a majority of items of reciprocating engine output, efficiency, and operating limitations.

The actual power output of any reciprocating engine is a direct function of the combination of engine torque and rotative speed. Thus, output brake horsepower can be related by the combination of *BMEP and RPM* or *torque pressure and RPM*. No other engine instruments can provide this immediate indication of output power.

If all other factors are constant, the engine power output is directly related to the engine airflow. Evidence of this fact could be appreciated from the equation for *BHP* in terms of *BMEP*.

$$BHP=\frac{(BMEP)(D)(N)}{792,000}$$

This equation relates that, for a given *BMEP*, the *BHP* is determined by the product of engine RPM, N, and displacement, D. In a sense, the reciprocating engine could be considered primarily as an air pump with the pump capacity directly affecting the power output. Thus, any engine instruments which relate factors affecting airflow can provide some indirect reflection of engine power. The pressure and temperature of the fuel-air mixture decide the density of the mixture entering the cylinder. The carburetor air temperature will provide the temperature of the inlet air at the carburetor. While this carburetor inlet air is not the same temperature as the air in the cylinder inlet manifold, the carburetor inlet temperature provides a stable indication independent of fuel flow and can be used as a standard of performance. Cylinder inlet manifold temperature is difficult to determine with the same degree of accuracy because of the normal variation of fuel-air mixture strength. The inlet manifold pressure provides an additional indication of the density of airflow entering the combustion chamber. The manifold absolute pressure, *MAP*, is affected by the carburetor inlet pressure, throttle position, and supercharger or impeller pressure ratio. Of course, the throttle is the principal control of manifold pressure and the throttling action controls the pressure of the fuel-air mixture delivered to the supercharger inlet. The pressure received by the supercharger is magnified by the supercharger in some proportion depending on impeller speed. Then the high pressure mixture is delivered to the manifold.

Of course, the engine airflow is a function of RPM for two reasons. A higher engine speed increases the pumping rate and the volume flow through the engine. Also, with the engine driven supercharger or impeller, an increase in engine speed increases the supercharger pressure ratio. With the exception of near closed throttle position, an increase in engine speed will produce an increase in manifold pressure.

The many variables affecting the character of the combustion process are an important subject of reciprocating engine operation. Uniform mixtures of fuel and air will support combustion between fuel-air ratios of approximately 0.04 and 0.20. The chemically correct proportions of air and hydrocarbon fuel would be 15 lbs. of air for each lb. of fuel, or a fuel-air ratio of 0.067. This chemically correct, or "stoichiometric," fuel-air ratio would provide the proportions of fuel and air to produce maximum release of heat during combustion of a given weight of mixture. If the fuel-air ratio were leaner than stoichiometric, the excess of air and deficiency of fuel would produce lower combustion temperatures and reduced heat release for a given weight of charge. If the fuel-air ratio were richer than stoichiometric, the excess of fuel and deficiency of air would produce lower combustion temperatures and reduced heat release for a given weight of charge.

The stoichiometric conditions would produce maximum heat release for ideal conditions of combustion and may apply quite closely for the individual cylinders of the low speed reciprocating engine. Because of the effects of flame propagation speed, fuel distribution, temperature variation, etc., the maximum power obtained with a fixed airflow occurs at fuel-air ratios of approximately 0.07 to 0.08. The first graph of figure 2.16 shows the variation of output power with fuel-air ratio for a a constant engine airflow, i.e., constant RPM, *MAP*, and *CAT* (carburetor air temperature). Combustion can be supported by fuel-air ratios just greater than 0.04 but the energy released is insufficient to overcome pumping losses and engine mechanical friction. Essentially, the same result is obtained for the rich fuel-air ratios just below 0.20. Fuel-air ratios between these limits produce varying amounts of output power and the maximum power output generally occurs at fuel-air ratios of approximately 0.07 to 0.08. Thus, this range of fuel-air ratios which produces maximum power for a given airflow is termed the "best power" range. At some lower range of fuel-air ratios, a maximum of power per fuel-air ratio is obtained and this the "best economy" range. The best economy range generally occurs between fuel-air ratios of 0.05 and 0.07. When maximum engine power is required for take-off, fuel-air ratios greater than 0.08 are necessary to suppress detonation. Hence, fuel-air ratios of 0.09 to 0.11 are typical during this operation.

The pattern of combustion in the cylinder is best illustrated by the second graph of figure 2.16. The normal combustion process begins by spark ignition toward the end of the compression stroke. The electric spark provides the beginning of combustion and a flame front is propagated smoothly through the compressed mixture. Such normal combustion is shown by the plot of cylinder pressure versus piston travel. Spark ignition begins a smooth rise of cylinder pressure to some peak value with subsequent expansion through the power stroke. The variation of pressure with piston travel must be controlled to achieve the greatest net work during the cycle of operation.

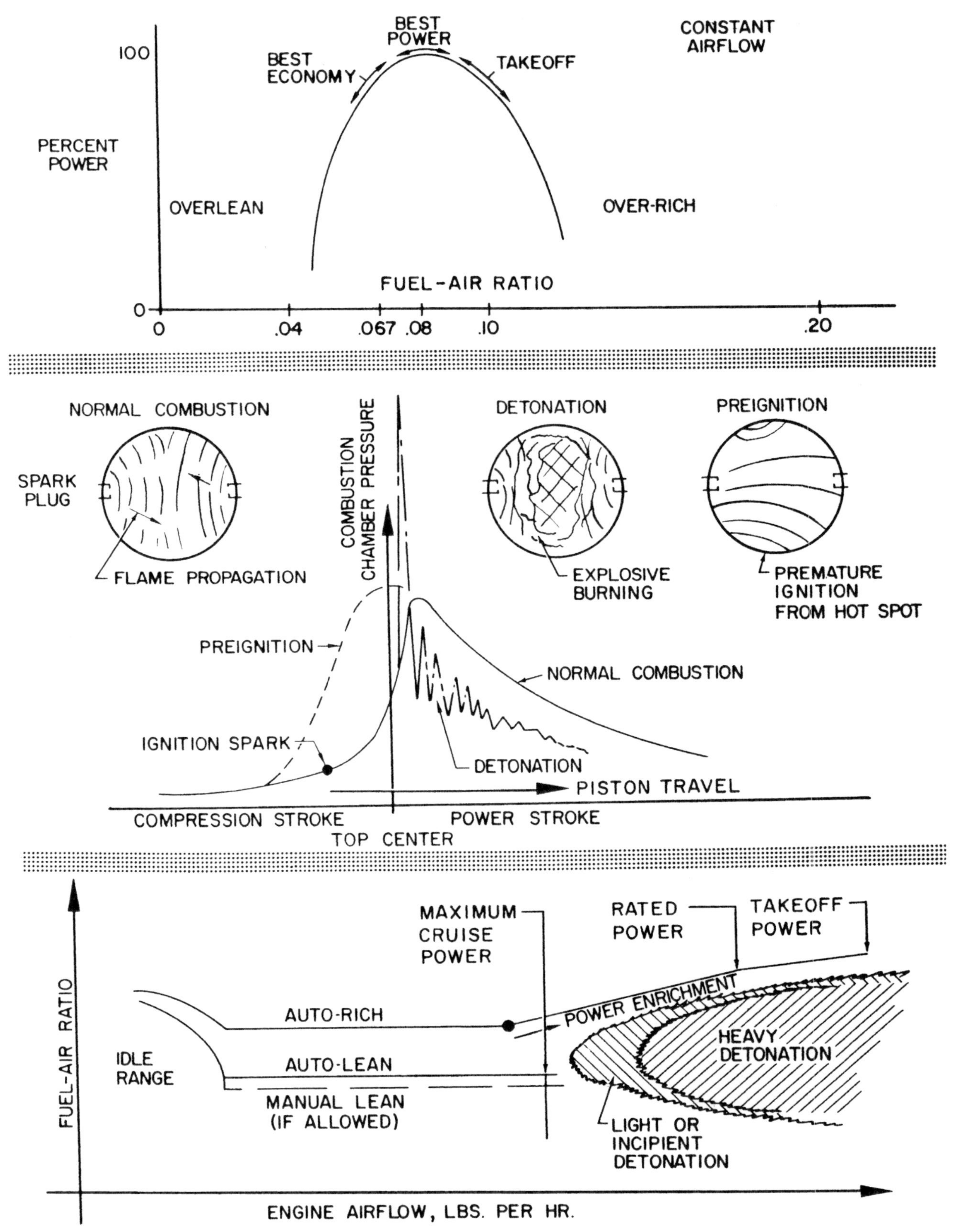

Figure 2.16. Reciprocating Engine Operation

Obviously, spark ignition timing is an important factor controlling the initial rise of pressure in the combustion chamber. The ignition of the fuel mixture must begin at the proper time to allow flame front propagation and the release of heat to build up peak pressure for the power stroke.

The speed of flame front propagation is a major factor affecting the power output of the reciprocating engine since this factor controls the rate of heat release and rate of pressure rise in the combustion chamber. For this reason, dual ignition is necessary for powerplants of high specific power output. Obviously, normal combustion can be accomplished more rapidly with the propagation of two flame fronts rather than one. The two sources of ignition are able to accomplish the combustion heat release and pressure rise in a shorter period of time. Fuel-air ratio is another factor affecting the flame propagation speed in the combustion chamber. The maximum flame propagation speed occurs near a fuel-air ratio of 0.08 and, thus, maximum power output for a given airflow will tend to occur at this value rather than the stoichiometric value.

Two aberrations of the combustion process are preignition and detonation. Preignition is simply a premature ignition and flame front propagation due to hot spots in the combustion chamber. Various lead and carbon deposits and feathered edges on metal surfaces can supply a glow ignition spot and begin a flame propagation prior to normal spark ignition. As shown on the graph of figure 2.16, preignition causes a premature rise of pressure during the piston travel. As a result, preignition combustion pressures and temperatures will exceed normal combustion values and are very likely to cause engine damage. Because of the premature rise of pressure toward the end of the compression stroke, the net work of the operating cycle is reduced. Preignition is evidenced by a rise in cylinder head temperature and drop in *BMEP* or torque pressure.

Denotation offers the possibility of immediate destruction of the powerplant. The normal combustion process is initiated by the spark and beginning of flame front propagation. As the flame front is propagated, the combustion chamber pressure and temperature begin to rise. Under certain conditions of high combustion pressure and temperature, the mixture ahead of the advancing flame front may suddenly explode with considerable violence and send strong detonation waves through the combustion chamber. The result is depicted by the graph of figure 2.16, where a sharp, explosive increase in pressure takes place with a subsequent reduction of the mean pressure during the power stroke. Detonation produces sharp explosive pressure peaks many times greater than normal combustion. Also, the exploding gases radiate considerable heat and cause excessive temperatures for many local parts of the engine. The effects of heavy detonation are so severe that structural damage is the immediate result. Rapid rise of cylinder head temperature, rapid drop in *BMEP*, and loud, expensive noises are evidence of detonation.

Detonation is not necessarily confined to a period after the beginning of normal flame front propagation. With extremely low grades of fuel, detonation can occur before normal ignition. In addition, the high temperatures and pressure caused by preignition will mean that detonation is usually a corollary of preignition. Detonation results from a sudden, unstable decomposition of fuel at some critical combination of high temperature and pressure. Thus, detonation is most likely to occur at any operating condition which produces high combustion pressures and temperatures. Generally, high engine airflow and fuel-air ratios for maximum heat release will produce the critical conditions. High engine airflow is common to high *MAP* and RPM and the engine is most sensitive to *CAT* and fuel-air ratio in this region.

The detonation properties of a fuel are determined by the basic molecular structure of the fuel and the various additives. The fuel detonation properties are generally specified by the antidetonation or antiknock qualities of an octane rating. Since the antiknock properties of a high quality fuel may depend on the mixture strength, provision must be made in the rating of fuels. Thus, a fuel grade of 115/145 would relate a lean mixture antiknock rating of 115 and a rich mixture antiknock rating of 145. One of the most common operational causes of detonation is fuel contamination. An extremely small contamination of high octane fuel with jet fuel can cause a serious decrease in the antiknock rating. Also, the contamination of a high grade fuel with the next lower grade will cause a noticeable loss of antiknock quality.

The fuel metering requirements for an engine are illustrated by the third graph of figure 2.16 which is a plot of fuel-air ratio versus engine airflow. The carburetor must provide specific fuel-air ratios throughout the range of engine airflow to accommodate certain output power. Most modern engines equipped with automatic mixture control provide a scheduling of fuel-air ratio for automatic rich or automatic lean operation. The auto-rich scheduling usually provides a fuel-air ratio at or near the maximum heat release value for the middle range of airflows. However, at high airflows a power enrichment must be provided to suppress detonation. The auto-rich schedule generally will provide an approximate fuel-air ratio of 0.08 which increases to 0.10 or 0.11 at the airflow for takeoff power. In addition, the low airflow and mixture dilution that occurs in the idle power range requires enrichment for satisfactory operation.

The schedule of fuel-air ratios with an automatic lean fuel-air ratio will automatically provide maximum usable economy. If manual leaning procedures are applicable a lower fuel-air ratio may be necessary for maximum possible efficiency. The maximum continuous cruise power is the upper limit of power that can be utilized for this operation. Higher airflows and higher power without a change in fuel-air ratio will intersect the knee of the detonation envelope.

The primary factor relating the efficiency of operation of the reciprocating engine is the brake specific fuel consumption, *BSFC*, or simply c.

Brake specific fuel consumption

$$=\frac{\text{engine fuel flow}}{\text{brake horsepower}}$$

$$c=\frac{\text{lbs. per hr.}}{\text{BHP}}$$

Typical minimum values for c range from 0.4 to 0.6 lbs. per hr. per *BHP* and most aircraft powerplants average 0.5. The turbocompound engine is generally the most efficient because of the power recovery turbines and can approach values of $c=0.38$ to 0.42. It should be noted that the minimum values of specific fuel consumption will be obtained only within the range of cruise power operation, 30 to 60 percent of the maximum power output. Generally, the conditions of minimum specific fuel consumption are achieved with auto-lean or manual lean scheduling of fuel-air ratios and high *BMEP* and low RPM. The low RPM is the usual requirement to minimize friction horsepower and improve output efficiency.

The effect of *altitude* is to reduce the engine airflow and power output and supercharging is necessary to maintain high power output at high altitude. Since the basic engine is able to process air only by the basic volume displacement, the function of the supercharger is to compress the inlet air and provide a greater weight of air for the engine to process. Of course, shaft power is necessary to operate the engine driven supercharger and a temperature rise occurs through the supercharger compression. The effect of various forms of supercharging on altitude performance is illustrated in figure 2.17.

The unsupercharged—or naturally aspirated—engine has no means of providing a

EFFECT OF SUPERCHARGING ON ALTITUDE PERFORMANCE

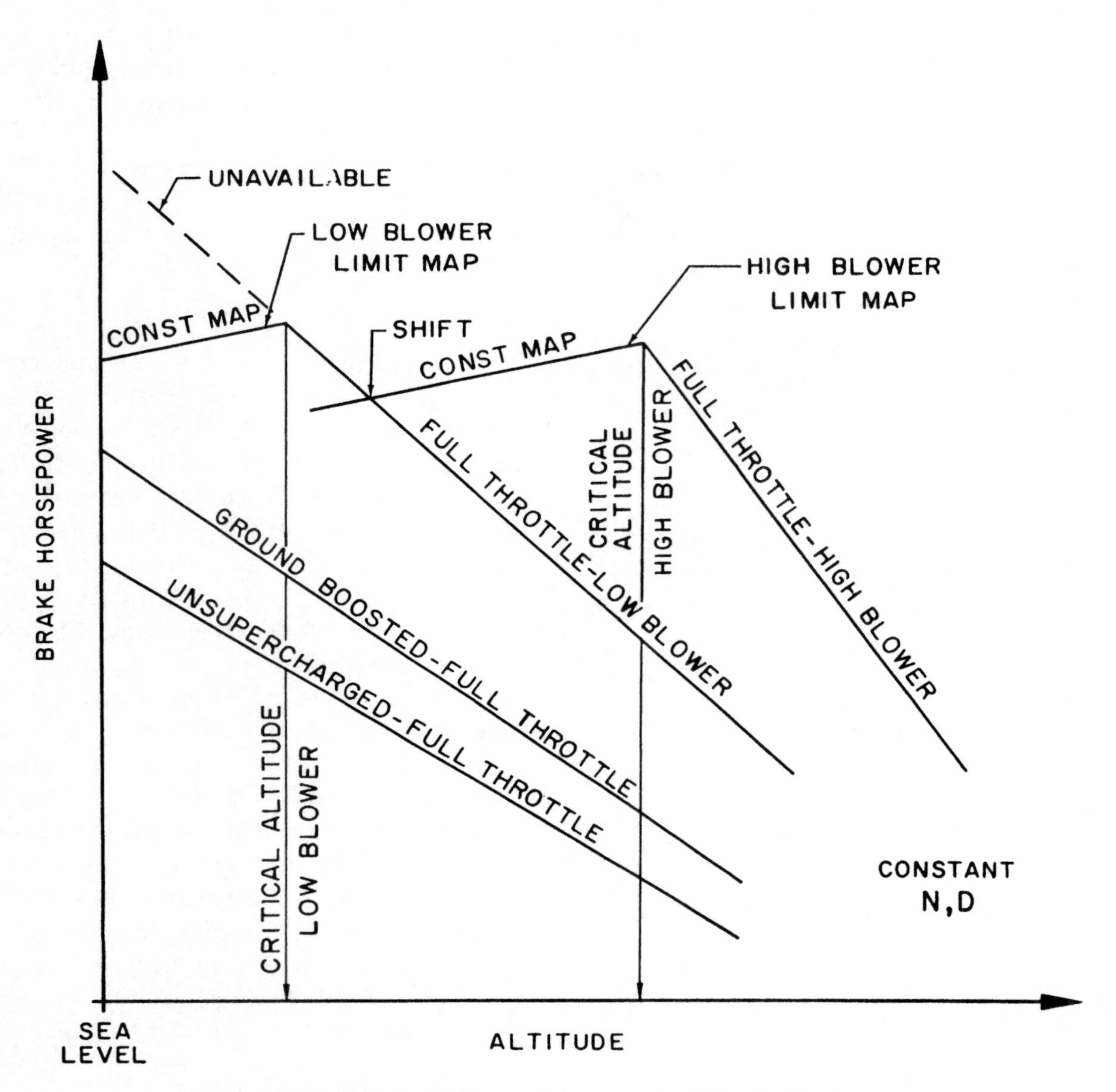

Figure 2.17. Effect of Supercharging on Altitude Performance

manifold pressure any greater than the induction system inlet pressure. As altitude is increased with full throttle and a governed RPM, the airflow through the engine is reduced and *BHP* decreases. The first forms of supercharging were of relatively low pressure ratio and the added airflow and power could be handled at full throttle within detonation limits. Such a "ground boosted" engine would achieve higher output power at all altitudes but an increase in altitude would produce a decrease in manifold pressure, airflow, and power output.

More advanced forms of supercharging with higher pressure ratios can produce very large engine airflow. In fact, the typical case of altitude supercharging will produce such high airflow at low altitude operation that full throttle operation cannot be utilized within detonation limits. Figure 2.17 illustrates this case for a typical two-speed engine driven altitude supercharging installation. At sea level, the limiting manifold pressure produces a certain amount of *BHP*. Full throttle operation could produce a higher *MAP* and *BHP* if detonation were not the problem. In this case full throttle operation is unavailable because of detonation limits. As altitude is increased with the supercharger or "blower" at low speed, the constant *MAP* is maintained by opening the throttle and the *BHP* increases above the sea level value because of the reduced exhaust back pressure. Opening the throttle allows the supercharger inlet to receive the same inlet pressure and produce the same *MAP*. Finally, the increase of altitude will require full throttle to produce the constant *MAP* with low blower and this point is termed the "critical altitude" or "full throttle height." If altitude is increased beyond the critical altitude, the engine *MAP*, airflow, and *BHP* decrease.

The critical altitude with a particular supercharger installation is specific to a given combination of *MAP* and RPM. Obviously, a lower *MAP* could be maintained to some higher altitude or a lower engine speed would produce less supercharging and a given *MAP* would require a greater throttle opening. Generally, the most important critical altitudes will be specified for maximum, rated, and maximum cruise power conditions.

A change of the blower to a high speed will provide greater supercharging but will require more shaft power and incur a greater temperature rise. Thus, the high blower speed can produce an increase in altitude performance within the detonation limitations. The variation of *BHP* with altitude for the blower at high speed shows an increase in critical altitude and greater *BHP* than is obtainable in low blower. Operation below the high blower critical altitude requires some limiting manifold pressure to remain within detonation limits. It is apparent that the shift to high blower is not required just past low blower critical altitude but at the point where the transition from low blower, full throttle to high blower, limit *MAP* will produce greater *BHP*. Of course, if the blower speed is increased without reducing the throttle opening, an "overboost" can occur.

Since the exhaust gases have considerable energy, exhaust turbines provide a source of supercharger power. The turbosupercharger (*TBS*) allows control of the supercharger speed and output to very high altitudes with a variable discharge exhaust turbine (*VDT*). The turbosupercharger is capable of providing the engine airflow with increasing altitude by increasing turbine and supercharger speed. Critical altitude for the turbosupercharger is usually defined by the altitude which produces the limiting exhaust turbine speed.

The minimum specific fuel consumption of the supercharged engine is not greatly affected by altitudes less than the critical altitude. At the maximum cruise power condition, specific fuel consumption will decrease slightly with an increase in altitude up to the critical altitude. Above critical altitude, maximum cruise power cannot be maintained but the specific fuel consumption is not adversely affected as long as auto-lean or manual lean power can be used at the cruise power setting.

One operating characteristic of the reciprocating engine is distinctly different from that of the turbojet. *Water vapor* in the air will cause a significant reduction in output power of the reciprocating engine but a negligible loss of thrust for the turbojet engine. This basic difference exists because the reciprocating engine operates with a fixed displacement and all air processed is directly associated with the combustion process. If water vapor enters the induction system of the reciprocating engine, the amount of air available for combustion is reduced and, since most carburetors do not distinguish water vapor from air, an enrichment of the fuel-air ratio takes place. The maximum power output at takeoff requires fuel-air ratios richer than that for maximum heat release so a further enrichment will take place with subsequent loss of power. The turbojet operates with such great excess of air that the combustion process essentially is unaffected and the reduction of air mass flow is the principal consideration. As an example, extreme conditions which would produce high specific humidity may cause a 3 percent thrust loss for a turbojet but a 12 percent loss of *BHP* for a reciprocating engine. Proper accounting of the loss due to humidity is essential in the operation of the reciprocating engine.

OPERATING LIMITATIONS. Reciprocating engines have achieved a great degree of refinement and development and are one of the most reliable of all types of aircraft powerplants. However, reliable operation of the reciprocating engine is obtained only by strict adherence to the specific operating limitations.

The most important operating limitations of the reciprocating engine are those provided to ensure that detonation and preignition do not take place. The pilot must ensure that proper fuel grades are used that limit *MAP*, *BMEP*, RPM, *CAT*, etc., are not exceeded. Since heavy detonation or preignition is common to the high airflow at maximum power, the most likely chance of detonation or preignition is at takeoff. In order to suppress detonation or allow greater power for takeoff, water injection is often used in the reciprocating engine. At high power settings, the injection of the water-alcohol mixture can replace the excess fuel required to suppress detonation, and derichment provisions can reduce the fuel-air ratio toward the value for maximum heat release. Thus, an increase in power will be obtained by the better fuel-air ratio. In some instances, a higher manifold pressure can be utilized to produce additional power. The injection fluid will require proportions of alcohol and water quite different from the injection fluid for jet engine thrust augmentation. Since derichment of the fuel-air ratio is desired, the anti-detonant injection (*ADI*) will contain alcohol in quantities to prevent residual fluid from fouling the plumbing.

When the fuel grades are altered during operation and the engine must be operated on a next lower fuel grade, proper account must be made for the change in the operating limitations. This accounting must be made for the maximum power for takeoff and the maximum cruise power since both of these operating conditions are near the detonation envelope. In addition, when the higher grade of fuel again becomes available, the higher operating limits cannot be used until it is sure that no contamination exists from the lower grade fuel remaining in the tanks.

Spark plug fouling can provide certain high as well as low limits of operating temperatures. When excessively low operating temperatures are encountered, rapid carbon fouling of the plugs will take place. On the other hand, excessively high operating temperatures will produce plug fouling from lead bromide deposits from the fuel additives.

Generally, the limited periods of time at various high power settings are set to minimize the accumulation of high rates of wear and fatigue damage. By minimizing the amount of total time spent at high power setting, greater overhaul life of the powerplant can be achieved. This should not imply that the takeoff rating of the engine should not be used. Actually, the use of the full maximum power at takeoff will accumulate less total engine wear than a reduced power setting at the same RPM because of less time required to climb to a given altitude or to accelerate to a given speed.

The most severe rate of wear and fatigue damage occurs at high RPM and low *MAP*. High RPM produces high centrifugal loads and reciprocating inertia loads. When the large reciprocating inertia loads are not cushioned by high compression pressures, critical resultant loads can be produced. Thus, operating time at maximum RPM and *MAP* must be held to a minimum and operation at maximum RPM and low *MAP* must be avoided.

AIRCRAFT PROPELLERS

The aircraft propeller functions to convert the powerplant shaft horsepower into propulsive horsepower. The basic principles of propulsion apply to the propeller in that thrust is produced by providing the airstream a momentum change. The propeller achieves high propulsive efficiency by processing a relatively large mass flow of air and imparting a relatively small velocity change. The momentum change created by propeller is shown by the illustration of figure 2.18.

The action of the propeller can be idealized by the assumption that the rotating propeller is simply an actuating disc. As shown in figure 2.18, the inflow approaching the propeller disc indicates converging streamlines with an increase in velocity and drop in pressure. The converging streamlines leaving the propeller disc indicate a drop in pressure and increase in velocity behind the propeller. The pressure change through the disc results from the distribution of thrust over the area of the propeller disc. In this idealized propeller disc, the pressure difference is uniformly distributed over the disc area but the actual case is rather different from this.

The final velocity of the propeller slipstream, V_2, is achieved some distance behind the propeller. Because of the nature of the flow pattern produced by the propeller, one half of the total velocity change is produced as the flow reaches the propeller disc. If the complete velocity increase amounts to $2a$, the flow velocity has increased by the amount a at the propeller disc. The ***propulsive efficiency***, η_p, of the ideal propeller could be expressed by the following relationship:

$$\eta_p = \frac{\text{output power}}{\text{input power}}$$

$$\eta_p = \frac{TV}{T(V+a)}$$

where

η_p = propulsive efficiency
T = thrust, lbs.
V = flight velocity, knots
a = velocity increment at the propeller disc, knots

Since the final velocity, V_2, is the sum of total velocity change $2a$ and the initial velocity, V_1, the propulsive efficiency rearranges to a form identical to that for the turbojet.

$$\eta_p = \frac{2}{1+\left(\frac{V_2}{V_1}\right)}$$

So, the same relationship exists as with the turbojet engine in that high efficiency is developed by producing thrust with the highest possible mass flow and smallest necessary velocity change.

The actual propeller must be evaluated in a more exact sense to appreciate the effect of nonuniform disc loading, propeller blade drag forces, interference flow between blades, etc. With these differences from the ideal propeller,

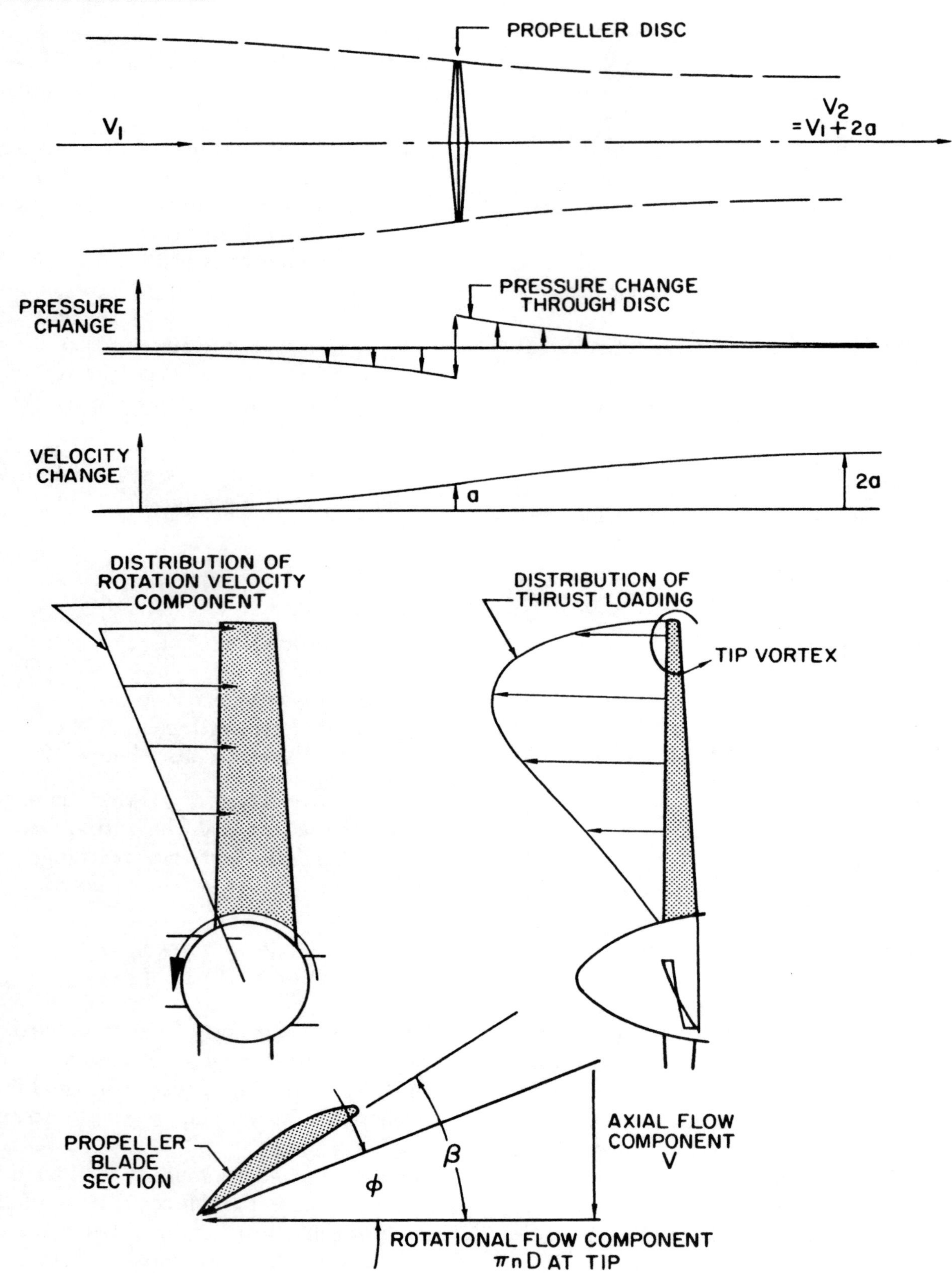

Figure 2.18. Principles of Propellers

it is more appropriate to define propeller efficiency in the following manner:

$$\eta_p = \frac{\text{output propulsive power}}{\text{input shaft horsepower}}$$

$$\eta_p = \frac{(T)(V)}{325\ BHP}$$

where

η_p = propeller efficiency
T = propeller thrust
V = flight velocity, knots
BHP = brake horsepower applied to the propeller

Many different factors govern the efficiency of a propeller. Generally, a large diameter propeller favors a high propeller efficiency from the standpoint of large mass flow. However, a powerful adverse effect on propeller efficiency is produced by high tip speeds and compressibility effects. Of course, small diameter propellers favor low tip speeds. In addition, the propeller and powerplant must be matched for compatibility of output and efficiency.

In order to appreciate some of the principal factors controlling the efficiency of a given propeller, figure 2.18 illustrates the distribution of rotative velocity along the rotating propeller blade. These rotative velocities add to the local inflow velocities to produce a variation of resultant velocity and direction along the blade. The typical distribution of thrust along the propeller blade is shown with the predominating thrust being located on the outer portions of the blade. Note that the propeller producing thrust develops a tip vortex similar to the wing producing lift. Evidence of this vortex can be seen by the condensation phenomenon occurring at this location under certain atmospheric conditions.

The component velocities at a given propeller blade section are shown by the diagram of figure 2.18. The inflow velocity adds vectorially to the velocity due to rotation to produce an inclination of the resultant wind with respect to the plane of rotation. This inclination is termed ϕ (phi), the effective pitch angle, and is a function of some proportion of the flight velocity, V, and the velocity due to rotation which is πnD at the tip. The proportions of these terms describe the propeller "advance ratio", J.

$$J = \frac{V}{nD}$$

where

J = propeller advance ratio
V = flight velocity, ft. per sec.
n = propeller rotative speed, revolutions per sec.
D = propeller diameter, ft.

The propeller blade angle, β (beta), varies throughout the length of the blade but a representative value is measured at 75 percent of the blade length from the hub.

Note that the difference between the effective pitch angle, ϕ, and the blade angle, β, determines an effective angle of attack for the propeller blade section. Since the angle of attack is the principal factor affecting the efficiency of an airfoil section, it is reasonable to make the analogy that the advance ratio, J, and blade angle, β, are the principal factors affecting propeller efficiency. The performance of a propeller is typified by the chart of figure 2.19 which illustrates the variation of propeller efficiency, η_p, with advance ratio, J, for various values of blade angle, β. The value of η_p for each β increases with J until a peak is reached, then decreases. It is apparent that a fixed pitch propeller may be selected to provide suitable performance in a narrow range of advance ratio but efficiency would suffer considerably outside this range.

In order to provide high propeller efficiency through a wide range of operation, the propeller blade angle must be controllable. The most convenient means of controlling the propeller is the provision of a constant speed governing apparatus. The constant speed governing feature is favorable from the standpoint of engine operation in that engine output and efficiency is positively controlled and governed.

The governing of the engine-propeller combination will allow operation throughout a wide range of power and speed while maintaining efficient operation.

If the envelope of maximum propeller efficiency is available, the propulsive horsepower available will appear as shown in the second chart of figure 2.19. The propulsive power available, Pa, is the product of the propeller efficiency and applied shaft horsepower.

$$Pa = \frac{TV}{325}$$

$$Pa = (\eta_p)\ (BHP)$$

The propellers used on most large reciprocating engines derive peak propeller efficiencies on the order of $\eta_p = 0.85$ to 0.88. Of course, the peak values are designed to occur at some specific design condition. For example, the selection of a propeller for a long range transport would require matching of the engine-propeller combination for peak efficiency at cruise condition. On the other hand, selection of a propeller for a utility or liaison type airplane would require matching of the engine-propeller combination to achieve high propulsive power at low speed and high power for good takeoff and climb performance.

Several special considerations must be made for the application of aircraft propellers. In the event of a powerplant malfunction or failure, provision must be made to streamline the propeller blades and reduce drag so that flight may be continued on the remaining operating engines. This is accomplished by feathering the propeller blades which stops rotation and incurs a minimum of drag for the inoperative engine. The necessity for feathering is illustrated in figure 2.19 by the change in equivalent parasite area, Δf, with propeller blade angle, β, of a typical installation. When the propeller blade angle is in the feathered position, the change in parasite drag is at a minimum and, in the case of a typical multiengine aircraft, the added parasite drag from a single feathered propeller is a relatively small contribution to the airplane total drag.

At smaller blade angles near the flat pitch position, the drag added by the propeller is very large. At these small blade angles, the propeller windmilling at high RPM can create such a tremendous amount of drag that the airplane may be uncontrollable. The propeller windmilling at high speed in the low range of blade angles can produce an increase in parasite drag which may be as great as the parasite drag of the basic airplane. An indication of this powerful drag is seen by the helicopter in autorotation. The windmilling rotor is capable of producing autorotation rates of descent which approach that of a parachute canopy with the identical disc area loading. Thus, the propeller windmilling at high speed and small blade angle can produce an effective drag coefficient of the disc area which compares with that of a parachute canopy. The drag and yawing moment caused by loss of power at high engine-propeller speed is considerable and the transient yawing displacement of the aircraft may produce critical loads for the vertical tail. For this reason, automatic feathering may be a necessity rather than a luxury.

The large drag which can be produced by the rotating propeller can be utilized to improve the stopping performance of the airplane. Rotation of the propeller blade to small positive values or negative values with applied power can produce large drag or reverse thrust. Since the thrust capability of the propeller is quite high at low speeds, very high deceleration can be provided by reverse thrust alone.

The *operating limitations* of the propeller are closely associated with those of the powerplant. Overspeed conditions are critical because of the large centrifugal loads and blade twisting moments produced by an excessive rotative speed. In addition, the propeller blades will have various vibratory modes and certain operating limitations may be necessary to prevent exciting resonant conditions.

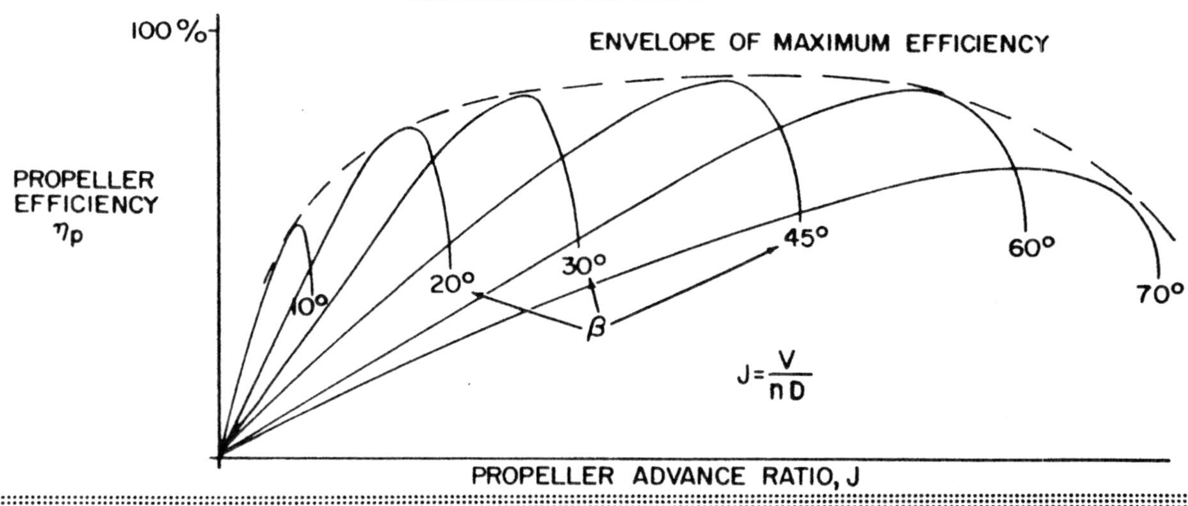

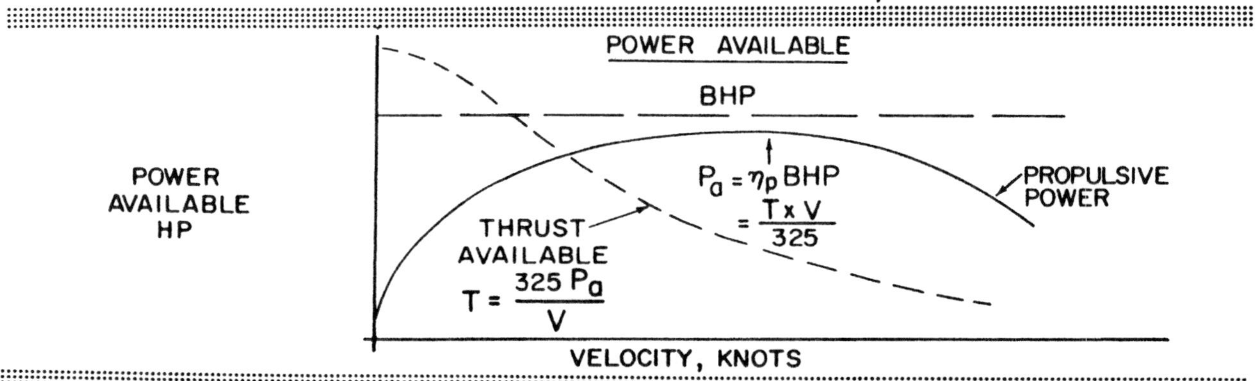

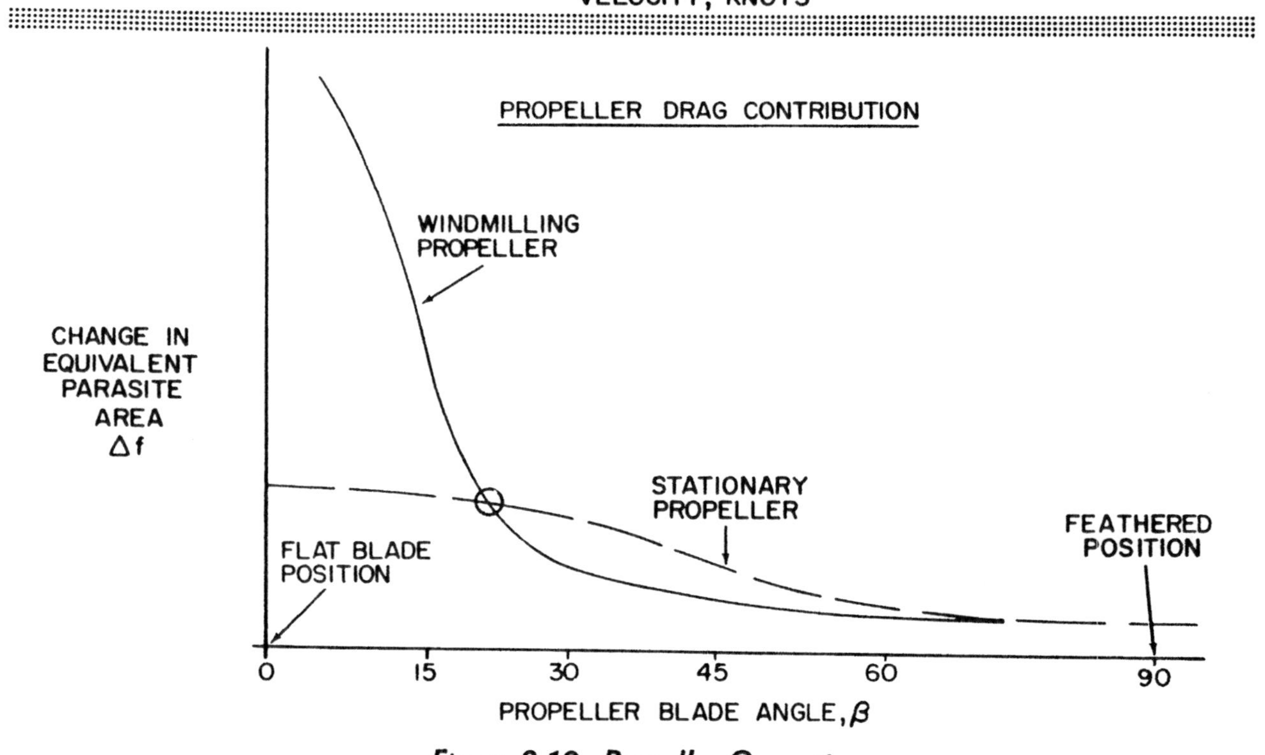

Figure 2.19. Propeller Operation

ITEMS OF AIRPLANE PERFORMANCE

The various items of airplane performance result from the combination of airplane and powerplant characteristics. The aerodynamic characteristics of the airplane generally define the power and thrust *requirements* at various conditions of flight while the powerplant characteristics generally define the power and thrust *available* at various conditions of flight. The matching of the aerodynamic configuration with the powerplant will be accomplished to provide maximum performance at the specific design condition, e.g., range, endurance, climb, etc.

STRAIGHT AND LEVEL FLIGHT

When the airplane is in steady, level flight, the condition of equilibrium must prevail. The unaccelerated condition of flight is achieved with the airplane trimmed for lift equal to weight and the powerplant set for a thrust to equal the airplane drag. In certain conditions of airplane performance it is convenient to consider the airplane requirements by the *thrust required* (or drag) while in other cases it is more applicable to consider the *power required*. Generally, the jet airplane will require consideration of the thrust required and the propeller airplane will require consideration of the power required. Hence, the airplane in steady level flight will require lift equal to weight and thrust available equal to thrust required (drag) or power available equal to power required.

The variation of power required and thrust required with velocity is illustrated in figure 2.20. Each specific curve of power or thrust required is valid for a particular aerodynamic configuration at a given weight and altitude. These curves define the power or thrust required to achieve equilibrium, lift-equal-weight, constant altitude flight at various airspeeds. As shown by the curves of figure 2.20, if it is desired to operate the airplane at the airspeed corresponding to point *A*, the power or thrust required curves define a particular value of thrust or power that must be made available from the powerplant to achieve equilibrium. Some different airspeed such as that corresponding to point *B* changes the value of thrust or power required to achieve equilibrium. Of course, the change of airspeed to point *B* also would require a change in angle of attack to maintain a constant lift equal to the airplane weight. Similarly, to establish airspeed and achieve equilibrium at point *C* will require a particular angle of attack and powerplant thrust or power. In this case, flight at point *C* would be in the vicinity of the minimum flying speed and a major portion of the thrust or power required would be due to induced drag.

The maximum level flight speed for the airplane will be obtained when the power or thrust required equals the maximum power or thrust available from the powerplant. The minimum level flight airspeed is not usually defined by thrust or power requirement since conditions of stall or stability and control problems generally predominate.

CLIMB PERFORMANCE

During climbing flight, the airplane gains potential energy by virtue of elevation. This increase in potential energy during a climb is provided by one, or a combination, of two means: (1) expenditure of propulsive energy above that required to maintain level flight or (2) expenditure of airplane kinetic energy, i.e., loss of velocity by a zoom. Zooming for altitude is a transient process of trading kinetic energy for potential energy and is of considerable importance for airplane configurations which can operate at very high levels of kinetic energy. However, the major portions of climb performance for most airplanes is a near steady process in which additional propulsive energy is converted into potential energy. The fundamental parts of airplane climb performance involve a flight condition where the airplane is in equilibrium but not at constant altitude.

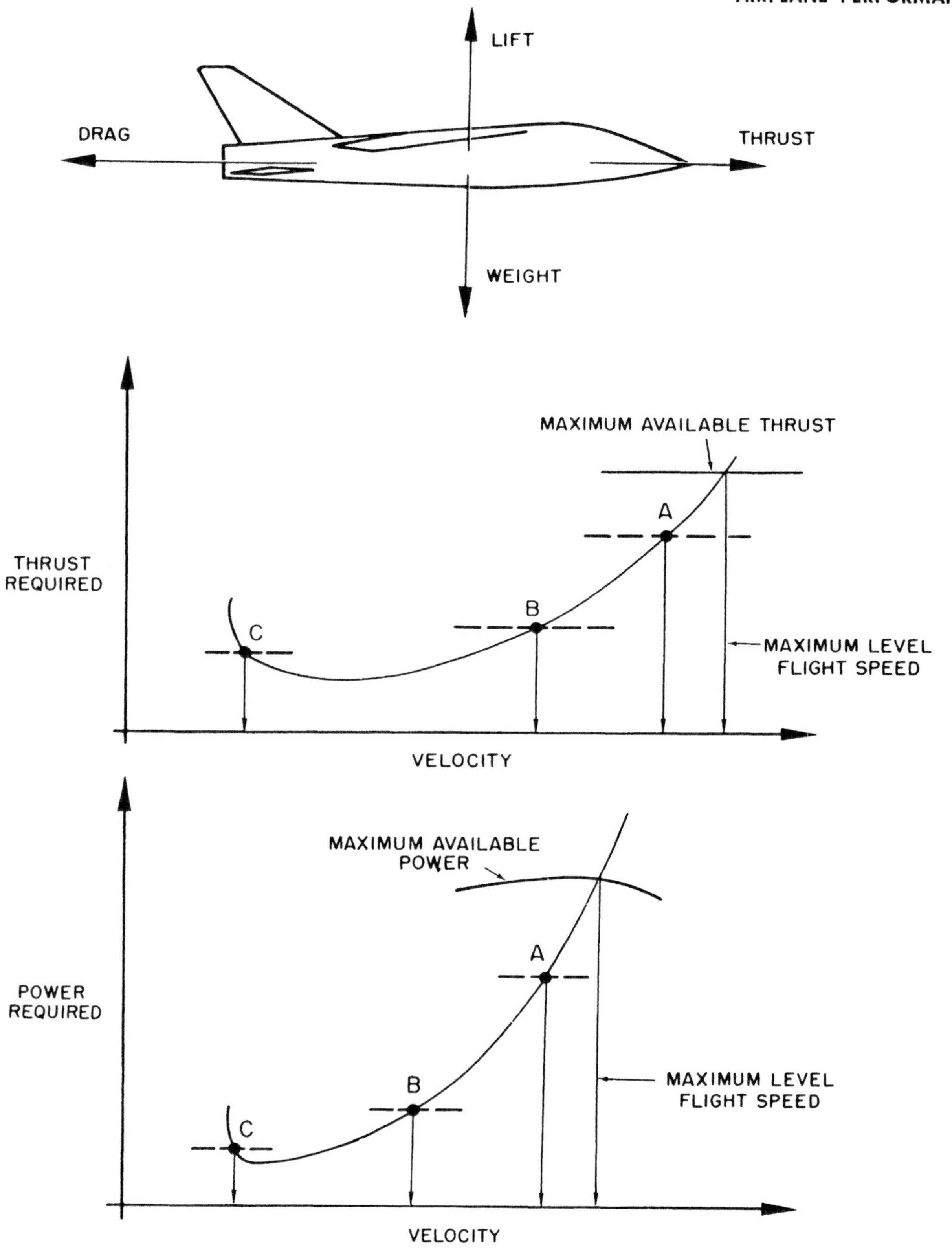

Figure 2.20. Level Flight Performance

The forces acting on the airplane during a climb are shown by the illustration of figure 2.21. When the airplane is in steady flight with moderate angle of climb, the vertical component of lift is very nearly the same as the actual lift. Such climbing flight would exist with the lift very nearly equal to the weight. The net thrust of the powerplant may be inclined relative to the flight path but this effect will be neglected for the sake of simplicity. Note that the weight of the aircraft is vertical but a component of weight will act aft along the flight path.

If it is assumed that the aircraft is in a steady climb with essentially small inclination of the flight path, the summation of forces along the flight path resolves to the following:

Forces forward = Forces aft

$$T = D + W \sin \gamma$$

where

T = thrust available, lbs.
D = drag, lbs.
W = weight, lbs.
γ = flight path inclination or angle of climb, degrees ("gamma")

This basic relationship neglects some of the factors which may be of importance for airplanes of very high climb performance. For example, a more detailed consideration would account for the inclination of thrust from the flight path, lift not equal to weight, subsequent change of induced drag, etc. However, this basic relationship will define the principal factors affecting climb performance. With this relationship established by the condition of equilibrium, the following relationship exists to express the trigonometric sine of the climb angle, γ:

$$\sin \gamma = \frac{T - D}{W}$$

This relationship simply states that, for a given weight airplane, the *angle of climb* (γ) depends on the difference between thrust and drag ($T-D$), or excess thrust. Of course, when the excess thrust is zero ($T-D=0$ or $T=D$), the inclination of the flight path is zero and the airplane is in steady, level flight. When the thrust is greater than the drag, the excess thrust will allow a climb angle depending on the value of excess thrust. Also, when the thrust is less than the drag, the deficiency of thrust will allow an angle of descent.

The most immediate interest in the climb angle performance involves obstacle clearance. The maximum angle of climb would occur where there exists the greatest difference between thrust available and thrust required, i.e., maximum ($T-D$). Figure 2.21 illustrates the climb angle performance with the curves of thrust available and thrust required versus velocity. The thrust required, or drag, curve is assumed to be representative of some typical airplane configuration which could be powered by either a turbojet or propeller type powerplant. The thrust available curves included are for a characteristic propeller powerplant and jet powerplant operating at maximum output.

The thrust curves for the representative propeller aircraft show the typical propeller thrust which is high at low velocities and decreases with an increase in velocity. For the propeller powered airplane, the maximum excess thrust and angle of climb will occur at some speed just above the stall speed. Thus, if it is necessary to clear an obstacle after takeoff, the propeller powered airplane will attain maximum angle of climb at an airspeed conveniently close to—if not at—the takeoff speed.

The thrust curves for the representative jet aircraft show the typical turbojet thrust which is very nearly constant with speed. If the thrust available is essentially constant with speed, the maximum excess thrust and angle of climb will occur where the thrust required

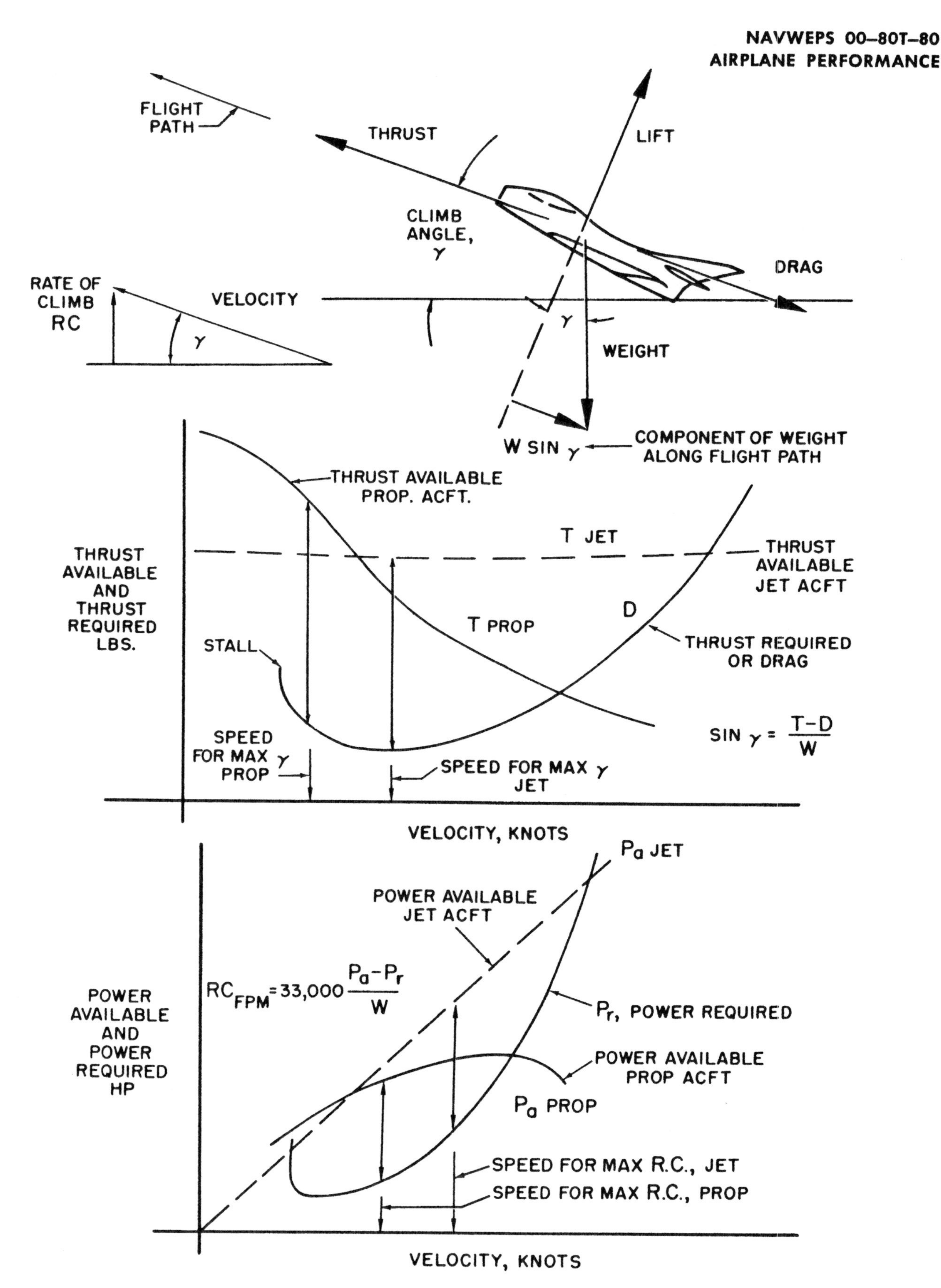

Figure 2.21. Climb Performance

is at a minimum, $(L/D)_{max}$. Thus, for maximum steady-state angle of climb, the turbojet aircraft would be operated at the speed for $(L/D)_{max}$. This poses somewhat of a problem in determining the proper procedure for obstacle clearance after takeoff. If the obstacle is a considerable distance from the takeoff point, the problem is essentially that of a long term gain and steady state conditions will predominate. That is, acceleration from the takeoff speed to $(L/D)_{max}$ speed will be favorable because the maximum steady climb angle can be attained. However, if the obstacle is a relatively short distance from the takeoff point, the additional distance required to accelerate to $(L/D)_{max}$ speed may be detrimental and the resulting situation may prove to be a short term gain problem. In this case, it may prove necessary to begin climb out at or near the takeoff speed or hold the aircraft on the runway for extra speed and a subsequent zoom. The problem is sufficiently varied that no general conclusion can be applied to all jet aircraft and particular procedures are specified for each aircraft in the Flight Handbook.

Of greater general interest in climb performance are the factors which affect the *rate of climb*. The vertical velocity of an airplane depends on the flight speed and the inclination of the flight path. In fact, the rate of climb is the vertical component of the flight path velocity. By the diagram of figure 2.21, the following relationship is developed:

$$RC = 101.3\ V \sin \gamma$$

since

$$\sin \gamma = \frac{T-D}{W}$$

then

$$RC = 101.3\ V \left(\frac{T-D}{W}\right)$$

and,

$$\text{with } Pa = \frac{TV}{325}$$

$$\text{and } Pr = \frac{DV}{325}$$

$$RC = 33{,}000\ \frac{Pa - Pr}{W}$$

where

RC = rate of climb, f.p.m.
Pa = power available, h.p.
Pr = power required, h.p.
W = weight, lbs
V = true airspeed, knots

and

33,000 is the factor converting horsepower to ft-lbs/min
101.3 is the factor converting knots to f.p.m.

The above relationship states that, for a given weight airplane, the *rate of climb* (RC) depends on the difference between the power available and the power required ($Pa - Pr$), or excess power. Of course, when the excess power is zero ($Pa - Pr = 0$ or $Pa = Pr$), the rate of climb is zero and the airplane is in steady level flight. When the power available is greater than the power required, the excess power will allow a rate of climb specific to the magnitude of excess power. Also, when the power available is less than the power required, the deficiency of power produces a rate of descent. This relationship provides the basis for an important axiom of flight technique: "For the conditions of steady flight, the power setting is the primary control of rate of climb or descent".

One of the most important items of climb performance is the maximum rate of climb. By the previous equation for rate of climb, maximum rate of climb would occur where there exists the greatest difference between power available and power required, i.e., maximum ($Pa - Pr$). Figure 2.21 illustrates the climb rate performance with the curves of power available and power required versus velocity. The power required curve is again a representative airplane which could be powered by either a turbojet or propeller type powerplant. The power available curves included are for a characteristic propeller powerplant and jet powerplant operating at maximum output.

The power curves for the representative propeller aircraft show a variation of propulsive power typical of a reciprocating engine-propeller combination. The maximum rate of climb for this aircraft will occur at some speed

US NAVY
US NAVY
US NAVY
US NAVY
US NAVY
US NAVY
US NAVY
US NAVY

near the speed for $(L/D)_{max}$. There is no direct relationship which establishes this situation since the variation of propeller efficiency is the principal factor accounting for the variation of power available with velocity. In an ideal sense, if the propeller efficiency were constant, maximum rate of climb would occur at the speed for minimum power required. However, in the actual case, the propeller efficiency of the ordinary airplane will produce lower power available at low velocity and cause the maximum rate of climb to occur at a speed greater than that for minimum power required.

The power curves for the representative jet aircraft show the near linear variation of power available with velocity. The maximum rate of climb for the typical jet airplane will occur at some speed much higher than that for maximum rate of climb of the equivalent propeller powered airplane. In part, this is accounted for by the continued increase in power available with speed. Note that a 50 percent increase in thrust by use of an afterburner may cause an increase in rate of climb of approximately 100 percent.

The climb performance of an airplane is affected by many various factors. The conditions of maximum climb angle or climb rate occur at specific speeds and variations in speed will produce variations in climb performance. Generally, there is sufficient latitude that small variations in speed from the optimum do not produce large changes in climb performance and certain operational items may require speeds slightly different from the optimum. Of course, climb performance would be most critical at high weight, high altitude, or during malfunction of a powerplant. Then, optimum climb speeds are necessary. A change in airplane weight produces a twofold effect on climb performance. First, the weight, W, appears directly in denominator of the equations for both climb angle and climb rate. In addition, a change in weight will alter the drag and power required. Generally, an increase in weight will reduce the maximum rate of climb but the airplane must be operated at some increase of speed to achieve the smaller peak climb rate (unless the airplane is compressibility limited).

The effect of altitude on climb performance is illustrated by the composite graphs of figure 2.22. Generally, an increase in altitude will increase the power required and decrease the power available. Hence, the climb performance of an airplane is expected to be greatly affected by altitude. The composite chart of climb performance depicts the variation with altitude of the speeds for maximum rate of climb, maximum angle of climb, and maximum and minimum level flight airspeeds. As altitude is increased, these various speeds finally converge at the absolute ceiling of the airplane. At the absolute ceiling, there is no excess of power or thrust and only one speed will allow steady level flight. The variation of rate of climb and maximum level flight speed with altitude for the typical propeller powered airplane give evidence of the effect of supercharging. Distinct aberrations in these curves take place at the supercharger critical altitudes and blower shift points. The curve of time to climb is the result of summing up the increments of time spent climbing through increments of altitude. Note that approach to the absolute ceiling produces tremendous increase in the time curve.

Specific reference points are established by these composite curves of climb performance. Of course, the absolute ceiling of the airplane produces zero rate of climb. The *service ceiling* is specified as the altitude which produces a rate of climb of 100 fpm. The altitude which produces a rate of climb of 500 fpm is termed the *combat ceiling*. Usually, these specific reference points are provided for the airplane at the combat configuration or a specific design configuration.

The composite curves of climb performance for the typical turbojet airplane are shown in figure 2.22. One particular point to note is the more rapid decay of climb performance

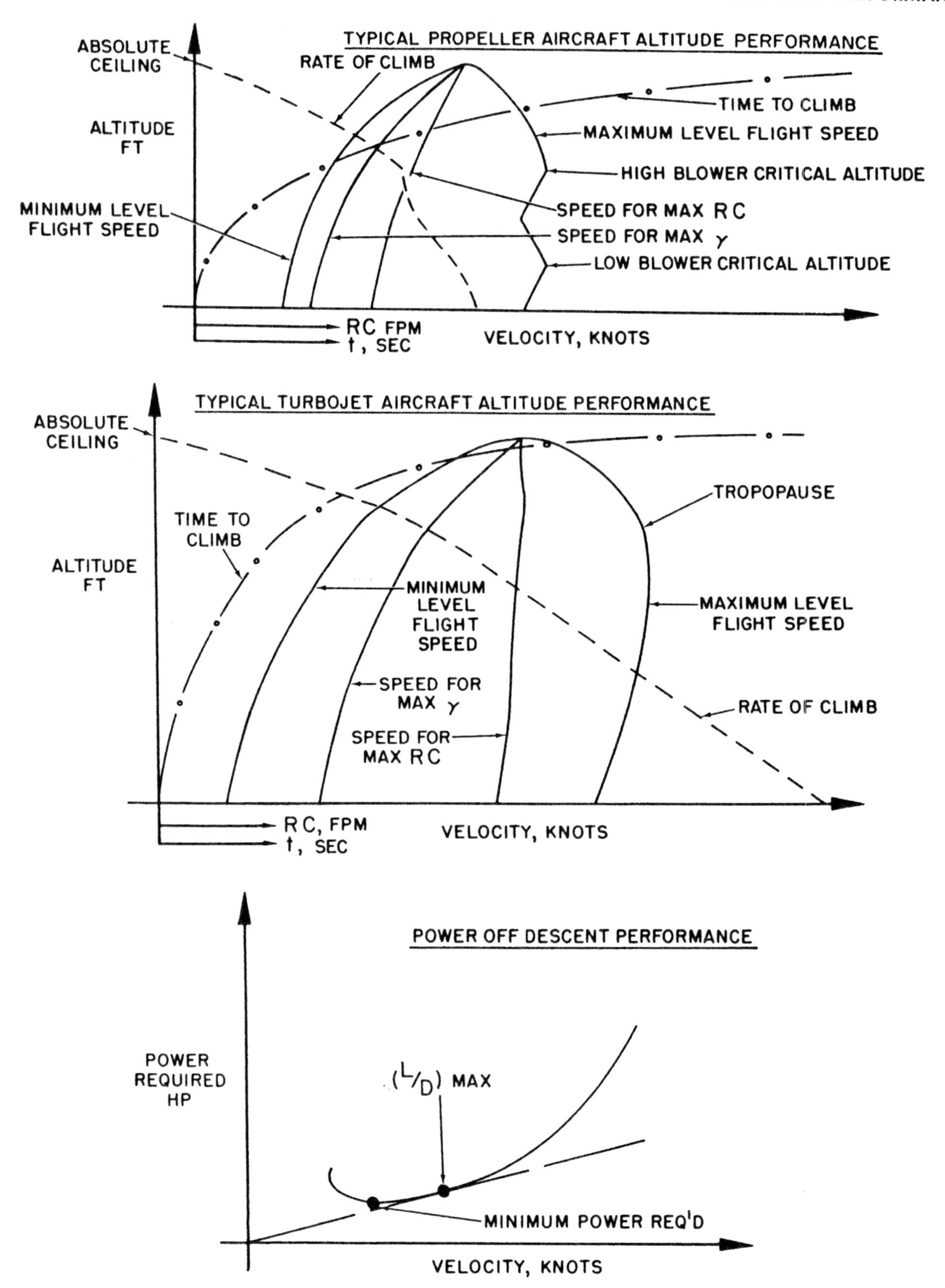

Figure 2.22. Climb and Descent Performance

with altitude above the tropopause. This is due in great part to the more rapid decay of engine thrust in the stratosphere.

During a power off descent the deficiency of thrust and power define the angle of descent and rate of descent. Two particular points are of interest during a power off descent: minimum angle of descent and minimum rate of descent. The minimum angle of descent would provide maximum glide distance through the air. Since no thrust is available from the power plant, minimum angle of descent would be obtained at $(L/D)_{max}$. At $(L/D)_{max}$ the deficiency of thrust is a minimum and, as shown by figure 2.22, the greatest proportion between velocity and power required is obtained. The minimum rate of descent in power off flight is obtained at the angle of attack and airspeed which produce minimum power required. For airplanes of moderate aspect ratio, the speed for minimum rate of descent is approximately 75 percent of the speed for minimum angle of descent

RANGE PERFORMANCE

The ability of an airplane to convert fuel energy into flying distance is one of the most important items of airplane performance. The problem of efficient range operation of an airplane appears of two general forms in flying operations: (1) to extract the maximum flying distance from a given fuel load or (2) to fly a specified distance with minimum expenditure of fuel. An obvious common denominator for each of these operating problems is the "specific range," nautical miles of flying distance per lb. of fuel. Cruise flight for maximum range conditions should be conducted so that the airplane obtains maximum specific range throughout the flight.

GENERAL RANGE PERFORMANCE. The principal items of range performance can be visualized by use of the illustrations of figure 2.23. From the characteristics of the aerodynamic configuration and the powerplant, the conditions of steady level flight will define various rates of fuel flow throughout the range of flight speed. The first graph of figure 2.23 illustrates a typical variation of fuel flow versus velocity. The specific range can be defined by the following relationship:

$$\text{specific range} = \frac{\text{nautical miles}}{\text{lbs. of fuel}}$$

or,

$$\text{specific range} = \frac{\text{nautical miles/hr.}}{\text{lbs. of fuel/hr.}}$$

thus,

$$\text{specific range} = \frac{\text{velocity, knots}}{\text{fuel flow, lbs. per hr.}}$$

If maximum specific range is desired, the flight condition must provide a maximum of velocity/fuel flow. This particular point would be located by drawing a straight line from the origin tangent to the curve of fuel flow versus velocity.

The general item of *range* must be clearly distinguished from the item of *endurance*. The item of *range* involves consideration of flying *distance* while *endurance* involves consideration of flying *time*. Thus, it is appropriate to define a separate term, "specific endurance."

$$\text{specific endurance} = \frac{\text{flight hours}}{\text{lb. of fuel}}$$

or,

$$\text{specific endurance} = \frac{\text{flight hours/hr.}}{\text{lbs. of fuel/hr.}}$$

then,

$$\text{specific endurance} = \frac{1}{\text{fuel flow, lbs. per hr.}}$$

By this definition, the specific endurance is simply the reciprocal of the fuel flow. Thus, if maximum endurance is desired, the flight condition must provide a minimum of fuel flow. This point is readily appreciated as the lowest point of the curve of fuel flow versus velocity. Generally, in subsonic performance, the speed at which maximum endurance is

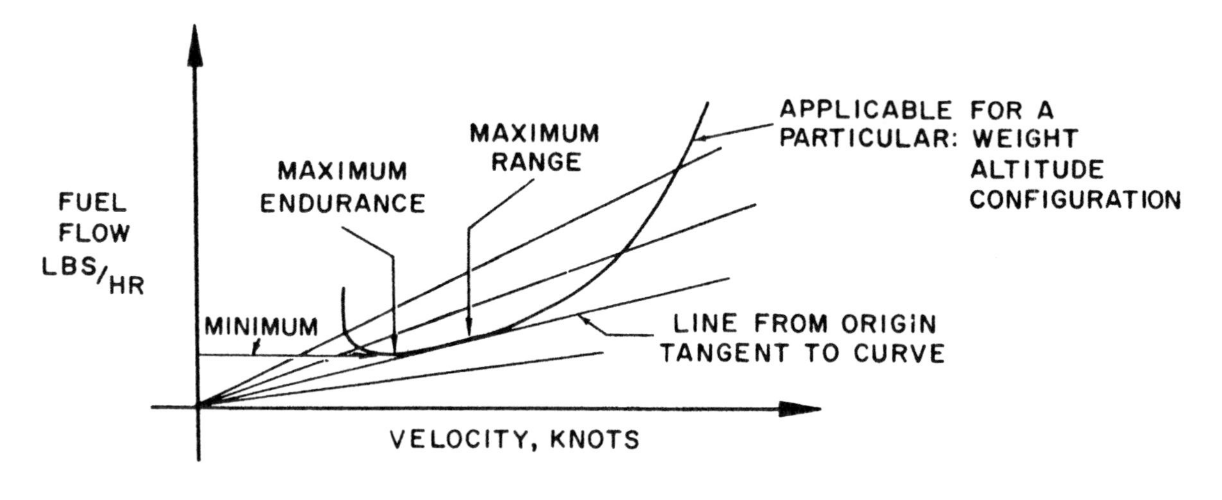

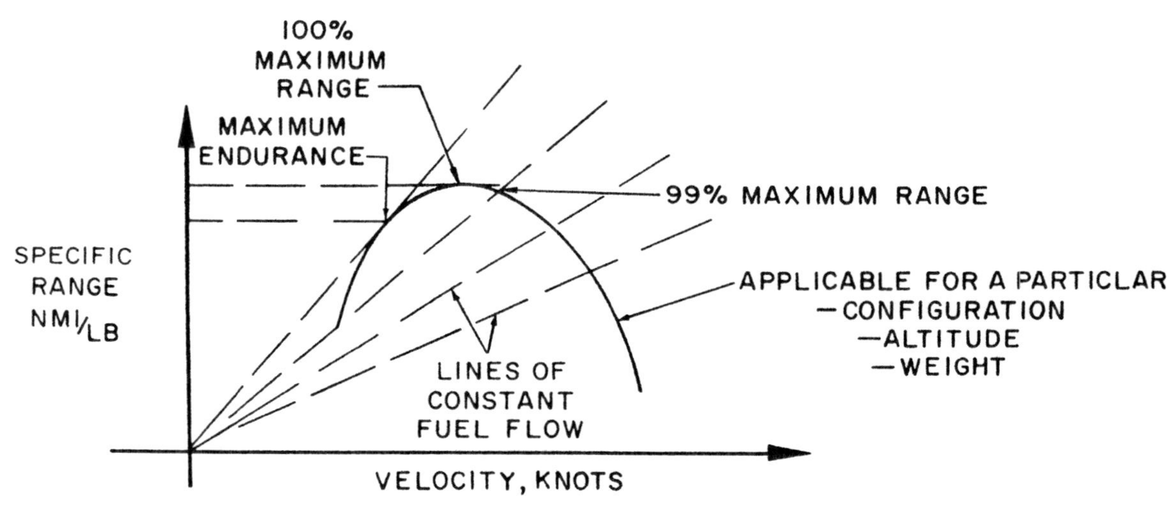

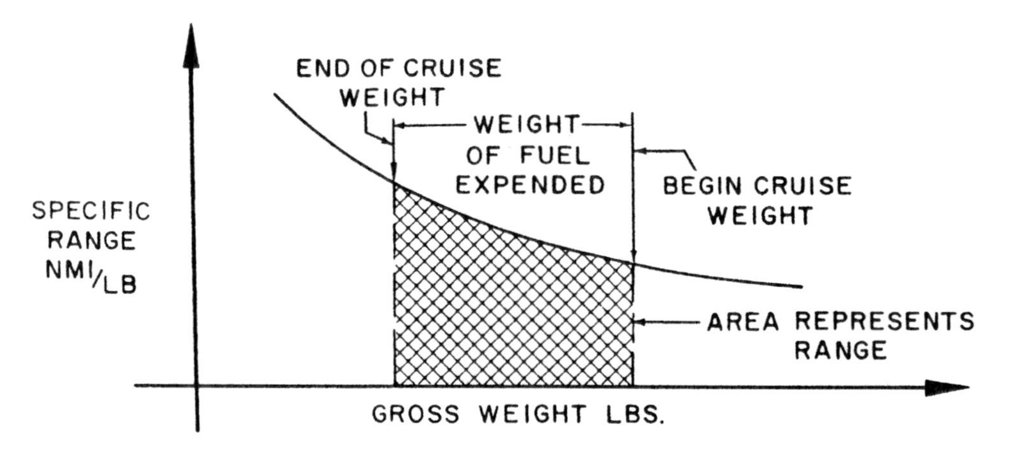

Figure 2.23. General Range Performance

obtained is approximately 75 percent of the speed for maximum range.

A more exact analysis of range may be obtained by a plot of specific range versus velocity similar to the second graph of figure 2.23. Of course, the source of these values of specific range is derived by the proportion of velocity and fuel flow from the previous curve of fuel flow versus velocity. The maximum specific range of the airplane is at the very peak of the curve. Maximum endurance point is located by a straight line from the origin tangent to the curve of specific range versus velocity. This tangency point defines a maximum of (nmi/lb.) per (nmi/hr.) or simply a maximum of (hrs./lb.).

While the very peak value of specific range would provide maximum range operation, long range cruise operation is generally recommended at some slightly higher airspeed. Most long range cruise operation is conducted at the flight condition which provides 99 percent of the absolute maximum specific range. The advantage of such operation is that 1 percent of range is traded for 3 to 5 percent higher cruise velocity. Since the higher cruise speed has a great number of advantages, the small sacrifice of range is a fair bargain. The curves of specific range versus velocity are affected by three principal variables: airplane gross weight, altitude, and the external aerodynamic configuration of the airplane. These curves are the source of range and endurance operating data and are included in the performance section of the flight handbook.

"Cruise control" of an airplane implies that the airplane is operated to maintain the recommended long range cruise condition throughout the flight. Since fuel is consumed during cruise, the gross weight of the airplane will vary and optimum airspeed, altitude, and power setting can vary. Generally, "cruise control" means the control of optimum airspeed, altitude, and power setting to maintain the 99 percent maximum specific range condition. At the beginning of cruise, the high initial weight of the airplane will require specific values of airspeed, altitude, and power setting to produce the recommended cruise condition. As fuel is consumed and the airplane gross weight decreases, the optimum airspeed and power setting may decrease or the optimum altitude may increase. Also, the optimum specific range will increase. The pilot must provide the proper cruise control technique to ensure that the optimum conditions are maintained.

The final graph of figure 2.23 shows a typical variation of specific range with gross weight for some particular cruise operation. At the beginning of cruise the gross weight is high and the specific range is low. As fuel is consumed, and the gross weight reduces, the specific range increases. This type of curve relates the range obtained by the expenditure of fuel by the crosshatched area between the gross weights at beginning and end of cruise. For example, if the airplane begins cruise at 18,500 lbs. and ends cruise at 13,000 lbs., 5,500 lbs. of fuel is expended. If the average specific range were 0.2 nmi/lb., the total range would be:

$$\text{range} = (0.2)\frac{\text{nmi}}{\text{lb.}}(5{,}500)\text{ lb.}$$

$$= 1{,}100 \text{ nmi.}$$

Thus, the total range is dependent on both the fuel available and the specific range. When range and economy of operation predominate, the pilot must ensure that the airplane will be operated at the recommended long range cruise condition. By this procedure, the airplane will be capable of its maximum design operating radius or flight distances less than the maximum can be achieved with a maximum of fuel reserve at the destination.

RANGE, PROPELLER DRIVEN AIRPLANES. The propeller driven airplane combines the propeller with the reciprocating engine or the gas turbine for propulsive power. In the case of either the reciprocating engine or the gas turbine combination, powerplant fuel flow is determined mainly by the shaft *power* put into the propeller rather than *thrust*. Thus, the powerplant fuel flow could be related directly to power required to maintain the airplane in steady, level flight. This fact allows study of the range of the propeller powered airplane by analysis of the curves of power required versus velocity.

Figure 2.24 illustrates a typical curve of power required versus velocity which, for the propeller powered airplane, would be analogous to the variation of fuel flow versus velocity. Maximum endurance condition would be obtained at the point of minimum power required since this would require the lowest fuel flow to keep the airplane in steady, level flight. Maximum range condition would occur where the proportion between velocity and power required is greatest and this point is located by a straight line from the origin tangent to the curve.

The maximum range condition is obtained at maximum lift-drag ratio and it is important to note that $(L/D)_{max}$ for a given airplane configuration occurs at a particular angle of attack and lift coefficient and is unaffected by weight or altitude (within compressibility limits). Since approximately 50 percent of the total drag at $(L/D)_{max}$ is induced drag, the propeller powered airplane which is designed specifically for long range will have a strong preference for the high aspect ratio planform.

The effect of the variation of airplane gross weight is illustrated by the second graph of figure 2.24. The flight condition of $(L/D)_{max}$ is achieved at one particular value of lift coefficient for a given airplane configuration. Hence, a variation of gross weight will alter the values of airspeed, power required, and specific range obtained at $(L/D)_{max}$. If a given configuration of airplane is operated at constant altitude and the lift coefficient for $(L/D)_{max}$, the following relationships will apply:

$$\frac{V_2}{V_1}=\sqrt{\frac{W_2}{W_1}}$$

$$\frac{Pr_2}{Pr_1}=\left(\frac{W_2}{W_1}\right)^{3/2}$$

$$\frac{SR_2}{SR_1}=\frac{W_1}{W_2}$$

where

condition (*1*) applies to some known condition of velocity, power required, and specific range for $(L/D)_{max}$ at some basic weight, W_1

condition (2) applies to some new values of velocity, power required, and specific range for $(L/D)_{max}$ at some different weight, W_2

and,

V = velocity, knots
W = gross weight, lbs.
Pr = power required, h.p.
SR = specific range, nmi/lb.

Thus a 10 percent increase in gross weight would create:

a 5 percent increase in velocity
a 15 percent increase in power required
a 9 percent decrease in specific range

when flight is maintained at the optimum conditions of $(L/D)_{max}$. The variations of velocity and power required must be monitored by the pilot as part of the cruise control to maintain $(L/D)_{max}$. When the airplane fuel weight is a small part of the gross weight and the range is small, the cruise control procedure can be simplified to essentially a constant speed and power setting throughout cruise. However, the long range airplane has a fuel weight which is a considerable part of the gross weight and cruise control procedure must employ scheduled airspeed and power changes to maintain optimum range conditions.

The effect of altitude on the range of the propeller powered airplane may be appreciated by inspection of the final graph of figure 2.24. If a given configuration of airplane is operated at constant gross weight and the lift coefficient

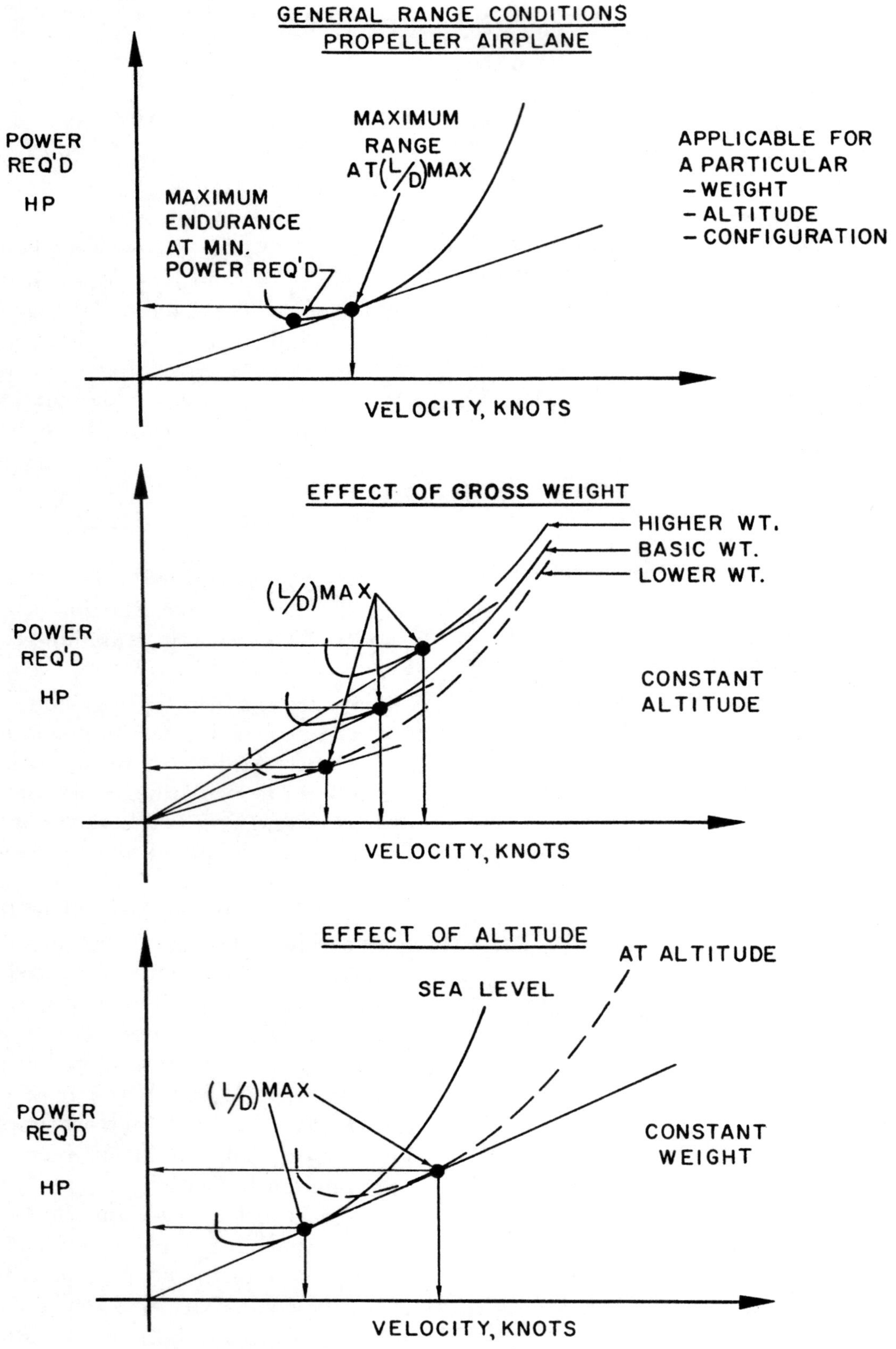

Figure 2.24. Range Performance, Propeller Aircraft

for $(L/D)_{max}$, a change in altitude will produce the following relationships:

$$\frac{V_2}{V_1}=\sqrt{\frac{\sigma_1}{\sigma_2}}$$

$$\frac{Pr_2}{Pr_1}=\sqrt{\frac{\sigma_1}{\sigma_2}}$$

where

condition (1) applies to some known condition of velocity and power required for $(L/D)_{max}$ at some original, basic altitude

condition (2) applies to some new values of velocity and power required for $(L/D)_{max}$ at some different altitude

and

V = velocity, knots (*TAS*, of course)
Pr = power required, h.p.
σ = altitude density ratio (sigma)

Thus, if flight is conducted at 22,000 ft. (σ=0.498), the airplane will have:

a 42 percent higher velocity
a 42 percent higher power required

than when operating at sea level. Of course, the greater velocity is a higher *TAS* since the airplane at a given weight and lift coefficient will require the same *EAS* independent of altitude. Also, the drag of the airplane at altitude is the same as the drag at sea level but the higher *TAS* causes a proportionately greater power required. Note that the same straight line from the origin tangent to the sea level power curve also is tangent to the altitude power curve.

The effect of altitude on specific range can be appreciated from the previous relationships. If a change in altitude causes identical changes in velocity and power required, the proportion of velocity to power required would be unchanged. This fact implies that the specific range of the propeller powered airplane would be unaffected by altitude. In the actual case, this is true to the extent that powerplant specific fuel consumption (c) and propeller efficiency (η_P) are the principal factors which could cause a variation of specific range with altitude. If compressibility effects are negligible, *any variation of specific range with altitude is strictly a function of engine-propeller performance.*

The airplane equipped with the reciprocating engine will experience very little, if any, variation of specific range with altitude at low altitudes. There is negligible variation of brake specific fuel consumption for values of *BHP* below the maximum cruise power rating of the powerplant which is the auto-lean or manual lean range of engine operation. Thus, an increase in altitude will produce a decrease in specific range only when the increased power requirement exceeds the maximum cruise power rating of the powerplants. One advantage of supercharging is that the cruise power may be maintained at high altitude and the airplane may achieve the range at high altitude with the corresponding increase in *TAS*. The principal differences in the high altitude cruise and low altitude cruise are the true airspeeds and climb fuel requirements.

The airplane equipped with the turboprop powerplant will exhibit a variation of specific range with altitude for two reasons. First, the specific fuel consumption (c) of the turbine engine improves with the lower inlet temperatures common to high altitudes. Also, the low power requirements to achieve optimum aerodynamic conditions at low altitude necessitate engine operation at low, inefficient output power. The increased power requirements at high altitudes allow the turbine powerplant to operate in an efficient output range. Thus, while the airplane has no particular preference for altitude, the powerplants prefer the higher altitudes and cause an increase in specific range with altitude. Generally, the upper limit of altitude for efficient cruise operation is defined by airplane gross weight (and power required) or compressibility effects.

The optimum climb and descent for the propeller powered airplane is affected by many different factors and no general, all-inclusive relationship is applicable. Handbook data for the specific airplane and various operational factors will define operating procedures.

RANGE, TURBOJET AIRPLANES. Many different factors influence the range of the turbojet airplane. In order to simplify the analysis of the overall range problem, it is convenient to separate airplane factors from powerplant factors and analyze each item independently. An analogy would be the study of "horsecart" performance by separating "cart" performance from "horse" performance to distinguish the principal factors which affect the overall performance.

In the case of the turbojet airplane, the fuel flow is determined mainly by the *thrust* rather than *power*. Thus, the fuel flow could be most directly related to the thrust required to maintain the airplane in steady, level flight. This fact allows study of the turbojet powered airplane by analysis of the curves of thrust required versus velocity. Figure 2.25 illustrates a typical curve of thrust required versus velocity which would be (somewhat) analogous to the variation of fuel flow versus velocity. Maximum endurance condition would be obtained at $(L/D)_{max}$ since this would incur the lowest fuel flow to keep the airplane in steady, level flight. Maximum range condition would occur where the proportion between velocity and thrust required is greatest and this point is located by a straight line from the origin tangent to the curve.

The maximum range is obtained at the aerodynamic condition which produces a maximum proportion between the square root of the lift coefficient (C_L) and the drag coefficient (C_D), or $(\sqrt{C_L}/C_D)_{max}$. In subsonic performance, $(\sqrt{C_L}/C_D)_{max}$ occurs at a particular value angle of attack and lift coefficient and is unaffected by weight or altitude (within compressibility limits). At this specific aerodynamic condition, induced drag is approximately 25 percent of the total drag so the turbojet airplane designed for long range does not have the strong preference for high aspect ratio planform like the propeller airplane. On the other hand, since approximately 75 percent of the total drag is parasite drag, the turbojet airplane designed specifically for long range has the special requirement for great aerodynamic cleanness.

The effect of the variation of airplane gross weight is illustrated by the second graph of figure 2.25. The flight condition of $(\sqrt{C_L}/C_D)_{max}$ is achieved at one value of lift coefficient for a given airplane in subsonic flight. Hence, a variation of gross weight will alter the values of airspeed, thrust required, and specific range obtained at $(\sqrt{C_L}/C_D)_{max}$. If a given configuration is operated at constant altitude and lift coefficient the following relationships will apply:

$$\frac{V_2}{V_1}=\sqrt{\frac{W_2}{W_1}}$$

$$\frac{Tr_2}{Tr_1}=\frac{W_2}{W_1}$$

$$\frac{SR_2}{SR_1}=\sqrt{\frac{W_1}{W_2}} \text{ (constant altitude)}$$

where

condition (1) applies to some known condition of velocity, thrust required, and specific range for $(\sqrt{C_L}/C_D)_{max}$ at some basic weight, W_1

condition (2) applies to some new values of velocity, thrust required, and specific range for $(\sqrt{C_L}/C_D)_{max}$ at some different weight, W_2

and

V = velocity, knots
W = gross weight, lbs.
Tr = thrust required, lbs.
SR = specific range, nmi/lb.

Thus, a 10 percent increase in gross weight would create:

a 5 percent increase in velocity
a 10 percent increase in thrust required
a 5 percent decrease in specific range

when flight is maintained at the optimum conditions of $(\sqrt{C_L}/C_D)_{max}$. Since most jet airplanes

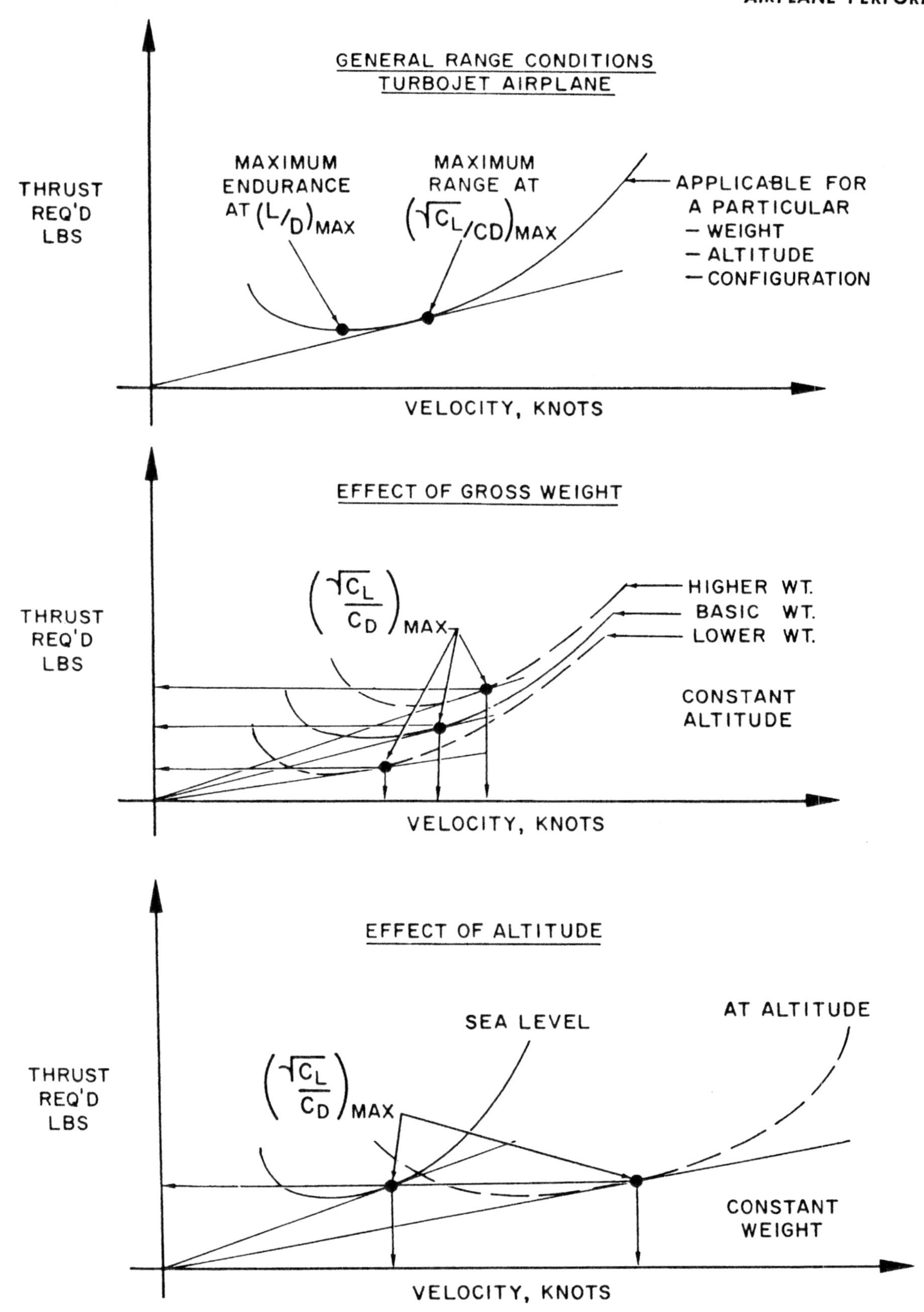

Figure 2.25. Range Performance, Jet Aircraft

have a fuel weight which is a large part of the gross weight, cruise control procedures will be necessary to account for the changes in optimum airspeeds and power settings as fuel is consumed.

The effect of altitude on the range of the turbojet airplane is of great importance because no other single item can cause such large variations of specific range. If a given configuration of airplane is operated at constant gross weight and the lift coefficient for $(\sqrt{C_L}/C_D)_{max}$, a change in altitude will produce the following relationships:

$$\frac{V_2}{V_1}=\sqrt{\frac{\sigma_1}{\sigma_2}}$$

Tr=constant (neglecting compressibility effects)

$$\frac{SR_2}{SR_1}=\sqrt{\frac{\sigma_1}{\sigma_2}}$$ (neglecting factors affecting engine performance)

where

condition (1) applies some known condition of velocity, thrust required, and specific range for $(\sqrt{C_L}/C_D)_{max}$ at some original, basic altitude.

condition (2) applies to some new values of velocity, thrust required, and specific range for $(\sqrt{C_L}/C_D)_{max}$ at some different altitude.

and

V=velocity, knots (TAS, of course)
Tr=thrust required, lbs.
SR=specific range, nmi/lb.
σ=altitude density ratio (sigma)

Thus, if flight is conducted at 40,000 ft. ($\sigma=0.246$), the airplane will have:

a 102 percent higher velocity
the same thrust required
a 102 percent higher specific range
(even when the beneficial effects of altitude on engine performance are neglected)

than when operating at sea level. Of course, the greater velocity is a higher TAS and the same thrust required must be obtained with a greater engine RPM.

At this point it is necessary to consider the effect of the operating condition on powerplant performance. An increase in altitude will improve powerplant performance in two respects. First, an increase in altitude when below the tropopause will provide lower inlet air temperatures which reduce the specific fuel consumption (c_t). Of course, above the tropopause the specific fuel consumption tends to increase. At low altitude, the engine RPM necessary to produce the required thrust is low and, generally, well below the normal rated value. Thus, a second benefit of altitude on engine performance is due to the increased RPM required to furnish cruise thrust. An increase in engine speed to the normal rated value will reduce the specific fuel consumption.

The increase in specific range with altitude of the turbojet airplane can be attributed to these three factors:

(1) An increase in altitude will increase the proportion of (V/Tr) and provide a *greater* TAS for the same Tr.

(2) An increase in altitude in the troposphere will produce *lower inlet air temperature* which reduces the specific fuel consumption.

(3) An increase in altitude requires *increased engine RPM* to provide cruise thrust and the specific fuel consumption reduces as normal rated RPM is approached.

The combined effect of these three factors defines altitude as the one most important item affecting the specific range of the turbojet airplane. As an example of this combined effect, the typical turbojet airplane obtains a specific range at 40,000 ft. which is approximately 150 percent greater than that obtained at sea level. The increased TAS accounts for approximately two-thirds of this benefit while increased engine performance (reduced c_t) accounts for the other one-third of the benefit. For example, at sea level the maximum specific range of a turbojet airplane may be 0.1 nmi/lb. but at 40,000 ft. the maximum specific range would be approximately 0.25 nmi/lb.

From the previous analysis, it is apparent that the cruise altitude of the turbojet should be as high as possible within compressibility or thrust limits. Generally, the optimum altitude to begin cruise is the highest altitude at which the maximum continuous thrust can provide the optimum aerodynamic conditions. Of course, the optimum altitude is determined mainly by the gross weight at the begin of cruise. For the majority of turbojet airplanes this altitude will be at or above the tropopause for normal cruise configurations.

Most turbojet airplanes which have transonic or moderate supersonic performance will obtain maximum range with a high subsonic cruise. However, the airplane designed specifically for high supersonic performance will obtain maximum range with a supersonic cruise and subsonic operation will cause low lift-drag ratios, poor inlet and engine performance and reduce the range capability.

The cruise control of the turbojet airplane is considerably different from that of the propeller driven airplane. Since the specific range is so greatly affected by altitude, the optimum altitude for begin of cruise should be attained as rapidly as is consistent with climb fuel requirements. The range-climb program varies considerably between airplanes and the performance section of the flight handbook will specify the appropriate procedure. The descent from cruise altitude will employ essentially the same feature, a rapid descent is necessary to minimize the time at low altitudes where specific range is low and fuel flow is high for a given engine speed.

During cruise flight of the turbojet airplane, the decrease of gross weight from expenditure of fuel can result in two types of cruise control. During a *constant altitude cruise*, a reduction in gross weight will require a reduction of airspeed and engine thrust to maintain the optimum lift coefficient of subsonic cruise. While such a cruise may be necessary to conform to the flow of traffic, it constitutes a certain inefficiency of operation. If the airplane were not restrained to a particular altitude, maintaining the same lift coefficient and engine speed would allow the airplane to climb as the gross weight decreases. Since altitude generally produces a beneficial effect on range, the *climbing cruise* implies a more efficient flight path.

The cruising flight of the turbojet airplane will begin usually at or above the tropopause in order to provide optimum range conditions. If flight is conducted at $(\sqrt{C_L}/C_D)_{max}$, optimum range will be obtained at specific values of lift coefficient and drag coefficient. When the airplane is fixed at these values of C_L and C_D and the *TAS* is held constant, both lift and drag are directly proportional to the density ratio, σ. Also, above the tropopause, the thrust is proportional to σ when the *TAS* and RPM are constant. As a result, a reduction of gross weight by the expenditure of fuel would allow the airplane to climb but the airplane would remain in equilibrium because lift, drag, and thrust all vary in the same fashion. This relationship is illustrated by figure 2.26.

The relationship of lift, drag, and thrust is convenient for, in part, it justifies the condition of a constant velocity. Above the tropopause, the speed of sound is constant hence a constant velocity during the cruise-climb would produce a constant Mach number. In this case, the optimum values of $(\sqrt{C_L}/C_D)$, C_L and C_D do not vary during the climb since the Mach number is constant. The specific fuel consumption is initially constant above the tropopause but begins to increase at altitudes much above the tropopause. If the specific fuel consumption is assumed to be constant during the cruise-climb, the following relationships will apply:

V, M, C_L and C_D are constant

$$\frac{\sigma_2}{\sigma_1}=\frac{W_2}{W_1}$$

$$\frac{FF_2}{FF_1}=\frac{\sigma_2}{\sigma_1}$$

$$\frac{SR_2}{SR_1}=\frac{W_1}{W_2}$$ (cruise climb above tropopause, constant M, c_t)

where

condition (*1*) applies to some known condition of weight, fuel flow, and specific range at some original basic altitude during cruise climb.

condition (2) applies to some new values of weight, fuel flow, and specific range at some different altitude along a particular cruise path.

and

V = velocity, knots
M = Mach number
W = gross weight, lbs.
FF = fuel flow, lbs./hr.
SR = specific range, nmi./lb.
σ = altitude density ratio

Thus, during a cruise-climb flight, a 10 percent decrease in gross weight from the consumption of fuel would create:

no change in Mach number or *TAS*
a 5 percent decrease in *EAS*
a 10 percent decrease in σ, i.e., higher altitude
a 10 percent decrease in fuel flow
an 11 percent increase in specific range

An important comparison can be made between the constant altitude cruise and the cruise-climb with respect to the variation of specific range. From the previous relationships, a 2 percent reduction in gross weight during

$$\frac{SR_2}{SR_1}=\sqrt{\frac{W_1}{W_2}} \qquad \textit{constant altitude}$$

$$\frac{SR_2}{SR_1}=\frac{W_1}{W_2} \qquad \textit{cruise-climb}$$

cruise would create a 1 percent increase in specific range in a constant altitude cruise but a 2 percent increase in specific range in a cruise-climb at constant Mach number. Thus, a higher average specific range can be maintained during the expenditure of a given increment of fuel. If an airplane begins a cruise at optimum conditions at or above the tropopause with a given weight of fuel, the following data provide a comparison of the total range available from a constant altitude or cruise-climb flight path.

Ratio of cruise fuel weight to airplane gross weight at beginning of cruise	*Ratio of cruise-climb range to constant altitude cruise range*
0.0	1.000
.1	1.026
.2	1.057
.3	1.092
.4	1.136
.5	1.182
.6	1.248
.7	1.331

For example, if the cruise fuel weight is 50 percent of the gross weight, the climbing cruise flight path will provide a range 18.2 percent greater than cruise at constant altitude. This comparison does not include consideration of any variation of specific fuel consumption during cruise or the effects of compressibility in defining the optimum aerodynamic conditions for cruising flight. However, the comparison is generally applicable for aircraft which have subsonic cruise.

When the airplane has a supersonic cruise for maximum range, the optimum flight path is generally one of a constant Mach number. The optimum flight path is generally—but not necessarily—a climbing cruise. In this case of subsonic or supersonic cruise, a Machmeter is of principal importance in cruise control of the jet airplane.

The *effect of wind on range* is of considerable importance in flying operations. Of course, a headwind will always reduce range and a tailwind will always increase range. The selection of a cruise altitude with the most favorable (or least unfavorable) winds is a relatively simple matter for the case of the propeller powered airplane. Since the range of the propeller powered airplane is relatively unaffected by altitude, the altitude with the most favorable winds is selected for range. However, the range of the turbojet airplane is greatly affected by altitude so the selection of an optimum altitude will involve considering the wind profile with the variation of range with altitude. Since the turbojet range increases

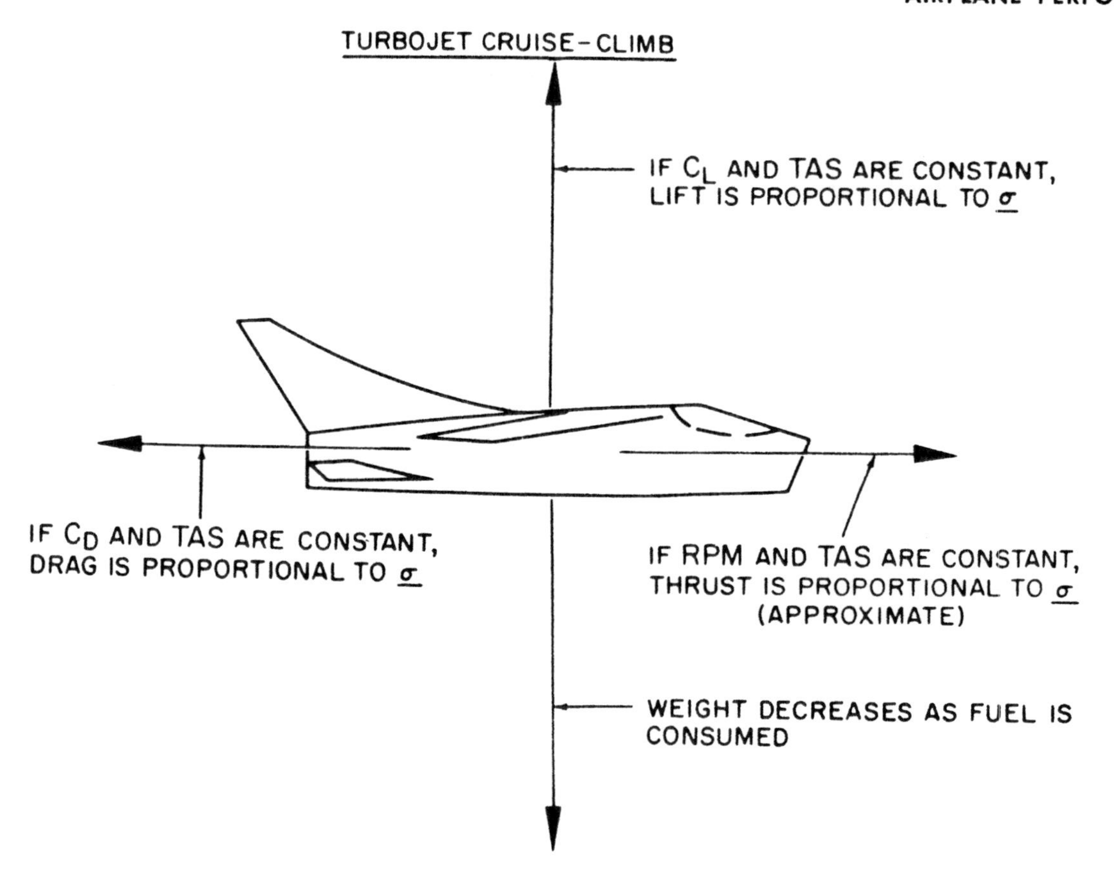

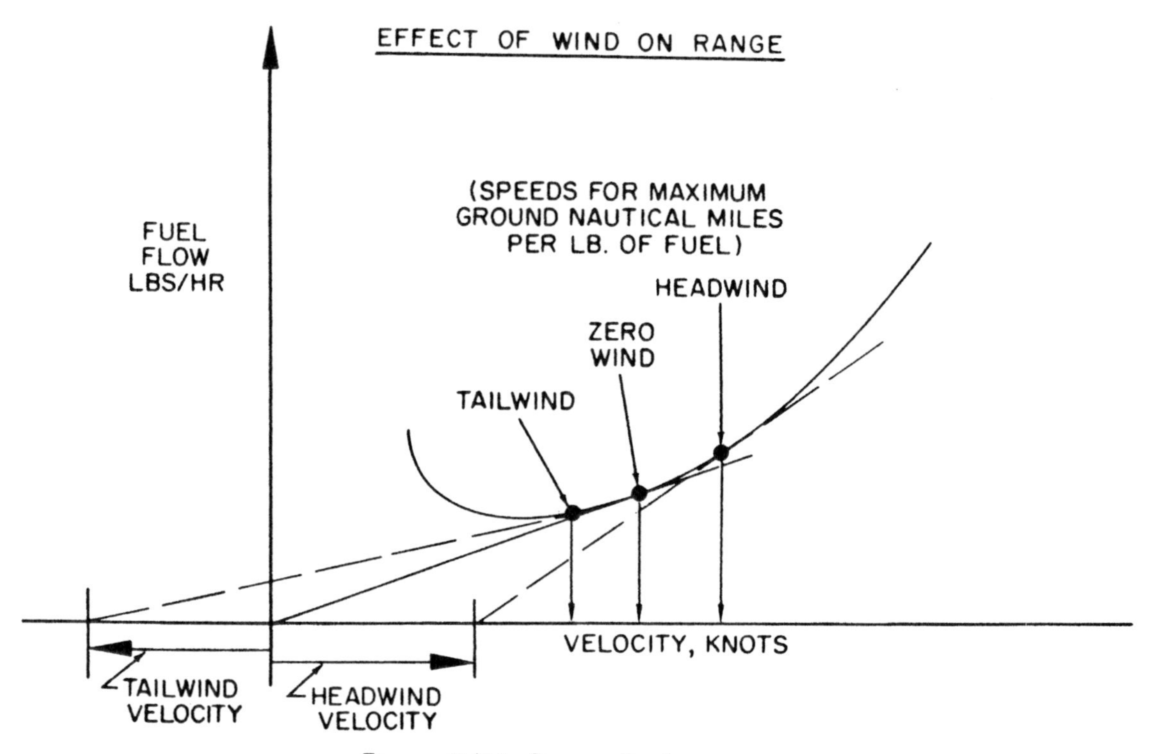

Figure 2.26. Range Performance

greatly with altitude, the turbojet can tolerate less favorable (or more unfavorable) winds with increased altitude.

In some cases, large values of wind may cause a significant change in cruise velocity to maintain maximum ground nautical miles per lb. of fuel. As an example of an extreme condition, consider an airplane flying into a headwind which equals the cruise velocity. In this case, *any* increase in velocity would improve range.

To appreciate the changes in optimum speeds with various winds, refer to the illustration of figure 2.26. When zero wind conditions exist, a straight line from the origin tangent to the curve of fuel flow versus velocity will locate maximum range conditions. When a headwind condition exists, the speed for maximum ground range is located by a line tangent drawn from a velocity offset equal to the headwind velocity. This will locate maximum range at some higher velocity and fuel flow. Of course, the range will be less than when at zero wind conditions but the higher velocity and fuel flow will minimize the range loss due to the headwind. In a similar sense, a tailwind will reduce the cruise velocity to maximize the benefit of the tailwind.

The procedure of employing different cruise velocities to account for the effects of wind is necessary only at extreme values of wind velocity. It is necessary to consider the change in optimum cruise airspeed when the wind velocities exceed 25 percent of the zero wind cruise velocity.

ENDURANCE PERFORMANCE

The ability of the airplane to convert fuel energy into flying time is an important factor in flying operations. The "specific endurance" of the airplane is defined as follows:

$$\text{specific endurance} = \frac{\text{flight hrs.}}{\text{lb. of fuel}}$$

$$\text{specific endurance} = \frac{1}{\text{fuel flow, lbs. per hr.}}$$

The specific endurance is simply the reciprocal of the fuel flow, hence maximum endurance conditions would be obtained at the lowest fuel flow required to hold the airplane in steady level flight. Obviously, minimum fuel flow will provide the maximum flying time from a given quantity of fuel. Generally, in subsonic performance, the speed at which maximum endurance is achieved is approximately 75 percent of the speed for maximum range.

While many different factors can affect the specific endurance, the most important factors at the control of the pilot are the configuration and operating altitude. Of course, for maximum endurance conditions the airplane must be in the clean configuration and operated at the proper aerodynamic conditions.

EFFECT OF ALTITUDE ON ENDURANCE, PROPELLER DRIVEN AIRPLANES. Since the fuel flow of the propeller driven airplane is proportional to power required, the propeller powered airplane will achieve maximum specific endurance when operated at minimum power required. The point of minimum power required is obtained at a specific value of lift coefficient for a particular airplane configuration and is essentially independent of weight or altitude. However, an increase in altitude will increase the value of the minimum power required as illustrated by figure 2.27. If the specific fuel consumption were not influenced by altitude or engine power, the specific endurance would be directly proportional to $\sqrt{\sigma}$, e.g., the specific endurance at 22,000 ft. ($\sigma = 0.498$) would be approximately 70 percent of the value at sea level. This example is very nearly the case of the airplane with the *reciprocating engine* since specific fuel consumption and propeller efficiency are not directly affected by altitude. The obvious conclusion is that maximum endurance of the reciprocating engine airplane is obtained at the lowest practical altitude.

The variation with altitude of the maximum endurance of the turboprop airplane requires consideration of powerplant factors in addition

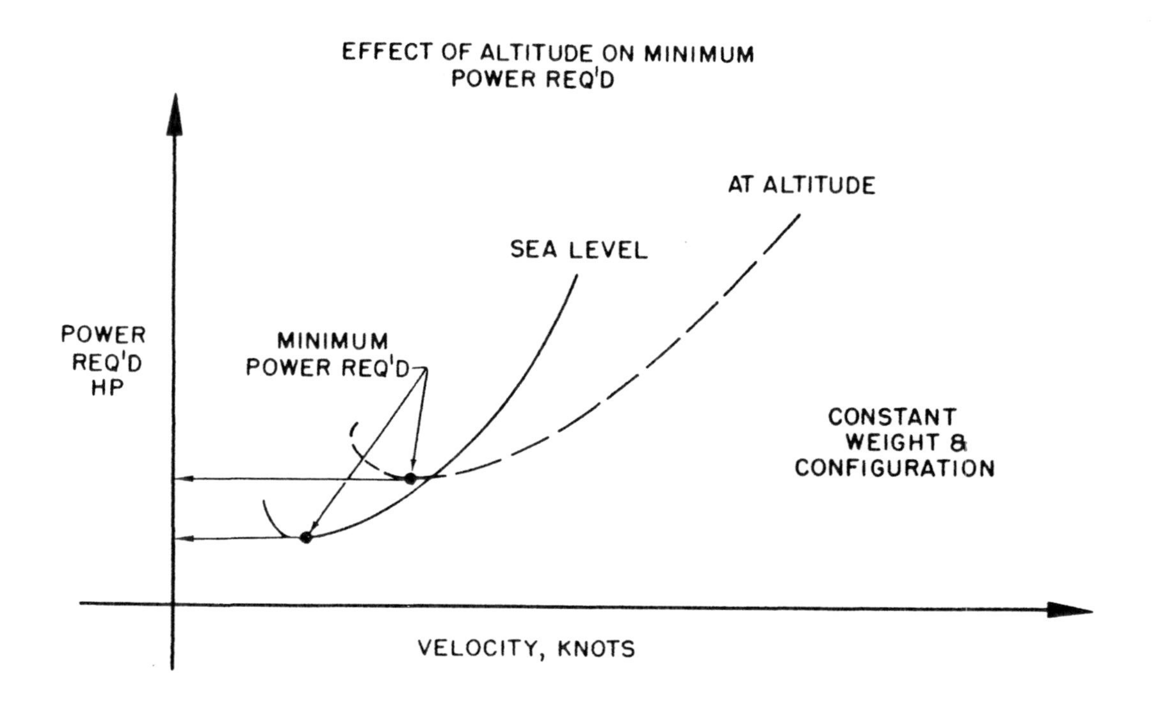

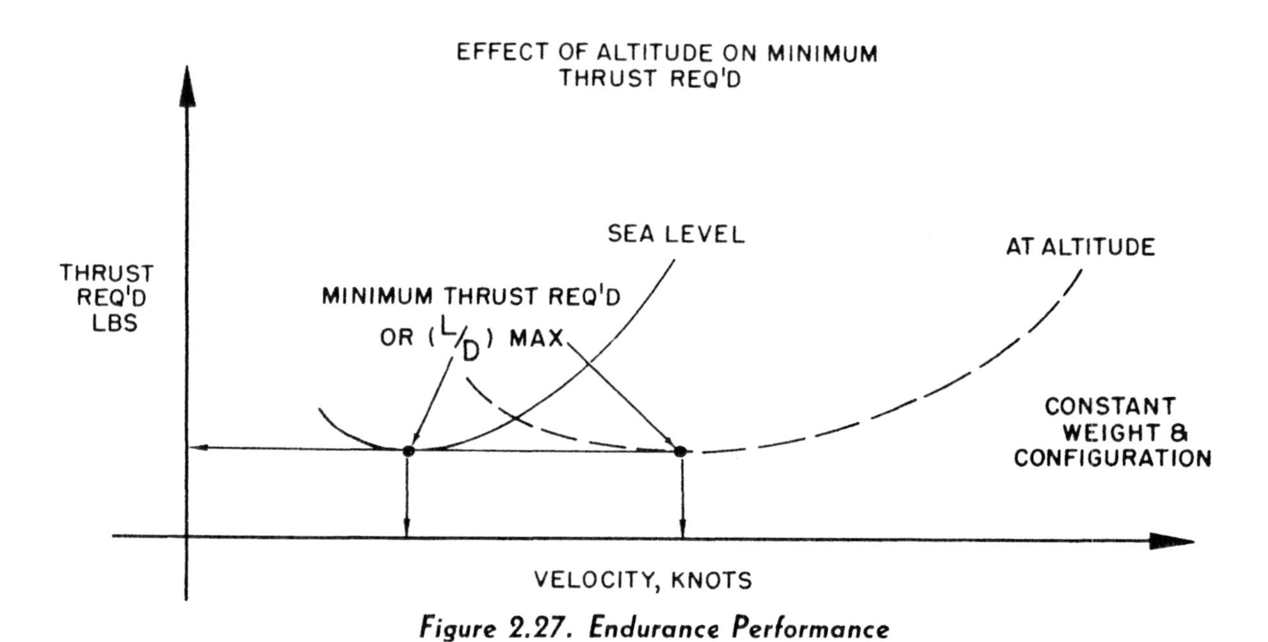

Figure 2.27. Endurance Performance

to airplane factors. The turboprop powerplant prefers operation at low inlet air temperatures and relatively high power setting to produce low specific fuel consumption. While an increase in altitude will increase the minimum power required for the airplane, the powerplant achieves more efficient operation. As a result of these differences, maximum endurance of the multiengine turboprop airplane at low altitudes may require shutting down some of the powerplants in order to operate the remaining powerplants at a higher, more efficient power setting.

EFFECT OF ALTITUDE ON ENDURANCE, TURBOJET AIRPLANES. Since the fuel flow of the turbojet powered airplane is proportional to thrust required, the turbojet airplane will achieve maximum specific endurance when operated at minimum thrust required or $(L/D)_{max}$. In subsonic flight, $(L/D)_{max}$ occurs at a specific value of lift coefficient for a given airplane and is essentially independent of weight or altitude. If a given weight and configuration of airplane is operated at various altitudes, the value of the minimum thrust required is unaffected by the curves of thrust required versus velocity shown in figure 2.27. Hence, it is apparent that the aerodynamic configuration has no preference for altitude (within compressibility limits) and specific endurance is a function only of engine performance.

The specific fuel consumption of the turbojet engine is strongly affected by operating RPM and altitude. Generally, the turbojet engine prefers the operating range near normal rated engine speed and the low temperatures of the stratosphere to produce low specific fuel consumption. Thus, increased altitude provides the favorable lower inlet air temperature and requires a greater engine speed to provide the thrust required at $(L/D)_{max}$. The typical turbojet airplane experiences an increase in specific endurance with altitude with the peak values occurring at or near the tropopause. For example, a typical single-engine turbojet airplane will have a maximum specific endurance at 35,000 ft. which is at least 40 percent greater than the maximum value at sea level. If the turbojet airplane is at low altitude and it is necessary to hold for a considerable time, maximum time in the air will be obtained by beginning a climb to some optimum altitude dependent upon the fuel quantity available. Even though fuel is expended during the climb, the higher altitude will provide greater total endurance. Of course, the use of afterburner for the climb would produce a prohibitive reduction in endurance.

OFF-OPTIMUM RANGE AND ENDURANCE

There are many conditions of flying operations in which optimum range or endurance conditions are not possible or practical. In many instances, the off-optimum conditions result from certain operational requirements or simplification of operating procedure. In addition, off-optimum performance may be the result of a powerplant malfunction or failure. The most important conditions are discussed for various airplanes by powerplant type.

RECIPROCATING POWERED AIRPLANE. In the majority of cases, the reciprocating powered airplane is operated at an engine dictated cruise. Service use will most probably define some continuous power setting which will give good service life and trouble-free operation of the powerplant. When range or endurance is of no special interest, the simple expedient is to operate the powerplant at the recommended power setting and accept whatever speed, range, or endurance that results. While such a procedure greatly simplifies the matter of cruise control, the practice does not provide the necessary knowledge required for operating a high performance, long range airplane.

The failure of an engine on the multiengine reciprocating powered airplane has interesting ramifications. The first problem appearing is to produce sufficient power from the remaining engines to keep the airplane airborne. The problem will be most critical if the airplane is at high altitude, high gross weight, and with flaps and gear extended. Lower altitude, jettisoning of weight items, and cleaning up the airplane will reduce the power required for flight. Of course, the propeller on the inoperative engine must be feathered or the power required may exceed that available from the remaining operating powerplants.

The effect on range is much dependent on the airplane configuration. When the propeller on the inoperative engine is feathered, the added drag is at a minimum, but there is added the trim drag required to balance the unsymmetrical power. When both these sources of added drag are accounted for, the $(L/D)_{max}$ is reduced but not by significant amounts. Generally, if the specific fuel consumption and propeller efficiency do not deteriorate, the maximum specific range is not greatly reduced. On the twin-engine airplane the power required must be furnished by the one remaining engine and this usually requires more than the maximum cruise-rating of the powerplant. As a result the powerplant cannot be operated in the auto-lean or manual lean power range and the specific fuel consumption increases greatly. Thus, noticeable loss of range must be anticipated when one engine fails on the twin-engine airplane. The failure of one engine on the four (or more) engine airplane may allow the required power to be developed by the three remaining powerplants operating in an economical power range. If the airplane is clean, at low altitude, and low gross weight, the failure of one engine is not likely to cause a loss of range. However, the loss of two engines is likely to cause a considerable loss of range.

When engine failure produces a critical power or range situation, improved performance is possible with the airplane in the clean configuration at low altitude. Also, jettisoning of expendable weight items will reduce the power required and improve the specific range.

TURBOPROP POWERED AIRPLANE. The turbine engine has the preference for relatively high power settings and high altitudes to provide low specific fuel consumption. Thus, the off-optimum conditions of range or endurance can be concerned with altitudes less than the optimum. Altitudes less than the optimum can reduce the range but the loss can be minimized on the multiengine airplane by shutting down some powerplants and operating the remaining powerplants at a higher, more efficient output. In this case the change of range is confined to the variation of specific fuel consumption with altitude.

Essentially the same situation exists in the case of engine failure when cruising at optimum altitude. If the propeller on the inoperative engine is feathered, the loss of range will be confined to the change in specific fuel consumption from the reduced cruise altitude. If a critical power situation exists due to engine failure, a reduction in altitude provides immediate benefit because of the reduction of power required and the increase in power available from the power plants. In addition, the jettisoning of expendable weight items will improve performance and, of course, the clean configuration provides minimum parasite drag.

Maximum specific endurance of the turboprop airplane does not vary as greatly with altitude as the turbojet airplane. While each configuration has its own particular operating requirements, low altitude endurance of the turboprop airplane requires special consideration. The single-engine turboprop will generally experience an increase in specific endurance with an increase in altitude from sea level. However, if the airplane is at low altitude and must hold or endure for a period of time, the decision to begin a climb or hold the existing altitude will depend on the quantity of fuel available. The decision depends primarily on the climb fuel requirements and the variation of specific endurance with altitude. A somewhat similar problem exists with the multiengine

NAVY

turboprop airplane but additional factors are available to influence the specific endurance at low altitude. In other words, low altitude endurance can be improved by shutting down some powerplants and operating the remaining powerplants at higher, more efficient power setting. Many operational factors could decide whether such procedure would be a suitable technique.

TURBOJET POWERED AIRPLANE. Increasing altitude has a powerful effect on both the range and endurance of the turbojet airplane. As a result of this powerful effect, the typical turbojet airplane will achieve maximum specific endurance at or near the tropopause. Also, the maximum specific range will be obtained at even higher altitudes since the peak specific range generally occurs at the highest altitude at which the normal rating of the engine can sustain the optimum aerodynamic conditions. At low altitude cruise conditions, the engine speed necessary to sustain optimum aerodynamic conditions is very low and the specific fuel consumption is relatively poor. Thus, at low altitude, the airplane prefers the low speeds to obtain $(\sqrt{C_L}/C_D)_{max}$ but the powerplant prefers the higher speeds common to higher engine efficiency. The compromise results in maximum specific range at flight speeds well above the optimum aerodynamic conditions. In a sense, low altitude cruise conditions are engine dictated.

Altitude is the one most important factor affecting the specific range of the turbojet airplane. Any operation below the optimum altitude will have a noticeable effect on the range capability and proper consideration must be given to the loss of range. In addition, turbojet airplanes designed specifically for long range will have a large percent of the gross weight as fuel. The large changes in gross weight during cruise will require particular methods of cruise control to extract the maximum flight range. A variation from the optimum flight path of cruise (constant Mach number, cruise-climb, or whatever the appropriate technique) will result in a loss of range capability.

The failure of an engine during the optimum cruise of a multiengine turbojet airplane will cause a noticeable loss of range. Since the optimum cruise of the turbojet is generally a thrust-limited cruise, the loss of part of the total thrust means that the airplane must descend to a lower altitude. For example, if a twin-engine jet begins an optimum cruise at 35,000 ft. ($\sigma=0.31$) and one powerplant fails, the airplane must descend to a lower altitude so that the operative engine can provide the cruise thrust. The resulting altitude would be approximately 16,000 ft. ($\sigma=0.61$). Thus, the airplane will experience a loss of the range remaining at the point of engine failure and loss could be accounted for by the reduced velocity (TAS) and the increase in specific fuel consumption (c_t) from the higher ambient air temperature. In the case of the example airplane, engine failure would cause a 30 to 40 percent loss of range from the point of engine failure. Of course, the jettisoning of expendable weight items would allow higher altitude and would increase the specific range.

Maximum endurance in the turbojet airplane varies with altitude but the variation is due to the changes in fuel flow necessary to provide the thrust required at $(L/D)_{max}$. The low inlet air temperature of the tropopause and the greater engine speed reduce the specific fuel consumption to a minimum. If the single-engine turbojet airplane is at low altitude and must hold or endure for a period of time, a climb should begin to take advantage of the higher specific endurance at higher altitude. The altitude to which to climb will be determined by the quantity of fuel remaining. In the case of the multiengine turbojet at low altitude, some slightly different procedure may be utilized. If all powerplants are operating, it is desirable to climb to a higher altitude which is a function of the remaining fuel quantity. An alternative at low altitude would be to provide the endurance thrust with some engine(s) shut down and the remaining engine(s) operating at a more efficient power output. This technique would cause a minimum loss of endurance if at low altitude. The feasibility of such a procedure is dependent on many operational factors.

In all cases, the airplane should be in the cleanest possible external configuration because the specific endurance is directly proportional to the (L/D).

MANEUVERING PERFORMANCE

When the airplane is in turning flight, the airplane is not in static equilibrium for there must be developed the unbalance of force to produce the acceleration of the turn. During a steady coordinated turn, the lift is inclined to produce a horizontal component of force to equal the centrifugal force of the turn. In addition, the steady turn is achieved by producing a vertical component of lift which is equal to the weight of the airplane. Figure 2.28 illustrates the forces which act on the airplane in a steady, coordinated turn.

For the case of the steady, coordinated turn, the vertical component of lift must equal the weight of the aircraft so that there will be no acceleration in the vertical direction. This requirement leads to the following relationship:

$$n=\frac{L}{W}$$

$$n=\frac{1}{\cos\phi}$$

$$n=\sec\phi$$

where

n = load factor or "G"
L = lift, lbs.
W = weight, lbs.
ϕ = bank angle, degrees (phi)

From this relationship, it is apparent that the steady, coordinated turn requires specific values of load factor, n, at various angles of bank, ϕ. For example, a bank angle of 60° requires a load factor of 2.0 ($\cos 60°=0.5$ or $\sec 60°=2.0$) to provide the steady, coordinated turn. If the airplane were at a 60° bank and lift were not provided to produce the exact load factor of 2.0, the aircraft would be accelerating in the vertical direction as well as the horizontal direction and the turn would not be steady. Also, any sideforce on the aircraft due to sideslip, etc., would place the resultant aerodynamic force out of the plane of symmetry perpendicular to the lateral axis and the turn would not be coordinated.

As a consequence of the increase lift required to produce the steady turn in a bank, the induced drag is increased above that incurred by steady, wing level, lift-equal-weight flight. In a sense, the increased lift required in a steady turn will increase the total drag or power required in the same manner as increased gross weight in level flight. The curves of figure 2.28 illustrate the general effect of turning flight on the total thrust and power required. Of course, the change in thrust required at any given speed is due to the change in induced drag and the magnitude of change depends on the value of induced drag in level flight and the angle of bank in turning flight. Since the induced drag generally varies as the square of C_L, the following data provide an illustration of the effect of various degrees of bank:

Bank angle, ϕ	Load factor, n	Percent increase in induced drag from level flight
0°	1.000	0 (of course)
15°	1.036	7.2
30°	1.154	33.3
45°	1.414	100.0
60°	2.000	300.0

Since the induced drag predominates at low speeds, steep turns at low speeds can produce significant increases in thrust or power required to maintain altitude. Thus, steep turns must be avoided after takeoff, during approach, and especially during a critical power situation from failure or malfunction of a powerplant. The greatly increased induced drag is just as

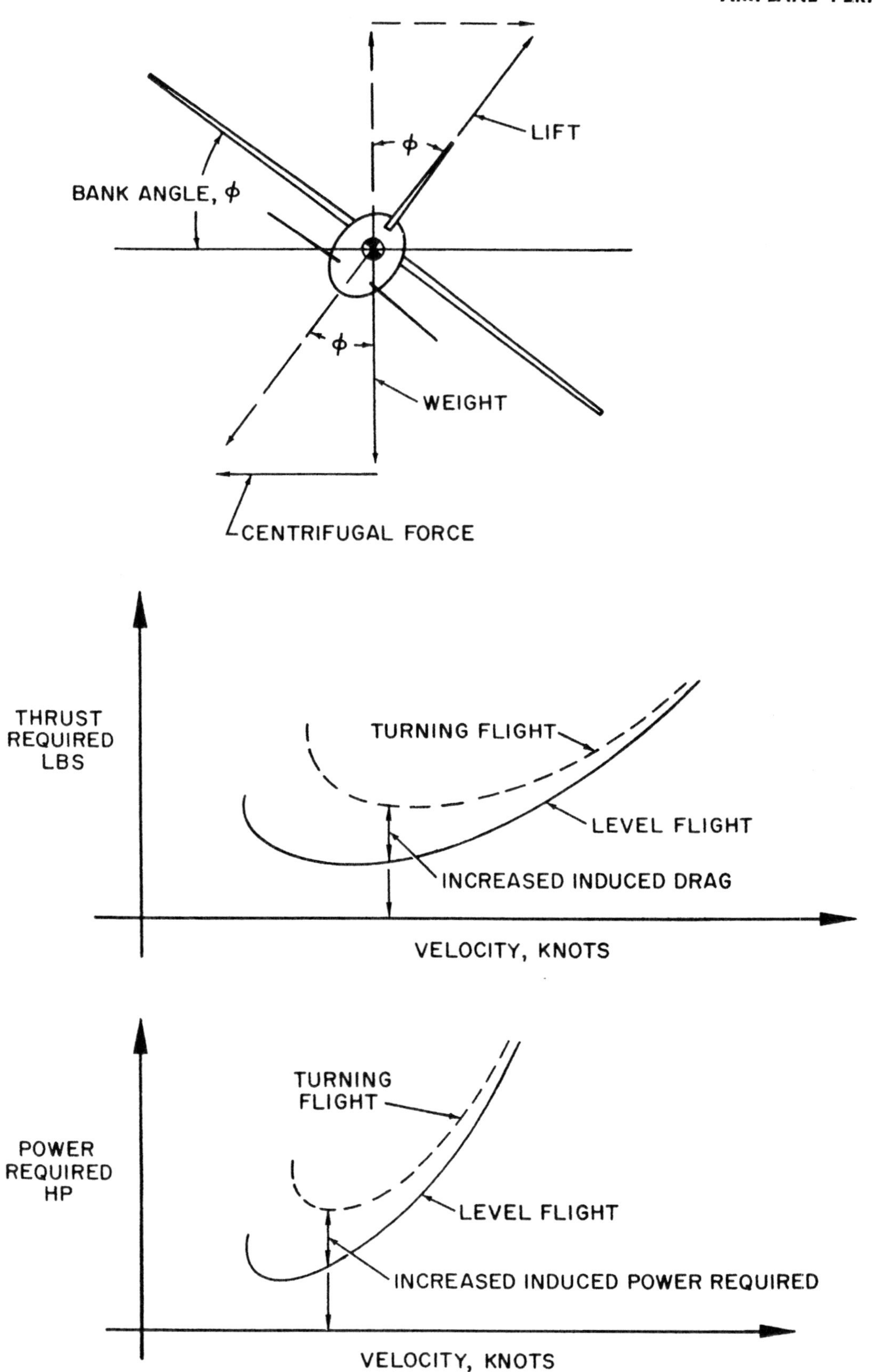

Figure 2.28. Effect of Turning Flight

important—if not more important—as the increased stall speed in turning flight. It is important also that any turn be well coordinated to prevent the increased drag attendant to a sideslip.

TURNING PERFORMANCE. The horizontal component of lift will equal the centrifugal force of steady, turning flight. This fact allows development of the following relationships of turning performance:

turn radius

$$r=\frac{V^2}{11.26 \tan \phi}$$

where

r = turn radius, ft.
V = velocity, knots (TAS)
ϕ = bank angle, degrees

turn rate

$$ROT=\frac{1{,}091 \tan \phi}{V}$$

where

ROT = rate of turn, degrees per sec.
ϕ = bank angle, degrees
V = velocity, knots, TAS

These relationships define the turn radius, r, and rate of turn, ROT, as functions of the two principal variables: bank angle, ϕ, and velocity, V (TAS). Thus, when the airplane is flown in the steady, coordinated turn at specific values of bank angle and velocity, the turn rate and turn radius are fixed and independent of the airplane type. As an example, an airplane in a steady, coordinated turn at a bank angle of 45° and a velocity of 250 knots (TAS) would have the following turn performance:

$$r=\frac{(250)^2}{(11.26)(1.000)} \quad (\tan 45°=1.000)$$

$$=5{,}550 \text{ ft.}$$

and

$$ROT=\frac{(1{,}091)(1.000)}{250}$$

$$=4.37 \text{ deg. per sec.}$$

If the airplane were to hold the same angle of bank at 500 knots (TAS), the turn radius would quadruple (r=22,200 ft.) and the turn rate would be one-half the original value (ROT=2.19 deg. per sec.).

Values of turn radius and turn rate versus velocity are shown in figure 2.29 for various angles of bank and the corresponding load factors. The conditions are for the steady, coordinated turn at constant altitude but the results are applicable for climbing or descending flight when the angle of climb or descent is relatively small. While the effect of altitude on turning performance is not immediately apparent from these curves, the principal effect must be appreciated as an increased true airspeed (TAS) for a given equivalent airspeed (EAS).

TACTICAL PERFORMANCE. Many tactical maneuvers require the use of the maximum turning capability of the airplane. The maximum turning capability of an airplane will be defined by three factors:

(1) *Maximum lift capability*. The combination of maximum lift coefficient, $C_{L_{max}}$, and wing loading, W/S, will define the ability of the airplane to develop aerodynamically the load factors of maneuvering flight.

(2) *Operating strength limits* will define the upper limits of maneuvering load factors which will not damage the primary structure of the airplane. These limits must not be exceeded in normal operations because of the possibility of structural damage or failure.

(3) *Thrust or power limits* will define the ability of the airplane to turn at constant altitude. The limiting condition would allow increased load factor and induced drag until the drag equals the maximum thrust available from the powerplant. Such a case would produce the maximum turning capability for maintaining constant altitude.

The first illustration of figure 2.30 shows how the aerodynamic and structural limits

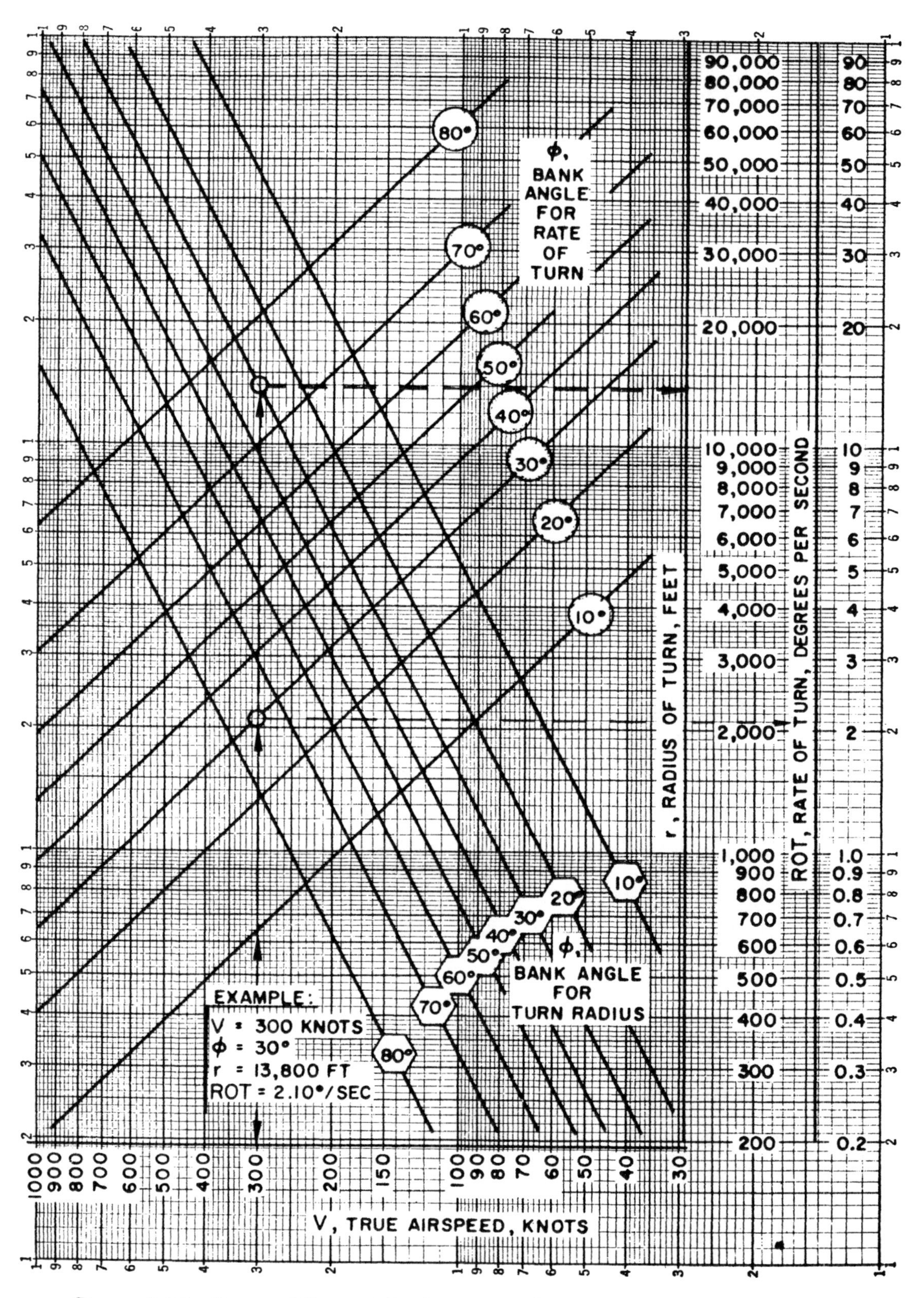

Figure 2.29. General Turning Performance (Constant Altitude, Steady Turn)

define the maximum turning performance. The *aerodynamic limit* describes the minimum turn radius available to the airplane when operated at $C_{L_{max}}$. When the airplane is at the stall speed in level flight, all the lift is necessary to sustain the aircraft in flight and none is available to produce a steady turn. Hence, the turn radius at the stall speed is infinite. As speed is increased above the stall speed, the airplane at $C_{L_{max}}$ is able to develop lift greater than weight and produce a finite turn radius. For example, at a speed twice the stall speed, the airplane at $C_{L_{max}}$ is able to develop a load factor of four and utilize a bank angle of 75.5° (cos 75.5°=0.25). Continued increase in speed increases the load factor and bank angle which is available aerodynamically but, because of the increase in velocity and the basic effect on turn radius, the turn radius approaches an absolute minimum value. When $C_{L_{max}}$ is unaffected by velocity, the aerodynamic minimum turn radius approaches this absolute value which is a function of $C_{L_{max}}$, W/S, and σ. Actually, the one common denominator of aerodynamic turning performance is the wing level stall speed.

The aerodynamic limit of turn radius requires that the increased velocity be utilized to produce increasing load factors and greater angles of bank. Obviously, very high speeds will require very high load factors and the absolute aerodynamic minimum turn radius will require an infinite load factor. Increasing speed above the stall speed will eventually produce the limit load factor and continued increase in speed above this point will require that load factor and bank angle be limited to prevent structural damage. When the load factor and bank angle are held constant at the structural limit, the turn radius varies as the square of the velocity and increases rapidly above the aerodynamic limit. The intersection of the aerodynamic limit and structural limit lines is the "maneuver speed." The maneuver speed is the minimum speed necessary to develop aerodynamically the limit load factor and it produces the minimum turn radius within aerodynamic and structural limitations. At speeds less than the maneuver speed, the limit load factor is not available aerodynamically and turning performance is aerodynamically limited. At speeds greater than the maneuver speed, $C_{L_{max}}$ and maximum aerodynamic load factor are not available and turning performance is structurally limited. When the stall speed and limit load factor are known for a particular configuration, the maneuver speed is related by the following expression:

$$V_p = V_s\sqrt{n \text{ limit}}$$

where

V_p = maneuver speed, knots
V_s = stall speed, knots
n limit = limit load factor

For example, an airplane with a limit load factor of 4.0 would have a maneuver speed which is twice the stall speed.

The aerodynamic limit line of the first illustration of figure 2.30 is typical of an airplane with a $C_{L_{max}}$ which is invariant with speed. While this is applicable for the majority of subsonic airplanes, considerable difference would be typical of the transonic or supersonic airplane at altitude. Compressibility effects and changes in longitudinal control power may produce a maximum available C_L which varies with velocity and an aerodynamic turn radius which is not an absolute minimum at the maximum of velocity.

The second illustration of figure 2.30 describes the constant altitude turning performance of an airplane. When an airplane is at high altitude, the turning performance at the high speed end of the flight speed range is more usually thrust limited rather than structurally limited. In flight at constant altitude, the thrust must equal the drag to maintain equilibrium and, thus, the constant altitude turn radius is infinite at the maximum level flight speed. Any bank or turn at maximum level flight speed would incur additional drag and

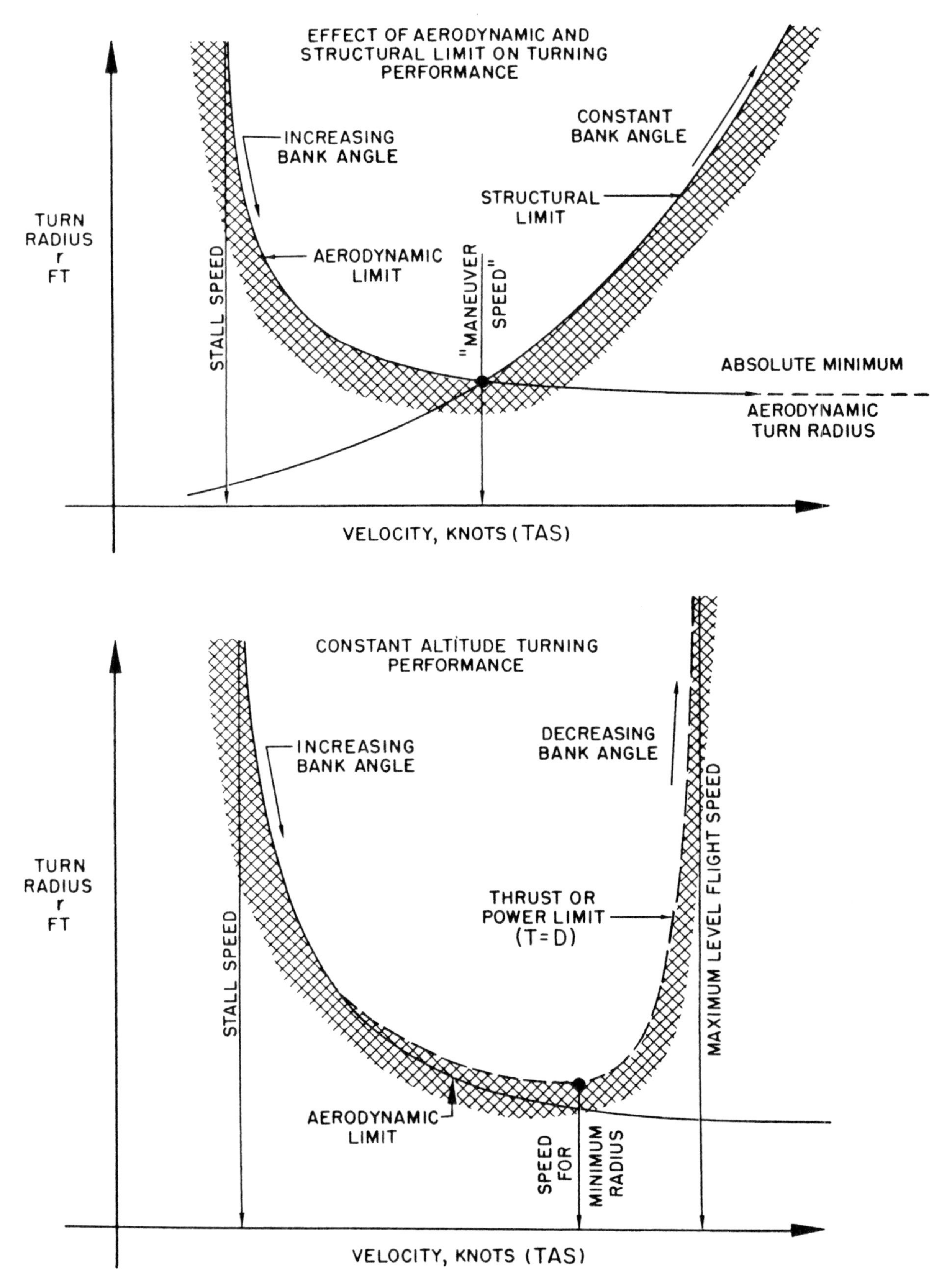

Figure 2.30. Maneuvering Performance

cause the airplane to descend. However, as speed is reduced below the maximum level flight speed, parasite drag reduces and allows increased load factors and bank angles and reduced radius of turn, i.e., decreased parasite drag allows increased induced drag to accommodate turns within the maximum thrust available. Thus, the considerations of constant altitude will increase the minimum turn radius above the aerodynamic limit and define a particular airspeed for minimum turn radius.

Each of the three limiting factors (aerodynamic, structural, and power) may combine to define the turning performance of an airplane. Generally, aerodynamic and structural limits predominate at low altitude while aerodynamic and power limits predominate at high altitude. The knowledge of this turning performance is particularly necessary for effective operation of fighter and interceptor types of airplanes.

TAKEOFF AND LANDING PERFORMANCE

The majority of pilot caused airplane accidents occur during the takeoff and landing phase of flight. Because of this fact, the Naval Aviator must be familiar with all the many variables which influence the takeoff and landing performance of an airplane and must strive for exacting, professional techniques of operation during these phases of flight.

Takeoff and landing performance is a condition of accelerated motion. For instance, during takeoff the airplane starts at zero velocity and accelerates to the takeoff velocity to become airborne. During landing, the airplane touches down at the landing speed and decelerates (or accelerates negatively) to the zero velocity of the stop. In fact, the landing performance could be considered as a takeoff in reverse for purposes of study. In either case, takeoff or landing, the airplane is accelerated between zero velocity and the takeoff or landing velocity. The important factors of takeoff or landing performance are:

(1) The takeoff or landing *velocity* which will generally be a function of the stall speed or minimum flying speed, e.g., 15 percent above the stall speed.

(2) The *acceleration* during the takeoff or landing roll. The acceleration experienced by any object varies directly with the unbalance of force and inversely as the mass of the object.

(3) The takeoff or landing roll *distance* is a function of both the acceleration and velocity.

In the actual case, the takeoff and landing distance is related to velocity and acceleration in a very complex fashion. The main source of the complexity is that the forces acting on the airplane during the takeoff or landing roll are difficult to define with simple relationships. Since the acceleration is a function of these forces, the acceleration is difficult to define in a simple fashion and it is a principal variable affecting distance. However, some simplification can be made to study the basic relationship of acceleration, velocity, and distance. While the acceleration is not necessarily constant or uniform throughout the takeoff or landing roll, the assumption of uniformly accelerated motion will facilitate study of the principal variables affecting takeoff and landing distance.

From basic physics, the relationship of velocity, acceleration, and distance for uniformly accelerated motion is defined by the following equation:

$$S=\frac{V^2}{2a}$$

where

S = acceleration distance, ft.

V = final velocity, ft. per sec., after accelerating uniformly from zero velocity

a = acceleration, ft. per sec.^{2}

This equation could relate the takeoff distance in terms of the takeoff velocity and acceleration when the airplane is accelerated uniformly from zero velocity to the final takeoff velocity. Also, this expression could relate the landing distance in terms of the landing velocity and deceleration when the airplane is accelerated (negatively) from the landing velocity to a complete stop. It is important to note that

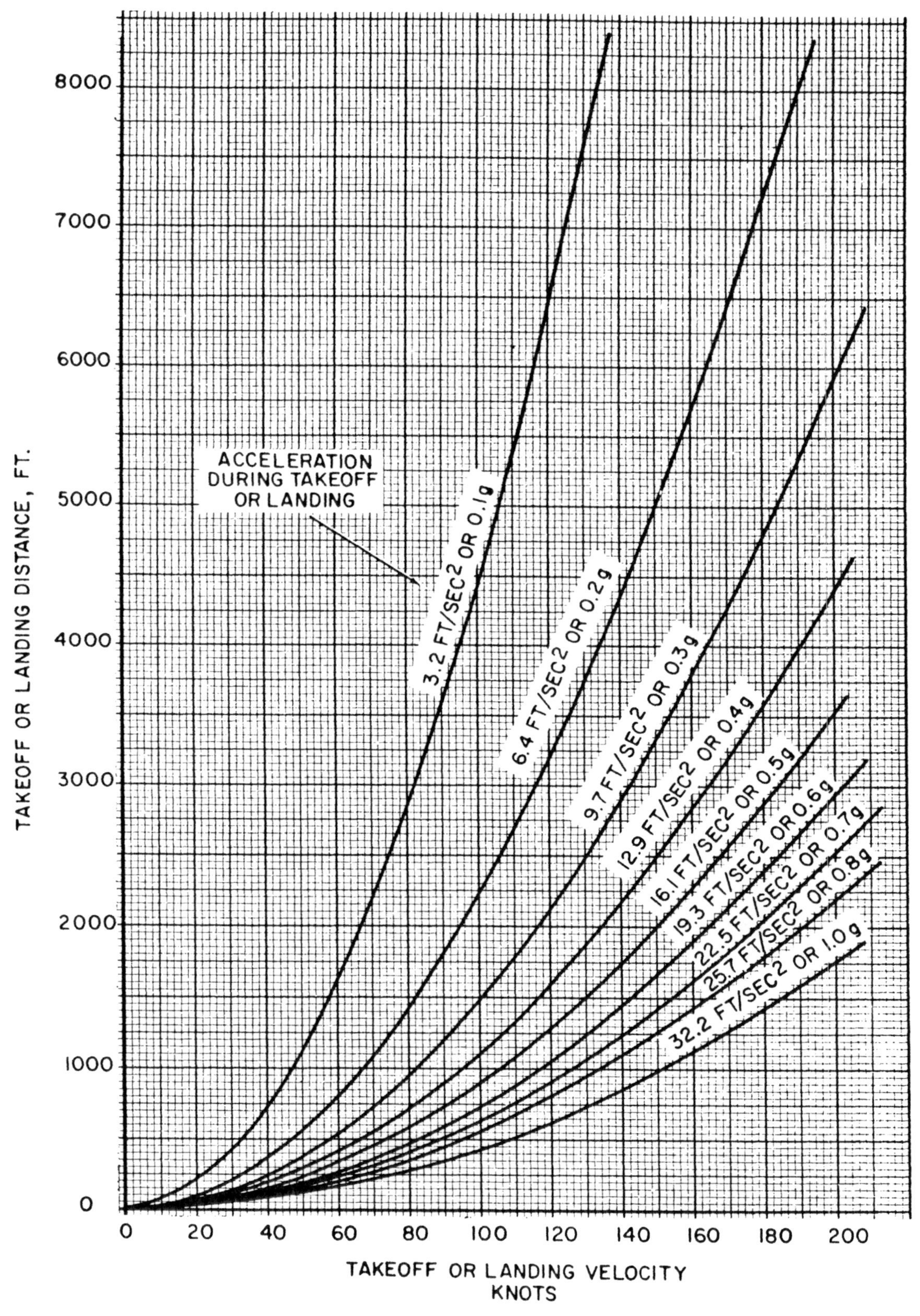

Figure 2.31. Relationship of Velocity, Acceleration, and Distance for Uniformly Accelerated Motion

the distance varies directly as the square of the velocity and inversely as the acceleration.

As an example of this relationship, assume that during takeoff an airplane is accelerated uniformly from zero velocity to a takeoff velocity of 150 knots (253.5 ft. per sec.) with an acceleration of 6.434 ft. per sec.^{2} (or, 0.2g, since g=32.17 ft. per sec.^{2}). The takeoff distance would be:

$$S=\frac{V^2}{2a}$$
$$=\frac{(253.5)^2}{(2)(6.434)}$$
$$=5,000 \text{ ft.}$$

If the acceleration during takeoff were reduced 10 percent, the takeoff distance would increase 11.1 percent; if the takeoff velocity were increased 10 percent, the takeoff distance would increase 21 percent. These relationships point to the fact that proper accounting must be made of altitude, temperature, gross weight, wind, etc. because any item affecting acceleration or takeoff velocity will have a definite effect on takeoff distance.

If an airplane were to land at a velocity of 150 knots and be decelerated uniformly to a stop with the same acceleration of 0.2g, the landing stop distance would be 5,000 ft. However, the case is not necessarily that an aircraft may have identical takeoff and landing performance but the principle illustrated is that distance is a function of velocity and acceleration. As before, a 10 percent lower acceleration increases stop distance 11.1 percent, and a 10 percent higher landing speed increases landing distance 21 percent.

The general relationship of velocity, acceleration, and distance for uniformly accelerated motion is illustrated by figure 2.31. In this illustration, acceleration distance is shown as a function of velocity for various values of acceleration.

TAKEOFF PERFORMANCE. The minimum takeoff distance is of primary interest in the operation of any aircraft because it defines the runway requirements. The minimum takeoff distance is obtained by takeoff at some minimum safe velocity which allows sufficient margin above stall and provides satisfactory control and initial rate of climb. Generally, the takeoff speed is some fixed percentage of the stall speed or minimum control speed for the airplane in the takeoff configuration. As such, the takeoff will be accomplished at some particular value of lift coefficient and angle of attack. Depending on the airplane characteristics, the takeoff speed will be anywhere from 1.05 to 1.25 times the stall speed or minimum control speed. If the takeoff speed is specified as 1.10 times the stall speed, the takeoff lift coefficient is 82.6 percent of $C_{L_{max}}$ and the angle of attack and lift coefficient for takeoff are fixed values independent of weight, altitude, wind, etc. Hence, an angle of attack indicator can be a valuable aid during takeoff.

To obtain minimum takeoff distance at the specified takeoff velocity, the forces which act on the aircraft must provide the maximum acceleration during the takeoff roll. The various forces acting on the aircraft may or may not be at the control of the pilot and various techniques may be necessary in certain airplanes to maintain takeoff acceleration at the highest value.

Figure 2.32 illustrates the various forces which act on the aircraft during takeoff roll. The powerplant *thrust* is the principal force to provide the acceleration and, for minimum takeoff distance, the output thrust should be at a maximum. *Lift and drag* are produced as soon as the airplane has speed and the values of lift and drag depend on the angle of attack and dynamic pressure. *Rolling friction* results when there is a normal force on the wheels and the friction force is the product of the normal force and the coefficient of rolling friction. The normal force pressing the wheels against the runway surface is the net of weight and lift while the rolling friction coefficient is a function of the tire type and runway surface texture.

The acceleration of the airplane at any instant during takeoff roll is a function of the net accelerating force and the airplane mass. From Newton's second law of motion:

$$a = Fn/M$$

or

$$a = g(Fn/W)$$

where

a = acceleration, ft. per sec.^{2}
Fn = net accelerating force, lbs.
W = weight, lbs.
g = gravitational acceleration
= 32.17 ft. per sec.^{2}
M = mass, slugs
= W/g

The net accelerating force on the airplane, F_n, is the net of thrust, T, drag, D, and rolling friction, F. Thus, the acceleration at any instant during takeoff roll is:

$$a = \frac{g}{W}(T - D - F)$$

Figure 2.32 illustrates the typical variation of the various forces acting on the aircraft throughout the takeoff roll. If it is assumed that the aircraft is at essentially constant angle of attack during takeoff roll, C_L and C_D are constant and the forces of lift and drag vary as the square of the speed. For the case of uniformly accelerated motion, distance along the takeoff roll is proportional also to the square of the velocity hence velocity squared and distance can be used almost synonomously. Thus, lift and drag will vary linearly with dynamic pressure (q) or V^2 from the point of beginning takeoff roll. As the rolling friction coefficient is essentially unaffected by velocity, the rolling friction will vary as the normal force on the wheels. At zero velocity, the normal force on the wheels is equal to the airplane weight but, at takeoff velocity, the lift is equal to the weight and the normal force is zero. Hence, rolling friction decreases linearly with q or V^2 from the beginning of takeoff roll and reaches zero at the point of takeoff.

The total retarding force on the aircraft is the sum of drag and rolling friction $(D+F)$ and, for the majority of configurations, this sum is nearly constant or changes only slightly during the takeoff roll. The net accelerating force is then the difference between the powerplant thrust and the total retarding force,

$$Fn = T - D - F$$

The variation of the net accelerating force throughout the takeoff roll is shown in figure 2.32. The typical propeller airplane demonstrates a net accelerating force which decreases with velocity and the resulting acceleration is initially high but decreases throughout the takeoff roll. The typical jet airplane demonstrates a net accelerating force which is essentially constant throughout the takeoff roll. As a result, the takeoff performance of the typical turbojet airplane will compare closely with the case for uniformly accelerated motion.

The pilot technique required to achieve peak acceleration throughout takeoff roll can vary considerably between airplane configurations. In some instances, maximum acceleration will be obtained by allowing the airplane to remain in the three-point attitude throughout the roll until the airplane simply reaches lift-equal-to-weight and flies off the ground. Other airplanes may require the three-point attitude until the takeoff speed is reached then rotation to the takeoff angle of attack to become airborne. Still other configurations may require partial or complete rotation to the takeoff angle of attack prior to reaching the takeoff speed. In this case, the procedure may be necessary to provide a smaller retarding force $(D+F)$ to achieve peak acceleration. Whenever any form of pitch rotation is necessary the pilot must provide the proper angle of attack since an excessive angle of attack will cause excessive drag and hinder (or possibly preclude) a successful takeoff. Also, insufficient rotation may provide added rolling resistance or require that the airplane accelerate to some excessive speed prior to becoming airborne.

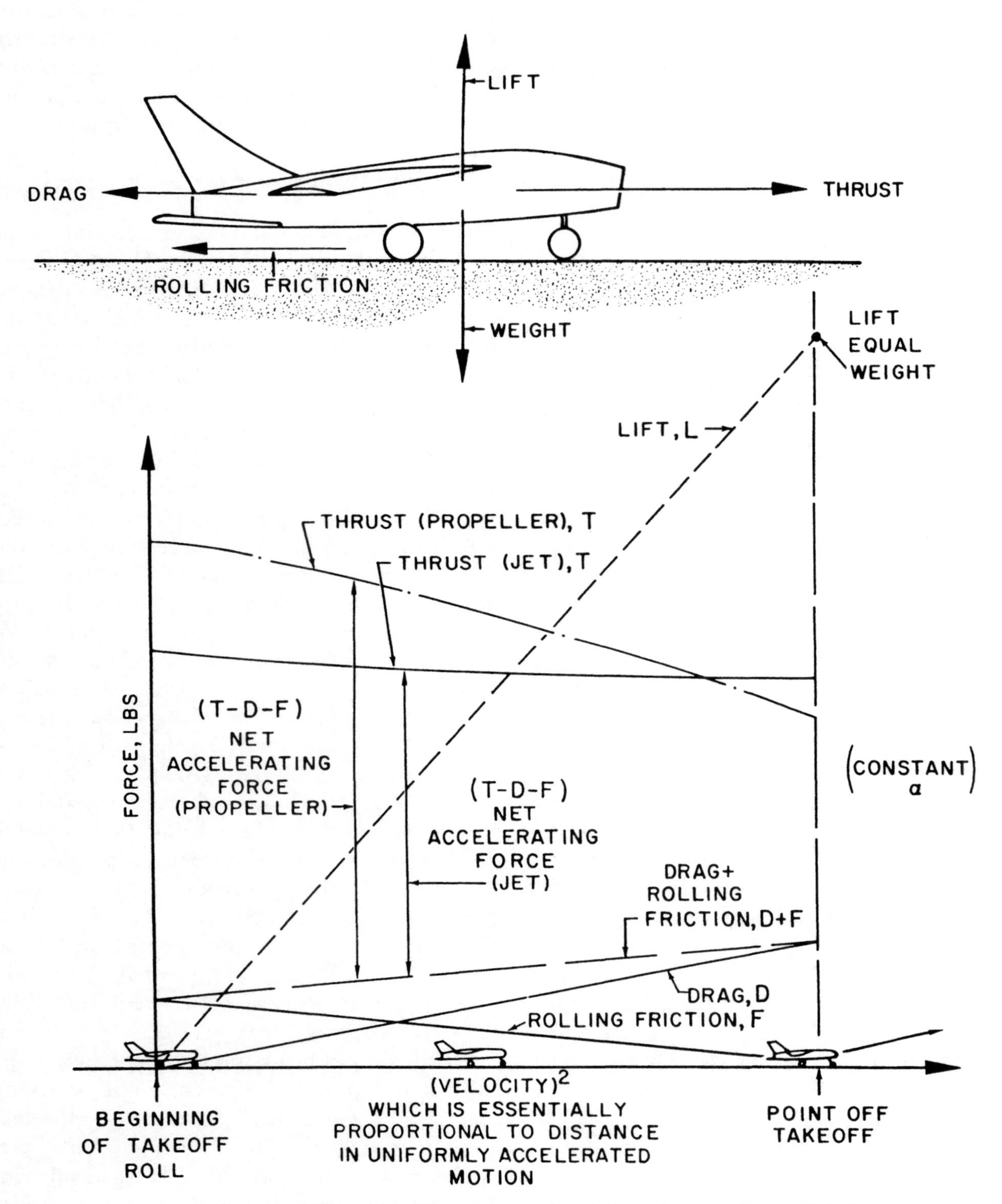

Figure 2.32. Forces Acting on the Airplane During Takeoff Roll

In this sense, an angle of attack indicator is especially useful for night or instrument takeoff conditions as well as the ordinary day VFR takeoff conditions. Acceleration errors of the attitude gyro usually preclude accurate pitch rotation under these conditions.

FACTORS AFFECTING TAKEOFF PERFORMANCE. In addition to the important factors of proper technique, many other variables affect the takeoff performance of an airplane. Any item which alters the takeoff velocity or acceleration during takeoff roll will affect the takeoff distance. In order to evaluate the effect of the many variables, the principal relationships of uniformly accelerated motion will be assumed and consideration will be given to those effects due to any nonuniformity of acceleration during the process of takeoff. Generally, in the case of uniformly accelerated motion, distance varies directly with the square of the takeoff velocity and inversely as the takeoff acceleration.

$$\frac{S_2}{S_1}=\left(\frac{V_2}{V_1}\right)^2\times\left(\frac{a_1}{a_2}\right)$$

where

S = distance
V = velocity
a = acceleration
condition (1) applies to some known takeoff distance, S_1, which was common to some original takeoff velocity, V_1, and acceleration, a_1.
condition (2) applies to some new takeoff distance, S_2, which is the result of some different value of takeoff velocity, V_2, or acceleration, a_2.

With this basic relationship, the effect of the many variables on takeoff distance can be approximated.

The *effect of gross weight* on takeoff distance is large and proper consideration of this item must be made in predicting takeoff distance. Increased gross weight can be considered to produce a threefold effect on takeoff performance: (1) increased takeoff velocity, (2) greater mass to accelerate, and (3) increased retarding force $(D+F)$. If the gross weight increases, a greater speed is necessary to produce the greater lift to get the airplane airborne at the takeoff lift coefficient. The relationship of takeoff speed and gross weight would be as follows:

$$\frac{V_2}{V_1}=\sqrt{\frac{W_2}{W_1}}\quad (EAS \text{ or } CAS)$$

where

V_1 = takeoff velocity corresponding to some original weight, W_1
V_2 = takeoff velocity corresponding to some different weight, W_2

Thus, a given airplane in the takeoff configuration at a given gross weight will have a specific takeoff speed (*EAS* or *CAS*) which is invariant with altitude, temperature, wind, etc. because a certain value of q is necessary to provide lift equal to weight at the takeoff C_L. As an example of the effect of a change in gross weight a 21 percent increase in takeoff weight will require a 10 percent increase in takeoff speed to support the greater weight.

A change in gross weight will change the net accelerating force, Fn, and change the mass, M, which is being accelerated. If the airplane has a relatively high thrust-to-weight ratio, the change in the net accelerating force is slight and the principal effect on acceleration is due to the change in mass.

To evaluate the effect of gross weight on takeoff distance, the following relationships are used:

the effect of weight on takeoff velocity is

$$\frac{V_2}{V_1}=\sqrt{\frac{W_2}{W_1}}\quad \text{or}\quad \left(\frac{V_2}{V_1}\right)^2=\frac{W_2}{W_1}$$

if the change in net accelerating force is neglected, the effect of weight on acceleration is

$$\frac{a_1}{a_2}=\frac{W_2}{W_1}\quad \text{or}\quad \frac{a_2}{a_1}=\frac{W_1}{W_2}$$

the effect of these items on takeoff distance is

$$\frac{S_2}{S_1}=\left(\frac{V_2}{V_1}\right)^2\times\left(\frac{a_1}{a_2}\right)$$

or

$$\frac{S_2}{S_1}=\left(\frac{W_2}{W_1}\right)\times\left(\frac{W_2}{W_1}\right)$$

$$\frac{S_2}{S_1}=\left(\frac{W_2}{W_1}\right)^2$$

(*at least* this effect because weight will alter the net accelerating force)

This result approximates the effect of gross weight on takeoff distance for airplanes with relatively high thrust-to-weight ratios. In effect, the takeoff distance will vary at least as the square of the gross weight. For example, a 10 percent increase in takeoff gross weight would cause:

a 5 percent increase in takeoff velocity
at least a 9 percent decrease in acceleration
at least a 21 percent increase in takeoff distance

For the airplane with a high thrust-to-weight ratio, the increase in takeoff distance would be approximately 21 to 22 percent but, for the airplane with a relatively low thrust-to-weight ratio, the increase in takeoff distance would be approximately 25 to 30 percent. Such a powerful effect requires proper consideration of gross weight in predicting takeoff distance.

The *effect of wind* on takeoff distance is large and proper consideration also must be provided when predicting takeoff distance. The effect of a headwind is to allow the airplane to reach the takeoff velocity at a lower ground velocity while the effect of a tailwind is to require the airplane to achieve a greater ground velocity to attain the takeoff velocity. The effect of the wind on acceleration is relatively small and, for the most part, can be neglected. To evaluate the effect of wind on takeoff distance, the following relationships are used:

the effect of a headwind is to reduce the takeoff ground velocity by the amount of the headwind velocity, V_w

$$V_2'=V_1-V_w$$

the effect of wind on acceleration is negligible,

$$a_2=a_1 \quad \text{or} \quad \frac{a_1}{a_2}=1$$

the effect of these items on takeoff distance is

$$\frac{S_2}{S_1}=\left(\frac{V_2}{V_1}\right)^2\times\left(\frac{a_1}{a_2}\right)$$

$$\frac{S_2}{S_1}=\left[\frac{V_1-V_w}{V_1}\right]^2$$

or

$$\frac{S_2}{S_1}=\left[1-\frac{V_w}{V_1}\right]^2$$

where

S_1 = zero wind takeoff distance
S_2 = takeoff distance into the headwind
V_w = headwind velocity
V_1 = takeoff ground velocity with zero wind, or, simply, the takeoff airspeed

As a result of this relationship, a headwind which is 10 percent of the takeoff airspeed will reduce the takeoff distance 19 percent. However, a tailwind (or negative headwind) which is 10 percent of the takeoff airspeed will increase the takeoff distance 21 percent. In the case where the headwind velocity is 50 percent of the takeoff speed, the takeoff distance would be approximately 25 percent of the zero wind takeoff distance (75 percent reduction).

The effect of wind on landing distance is identical to the effect on takeoff distance. Figure 2.33 illustrates the general effect of wind by the percent change in takeoff or landing distance as a function of the ratio of wind velocity to takeoff or landing speed.

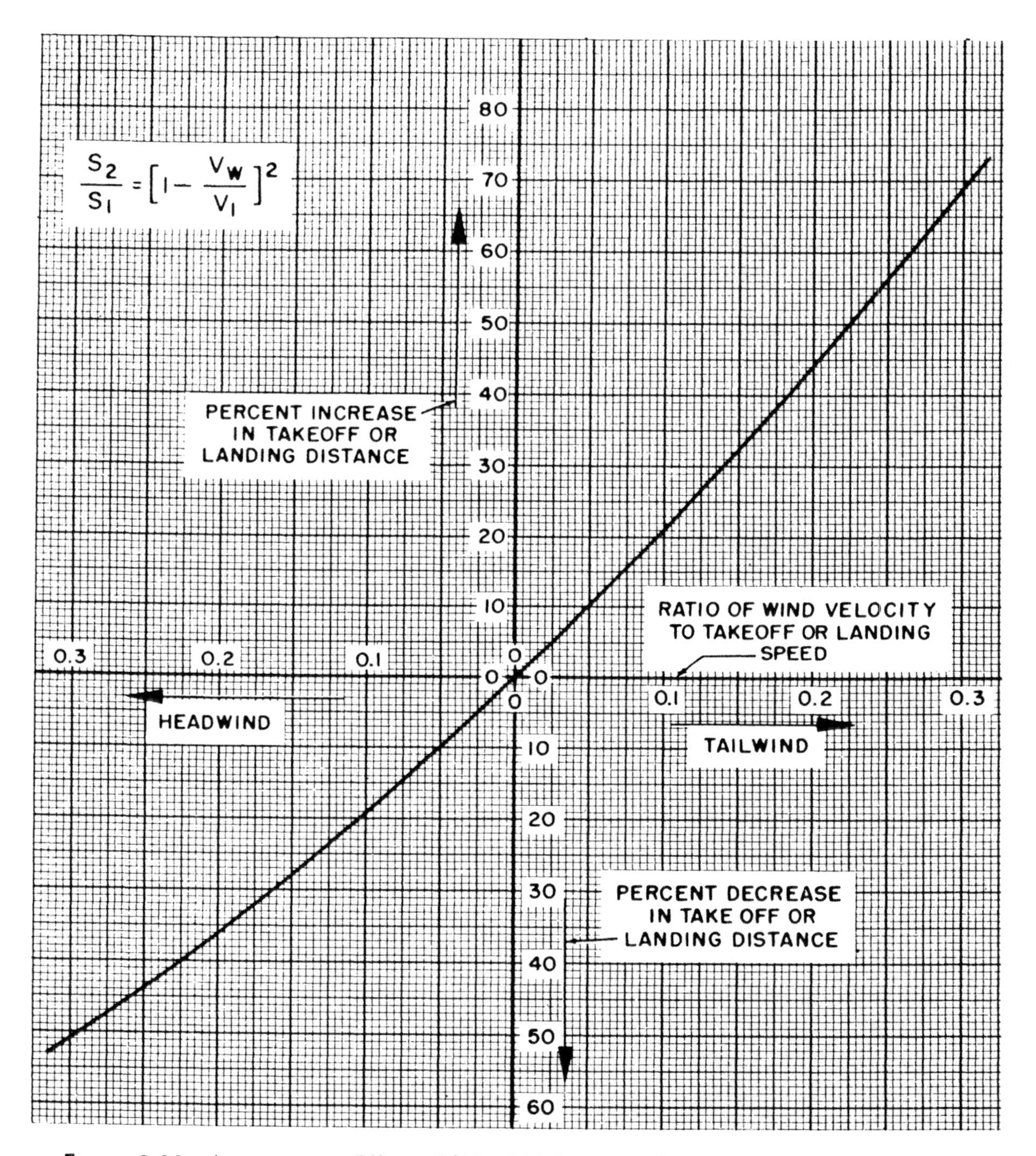

Figure 2.33. Approximate Effect of Wind Velocity on Takeoff or Landing Distance

The *effect of runway slope* on takeoff distance is due to the component of weight along the inclined path of the airplane. A runway slope of 1 percent would provide a force component along the path of the airplane which is 1 percent of the gross weight. Of course, an upslope would contribute a retarding force component while a downslope would contribute an accelerating force component. For the case of the upslope, the retarding force component adds to drag and rolling friction to reduce the net accelerating force. Ordinarily, a 1 percent runway slope can cause a 2 to 4 percent change in takeoff distance depending on the airplane characteristics. The airplane with the high thrust-to-weight ratio is least affected while the airplane with the low thrust-to-weight ratio is most affected because the slope force component causes a relatively greater change in the net accelerating force.

The effect of runway slope must be considered when predicting the takeoff distance but the effect is usually minor for the ordinary runway slopes and airplanes with moderate thrust-to-weight ratios. In fact, runway slope considerations are of great significance only when the runway slope is large and the airplane has an intrinsic low acceleration, i.e., low thrust-to-weight ratio. In the ordinary case, the selection of the takeoff runway will favor the direction with an upslope and headwind rather than the direction with a downslope and tailwind.

The effect of *proper takeoff velocity* is important when runway lengths and takeoff distances are critical. The takeoff speeds specified in the flight handbook are generally the minimum safe speeds at which the airplane can become airborne. Any attempt to take off below the recommended speed may mean that the aircraft may stall, be difficult to control, or have very low initial rate of climb. In some cases, an excessive angle of attack may not allow the airplane to climb out of ground effect. On the other hand, an excessive airspeed at takeoff may improve the initial rate of climb and "feel" of the airplane but will produce an undesirable increase in takeoff distance. Assuming that the acceleration is essentially unaffected, the takeoff distance varies as the square of the takeoff velocity,

$$\frac{S_2}{S_1}=\left(\frac{V_2}{V_1}\right)^2$$

Thus, 10 percent excess airspeed would increase the takeoff distance 21 percent. In most critical takeoff conditions, such an increase in takeoff distance would be prohibitive and the pilot must adhere to the recommended takeoff speeds.

The *effect of pressure altitude and ambient temperature* is to define primarily the *density altitude* and its effect on takeoff performance. While subsequent corrections are appropriate for the effect of temperature on certain items of powerplant performance, density altitude defines certain effects on takeoff performance. An increase in density altitude can produce a two-fold effect on takeoff performance: (1) increased takeoff velocity and (2) decreased thrust and reduced net accelerating force. If a given weight and configuration of airplane is taken to altitude above standard sea level, the airplane will still require the same dynamic pressure to become airborne at the takeoff lift coefficient. Thus, the airplane at altitude will take off at the same equivalent airspeed (EAS) as at sea level, but because of the reduced density, the true airspeed (TAS) will be greater. From basic aerodynamics, the relationship between true airspeed and equivalent airspeed is as follows:

$$\frac{TAS}{EAS}=\frac{1}{\sqrt{\sigma}}$$

where

TAS = true airspeed
EAS = equivalent airspeed
σ = altitude density ratio
$\sigma=\rho/\rho_0$

The effect of density altitude on powerplant thrust depends much on the type of powerplant. An increase in altitude above standard sea level will bring an immediate decrease in power output for the unsupercharged or ground boosted reciprocating engine or the turbojet and turboprop engines. However, an increase in altitude above standard sea level will not cause a decrease in power output for the supercharged reciprocating engine until the altitude exceeds the critical altitude. For those powerplants which experience a decay in thrust with an increase in altitude, the effect on the net accelerating force and acceleration can be approximated by assuming a direct variation with density. Actually, this assumed variation would closely approximate the effect on airplanes with high thrust-to-weight ratios. This relationship would be as follows:

$$\frac{a_2}{a_1}=\frac{Fn_2}{Fn_1}=\frac{\rho}{\rho_0}=\sigma$$

where

a_1, Fn_1 = acceleration and net accelerating force corresponding to sea level
a_2, Fn_2 = acceleration and net accelerating force corresponding to altitude
σ = altitude density ratio

In order to evaluate the effect of these items on takeoff distance, the following relationships are used:

if an increase in altitude does not alter acceleration, the principal effect would be due to the greater *TAS*

$$\frac{S_2}{S_1}=\left(\frac{V_2}{V_1}\right)^2\times\left(\frac{a_1}{a_2}\right)$$

$$\frac{S_2}{S_1}=\frac{1}{\sigma}$$

where

S_1 = standard sea level takeoff distance
S_2 = takeoff distance at altitude
σ = altitude density ratio

if an increase in altitude reduces acceleration in addition to the increase in *TAS*, the combined effects would be approximated for the case of the airplane with high intrinsic acceleration by the following:

$$\frac{S_2}{S_1}=\left(\frac{V_2}{V_1}\right)^2\times\left(\frac{a_1}{a_2}\right)$$

$$\frac{S_2}{S_1}=\left(\frac{1}{\sigma}\right)\times\left(\frac{1}{\sigma}\right)$$

$$\frac{S_2}{S_1}=\left(\frac{1}{\sigma}\right)^2$$

where

S_1 = standard sea level takeoff distance
S_2 = takeoff distance at altitude
σ = altitude density ratio

As a result of these relationships, it should be appreciated that density altitude will affect takeoff performance in a fashion depending much on the powerplant type. The effect of density altitude on takeoff distance can be appreciated by the following comparison:

TABLE 2-1. Approximate Effect of Altitude on Takeoff Distance

Density altitude	σ	$\frac{1}{\sigma}$	$\left(\frac{1}{\sigma}\right)^2$	Percent increase in takeoff distance from standard sea level		
				Supercharged reciprocating airplane below critical altitude	Turbojet high (T/W)	Turbojet low (T/W)
Sea level	1.000	1.000	1.000	0	0	0
1,000 ft	.9711	1.0298	1.0605	2.98	6.05	9.8
2,000 ft	.9428	1.0605	1.125	6.05	12.5	19.9
3,000 ft	.9151	1.0928	1.195	9.28	19.5	30.1
4,000 ft	.8881	1.126	1.264	12.6	26.4	40.6
5,000 ft	.8617	1.1605	1.347	16.05	34.7	52.3
6,000 ft	.8359	1.1965	1.432	19.65	43.2	65.8

From the previous table, some approximate rules of thumb may be derived to illustrated the differences between the various airplane types. A 1,000-ft. increase in density altitude will cause these approximate increases in takeoff distance:

- 3½ percent for the supercharged reciprocating airplane when below critical altitude
- 7 percent for the turbojet with high thrust-to-weight ratio
- 10 percent for the turbojet with low thrust-to-weight ratio

These approximate relationships show the turbojet airplane to be much more sensitive to density altitude than the reciprocating powered airplane. This is an important fact which must be appreciated by pilots in transition from propeller type to jet type airplanes. Proper accounting of pressure altitude (field elevation is a poor substitute) and temperature is mandatory for accurate prediction of takeoff roll distance.

The most critical conditions of takeoff performance are the result of some combination of high gross weight, altitude, temperature and unfavorable wind. In all cases, it behooves the pilot to make an accurate prediction of takeoff distance from the performance data of the *Flight Handbook*, regardless of the runway available, and to strive for a polished, professional takeoff technique.

In the prediction of takeoff distance from the handbook data, the following primary considerations must be given:

Reciprocating powered airplane

(1) Pressure altitude and temperature—to define the effect of density altitude on distance.

(2) Gross weight—a large effect on distance.

(3) Specific humidity—to correct takeoff distance for the power loss associated with water vapor.

(4) Wind—a large effect due to the wind or wind component along the runway.

Turbine powered airplane

(1) Pressure altitude and temperature—to define the effect of density altitude.

(2) Gross weight.

(3) Temperature—an additional correction for nonstandard temperatures to account for the thrust loss associated with high compressor inlet air temperature. For this correction the ambient temperature at the runway conditions is appropriate rather than the ambient temperature at some distant location.

(4) Wind.

In addition, corrections are necessary to account for runway slope, engine power deficiencies, etc.

LANDING PERFORMANCE. In many cases, the landing distance of an airplane will define the runway requirements for flying operations. This is particularly the case of high speed jet airplanes at low altitudes where landing distance is the problem rather than takeoff performance. The minimum landing distance is obtained by landing at some minimum safe velocity which allows sufficient margin above stall and provides satisfactory control and capability for waveoff. Generally, the landing speed is some fixed percentage of the stall speed or minimum control speed for the airplane in the landing configuration. As such, the landing will be accomplished at some particular value of lift coefficient and angle of attack. The exact value of C_L and α for landing will depend on the airplane characteristics but, once defined, the values are independent of weight, altitude, wind, etc. Thus, an angle of attack indicator can be a valuable aid during approach and landing.

To obtain minimum landing distance at the specified landing velocity, the forces which act on the airplane must provide maximum deceleration (or negative acceleration) during the landing roll. The various forces acting on the airplane during the landing roll may require various techniques to maintain landing deceleration at the peak value.

Figure 2.34 illustrates the forces acting on the aircraft during landing roll. The powerplant *thrust* should be a minimum positive value, or, if reverse thrust is available, a maximum negative value for minimum landing distance. *Lift and drag* are produced as long as the airplane has speed and the values of lift and drag depend on dynamic pressure and angle of attack. *Braking friction* results when there is a normal force on the braking wheel surfaces and the friction force is the product of the normal force and the coefficient of braking friction. The normal force on the braking surfaces is some part of the net of weight and lift, i.e., some other part of this net may be distributed to wheels which have no brakes. The maximum coefficient of braking friction is primarily a function of the runway surface condition (dry, wet, icy, etc.) and rather independent of the type of tire for ordinary conditions (dry, hard surface runway). However, the operating coefficient of braking friction is controlled by the pilot by the use of brakes.

The acceleration of the airplane during the landing roll is negative (deceleration) and will be considered to be in that sense. At any instant during the landing roll the acceleration is a function of the net retarding force and the airplane mass. From Newton's second law of motion:

$$a=Fr/M$$

or

$$a=g\,(Fr/W)$$

where

a = acceleration, ft. per sec.^{2} (negative)
Fr = net retarding force, lbs.
g = gravitational acceleration, ft. per sec.^{2}
W = weight, lbs.
M = mass, slugs
$= W/g$

The net retarding force on the airplane, Fr, is the net of drag, D, braking friction, F, and thrust, T. Thus, the acceleration (negative) at any instant during the landing roll is:

$$a=\frac{g}{w}(D+F-T)$$

Figure 2.34 illustrates the typical variation of the various forces acting on the aircraft throughout the landing roll. If it is assumed that the aircraft is at essentially constant angle of attack from the point of touchdown, C_L and C_D are constant and the forces of lift and drag vary as the square of the velocity. Thus, lift and drag will decrease linearly with q or V^2 from the point of touchdown. If the braking coefficient is maintained at the maximum value, this maximum value of coefficient of friction is essentially constant with speed and the braking friction force will vary as the normal force on the braking surfaces. As the airplane nears a complete stop, the velocity and lift approach zero and the normal force on the wheels approaches the weight of the airplane. At this point, the braking friction force is at a maximum. Immediately after touchdown, the lift is quite large and the normal force on the wheels is small. As a result, the braking friction force is small. A common error at this point is to apply excessive brake pressure without sufficient normal force on the wheels. This may develop a skid with a locked wheel and cause the tire to blow out so suddenly that judicious use of the brakes is necessary.

The coefficient of braking friction can reach peak values of 0.8 but ordinarily values near 0.5 are typical for the dry hard surface runway. Of course, a slick, icy runway can reduce the maximum braking friction coefficient to values as low as 0.2 or 0.1. If the entire weight of the airplane were the normal force on the braking surfaces, a coefficient of braking friction of 0.5 would produce a deceleration of ½g, 16.1 ft. per sec.^{2} Most airplanes in ground effect rarely produce lift-drag ratios lower than 3 or 4. If the lift of the airplane were equal to the weight, an $L/D=4$ would produce a deceleration of ¼g, 8 ft. per sec.^{2} By this comparison it should be apparent that friction braking offers the possibility of greater deceleration than airplane aerodynamic braking. To this end, the majority of airplanes operating from

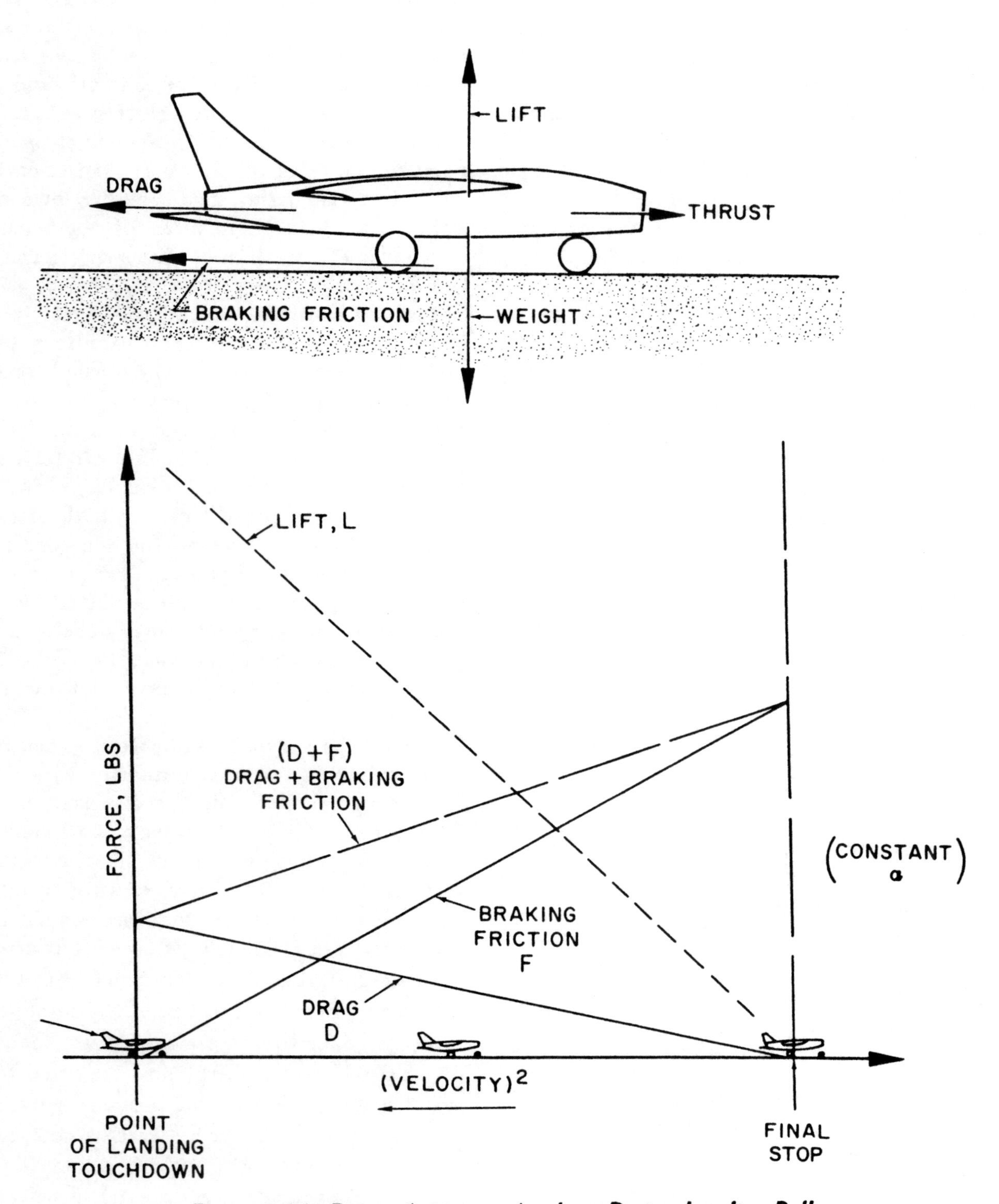

Figure 2.34. Forces Acting on Airplane During Landing Roll

dry hard surface runways will require particular techniques to obtain minimum landing distance. Generally, the technique involves lowering the nose wheel to the runway and retracting the flaps to increase the normal force on the braking surfaces. While the airplane drag is reduced, the greater normal force can provide greater braking friction force to compensate for the reduced drag and the net retarding force is increased.

The technique necessary for minimum landing distance can be altered to some extent in certain situations. For example, low aspect ratio airplanes with high longitudinal control power can create very high drag at the high speeds immediate to landing touchdown. If the landing gear configuration or flap or incidence setting precludes a large reduction of C_L, the normal force on the braking surfaces and braking friction force capability are relatively small. Thus, in the initial high speed part of the landing roll, maximum deceleration would be obtained by creating the greatest possible aerodynamic drag. By the time the aircraft has slowed to 70 or 80 percent of the touchdown speed, aerodynamic drag decays but braking action will then be effective. Some form of this technique may be necessary to achieve minimum distance for some configurations when the coefficient of braking friction is low (wet, icy runway) and the braking friction force capability is reduced relative to airplane aerodynamic drag.

A distinction should be made between the techniques for minimum landing distance and an ordinary landing roll with considerable excess runway available. Minimum landing distance will be obtained from the landing speed by creating a continuous peak deceleration of the airplane. This condition usually requires extensive use of the brakes for maximum deceleration. On the other hand, an ordinary landing roll with considerable excess runway may allow extensive use of aerodynamic drag to minimize wear and tear on the tires and brakes. If aerodynamic drag is sufficient to cause deceleration of the airplane it can be used in deference to the brakes in the early stages of the landing roll, i.e., brakes and tires suffer from continuous, hard use but airplane aerodynamic drag is free and does not wear out with use. The use of aerodynamic drag is applicable only for deceleration to 60 or 70 percent of the touchdown speed. At speeds less than 60 to 70 percent of the touchdown speed, aerodynamic drag is so slight as to be of little use and braking must be utilized to produce continued deceleration of the airplane.

Powerplant thrust is not illustrated on figure 2.34 for there are so many possible variations. Since the objective during the landing roll is to decelerate, the powerplant thrust should be the smallest possible positive value or largest possible negative value. In the case of the turbojet aircraft, the idle thrust of the engine is nearly constant with speed throughout the landing roll. The idle thrust is of significant magnitude on cold days because of the low compressor inlet air temperature and low density altitude. Unfortunately, such atmospheric conditions usually have the corollary of poor braking action because of ice or water on the runway. The thrust from a windmilling propeller with the engine at idle can produce large negative thrust early in the landing roll but the negative force decreases with speed. The large negative thrust at high speed is valuable in adding to drag and braking friction to increase the net retarding force.

Various devices can be utilized to provide greater deceleration of the airplane or to minimize the wear and tear on tires and brakes. The drag parachute can provide a large retarding force at high q and greatly increase the deceleration during the initial phase of landing roll. It should be noted that the contribution of the drag chute is important only during the high speed portion of the landing roll. For maximum effectiveness, the drag chute must be deployed immediately after the airplane is in contact with the runway. Reverse thrust of propellers is obtained by rotating the blade angle well below the low pitch stop and applying engine power. The action is to extract a large amount of momentum from the airstream and thereby create negative thrust. The magnitude of the reverse thrust from propellers is very large, especially in the case of the turboprop where a very large shaft power can be fed into the propeller. In the case of reverse propeller thrust, maximum effectiveness is achieved by use immediately after the airplane is in contact with the runway. The reverse thrust capability is greatest at the high speed and, obviously, any delay in producing deceleration allows runway to pass by at a rapid rate. Reverse thrust of turbojet engines will usually employ some form of vanes, buckets, or clamshells in the exhaust to turn or direct the exhaust gases forward. Whenever the exit velocity is less than the inlet velocity (or negative), a negative momentum change occurs and negative thrust is produced. The reverse jet thrust is valuable and effective but it should not be compared with the reverse thrust capability of a comparable propeller powerplant which has the high intrinsic thrust at low velocities. As with the propeller reverse thrust, jet reverse thrust must be applied immediately after ground contact for maximum effectiveness in reducing landing distance.

FACTORS AFFECTING LANDING PERFORMANCE. In addition to the important factors of proper technique, many other variables affect the landing performance of an airplane. Any item which alters the landing velocity or deceleration during landing roll will affect the landing distance. As with takeoff performance, the relationships of uniformly accelerated motion will be assumed applicable for studying the principal effects on landing distance. The case of uniformly accelerated motion defines landing distance as varying directly as the square of the landing velocity and inversely as the acceleration during landing roll.

$$\frac{S_2}{S_1}=\left(\frac{V_2}{V_1}\right)^2\times\left(\frac{a_1}{a_2}\right)$$

where

S_1 = landing distance resulting from certain values of landing velocity, V_1, and acceleration, a_1

S_2 = landing distance resulting from some different values of landing velocity, V_2, or acceleration, a_2

With this relationship, the effect of the many variables on landing distance can be approximated.

The *effect of gross weight* on landing distance is one of the principal items determining the landing distance of an airplane. One effect of an increased gross weight is that the airplane will require a greater speed to support the airplane at the landing angle of attack and lift coefficient. The relationship of landing speed and gross weight would be as follows:

$$\frac{V_2}{V_1}=\sqrt{\frac{W_2}{W_1}} \quad (EAS \text{ or } CAS)$$

where

V_1 = landing velocity corresponding to some original weight, W_1

V_2 = landing velocity corresponding to some different weight, W_2

Thus, a given airplane in the landing configuration at a given gross weight will have a specific landing speed (EAS or CAS) which is invariant with altitude, temperature, wind, etc., because a certain value of q is necessary to provide lift equal to weight at the landing C_L. As an example of the effect of a change in gross weight, a 21 percent increase in landing weight will require a 10 percent increase in landing speed to support the greater weight.

When minimum landing distances are considered, braking friction forces predominate during the landing roll and, for the majority of airplane configurations, braking friction is the main source of deceleration. In this case, an increase in gross weight provides a greater

normal force and increased braking friction force to cope with the increased mass. Also, the higher landing speed at the same C_L and C_D produce an average drag which increased in the same proportion as the increased weight. Thus, increased gross weight causes like increases in the sum of drag plus braking friction and the acceleration is essentially unaffected.

To evaluate the effect of gross weight on landing distance, the following relationships are used:

the effect of weight on landing velocity is

$$\frac{V_2}{V_1}=\sqrt{\frac{W_2}{W_1}} \quad \text{or} \quad \left(\frac{V_2}{V_1}\right)^2=\frac{W_2}{W_1}$$

if the net retarding force increases in the same proportion as the weight, the acceleration is unaffected.

the effect of these items on landing distance is,

$$\frac{S_2}{S_1}=\left(\frac{V_2}{V_1}\right)^2\times\left(\frac{a_1}{a_2}\right)$$

or

$$\frac{S_2}{S_1}=\frac{W_2}{W_1}$$

In effect, the minimum landing distance will vary directly as the gross weight. For example, a 10 percent increase in gross weight at landing would cause:

a 5 percent increase in landing velocity
a 10 percent increase in landing distance

A contingency of the previous analysis is the relationship between weight and braking friction force. The maximum coefficient of braking friction is relatively independent of the usual range of normal forces and rolling speeds, e.g., a 10 percent increase in normal force would create a like 10 percent increase in braking friction force. Consider the case of two airplanes of the same type and c.g. position but of different gross weights. If these two airplanes are rolling along the runway at some speed at which aerodynamic forces are negligible, the use of the maximum coefficient of braking friction will bring both airplanes to a stop in the same distance. The heavier airplane will have the greater mass to decelerate but the greater normal force will provide a greater retarding friction force. As a result, both airplanes would have identical acceleration and identical stop distances from a given velocity. However, the heavier airplane would have a greater kinetic energy to be dissipated by the brakes and the principal difference between the two airplanes as they reach a stop would be that the heavier airplane would have the hotter brakes. Therefore, one of the factors of braking performance is the ability of the brakes to dissipate energy without developing excessive temperatures and losing effectiveness.

To appreciate the effectiveness of modern brakes, a 30,000-lb. aircraft landing at 175 knots has a kinetic energy of 41 million ft.-lbs. at the instant of touchdown. In a minimum distance landing, the brakes must dissipate most of this kinetic energy and *each* brake must absorb an input power of approximately 1,200 h.p. for 25 seconds. Such requirements for brakes are extreme but the example serves to illustrate the problems of brakes for high performance airplanes.

While a 10 percent increase in landing weight causes:

a 5 percent higher landing speed
a 10 percent greater landing distance,

it also produces a 21 percent increase in the *kinetic energy* of the airplane to be dissipated during the landing roll. Hence, high landing weights may approach the energy dissipating capability of the brakes.

The *effect of wind* on landing distance is large and deserves proper consideration when predicting landing distance. Since the airplane will land at a particular airspeed independent of the wind, the principal effect of wind on landing distance is due to the change in the ground velocity at which the airplane touches down. The effect of wind on acceleration during the landing distance is identical to the effect on takeoff distance and is approximated by the following relationship:

$$\frac{S_2}{S_1}=\left[1-\frac{V_w}{V_1}\right]^2$$

where

S_1 = zero wind landing distance
S_2 = landing distance into a headwind
V_w = headwind velocity
V_1 = landing ground velocity with zero wind or, simply, the landing airspeed

As a result of this relationship, a headwind which is 10 percent of the landing airspeed will reduce the landing distance 19 percent but a tailwind (or negative headwind) which is 10 percent of the landing speed will increase the landing distance 21 percent. Figure 2.33 illustrates this general effect.

The *effect of runway slope* on landing distance is due to the component of weight along the inclined path of the airplane. The relationship is identical to the case of takeoff performance but the magnitude of the effect is not as great. While account must be made for the effect, the ordinary values of runway slope do not contribute a large effect on landing distance. For this reason, the selection of the landing runway will ordinarily favor the direction with a downslope and headwind rather than an upslope and tailwind.

The *effect of pressure altitude and ambient temperature* is to define *density altitude* and its effect on landing performance. An increase in density altitude will increase the landing velocity but will not alter the net retarding force. If a given weight and configuration of airplane is taken to altitude above standard sea level, the airplane will still require the same q to provide lift equal to weight at the landing C_L. Thus, the airplane at altitude will land at the same equivalent airspeed (EAS) as at sea level but, because of the reduced density, the true airspeed (TAS) will be greater. The relationship between true airspeed and equivalent airspeed is as follows:

$$\frac{TAS}{EAS}=\frac{1}{\sqrt{\sigma}}$$

where

TAS = true airspeed
EAS = equivalent airspeed
σ = altitude density ratio

Since the airplane lands at altitude with the same weight and dynamic pressure, the drag and braking friction throughout the landing roll have the same values as at sea level. As long as the condition is within the capability of the brakes, the net retarding force is unchanged and the acceleration is the same as with the landing at sea level.

To evaluate the effect of density altitude on landing distance, the following relationships are used:

since an increase in altitude does not alter acceleration, the effect would be due to the greater TAS

$$\frac{S_2}{S_1}=\left(\frac{V_2}{V_1}\right)^2\times\left(\frac{a_1}{a_2}\right)$$

$$\frac{S_2}{S_1}=\frac{1}{\sigma}$$

where

S_1 = standard sea level landing distance
S_2 = landing distance at altitude
σ = altitude density ratio

From this relationship, the minimum landing distance at 5,000 ft. ($\sigma=0.8617$) would be 16 percent greater than the minimum landing distance at sea level. The approximate increase in landing distance with altitude is approximately 3½ percent for each 1,000 ft. of altitude. Proper accounting of density altitude is necessary to accurately predict landing distance.

The effect of *proper landing velocity* is important when runway lengths and landing distances are critical. The landing speeds specified in the flight handbook are generally the minimum safe speeds at which the airplane can be landed. Any attempt to land at below the specified speed may mean that the airplane may stall, be difficult to control, or develop high rates of descent. On the other hand, an excessive speed at landing may improve the controllability (especially in crosswinds) but will cause an undesirable increase in landing distance. The principal effect of excess landing speed is described by:

$$\frac{S_2}{S_1}=\left(\frac{V_2}{V_1}\right)^2$$

Thus, a 10 percent excess landing speed would cause a 21 percent increase in landing distance. The excess speed places a greater working load on the brakes because of the additional kinetic energy to be dissipated. Also, the additional speed causes increased drag and lift in the normal ground attitude and the increased lift will reduce the normal force on the braking surfaces. The acceleration during this range of speed immediately after touchdown may suffer and it will be more likely that a tire can be blown out from braking at this point. As a result, 10 percent excess landing speed will cause *at least* 21 percent greater landing distance.

The most critical conditions of landing performance are the result of some combination of high gross weight, density altitude, and unfavorable wind. These conditions produce the greatest landing distance and provide critical levels of energy dissipation required of the brakes. In all cases, it is necessary to make an accurate prediction of minimum landing distance to compare with the available runway. A polished, professional landing technique is necessary because the landing phase of flight accounts for more pilot caused aircraft accidents than any other single phase of flight.

In the prediction of minimum landing distance from the handbook data, the following considerations must be given:

(1) Pressure altitude and temperature—to define the effect of density altitude.

(2) Gross weight—which define the *CAS* or *EAS* for landing.

(3) Wind—a large effect due to wind or wind component along the runway.

(4) Runway slope—a relatively small correction for ordinary values of runway slope.

IMPORTANCE OF HANDBOOK PERFORMANCE DATA. The performance section or supplement of the flight handbook contains all the operating data for the airplane. For example, all data specific to takeoff, climb, range, endurance, descent and landing are included in this section. The ordinary use of these data in flying operations is mandatory and great knowledge and familiarity of the airplane can be gained through study of this material. A complete familiarity of an airplane's characteristics can be obtained only through extensive analysis and study of the handbook data.

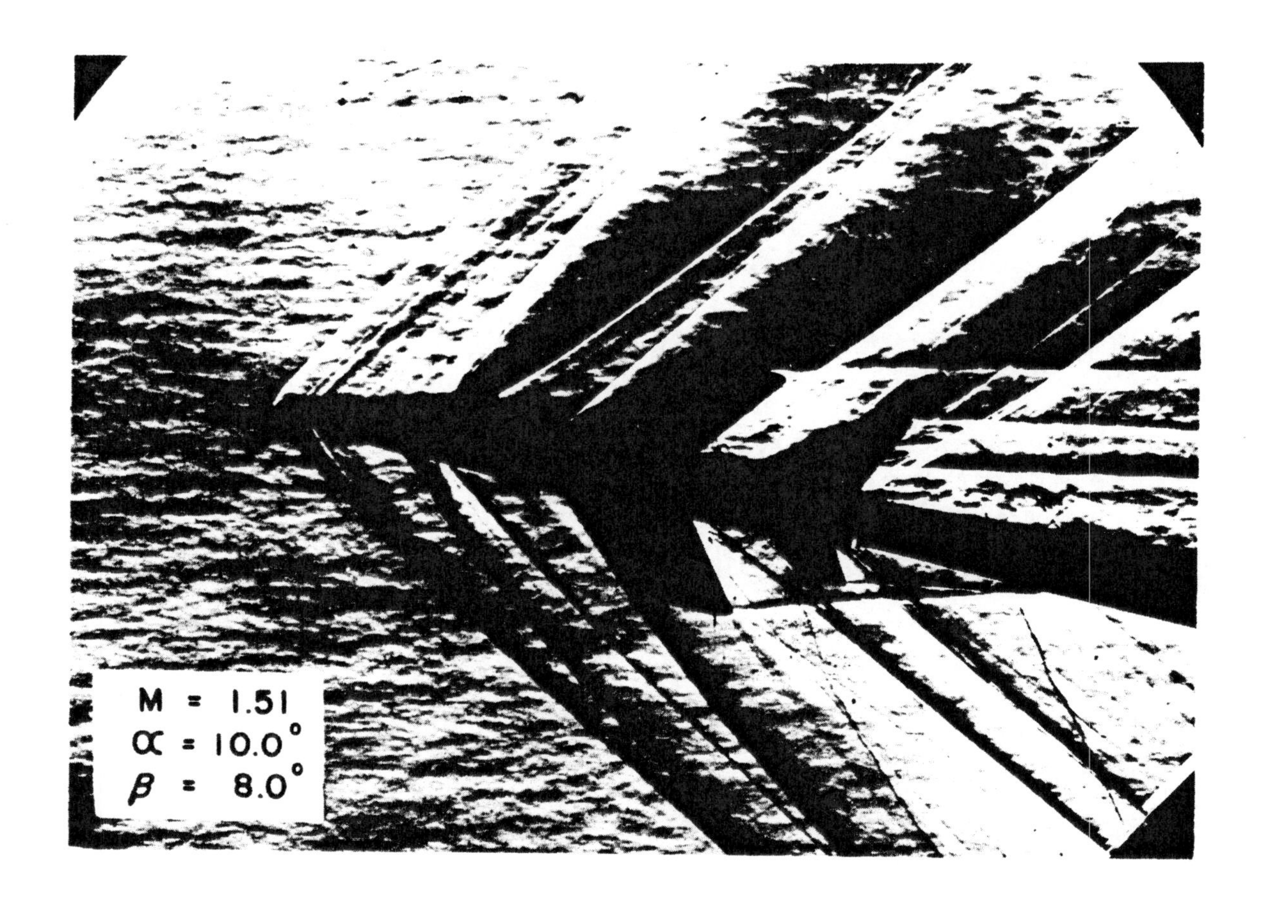

Chapter 3

HIGH SPEED AERODYNAMICS

Developments in aircraft and powerplants have produced high performance airplanes with capabilities for very high speed flight. The study of aerodynamics at these very high flight speeds has many significant differences from the study of classical low speed aerodynamics. Therefore, it is quite necessary that the Naval Aviator be familiar with the nature of high speed airflow and the characteristics of high performance airplane configurations.

GENERAL CONCEPTS AND SUPERSONIC FLOW PATTERNS

NATURE OF COMPRESSIBILITY

At low flight speeds the study of aerodynamics is greatly simplified by the fact that air may experience relatively small changes in pressure with only negligible changes in density. This airflow is termed *incompressible* since the air may undergo changes in pressure without apparent changes in density. Such a condition of airflow is analogous to the flow of water, hydraulic fluid, or any other incompressible fluid. However, at high flight speeds the pressure changes that take place are quite large and significant changes in air density occur. The study of airflow at high speeds must account for these changes in air density and must consider that the air is compressible and that there will be "compressibility effects."

A factor of great importance in the study of high speed airflow is the speed of sound. The speed of sound is the rate at which small pressure disturbances will be propagated through the air and this propagation speed is solely a function of air temperature. The accompanying table illustrates the variation of the speed of sound in the standard atmosphere.

TABLE 3-1. ***Variation of Temperature and Speed of Sound with Altitude in the Standard Atmosphere***

Altitude	Temperature		Speed of sound
Ft.	° *F.*	° *C.*	*Knots*
Sea level	59.0	15.0	661.7
5,000	41.2	5.1	650.3
10,000	23.3	−4.8	638.6
15,000	5.5	−14.7	626.7
20,000	−12.3	−24.6	614.6
25,000	−30.2	−34.5	602.2
30,000	−48.0	−44.4	589.6
35,000	−65.8	−54.3	576.6
40,000	−69.7	−56.5	573.8
50,000	−69.7	−56.5	573.8
60,000	−69.7	−56.5	573.8

As an object moves through the air mass, velocity and pressure changes occur which create pressure disturbances in the airflow surrounding the object. Of course, these pressure disturbances are propagated through the air at the speed of sound. If the object is travelling at low speed the pressure disturbances are propagated ahead of the object and the airflow immediately ahead of the object is influenced by the pressure field on the object. Actually, these pressure disturbances are transmitted in all directions and extend indefinitely in all directions. Evidence of this "pressure warning" is seen in the typical subsonic flow pattern of figure 3.1 where there is upwash and flow direction change well ahead of the leading edge. If the object is travelling at some speed above the speed of sound the airflow ahead of the object will not be influenced by the pressure field on the object since pressure disturbances cannot be propagated ahead of the object. Thus, as the flight speed nears the speed of sound a compression wave will form at the leading edge and all changes in velocity and pressure will take place quite sharply and suddenly. The airflow ahead of the object is not influenced until the air particles are suddenly forced out of the way by the concentrated pressure wave set up by the object. Evidence of this phenomenon is seen in the typical supersonic flow pattern of figure 3.1.

The analogy of surface waves on the water may help clarify these phenomena. Since a surface wave is simply the propagation of a pressure disturbance, a ship moving at a speed much less than the wave speed will *not* form a "bow wave." As the ship's speed nears the wave propagation speed the bow wave will form and become stronger as speed is increased beyond the wave speed.

At this point it should become apparent that all compressibility effects depend upon the relationship of airspeed to the speed of sound. The term used to describe this relationship is the Mach number, M, and this term is the ratio of the true airspeed to the speed of sound.

$$M=\frac{V}{a}$$

where

M = Mach number
V = true airspeed, knots
a = speed of sound, knots
$= a_0\sqrt{\theta}$
a_0 = speed of sound at standard sea level conditions, 661 knots
θ = temperature ratio
$= T/T_o$

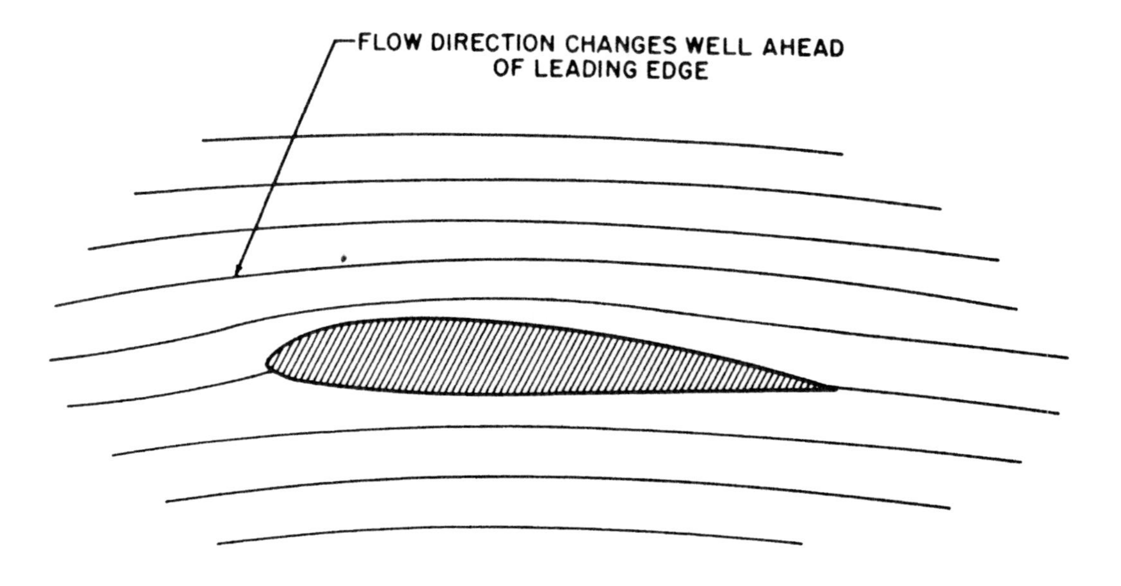

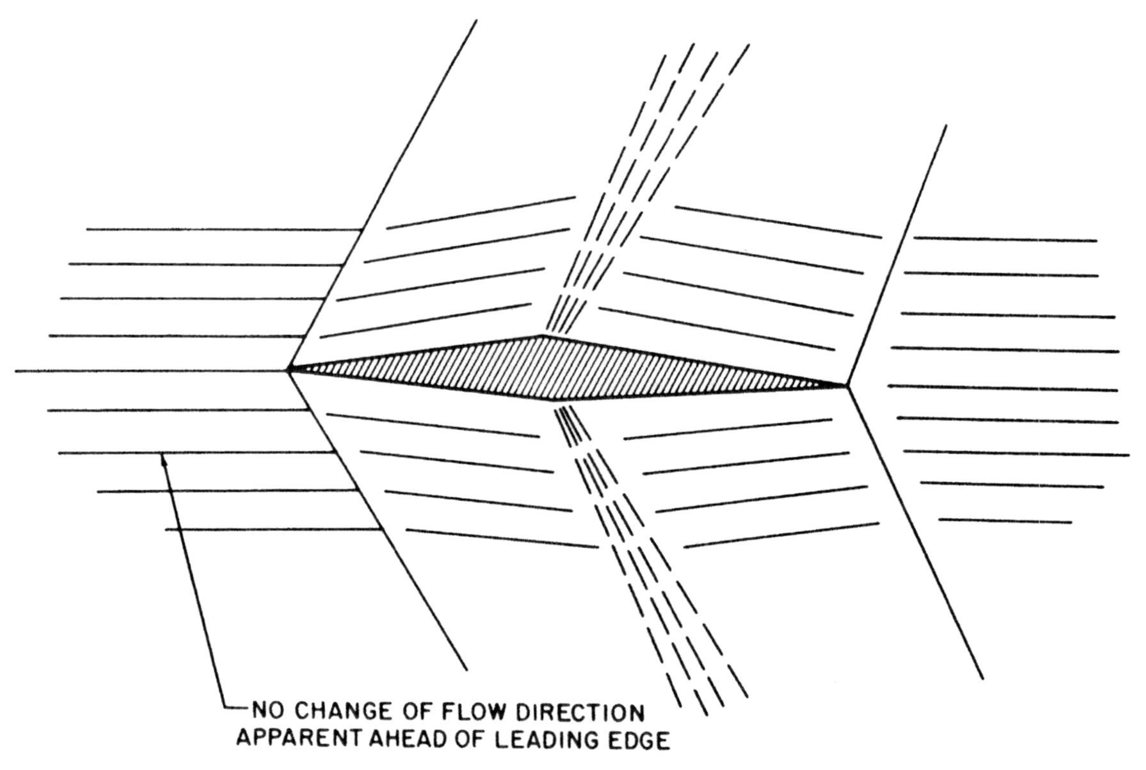

Figure 3.1. Comparison of Subsonic and Supersonic Flow Patterns

It is important to note that compressibility effects are not limited to flight speeds at and above the speed of sound. Since any aircraft will have some aerodynamic shape and will be developing lift there will be local flow velocities on the surfaces which are greater than the flight speed. Thus, an aircraft can experience compressibility effects at flight speeds well below the speed of sound. Since there is the possibility of having *both* subsonic and supersonic flows existing on the aircraft it is convenient to define certain regimes of flight. These regimes are defined approximately as follows:

Subsonic—Mach numbers below 0.75
Transonic—Mach numbers from 0.75 to 1.20
Supersonic—Mach numbers from 1.20 to 5.00
Hypersonic—Mach numbers above 5.00

While the flight Mach numbers used to define these regimes of flight are quite approximate, it is important to appreciate the types of flow existing in each area. In the subsonic regime it is most likely that pure subsonic airflow exists on all parts of the aircraft. In the transonic regime it is very probable that flow on the aircraft components may be partly subsonic and partly supersonic. The supersonic and hypersonic flight regimes will provide definite supersonic flow velocities on all parts of the aircraft. Of course, in supersonic flight there will be some portions of the boundary layer which are subsonic but the predominating flow is still supersonic.

The principal differences between subsonic and supersonic flow are due to the *compressibility* of the supersonic flow. Thus, any change of velocity or pressure of a supersonic flow will produce a related change of density which must be considered and accounted for. Figure 3.2 provides a comparison of incompressible and compressible flow through a closed tube. Of course, the condition of continuity must exist in the flow through the closed tube; the mass flow at any station along the tube is constant. This qualification must exist in both compressible and incompressible cases.

The example of subsonic incompressible flow is simplified by the fact that the density of flow is constant throughout the tube. Thus, as the flow approaches a constriction and the streamlines converge, velocity increases and static pressure decreases. In other words, a convergence of the tube requires an increasing velocity to accommodate the continuity of flow. Also, as the subsonic incompressible flow enters a diverging section of the tube, velocity decreases and static pressure increases but density remains unchanged. The behavior of subsonic incompressible flow is that a convergence causes expansion (decreasing pressure) while a divergence causes compression (increasing pressure).

The example of supersonic compressible flow is complicated by the fact that the variations of flow density are related to the changes in velocity and static pressure. The behavior of supersonic compressible flow is that a convergence causes compression while a divergence causes expansion. Thus, as the supersonic compressible flow approaches a constriction and the streamlines converge, velocity decreases and static pressure increases. Continuity of mass flow is maintained by the increase in flow density which accompanies the decrease in velocity. As the supersonic compressible flow enters a diverging section of the tube, velocity increases, static pressure decreases, and density decreases to accommodate the condition of continuity.

The previous comparison points out three significant differences between supersonic compressible and subsonic incompressible flow.

(*a*) Compressible flow includes the additional variable of flow density.

(*b*) Convergence of flow causes expansion of incompressible flow but compression of compressible flow.

(*c*) Divergence of flow causes compression of incompressible flow but expansion of compressible flow.

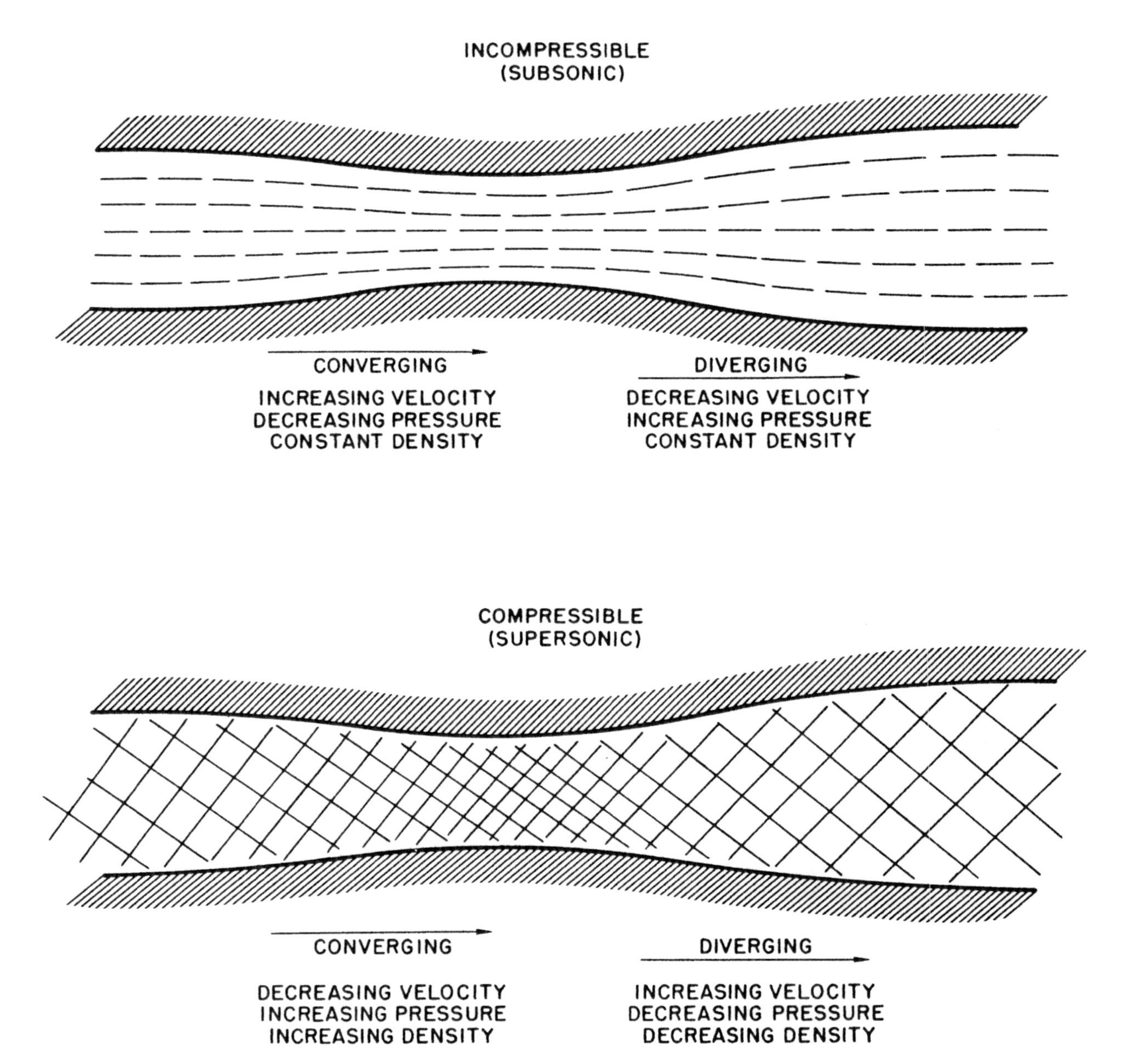

Figure 3.2. Comparison of Compressible and Incompressible Flow Through a Closed Tube

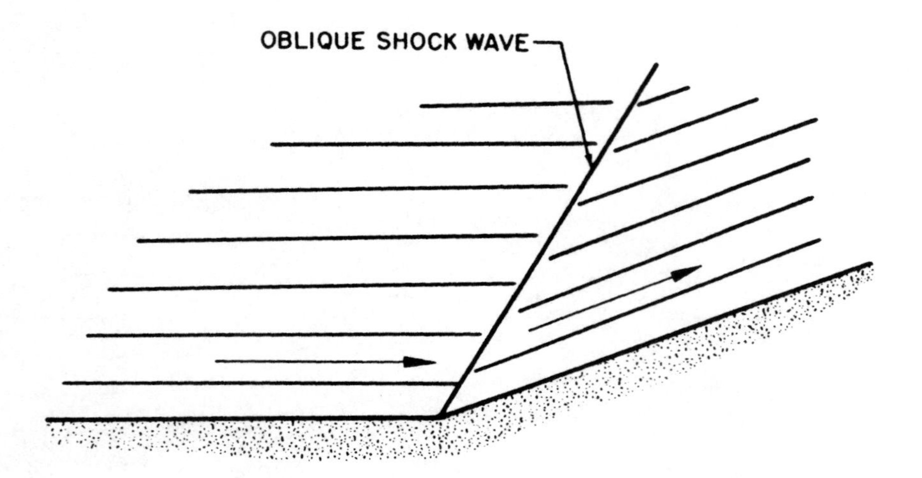

SUPERSONIC FLOW INTO A CORNER

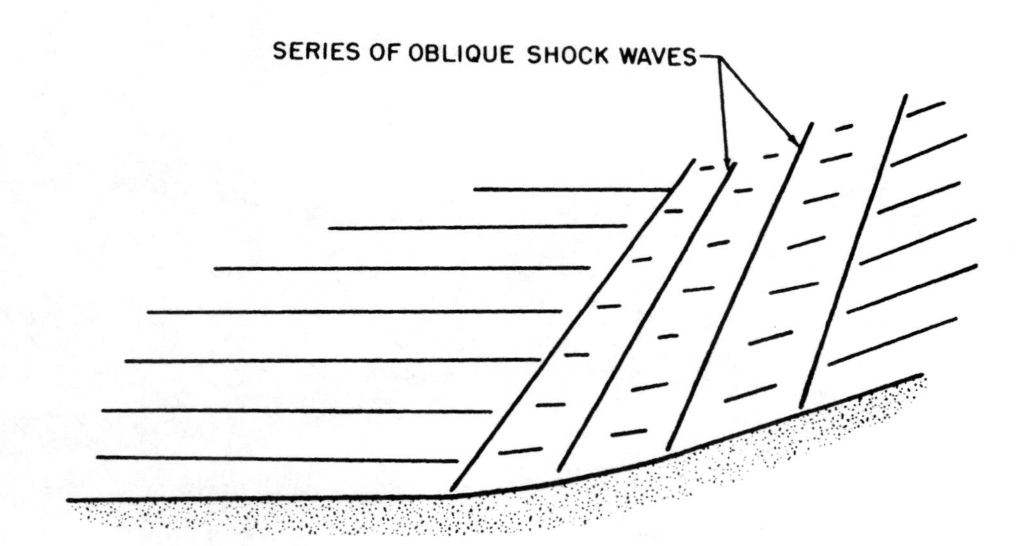

SUPERSONIC FLOW INTO A ROUNDED CORNER

Figure 3.3. Oblique Shock Wave Formation

TYPICAL SUPERSONIC FLOW PATTERNS

When supersonic flow is clearly established, all changes in velocity, pressure, density, flow direction, etc., take place quite suddenly and in relatively confined areas. The areas of flow change are generally distinct and the phenomena are referred to as "wave" formations. All compression waves occur suddenly and are wasteful of energy. Hence, the compression waves are distinguished by the sudden "shock" type of behavior. All expansion waves are not so sudden in their occurrence and are not wasteful of energy like the compression shock waves. Various types of waves can occur in supersonic flow and the nature of the wave formed depends upon the airstream and the shape of the object causing the flow change. Essentially, there are three fundamental types of waves formed in supersonic flow: (1) the *oblique* shock wave (compression), (2) the *normal* shock wave (compression), (3) the *expansion* wave (no shock).

OBLIQUE SHOCK WAVE. Consider the case where a supersonic airstream is turned into the preceding airflow. Such would be the case of a supersonic flow "into a corner" as shown in figure 3.3. A supersonic airstream passing through the oblique shock wave will experience these changes:

(1) The airstream is slowed down; the velocity and Mach number behind the wave are reduced but the flow is still supersonic

(2) The flow direction is changed to flow along the surface

(3) The static pressure of the airstream behind the wave is increased

(4) The density of the airstream behind the wave is increased

(5) Some of the available energy of the airstream (indicated by the sum of dynamic and static pressure) is dissipated and turned into unavailable heat energy. Hence, the shock wave is wasteful of energy.

A typical case of oblique shock wave formation is that of a wedge pointed into a supersonic airstream. The oblique shock wave will form on each surface of the wedge and the inclination of the shock wave will be a function of the free stream Mach number and the wedge angle. As the free stream Mach number increases, the shock wave angle decreases; as the wedge angle increases the shock wave angle increases, and, if the wedge angle is increased to some critical amount, the shock wave will detach from the leading edge of the wedge. It is important to note that detachment of the shock wave will produce *subsonic* flow immediately after the central portion of the shock wave. Figure 3.4 illustrates these typical flow patterns and the effect of Mach number and wedge angle.

The previous flow across a wedge in a supersonic airstream would allow flow in *two* dimensions. If a cone were placed in a supersonic airstream the airflow would occur in *three* dimensions and there would be some noticeable differences in flow characteristics. Three-dimensional flow for the same Mach number and flow direction change would produce a weaker shock wave with less change in pressure and density. Also, this conical wave formation allows changes in airflow that continue to occur past the wave front and the wave strength varies with distance away from the surface. Figure 3.5 depicts the typical three-dimensional flow past a cone.

Oblique shock waves can be reflected like any pressure wave and this effect is shown in figure 3.5. This reflection appears logical and necessary since the original wave changes the flow direction toward the wall and the reflected wave creates the subsequent flow change to cause the flow to remain parallel to the wall surface. This reflection phenomenon places definite restrictions on the size of a model in a wind tunnel since a wave reflected back to the model would cause a pressure distribution not typical of free flight.

NORMAL SHOCK WAVE. If a blunt-nosed object is placed in a supersonic airstream the shock wave which is formed will be detached from the leading edge. This detached

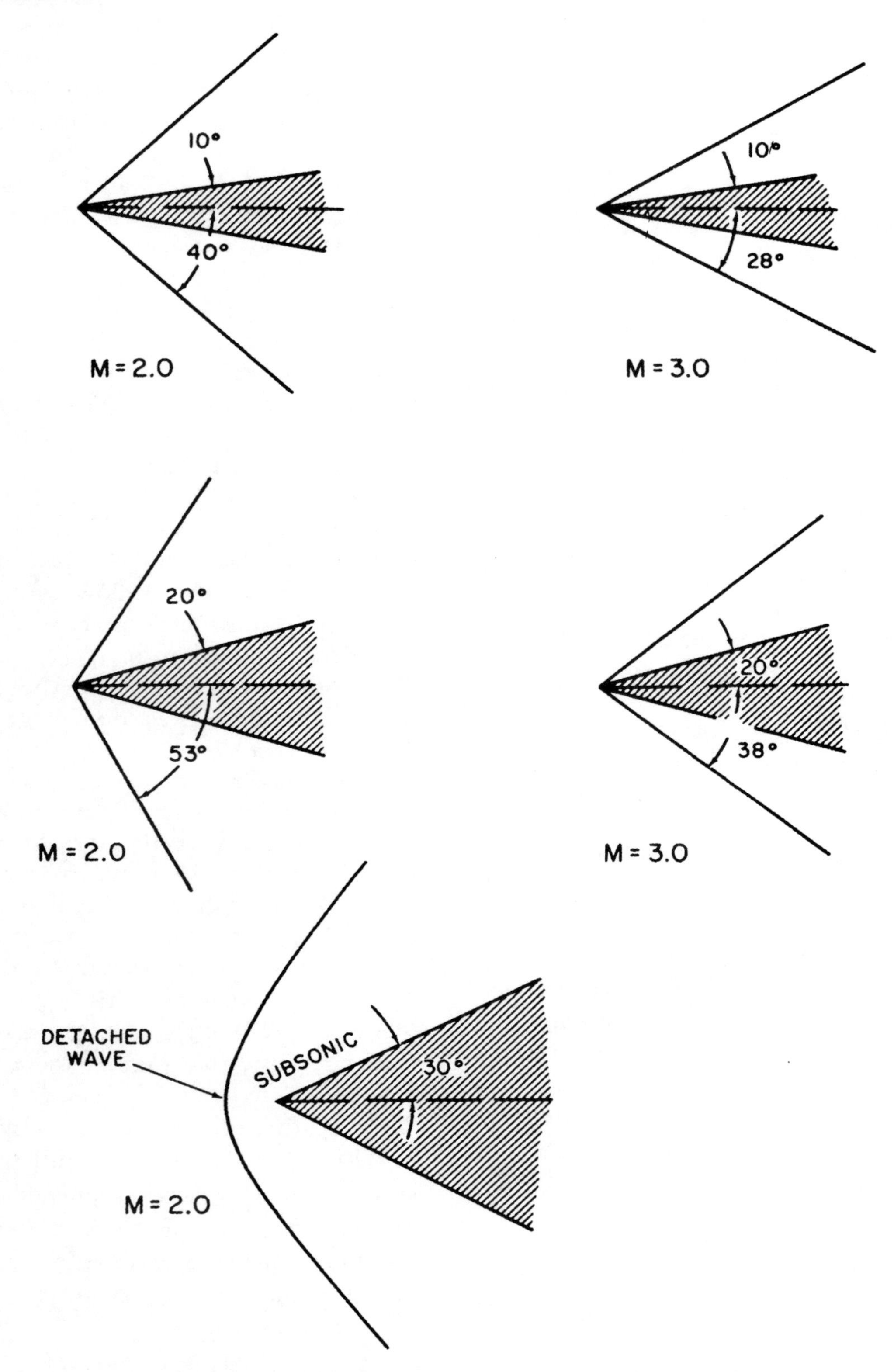

Figure 3.4. Shock Waves Formed by Various Wedge Shapes

CONE IN SUPERSONIC FLOW

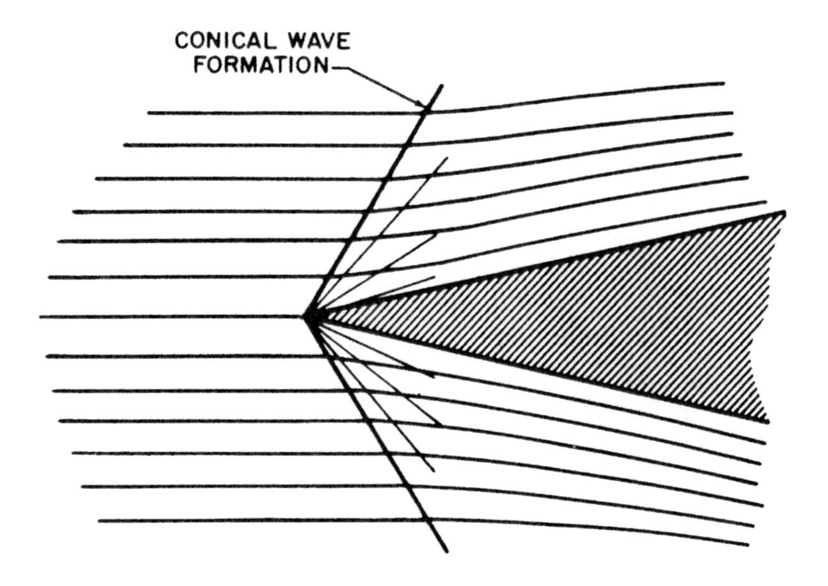

REFLECTED OBLIQUE WAVES

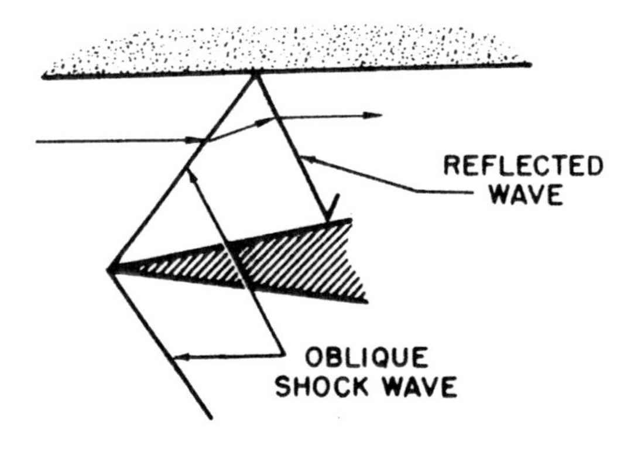

MODEL IN WIND
TUNNEL WITH WAVES
REFLECTED FROM
WALLS

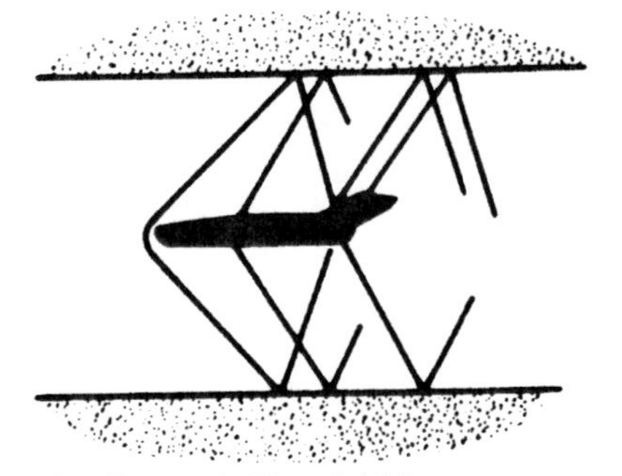

Figure 3.5. Three Dimensional and Reflected Shock Waves

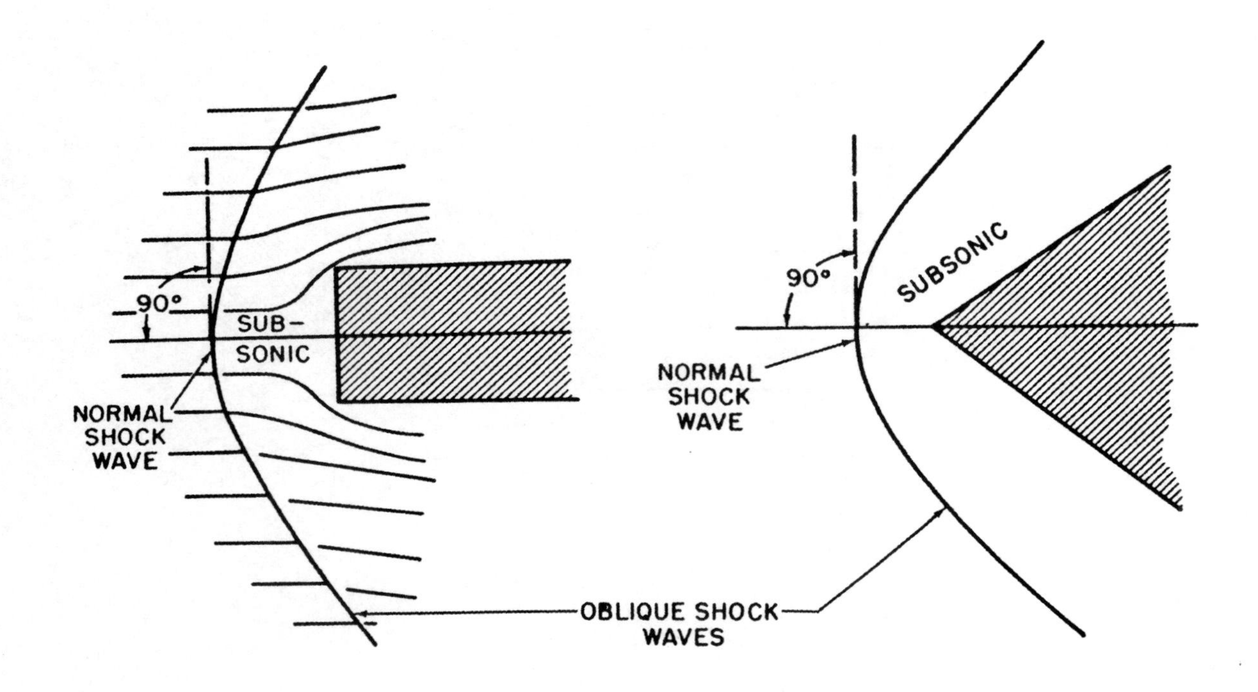

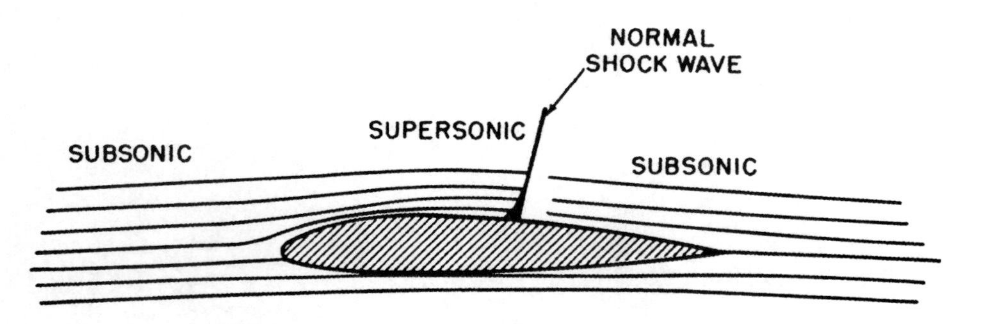

Figure 3.6. Normal Shock Wave Formation

wave also occurs when a wedge or cone angle exceeds some critical value. Whenever the shock wave forms perpendicular to the upstream flow, the shock wave is termed a "normal" shock wave and the flow immediately behind the wave is *subsonic*. Any relatively blunt object in a supersonic airstream will form a normal shock wave immediately ahead of the leading edge slowing the airstream to subsonic so the airstream may feel the presence of the blunt nose and flow around it. Once past the blunt nose the airstream may remain subsonic or accelerate back to supersonic depending on the shape of the nose and the Mach number of the free stream.

In addition to the formation of normal shock waves described above, this same type of wave may be formed in an entirely different manner when there is no object in the supersonic airstream. It is particular that whenever a supersonic airstream is slowed to subsonic without a change in direction a normal shock wave will form as a boundary between the supersonic and subsonic regions. This is an important fact since aircraft usually encounter some "compressibility effects" before the flight speed is sonic. Figure 3.6 illustrates the manner in which an airfoil at high subsonic speeds has local flow velocities which are supersonic. As the local supersonic flow moves aft, a normal shock wave forms slowing the flow to subsonic. The transition of flow from subsonic to supersonic is smooth and is not accompanied by shock waves if the transition is made gradually with a smooth surface. The transition of flow from supersonic to subsonic without direction change *always* forms a normal shock wave.

A supersonic airstream passing through a normal shock wave will experience these changes:

(1) The airstream is slowed to subsonic; the local Mach number behind the wave is approximately equal to the reciprocal of the Mach number ahead of the wave—e.g., if Mach number ahead of the wave is 1.25, the Mach number of the flow behind the wave is approximately 0.80.

(2) The airflow direction immediately behind the wave is unchanged.

(3) The static pressure of the airstream behind the wave is increased greatly.

(4) The density of the airstream behind the wave is increased greatly.

(5) The energy of the airstream (indicated by total pressure—dynamic plus static) is greatly reduced. The normal shock wave is very wasteful of energy.

EXPANSION WAVE. If a supersonic airstream were turned away from the preceding flow an expansion wave would form. The flow "around a corner" shown in figure 3.7 will not cause sharp, sudden changes in the airflow except at the corner itself and thus is not actually a "shock" wave. A supersonic airstream passing through an expansion wave will experience these changes:

(1) The airstream is accelerated; the velocity and Mach number behind the wave are greater.

(2) The flow direction is changed to flow along the surface—provided separation does not occur.

(3) The static pressure of the airstream behind the wave is decreased.

(4) The density of the airstream behind the wave is decreased.

(5) Since the flow changes in a rather gradual manner there is no "shock" and no loss of energy in the airstream. The expansion wave does not dissipate airstream energy.

The expansion wave in three dimensions is a slightly different case and the principal difference is the tendency for the static pressure to continue to increase past the wave.

The following table is provided to summarize the characteristics of the three principal wave forms encountered with supersonic flow.

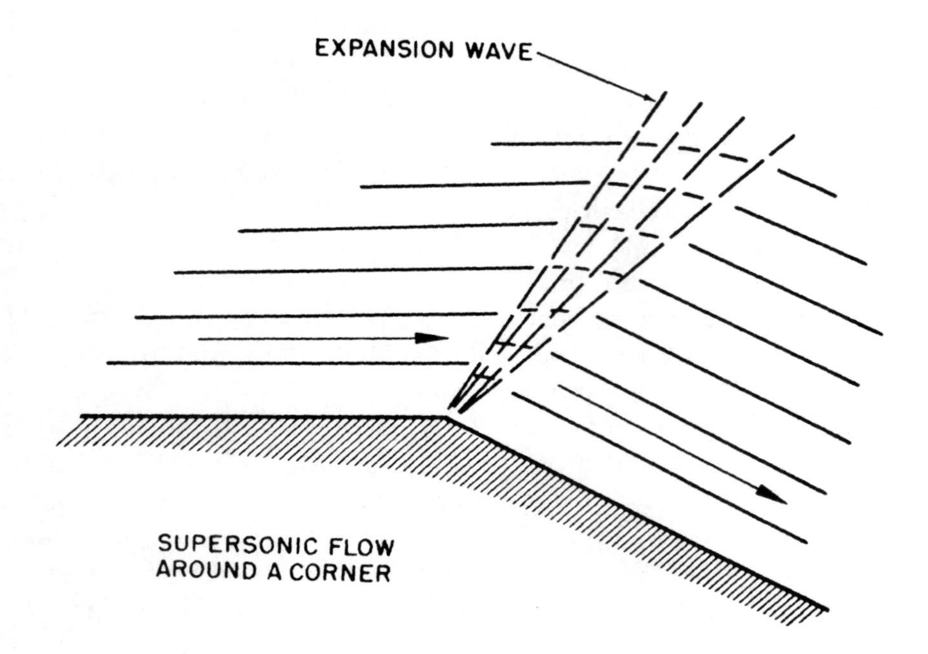

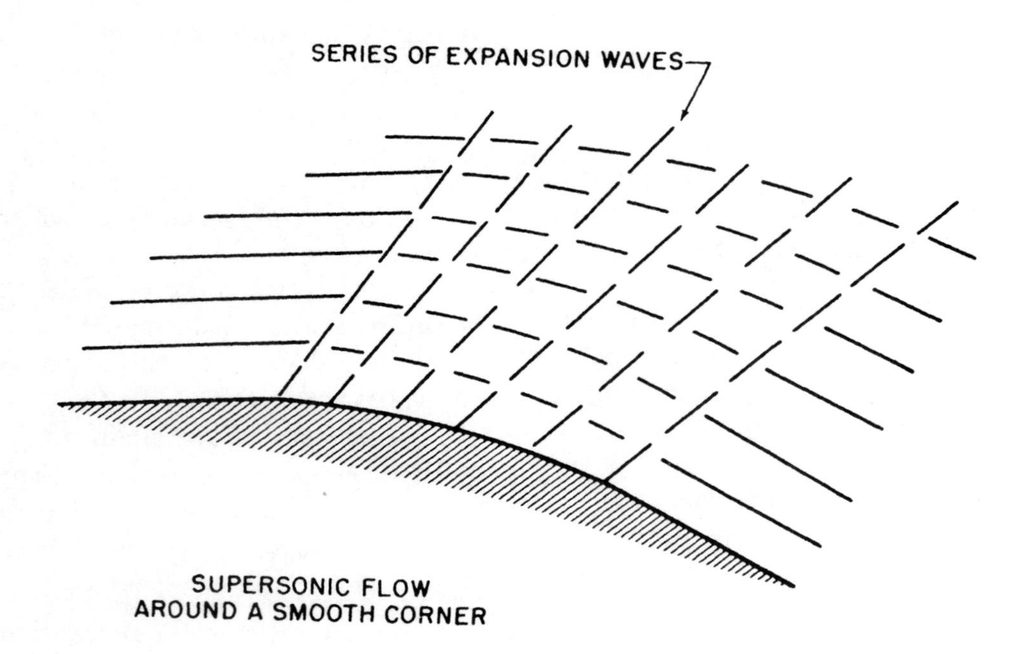

Figure 3.7. Expansion Wave Formation

TABLE 3-2. Supersonic Wave Characteristics

Type of wave formation	Oblique shock wave 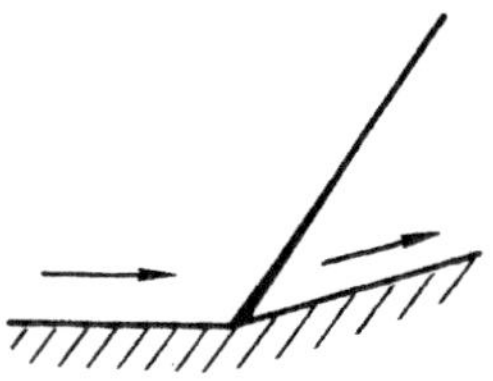	Normal shock wave 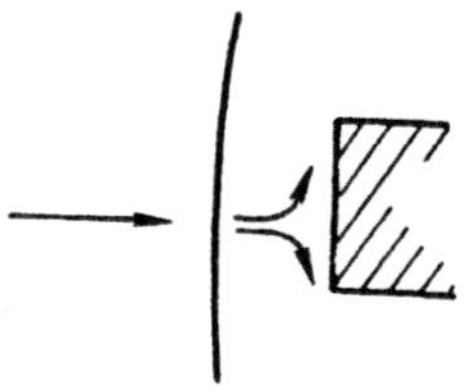	Expansion wave. 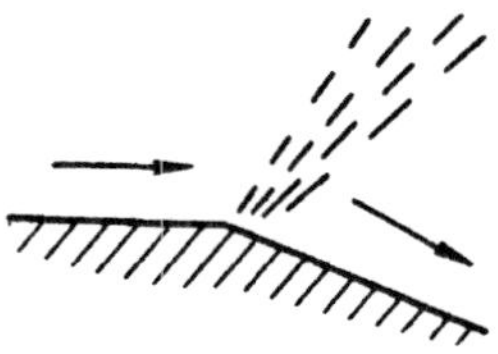
Flow direction change	"Flow into a corner," turned into preceding flow.	No change	"Flow around a corner," turned away from preceding flow.
Effect on velocity and Mach number.	Decreased but still supersonic.	Decreased to subsonic	Increased to higher supersonic.
Effect on static pressure and density.	Increase	Great increase	Decrease.
Effect on energy or total pressure.	Decrease	Great decrease	No change (no shock).

SECTIONS IN SUPERSONIC FLOW

In order to appreciate the effect of these various wave forms on the aerodynamic characteristics in supersonic flow, inspect figure 3.8. Parts (a) and (b) show the wave pattern and resulting pressure distribution for a thin flat plate at a positive angle of attack. The airstream moving over the upper surface passes through an expansion wave at the leading edge and then an oblique shock wave at the trailing edge. Thus, a uniform suction pressure exists over the upper surface. The airstream moving underneath the flat plate passes through an oblique shock wave at the leading edge then an expansion wave at the trailing edge. This produces a uniform positive pressure on the underside of the section. This distribution of pressure on the surface will produce a net lift and incur a subsequent drag due to lift from the inclination of the resultant lift from a perpendicular to the free stream.

Parts (c) and (d) of figure 3.8 show the wave pattern and resulting pressure distribution for a double wedge airfoil at zero lift. The airstream moving over the surface passes through an oblique shock, an expansion wave, and another oblique shock. The resulting pressure distribution on the surfaces produces no net lift, but the increased pressure on the forward half of the chord along with the decreased pressure on the aft half of the chord produces a "wave" drag. This wave drag is caused by the components of pressure forces which are parallel to the free stream direction. The wave drag is in addition to the drag due to friction, separation, lift, etc., and can be a very considerable part of the total drag at high supersonic speeds.

Parts (e) and (f) of figure 3.8 illustrate the wave pattern and resulting pressure distribution for the double wedge airfoil at a small positive angle of attack. The net pressure

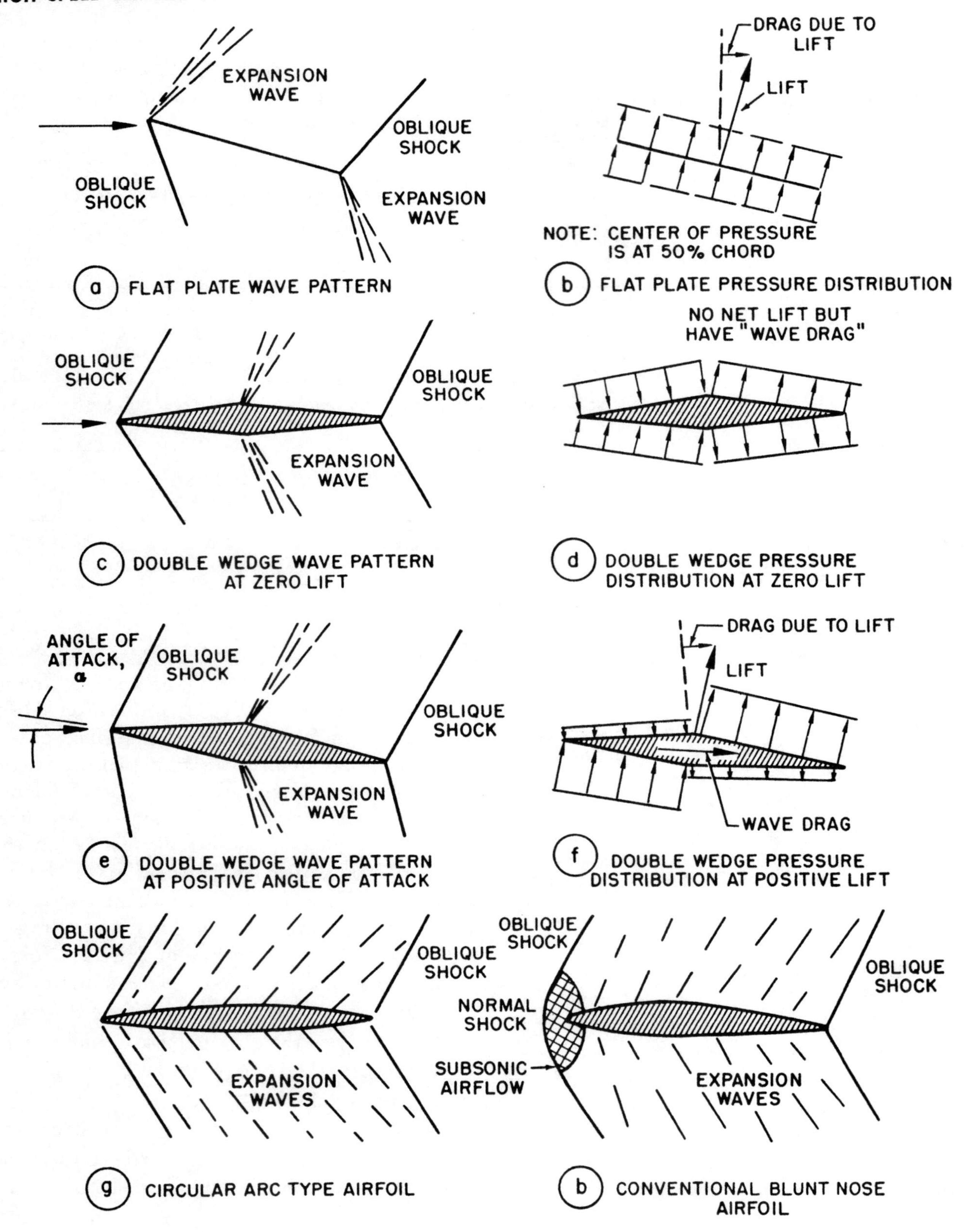

Figure 3.8. Typical Supersonic Flow Patterns and Distribution of Pressure

distribution produces an inclined lift with drag due to lift which is in addition to the wave drag at zero lift. Part (g) of figure 3.8 shows the wave pattern for a circular arc airfoil. After the airflow traverses the oblique shock wave at the leading edge, the airflow undergoes a gradual but continual expansion until the trailing edge shock wave is encountered. Part (h) of figure 3.8 illustrates the wave pattern on a conventional blunt nose airfoil in supersonic flow. When the nose is blunt the wave must detach and become a normal shock wave immediately ahead of the leading edge. Of course, this wave form produces an area of subsonic airflow at the leading edge with very high pressure and density behind the detached wave.

The drawings of figure 3.8 illustrate the typical patterns of supersonic flow and point out these facts concerning aerodynamic surfaces in two dimensional supersonic flow:

(1) All changes in velocity, pressure, density and flow direction will take place quite suddenly through the various wave forms. The shape of the object and the required flow direction change dictate the type and strength of the wave formed.

(2) As always, lift results from the distribution of pressure on a surface and is the net force perpendicular to the free stream direction. Any component of the lift in a direction parallel to the windstream will be drag due to lift.

(3) In supersonic flight, the zero lift drag of an airfoil of some finite thickness will include a "wave drag." The thickness of the airfoil will have an extremely powerful effect on this wave drag since the wave drag varies as the square of the thickness ratio—if the thickness is reduced 50 percent, the wave drag is reduced 75 percent. The leading edges of supersonic shapes must be sharp or the wave formed at the leading edge will be a strong detached shock wave.

(4) Once the flow on the airfoil is supersonic, the aerodynamic center of the surface will be located approximately at the 50 percent chord position. As this contrasts with the subsonic location for the aerodynamic center of the 25 percent chord position, significant changes in aerodynamic trim and stability may be encountered in transonic flight.

CONFIGURATION EFFECTS

TRANSONIC AND SUPERSONIC FLIGHT

Any object in subsonic flight which has some finite thickness or is producing lift will have local velocities on the surface which are greater than the free stream velocity. Hence, compressibility effects can be expected to occur at flight speeds less than the speed of sound. The transonic regime of flight provides the opportunity for mixed subsonic and supersonic flow and accounts for the first significant effects of compressibility.

Consider a conventional airfoil shape as shown in figure 3.9. If this airfoil is at a flight Mach number of 0.50 and a slight positive angle of attack, the maximum local velocity on the surface will be greater than the flight speed but most likely less than sonic speed. Assume that an increase in flight Mach number to 0.72 would produce *first evidence of local sonic flow*. This condition of flight would be the highest flight speed possible without *supersonic* flow and would be termed the "critical Mach number." Thus, critical Mach number is the boundary between subsonic and transonic flight and is an important point of reference for all compressibility effects encountered in transonic flight. By definition, critical Mach number is the "free stream Mach number which produces first evidence of local sonic flow." Therefore, shock waves, buffet, airflow separation, etc., take place above critical Mach number.

As critical Mach number is exceeded an area of *supersonic* airflow is created and a normal

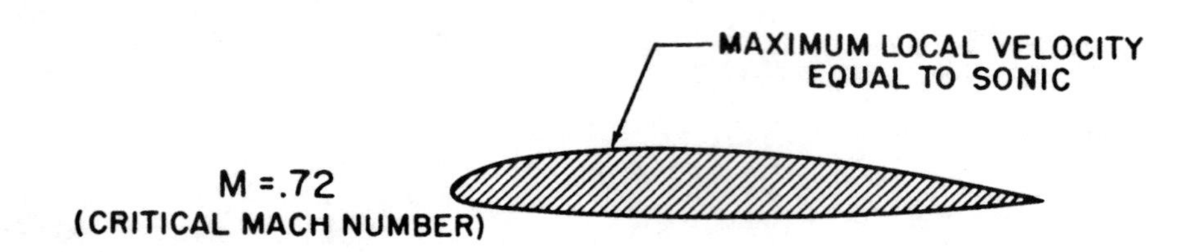

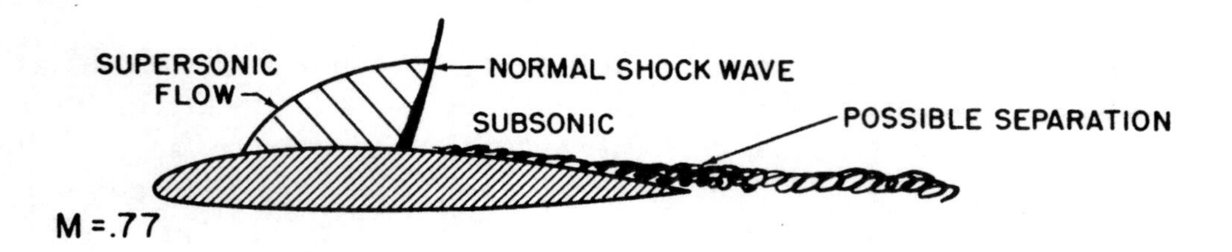

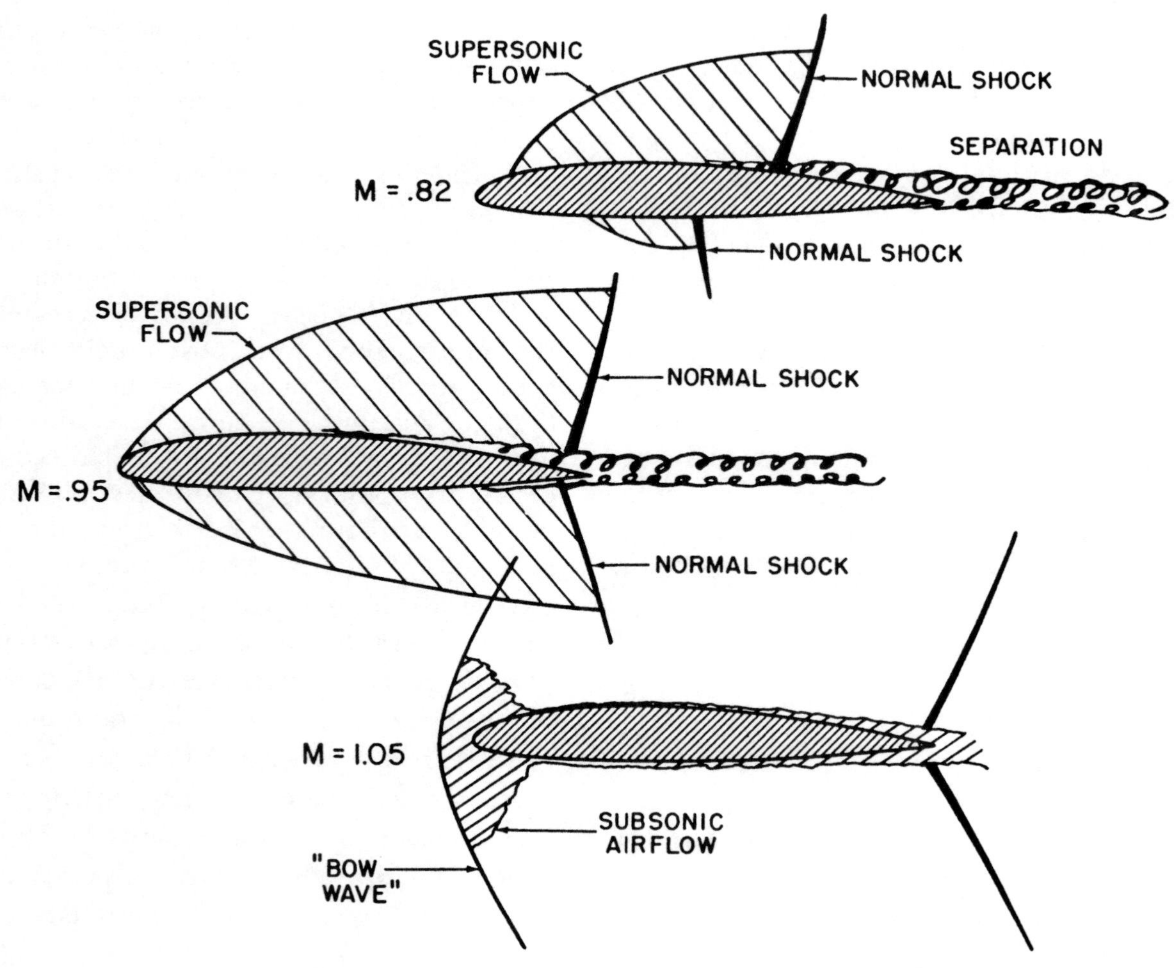

Figure 3.9. Transonic Flow Patterns (sheet 1 of 2)

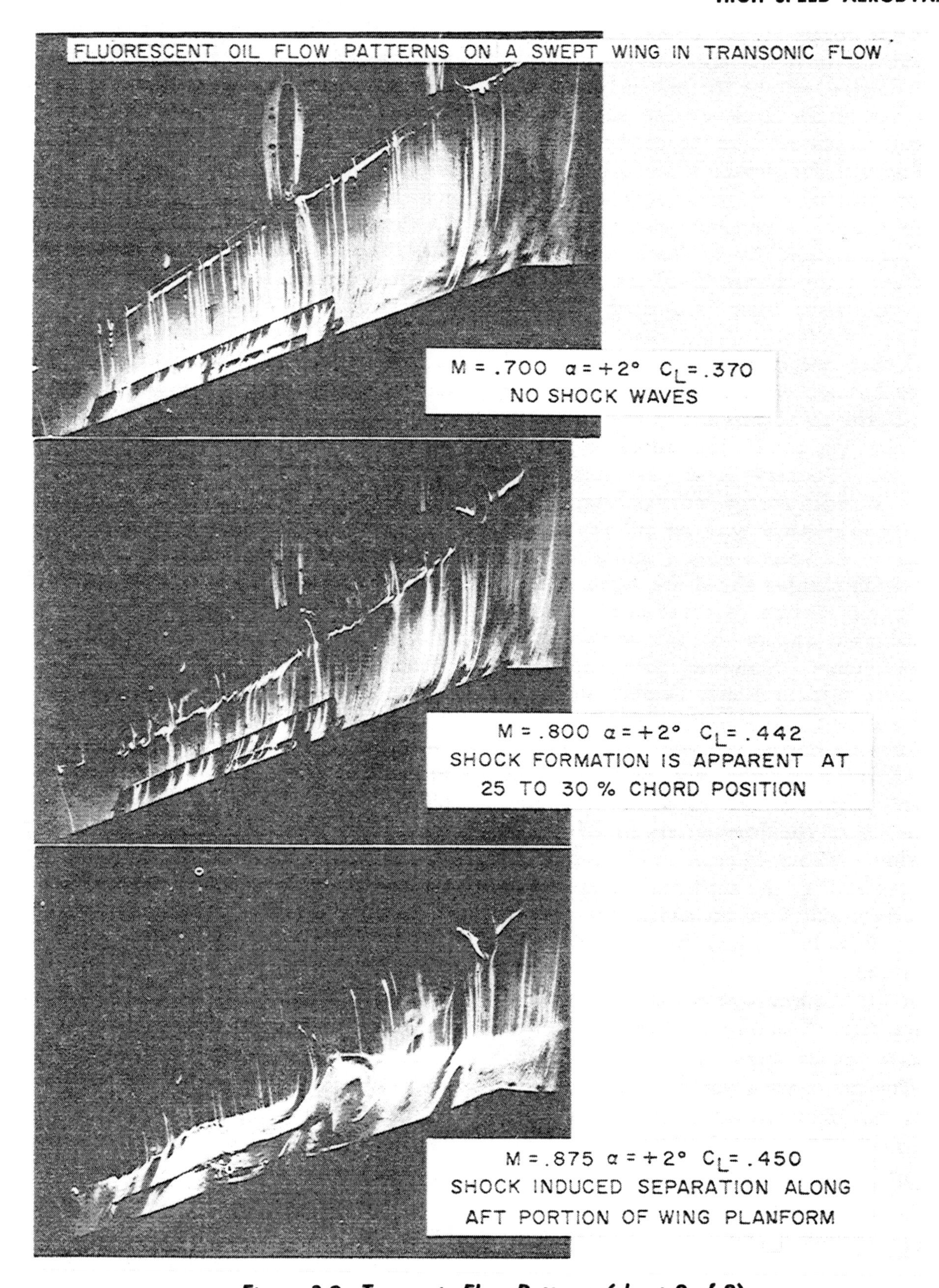

Figure 3.9. Transonic Flow Patterns (sheet 2 of 2)

shock wave forms as the boundary between the supersonic flow and the subsonic flow on the aft portion of the airfoil surface. The acceleration of the airflow from subsonic to supersonic is smooth and unaccompanied by shock waves if the surface is smooth and the transition gradual. However, the transition of airflow from supersonic to subsonic is always accompanied by a shock wave and, when there is no change in direction of the airflow, the wave form is a normal shock wave.

Recall that one of the principal effects of the normal shock wave is to produce a large increase in the static pressure of the airstream behind the wave. If the shock wave is strong, the boundary layer may not have sufficient kinetic energy to withstand the large, adverse pressure gradient and separation will occur. At speeds only slightly beyond critical Mach number the shock wave formed is not strong enough to cause spearation or any noticeable change in the aerodynamic force coefficients. However, an increase in speed above critical Mach number sufficient to form a strong shock wave can cause separation of the boundary layer and produce sudden changes in the aerodynamic force coefficients. Such a flow condition is shown in figure 3.9 by the flow pattern for $M=0.77$. Notice that a further increase in Mach number to 0.82 can enlarge the supersonic area on the upper surface and form an additional area of supersonic flow and normal shock wave on the lower surface.

As the flight speed approaches the speed of sound the areas of supersonic flow enlarge and the shock waves move nearer the trailing edge. The boundary layer may remain separated or may reattach depending much upon the airfoil shape and angle of attack. When the flight speed exceeds the speed of sound the "bow" wave forms at the leading edge and this typical flow pattern is illustrated in figure 3.9 by the drawing for $M=1.05$. If the speed is increased to some higher supersonic value all oblique portions of the waves incline more greatly and the detached normal shock portion of the bow wave moves closer to the leading edge.

Of course, all components of the aircraft are affected by compressibility in a manner somewhat similar to that of basic airfoil. The tail, fuselage, nacelles, canopy, etc. and the effect of the interference between the various surfaces of the aircraft must be considered.

FORCE DIVERGENCE. The airflow separation induced by shock wave formation can create significant variations in the aerodynamic force coefficients. When the free stream speed is greater than critical Mach number some typical effects on an airfoil section are as follows:

(1) An increase in the section drag coefficient for a given section lift coefficient.

(2) A decrease in section lift coefficient for a given section angle of attack.

(3) A change in section pitching moment coefficient.

A reference point is usually taken by a plot of drag coefficient versus Mach number for a constant lift coefficient. Such a graph is shown in figure 3.10. The Mach number which produces a sharp change in the drag coefficient is termed the "force divergence" Mach number and, for most airfoils, usually exceeds the critical Mach number at least 5 to 10 percent. This condition is also referred to as the "drag divergence" or "drag rise."

PHENOMENA OF TRANSONIC FLIGHT. Associated with the "drag rise" are buffet, trim and stability changes, and a decrease in control surface effectiveness. Conventional aileron, rudder, and elevator surfaces subjected to this high frequency buffet may "buzz," and changes in hinge moments may produce undesirable control forces. Of course, if the buffet is quite severe and prolonged, structural damage may occur if this operation is in violation of operating limitations. When airflow separation occurs on the wing due to

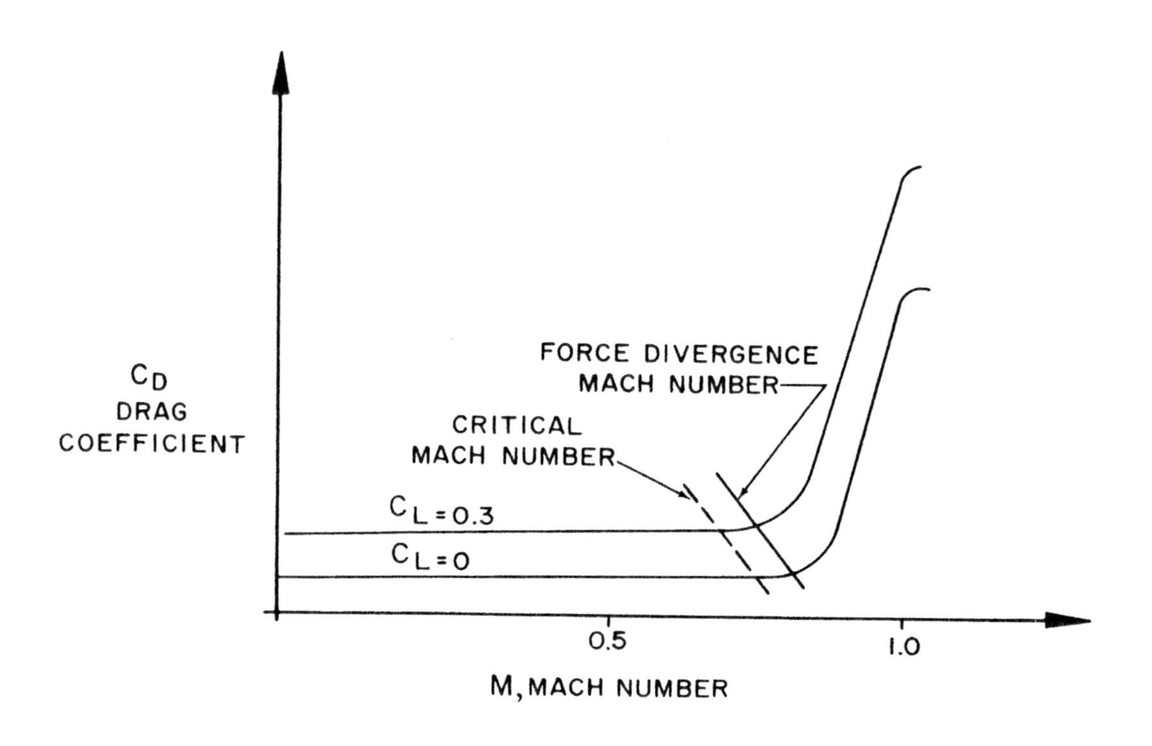

Figure 3.10. Compressibility Drag Rise

shock wave formation, there will be a loss of lift and subsequent loss of downwash aft of the affected area. If the wings shock unevenly due to physical shape differences or sideslip, a rolling moment will be created in the direction of the initial loss of lift and contribute to control difficulty ("wing drop"). If the shock induced separation occurs symmetrically near the wing root, a decrease in downwash behind this area is a corollary of the loss of lift. A decrease in downwash on the horizontal tail will create a diving moment and the aircraft will "tuck under." If these conditions occur on a swept wing planform, the wing center of pressure shift contributes to the trim change—root shock first moves the wing center of pressure aft and adds to the diving moment; shock formation at the wing tips first moves the center of pressure forward and the resulting climbing moment and tail downwash change can contribute to "pitch up."

Since most of the difficulties of transonic flight are associated with shock wave induced flow separation, any means of delaying or alleviating the shock induced separation will improve the aerodynamic characteristics. An aircraft configuration may utilize thin surfaces of low aspect ratio with sweepback to delay and reduce the magnitude of transonic force divergence. In addition, various methods of boundary layer control, high lift devices, vortex generators, etc., may be applied to improve transonic characteristics. For example, the application of vortex generators to a surface can produce higher local surface velocities and increase the kinetic energy of the boundary layer. Thus, a more severe pressure gradient (stronger shock wave) will be necessary to produce airflow separation.

Once the configuration of a transonic aircraft is fixed, the pilot must respect the effect of angle of attack and altitude. The local flow velocities on any upper surface increase with an increase in angle of attack. Hence, local sonic flow and subsequent shock wave formation can occur at lower free stream Mach numbers. A pilot must appreciate this reduction of force divergence Mach number with lift coefficient since maneuvers at high speed may produce compressibility effects which may not be encountered in unaccelerated flight. The effect of altitude is important since the *magnitude* of any force or moment change due to compressibility will depend upon the dynamic pressure of the airstream. Compressibility effects encountered at high altitude and low dynamic pressure may be of little consequence in the operation of a transonic aircraft. However, the same compressibility effects encountered at low altitudes and high dynamic pressures will create greater trim changes, heavier buffet, etc., and perhaps transonic flight restrictions which are of principal interest only to low altitude.

PHENOMENA OF SUPERSONIC FLIGHT. While many of the particular effects of supersonic flight will be presented in the detail of later discussion, many general effects may be anticipated. The airplane configuration must have aerodynamic shapes which will have low drag in compressible flow. Generally, this will require airfoil sections of low thickness ratio and sharp leading edges and body shapes of high fineness ratio to minimize the supersonic wave drag. Because of the aft movement of the aerodynamic center with supersonic flow, the increase in static longitudinal stability will demand effective, powerful control surfaces to achieve adequate controllability for supersonic maneuvering.

As a corollary of supersonic flight the shock wave formation on the airplane may create special problems outside the immediate vicinity of the airplane surfaces. While the shock waves a great distance away from the airplane can be quite weak, the pressure waves can be of sufficient magnitude to create an audible disturbance. Thus, "sonic booms" will be a simple consequence of supersonic flight.

The aircraft powerplants for supersonic flight must be of relatively high thrust output. Also, in many cases it may be necessary to provide the air breathing powerplant with special inlet configurations which will slow the airflow to subsonic prior to reaching the compressor face or combustion chamber. Aerodynamic heating of supersonic flight can provide critical inlet temperatures for the gas turbine engine as well as critical structural temperatures.

The density variations in airflow may be shown by certain optical techniques. Schlieren photographs and shadowgraphs can define the various wave patterns and their effect on the airflow. The Schlieren photographs presented in figure 3.11 define the flow conditions on an aircraft in supersonic flight.

TRANSONIC AND SUPERSONIC CONFIGURATIONS

Aircraft configurations developed for high speed flight will have significant differences in shape and planform when compared with aircraft designed for low speed flight. One of the outstanding differences will be in the selection of airfoil profiles for transonic or supersonic flight.

AIRFOIL SECTIONS. It should be obvious that airfoils for high speed subsonic flight should have high critical Mach numbers since critical Mach number defines the lower limit for shock wave formation and subsequent force divergence. An additional complication to airfoil selection in this speed range is that the airfoil should have a high maximum lift coefficient and sufficient thickness to allow application of high lift devices. Otherwise an excessive wing area would be required to provide maneuverability and reasonable takeoff and landing speeds.

F8U MODEL AT VARIOUS
MACH NUMBERS
$\alpha = 0°$ $\beta = 0°$

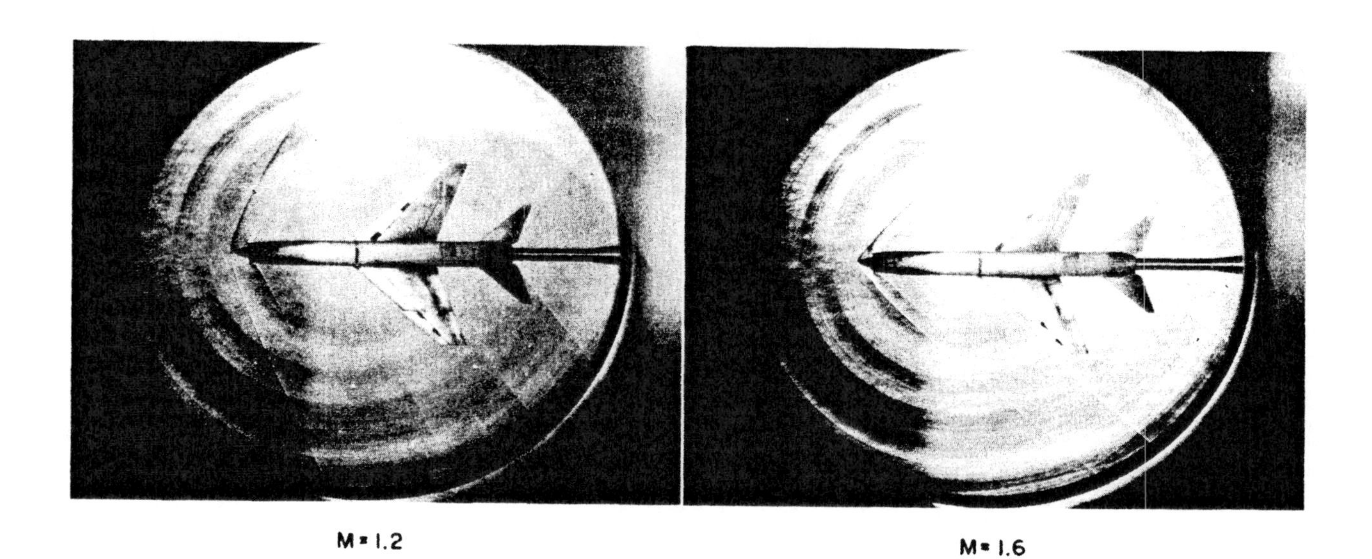

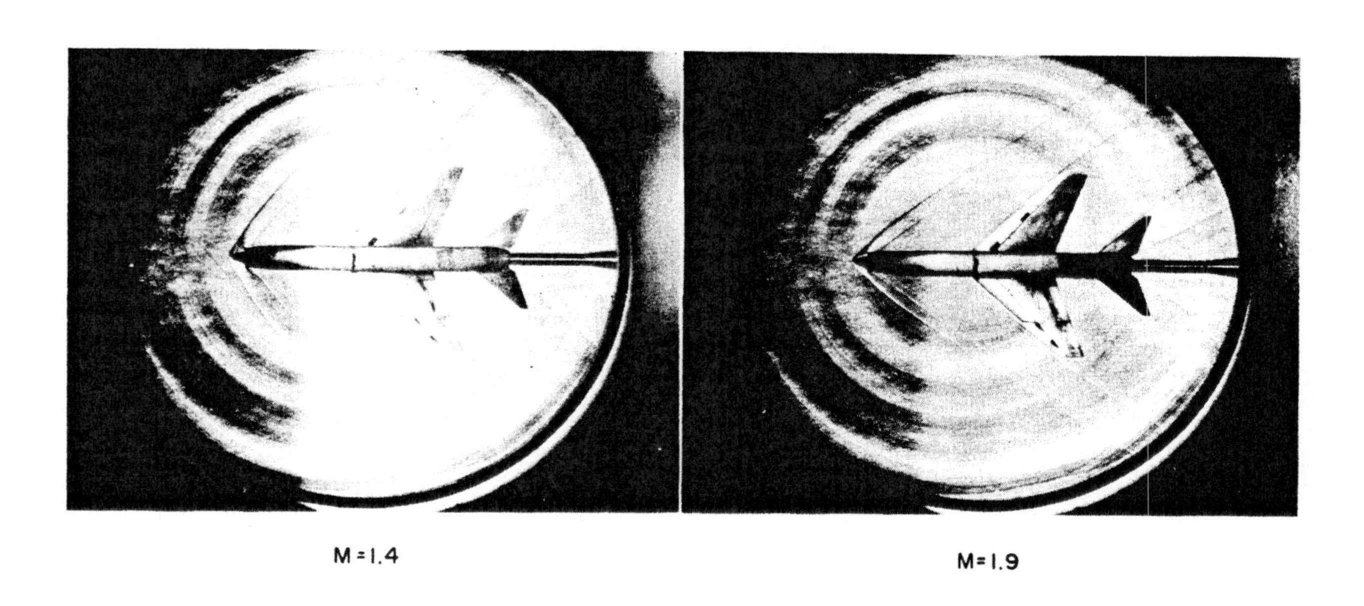

Figure 3.11. Schlieren Photographs of Supersonic Flight (sheet 1 of 2)

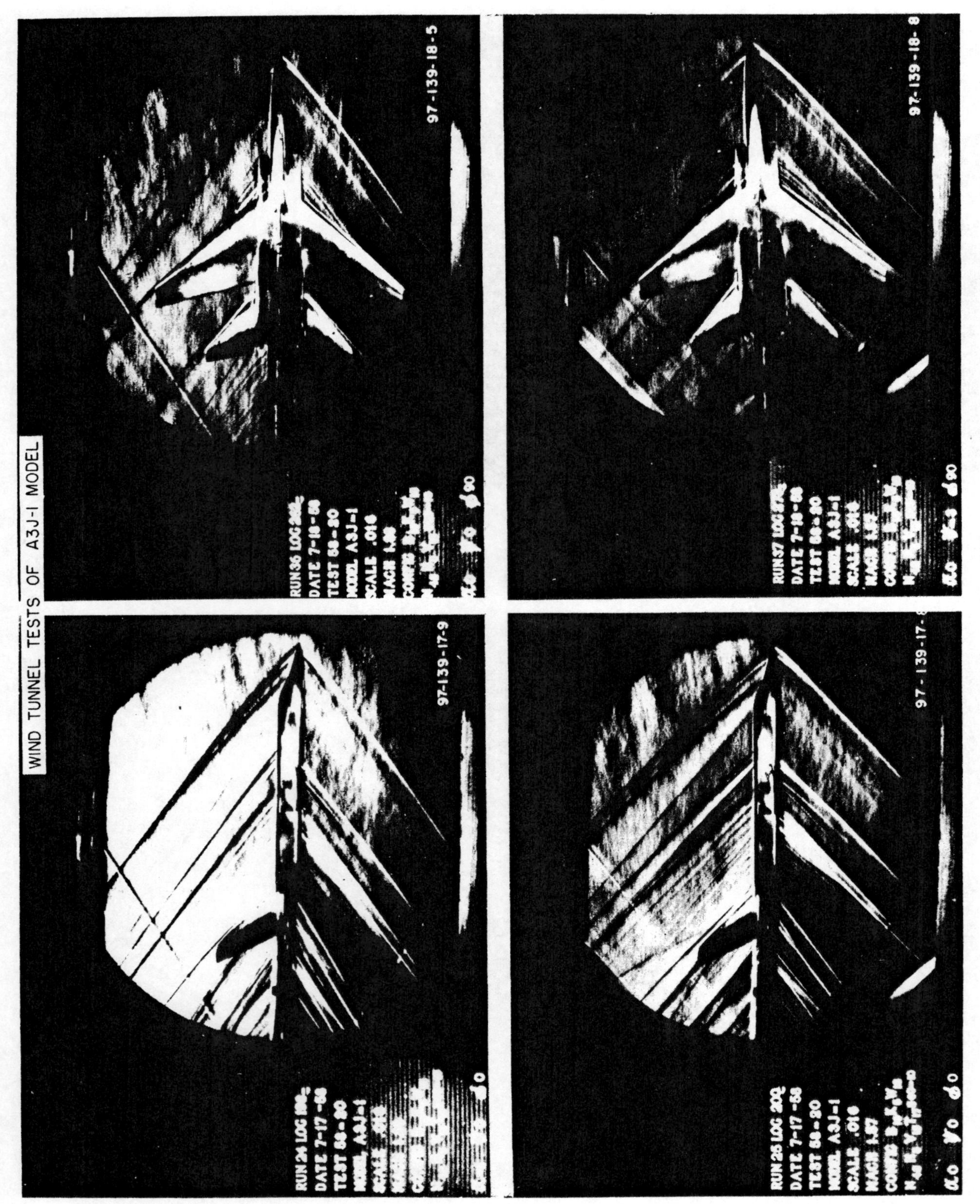

Figure 3.11. Schlieren Photographs of Supersonic Flight (sheet 2 of 2)

However, if high speed flight is the primary consideration, the airfoil must be chosen to have the highest practical critical Mach number.

Critical Mach number has been defined as the flight Mach number which produces first evidence of local sonic flow. Thus, the airfoil shape and lift coefficient—which determine the pressure and velocity distribution—will have a profound effect on critical Mach number. Conventional, low speed airfoil shapes have relatively poor compressibility characteristics because of the high local velocities near the leading edge. These high local velocities are inevitable if both the maximum thickness and camber are well forward on the chord. An improvement of the compressibility characteristics can be obtained by moving the points of maximum camber and thickness aft on the chord. This would distribute the pressure and velocity more evenly along the chord and produce a lower peak velocity for the same lift coefficient. Fortunately, the airfoil shape to provide extensive laminar flow and low profile drag in low speed, subsonic flight will provide a pressure distribution which is favorable for high speed flight. Figure 3.12 illustrates the pressure distributions and variation of critical Mach number with lift coefficient for a conventional low speed airfoil and a high speed section.

In order to obtain a high critical Mach number from an airfoil at some low lift coefficient the section must have:

(*a*) Low thickness ratio. The point of maximum thickness should be aft to smooth the pressure distribution.

(*b*) Low camber. The mean camber line should be shaped to help minimize the local velocity peaks.

In addition, the higher the required lift coefficient the lower the critical Mach number and more camber is required of the airfoil. If supersonic flight is a possibility the thickness ratio and leading edge radius must be small to decrease wave drag.

Figure 3.13 shows the flow patterns for two basic supersonic airfoil sections and provides the approximate equations for lift,drag, and lift curve slope. Since the wave drag is the only factor of difference between the two airfoil sections, notice the configuration factors which affect the wave drag. For the same thickness ratio, the circular arc airfoil would have a larger wedge angle formed between the upper and lower surfaces at the leading edge. At the same flight Mach number the larger angle at the leading edge would form the stronger shock wave at the nose and cause a greater pressure change on the circular arc airfoil. This same principle applies when investigating the effect of airfoil thickness. Notice that the wave drag coefficients for both airfoils vary as the SQUARE of the thickness ratio, e.g., if the thickness ratio were doubled, the wave drag coefficient would be four times as great. If the thickness were increased, the airflow at the leading edge will experience a greater change in direction and a stronger shock wave will be formed. This powerful variation of wave drag with thickness ratio necessitates the use of very thin airfoils with sharp leading edges for supersonic flight. An additional consideration is that thin airfoil sections favor the use of low aspect ratios and high taper to obtain lightweight structures and preserve stiffness and rigidity.

The parameter $\sqrt{M^2-1}$ appears in the denominator of each of the equations for the aerodynamic coefficients and indicates a decrease in each of these coefficients with an increase in Mach number. Essentially, this means that any aerodynamic surface becomes less sensitive to changes in angle of attack at higher Mach numbers. The decrease in lift curve slope with Mach number has tremendous implications in the stability and control of high speed aircraft. The vertical tail becomes less sensitive to angles of sideslip and the directional stability of the aircraft will deteriorate with Mach number. The horizontal tail of the airplane experiences the same

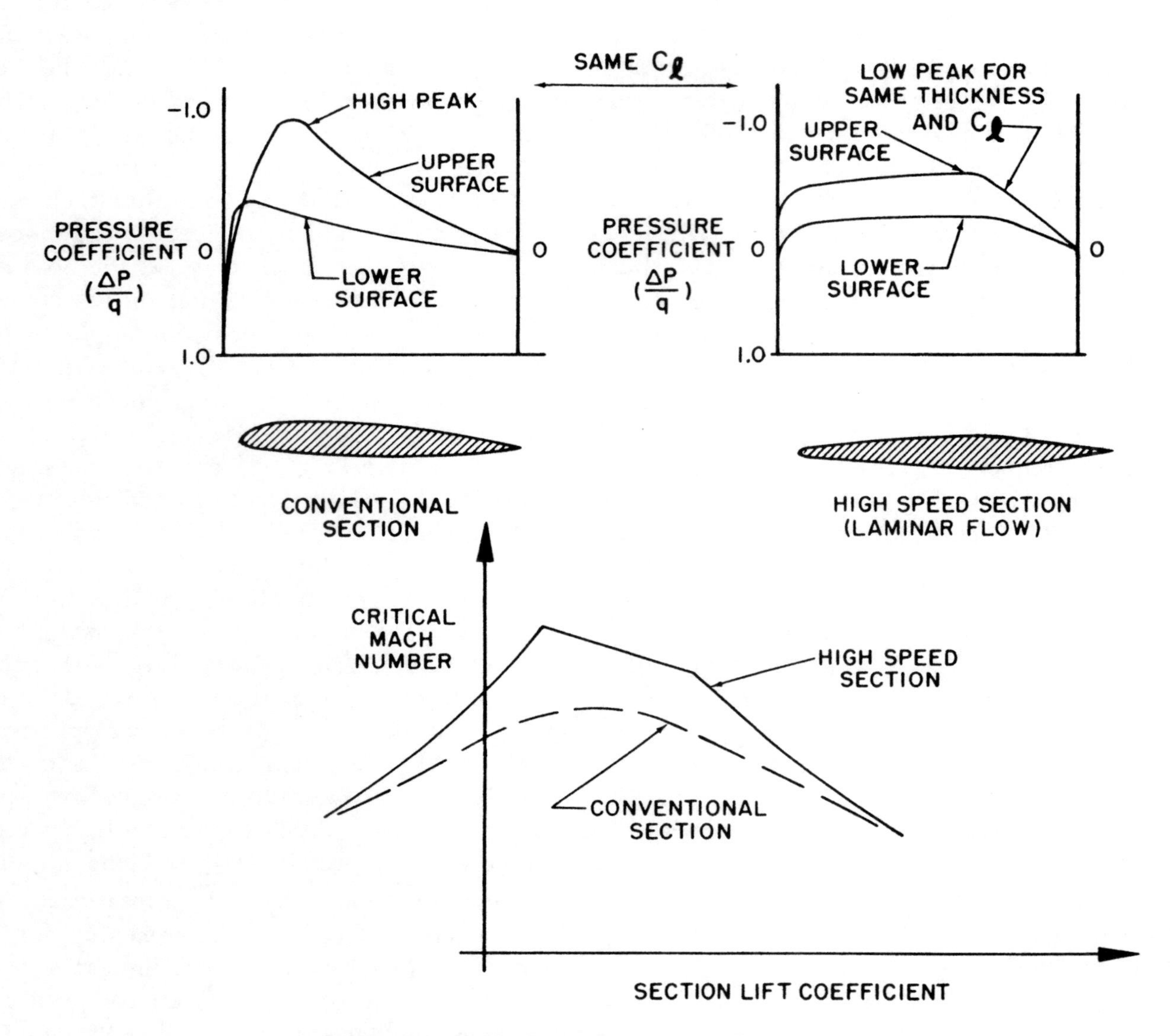

Figure 3.12. High Speed Section Characteristics

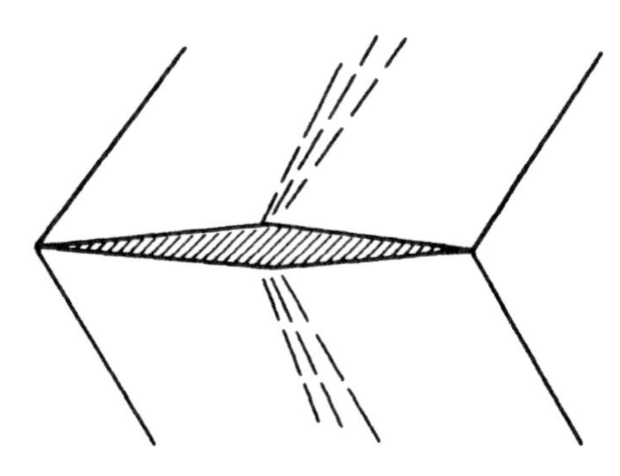

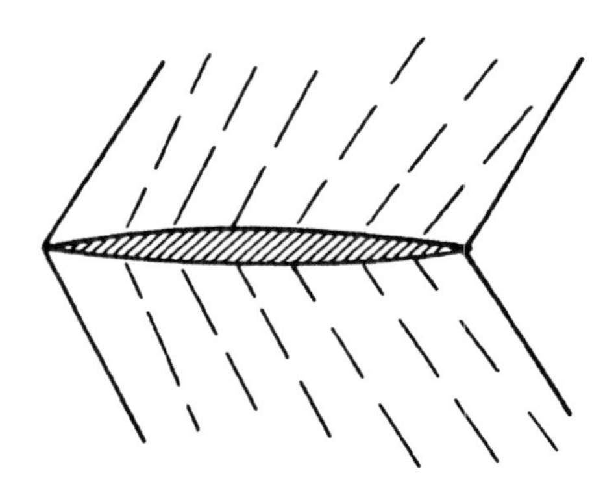

WAVE DRAG COEFFICIENT:

$$C_D = \frac{4\,(t/c)^2}{\sqrt{M^2-1}} \qquad\qquad C_D = \frac{5.33\,(t/c)^2}{\sqrt{M^2-1}}$$

LIFT COEFFICIENT:

$$C_L = \frac{4\alpha}{\sqrt{M^2-1}} \qquad\qquad C_L = \frac{4\alpha}{\sqrt{M^2-1}}$$

DRAG DUE TO LIFT:

$$C_D = \frac{4\alpha^2}{\sqrt{M^2-1}} \qquad\qquad C_D = \frac{4\alpha^2}{\sqrt{M^2-1}}$$

LIFT CURVE SLOPE:

$$C_{L_\alpha} = \frac{4}{\sqrt{M^2-1}} \qquad\qquad C_{L_\alpha} = \frac{4}{\sqrt{M^2-1}}$$

WHERE

(t/c) = AIRFOIL THICKNESS RATIO

α = ANGLE OF ATTACK (IN RADIANS)

M = MACH NUMBER

Figure 3.13. Approximate Equations for Supersonic Section Characteristics

general effect and contributes less damping to longitudinal pitching oscillations. These effects can become so significant at high Mach numbers that the aircraft might require complete synthetic stabilization.

PLANFORM EFFECTS. The development of surfaces for high speed involves consideration of many items in addition to the airfoil sections. Taper, aspect ratio, and sweepback can produce major effects on the aerodynamic characteristics of a surface in high speed flight. Sweepback produces an unusual effect on the high speed characteristics of a surface and has basis in a very fundamental concept of aerodynamics. A grossly simplified method of visualizing the effect of sweepback is shown in figure 3.14. The swept wing shown has the streamwise velocity broken down to a component of velocity perpendicular to the leading edge and a component parallel to the leading edge. The component of speed perpendicular to the leading edge is less than the free stream speed (by the cosine of the sweep angle) and it is this velocity component which determines the magnitude of the pressure distribution.

The component of speed parallel to the leading edge could be visualized as moving across constant sections and, in doing so, does not contribute to the pressure distribution on the swept wing. Hence, sweep of a surface produces a beneficial effect in high speed flight since higher flight speeds may be obtained before components of speed perpendicular to the leading edge produce critical conditions on the wing. This is one of the most important advantage of sweep since there is an increase in critical Mach number, force divergence Mach number, and the Mach number at which the drag rise will peak. In other words, sweep will *delay* the onset of compressibility effects.

Generally, the effect of wing sweep will apply to either sweep back or sweep forward. While the swept forward wing has been used in rare instances, the aeroelastic instability of such a wing creates such a problem that sweep back is more practical for ordinary applications.

In addition to the *delay* of the onset of compressibility effects, sweepback will *reduce* the magnitude of the changes in force coefficients due to compressibility. Since the component of velocity perpendicular to the leading edge is less than the free stream velocity, the magnitude of all pressure forces on the wing will be reduced (approximately by the square of the cosine of the sweep angle). Since compressibility force divergence occurs due to changes in pressure distribution, the use of sweepback will "soften" the force divergence. This effect is illustrated by the graph of figure 3.14 which shows the typical variation of drag coefficient with Mach number for various sweepback angles. The straight wing shown begins drag rise at $M=0.70$, reaches a peak near $M=1.0$, and begins a continual drop past $M=1.0$. Note that the use of sweepback then *delays* the drag rise to some higher Mach number and *reduces* the magnitude of the drag rise.

In view of the preceding discussion, sweepback will have the following principal advantages:

(1) Sweepback will *delay* the onset of all compressibility effects. Critical Mach number and force divergence Mach number will increase since the velocity component affecting the pressure distribution is less than the free stream velocity. Also, the peak of drag rise is delayed to some higher supersonic speed—approximately the speed which produces sonic flow perpendicular to the leading edge. Various sweeps applied to wings of moderate aspect ratio will produce these *approximate* effects in transonic flight:

Sweep angle (Λ)	Percent increase in critical Mach number	Percent increase in drag peak Mach number
0°	0	0
15°	2	4
30°	8	15
45°	20	41
60°	41	100

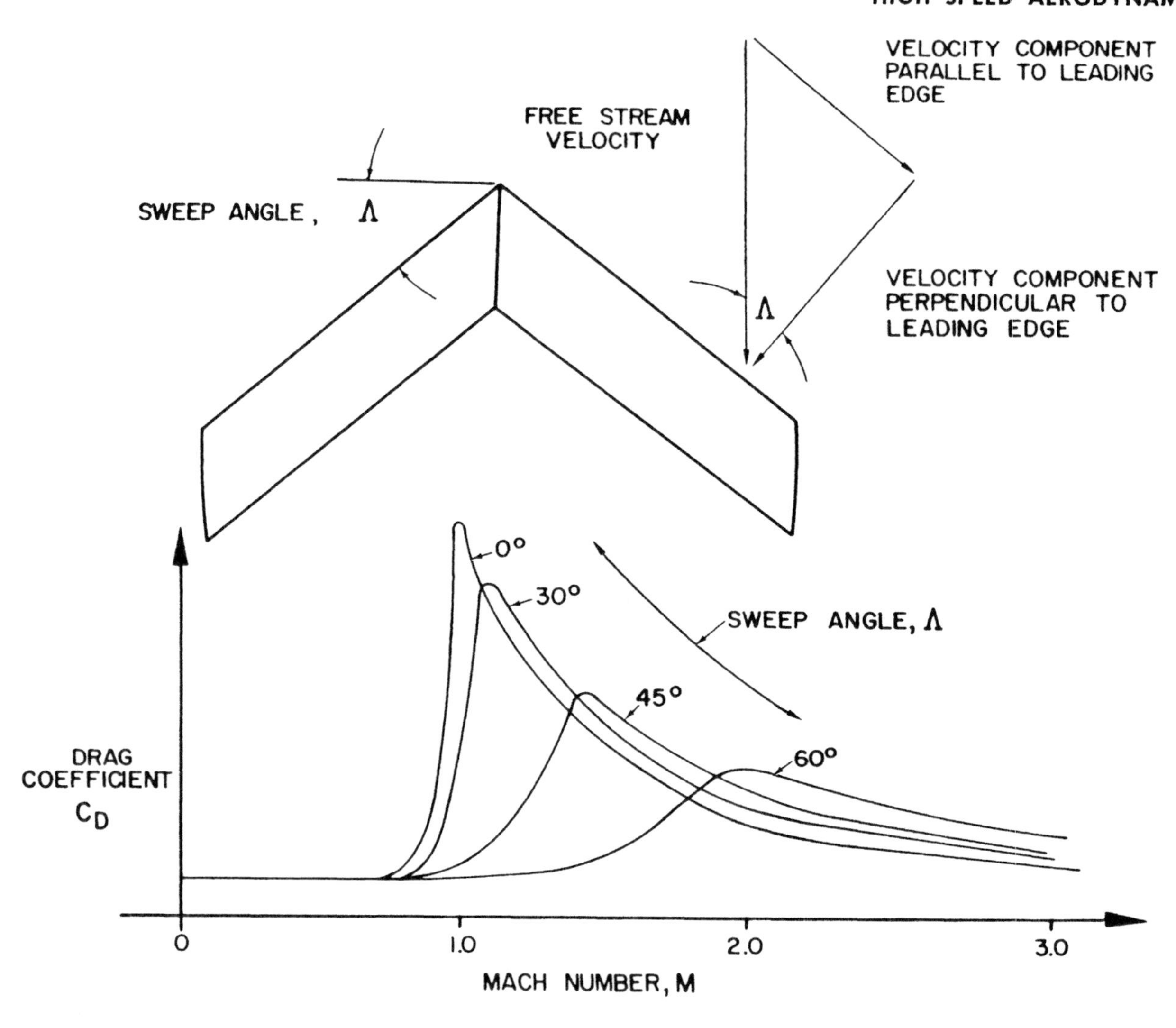

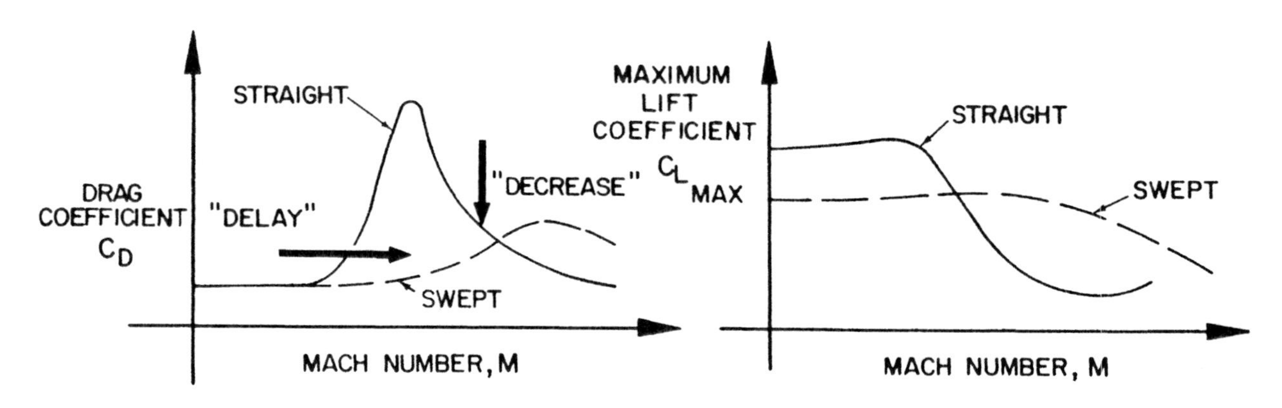

Figure 3.14. General Effects of Sweepback

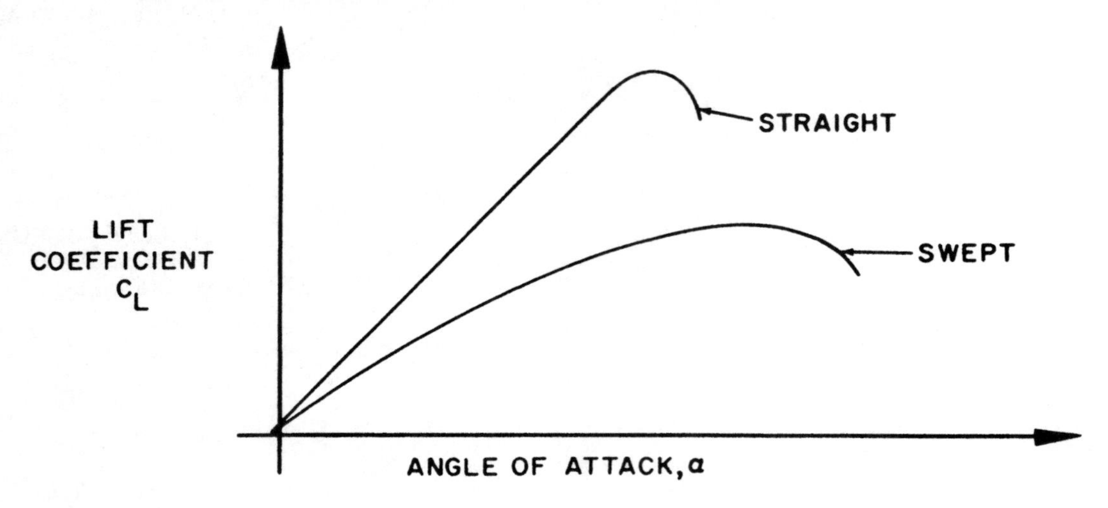

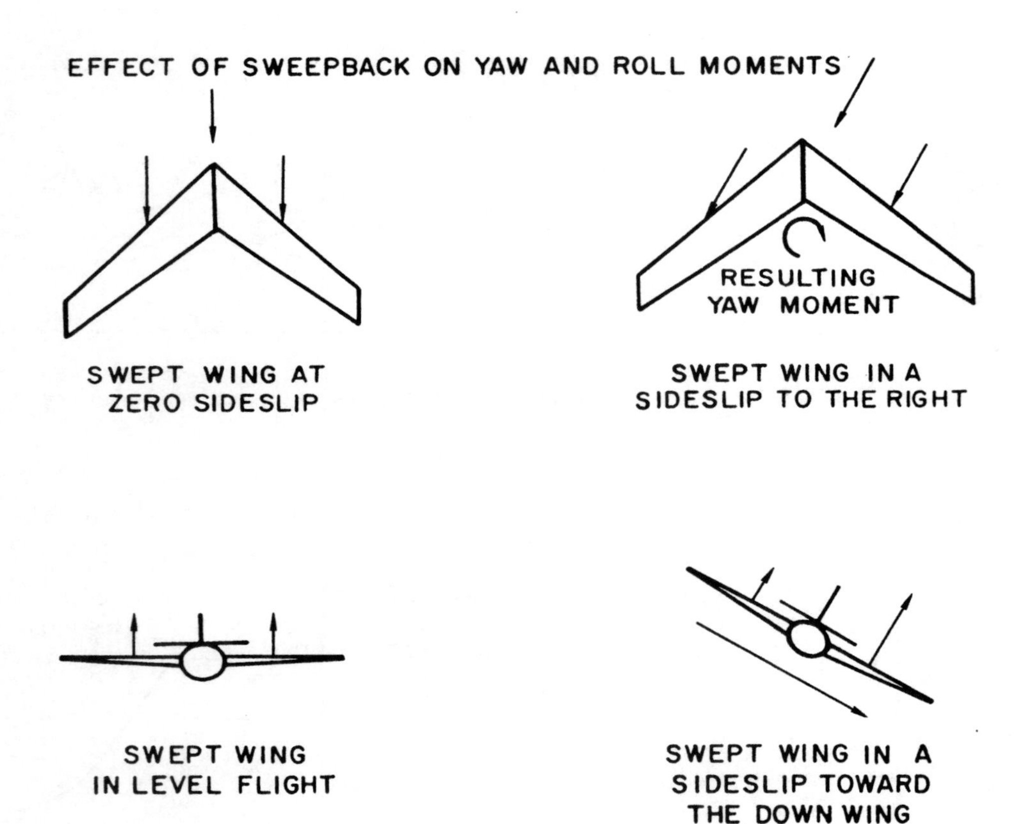

Figure 3.15. Aerodynamic Effects Due to Sweepback

(2) Sweepback will reduce the magnitude of change in the aerodynamic force coefficients due to compressibility. Any change in drag, lift, or moment coefficients will be reduced by the use of sweepback. Various sweep angles applied to wings of moderate aspect ratio will produce these *approximate* effects in transonic flight.

Sweep angle (Λ)	Percent reduction in drag rise	Percent reduction in *loss* of $C_{l_{max}}$
0°	0	0
15°	5	3
30°	15	13
45°	35	30
60°	60	50

These advantages of drag reduction and preservation of the transonic maximum lift coefficient are illustrated in figure 3.14.

Thus, the use of sweepback on a transonic aircraft will reduce and delay the drag rise and preserve the maneuverability of the aircraft in transonic flight. It should be noted that a small amount of sweepback produces very little benefit. If sweepback is to be used at all, at least 30° to 35° must be used to produce any significant benefit. Also note from figure 3.14 that the amount of sweepback required to *delay* drag rise in supersonic flight is very large, e.g., more than 60° necessary at $M=2.0$. By comparison of the drag curves at high Mach numbers it will be appreciated that extremely high (and possibly impractical) sweepback is necessary to delay drag rise and that the lowest drag is obtained with zero sweepback. Therefore, the planform of a wing designed to operate continuously at high Mach numbers will tend to be very thin, low aspect ratio, and unswept. An immediate conclusion is that sweepback is a device of greatest application in the regime of transonic flight.

A few of the less significant advantages of sweepback are as follows:

(1) The wing lift curve slope is reduced for a given aspect ratio. This is illustrated by the lift curve comparison of figure 3.15 for the straight and swept wing. Any reduction of lift curve slope implies the wing is less sensitive to changes in angle of attack. This is a beneficial effect only when the effect of gusts and turbulence is considered. Since the swept wing has the lower lift curve slope it will be less sensitive to gusts and experience less "bump" due to gust for a given aspect ratio and wing loading. This is a consideration particular to the aircraft whose structural design shows a predominating effect of the gust load spectrum, e.g., transport, cargo, and patrol types.

(2) "Divergence" of a surface is an aeroelastic problem which can occur at high dynamic pressures. Combined bending and twisting deflections interact with aerodynamic forces to produce sudden failure of the surface at high speeds. Sweep forward will aggravate this situation by "leading" the wing into the windstream and tends to lower the divergence speed. On the other hand, sweepback tends to stabilize the surface by "trailing" and tends to raise the divergence speed. By this tendency, sweepback may be beneficial in preventing divergence within the anticipated speed range.

(3) Sweepback contributes slightly to the static directional—or weathercock—stability of an aircraft. This effect may be appreciated by inspection of figure 3.15 which shows the swept wing in a yaw or sideslip. The wing into the wind has less sweep and a slight increase in drag; the wing away from the wind has more sweep and less drag. The net effect of these force changes is to produce a yawing moment tending to return the nose into the relative wind. This directional stability contribution is usually small and of importance in tailless aircraft only.

1
2
3
4

(4) Sweepback contributes to lateral stability in the same sense as dihedral. When the swept wing aircraft is placed in a sideslip, the wing into the wind experiences an increase in lift since the sweep is less and the wing away from the wind produces less lift since the sweep is greater. As shown in figure 3.15, the swept wing aircraft in a sideslip experiences lift changes and a subsequent rolling moment which tends to right the aircraft. This lateral stability contribution depends on the sweepback and the lift coefficient of the wing. A highly swept wing operating at high lift coefficient usually experiences such an excess of this lateral stability contribution that adequate controllability may be a significant problem.

As shown, the swept wing has certain important advantages. However, the use of sweepback produces certain inevitable disadvantages which are important from the standpoint of both airplane design and flight operations. The most important of these disadvantages are as follows:

(1) When sweepback is combined with taper there is an extremely powerful tendency for the wing to stall tip first. This pattern of stall is very undesirable since there would be little stall warning, a serious reduction in lateral control effectiveness, and the forward shift of the center of pressure would contribute to a nose up moment ("pitch up" or "stick force lightening"). Taper has its own effect of producing higher local lift coefficients toward the tip and one of the effects of sweepback is very similar. All outboard wing sections are affected by the upwash of the preceding inboard sections and the lift distribution resulting from sweepback alone is similar to that of high taper.

An additional effect is the tendency to develop a strong spanwise flow of the boundary layer toward the tip when the wing is at high lift coefficients. This spanwise flow produces a relatively low energy boundary layer near the tip which can be easily separated. The combined effect of taper and sweep present a considerable problem of tip stall and this is illustrated by the flow patterns of figure 3.16. Design for high speed performance may dictate high sweepback, while structural efficiency may demand a highly tapered planform. When such is the case, the wing may require extensive aerodynamic tailoring to provide a suitable stall pattern and a lift distribution at cruise condition which reduces drag due to lift. Washout of the tip, variation of section camber throughout span, flow fences, slats, leading edge extension, etc., are typical devices used to modify the stall pattern and minimize drag due to lift at cruise condition.

(2) As shown by the lift curve of figure 3.15 the use of sweepback will reduce the lift curve slope and the subsonic maximum lift coefficient. It is important to note this case is definitely subsonic since sweepback may be used to improve the transonic maneuvering capability. Various sweep angles applied to wings of moderate aspect ratio produce these *approximate* effects on the subsonic lift characteristics:

Sweep Angle (Λ):	*Percent reduction of subsonic maximum lift coefficient and lift curve slope*
0°	0
15°	4
30°	14
45°	30
60°	50

The reduction of the low speed maximum lift coefficient (which is in addition to that lost due to tip stall) has very important implications in design. If wing loading is not reduced, stall speeds increase and subsonic maneuverability decreases. On the other hand, if wing loading is reduced, the increase in wing surface area may reduce the anticipated benefit of sweepback in the transonic flight regime. Since the requirements of performance predominate, certain increases of stall speeds, takeoff speeds,

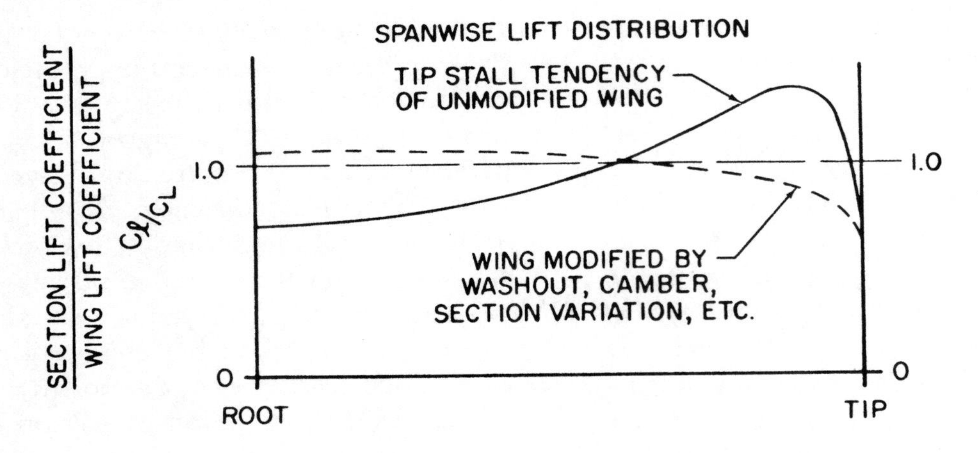

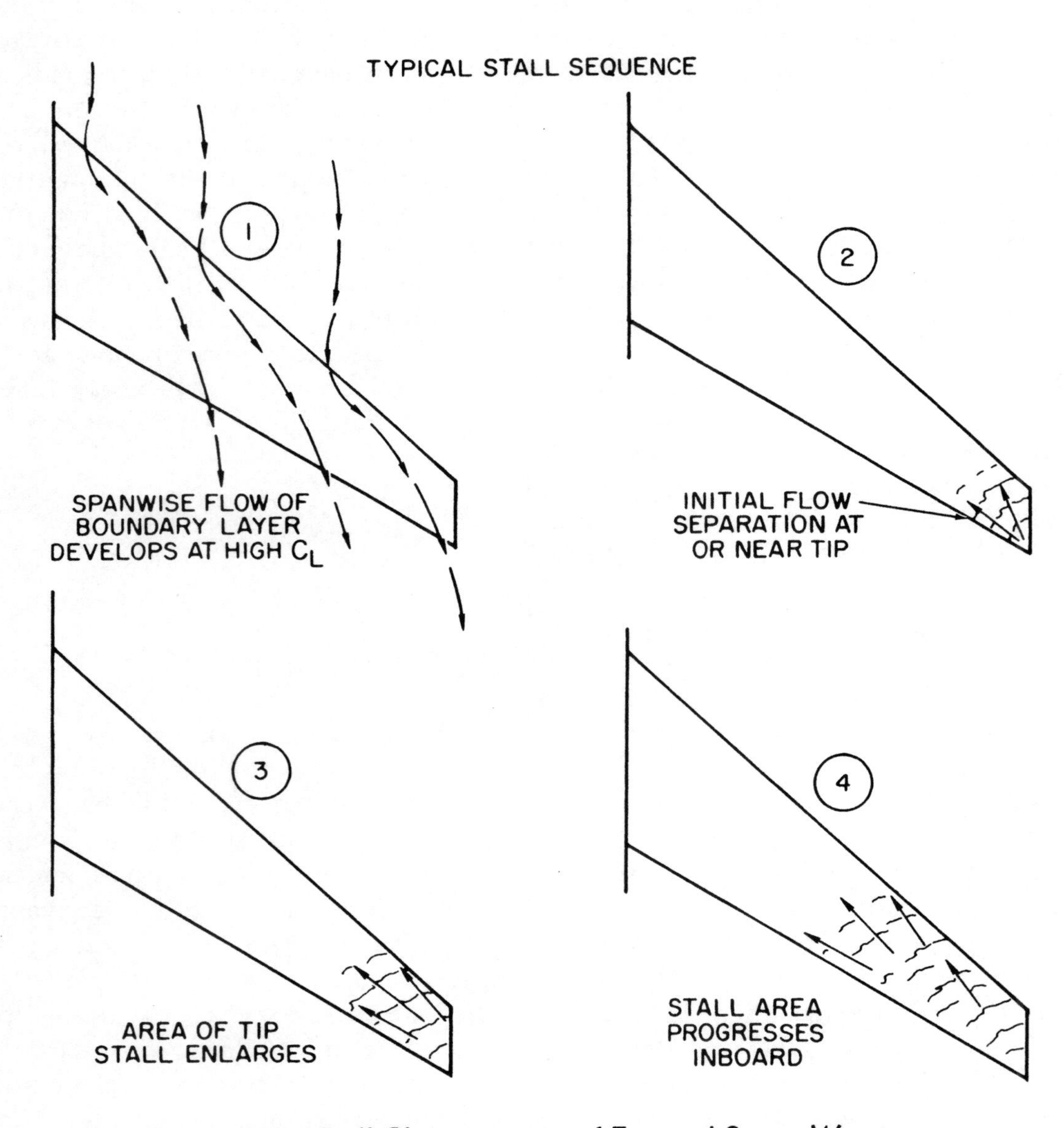

Figure 3.16. Stall Characteristics of Tapered Swept Wing

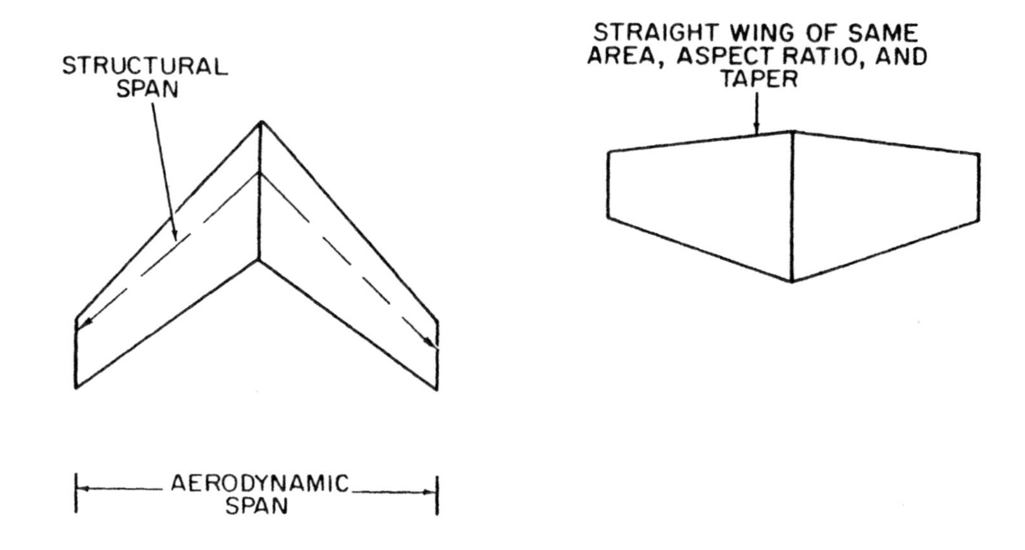

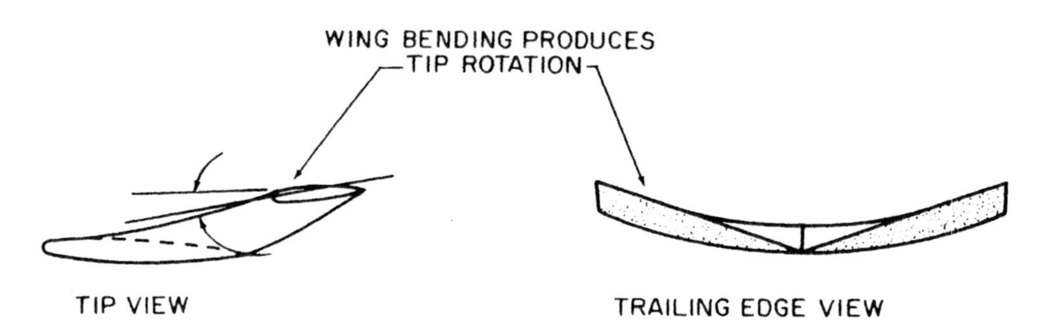

Figure 3.17. Structural Complications Due to Sweepback

and landing speeds usually will be accepted. While the reduction of lift curve slope may be an advantage for gust considerations, the reduced sensitivity to changes in angle of attack has certain undesirable effects in subsonic flight. The reduced wing lift curve slope tends to increase maximum lift angles of attack and complicate the problem of landing gear design and cockpit visibility. Also, the lower lift curve slope would reduce the contribution to stability of a given tail surface area.

(3) The use of sweepback will reduce the effectiveness of trailing edge control surfaces and high lift devices. A typical example of this effect is the application of a single slotted flap over the inboard 60 percent span to both a straight wing and a wing with 35° sweepback. The flap applied to the straight wing produces an increase in maximum lift coefficient of approximately 50 percent. The same type flap applied to the swept wing produces an increase in maximum lift coefficient of approximately 20 percent. To produce some reasonable maximum lift coefficient on a swept wing may require unsweeping the flap hinge line, application of leading edge high lift devices such as slots or slats, and possibly boundary layer control.

(4) As described previously, sweepback contributes to lateral stability by producing stable rolling moments with sideslip. The lateral stability contribution of sweepback varies with the amount of wing sweepback and wing lift coefficient—large sweepback and high lift coefficients producing large contribution to lateral stability. While stability is desirable, any excess of stability will reduce controllability. For the majority of airplane configurations, high lateral stability is neither necessary nor desirable, but adequate control in roll is absolutely necessary for good flying qualities. An excess of lateral stability from sweepback can aggravate "Dutch roll" problems and produce marginal control during crosswind takeoff and landing where the aircraft must move in a controlled sideslip. Therefore, it is not unusual to find swept wing aircraft with negative dihedral and lateral control devices designed principally to meet cross wind takeoff and landing requirements.

(5) The structural complexity and aeroelastic problems created by sweepback are of great importance. First, there is the effect shown in figure 3.17 that swept wing has a greater structural span than a straight wing of the same area and aspect ratio. This effect increases wing structural weight since greater bending and shear material must be distributed in the wing to produce the same design strength. An additional problem is created near the wing root and "carry-through" structure due to the large twisting loads and the tendency of the bending stress distribution to concentrate toward the trailing edge. Also shown in figure 3.17 is the influence of wing deflection on the spanwise lift distribution. Wing bending produces tip rotation which tends to unload the tip and move the center of pressure forward. Thus, the same effect which tends to allay divergence can make an undesirable contribution to longitudinal stability.

EFFECT OF ASPECT RATIO AND TIP SHAPE. In addition to wing sweep, planform properties such as aspect ratio, and tip shape, can produce significant effects on the aerodynamic characteristics at high speeds. There is no particular effect of aspect ratio on critical Mach number at high or medium aspect ratios. The aspect ratio must be less than four or five to produce any apparent change in critical Mach number. This effect is shown for a typical 9 percent thick symmetrical airfoil in the graph of figure 3.18. Note that very low aspect ratios are required to cause a significant increase in critical Mach number. Very low aspect ratios create the extremes of three dimensional flow and subsequent increase in free stream speed to create

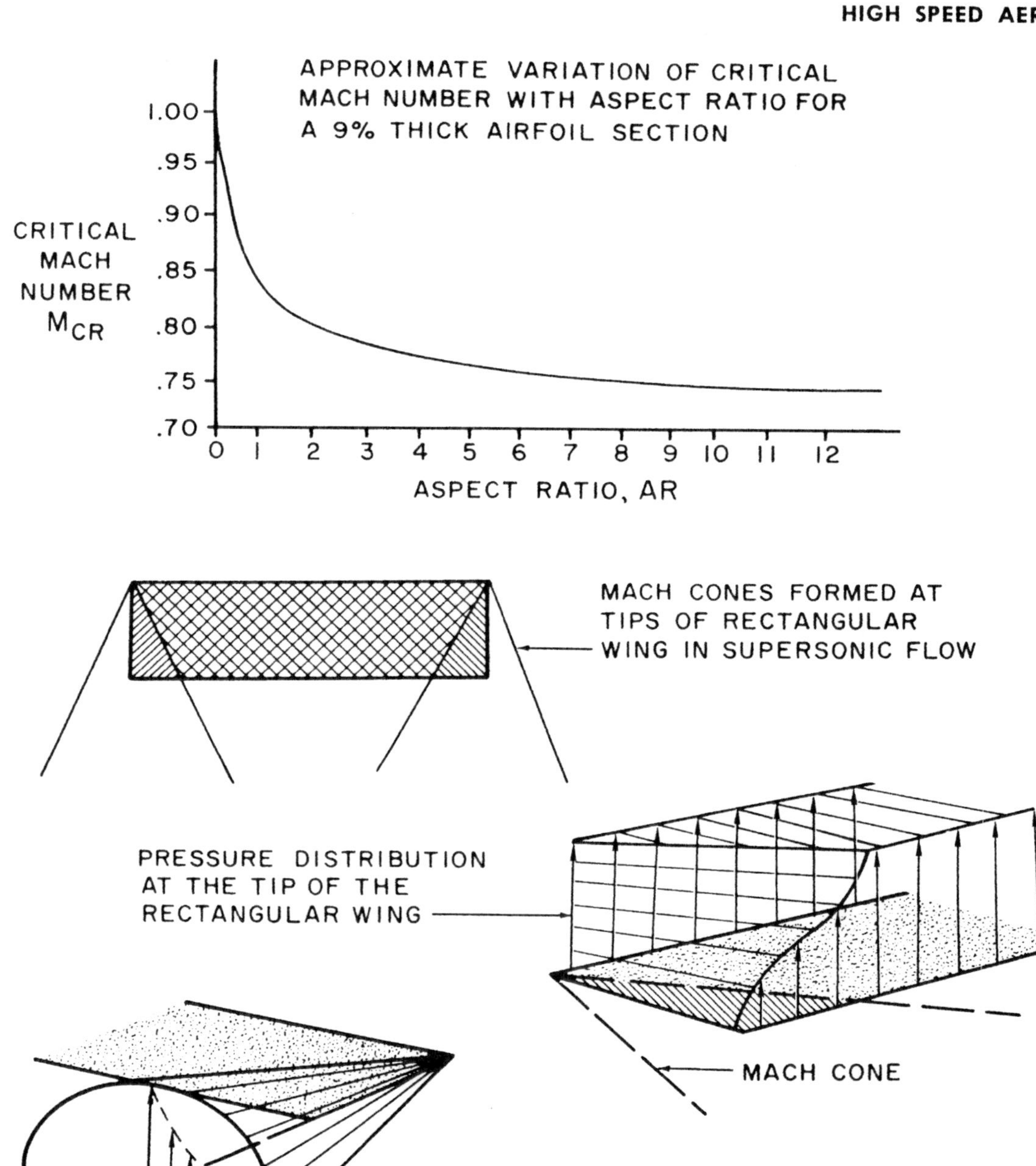

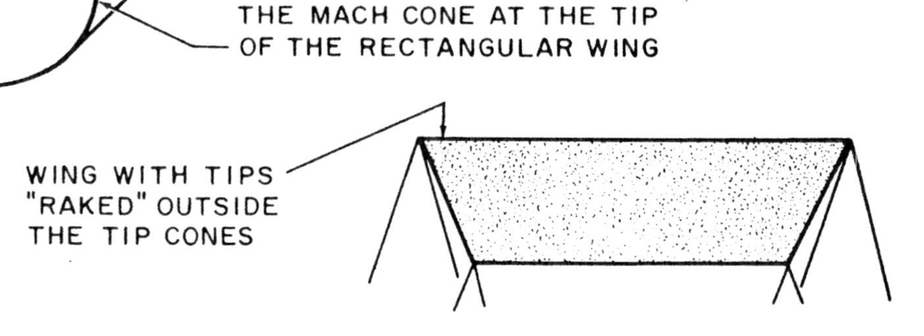

Figure 3.18. General Planform Effects

local sonic flow. Actually, the extremely low aspect ratios required to produce high critical Mach number are not too practical. Generally, the advantage of low aspect ratio must be combined with sweepback and high speed airfoil sections.

The thin rectangular wing in supersonic flow illustrates several important facts. As shown in figure 3.18, Mach cones form at the tips of the rectangular wing and affect the pressure distribution on the area within the cone. The vortex develops within the tip cone due to the pressure differential and the resulting average pressure on the area within the cone is approximately one-half the pressure between the cones. Three-dimensional flow on the wing is then confined to the area within the tip cones, while the area between the cones experiences pure two-dimensional flow.

It is important to realize that the three-dimensional flow on the rectangular wing in supersonic flight differs greatly from that of subsonic flight. A wing of finite aspect ratio in subsonic flight experiences a three-dimensional flow which includes the tip vortices, downwash behind the wing, upwash ahead of the wing, and local induced velocities along the span. Recall that the local induced velocities along the span of the wing would incline the section lift aft relative to the free stream and result in "induced drag." Such a flow condition cannot be directly correlated with the wing in supersonic flow. The flow pattern for the rectangular wing of figure 3.18 demonstrates that the three-dimensional flow is confined to the tip, and pure two-dimensional flow exists on the wing area between the tip cones. If the wing tips were to be "raked" outside the tip cones, the entire wing flow would correspond to the two-dimensional (or section) conditions.

Therefore, for the wing in supersonic flow, no upwash exists ahead of the wing, three-dimensional effects are confined to the tip cones, and no local induced velocities occur along the span between the tip cones. The supersonic drag due to lift is a function of the section and angle of attack while the subsonic induced drag is a function of lift coefficient and aspect ratio. This comparison makes it obvious that supersonic flight does not demand the use of high aspect ratio planforms typical of low speed aircraft. In fact, low aspect ratios and high taper are favorable from the standpoint of structural considerations if very thin sections are used to minimize wave drag.

If sweepback is applied to the supersonic wing, the pressure distribution will be affected by the location of the Mach cone with respect to the leading edge. Figure 3.19 illustrates the pressure distribution for the delta wing planform in supersonic flight with the leading edge behind or ahead of the Mach cone. When the leading edge is behind the Mach cone the components of velocity perpendicular to the leading edge are still subsonic even though the free stream flow is supersonic and the resulting pressure distribution will greatly resemble the subsonic pressure distribution for such a planform. Tailoring the leading edge shape and camber can minimize the components of the high leading edge suction pressure which are inclined in the drag direction and the drag due to lift can be reduced. If the leading edge is ahead of the Mach cone, the flow over this area will correspond to the two-dimensional supersonic flow and produce constant pressure for that portion of the surface between the leading edge and the Mach cone.

CONTROL SURFACES. The design of control surfaces for transonic and supersonic flight involves many important considerations. This fact is illustrated by the typical transonic and supersonic flow patterns of figure 3.19. Trailing edge control surfaces can be affected adversely by the shock waves formed in flight above critical Mach number. If the airflow is separated by the shock wave the resulting buffet of the control surface can be very objectionable. In addition to the buffet of the surface, the change in the pressure distribution due to separation and the shock wave location can

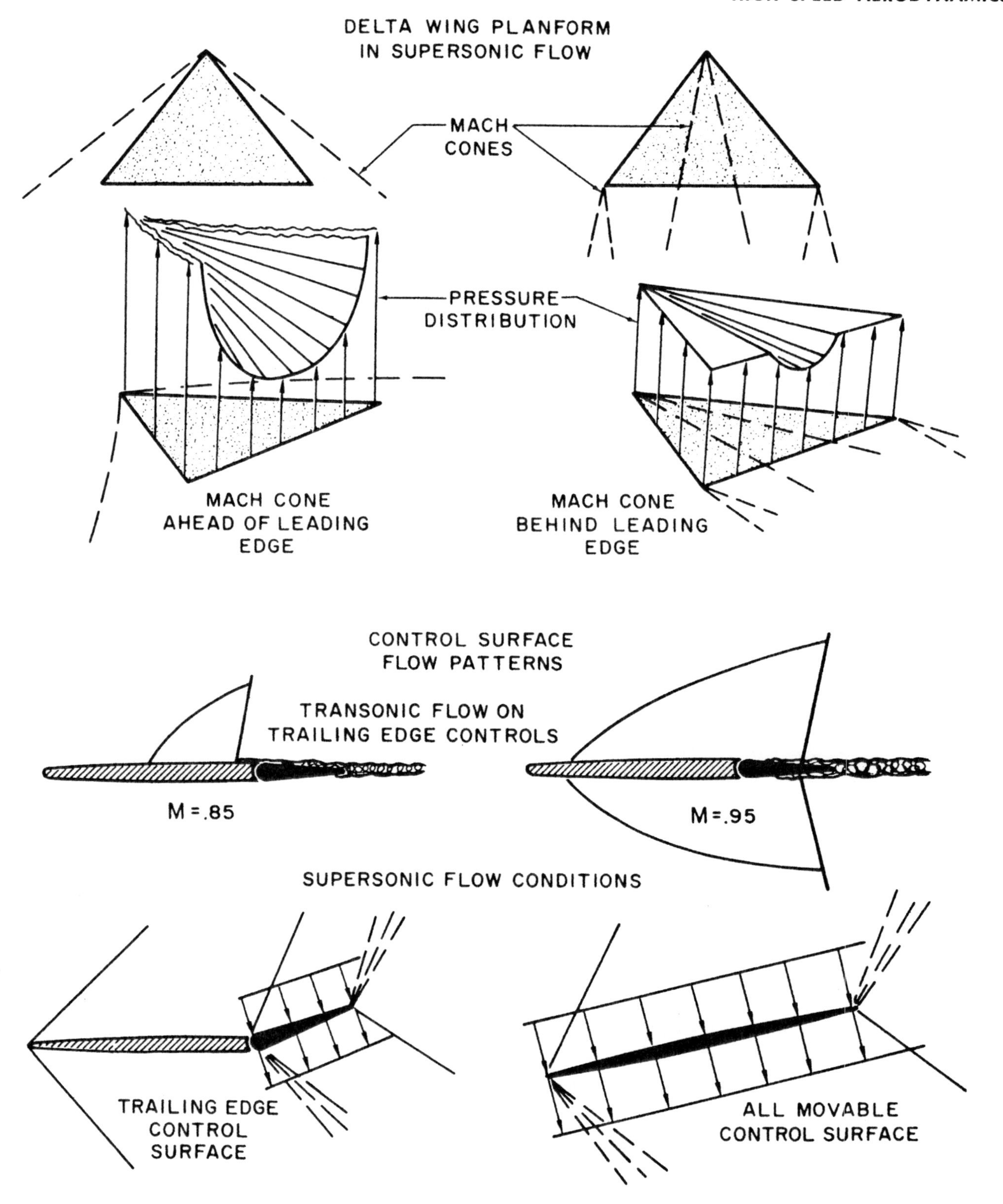

Figure 3.19. Planform Effects and Control Surfaces

create very large changes in control surface hinge moments. Such large changes in hinge moments create very undesirable control forces and present the need for an "irreversible" control system. An irreversible control system would employ powerful hydraulic or electric actuators to move the surfaces upon control by the pilot and the airloads developed on the surface could not feed back to the pilot. Of course, suitable control forces would be synthesized by bungees, "*q*" springs, bobweights, etc.

Transonic and supersonic flight can cause a noticeable reduction in the effectiveness of trailing edge control surfaces. The deflection of a trailing edge control surface at low subsonic speeds alters the pressure distribution on the fixed portion as well as the movable portion of the surface. This is true to the extent that a 1-degree deflection of a 40 percent chord elevator produces a lift change very nearly the equivalent of a 1-degree change in stabilizer setting. However, if supersonic flow exists on the surface, a deflection of the trailing edge control surface cannot influence the pressure distribution in the supersonic area ahead of the movable control surface. This is especially true in high supersonic flight where supersonic flow exists over the entire chord and the change in pressure distribution is limited to the area of the control surface. The reduction in effectiveness of the trailing edge control surface at transonic and supersonic speeds necessitates the use of an all movable surface. Application of the all movable control surface to the horizontal tail is most usual since the increase in longitudinal stability in supersonic flight requires a high degree of control effectiveness to achieve required controllability for supersonic maneuvering.

SUPERSONIC ENGINE INLETS. Air which enters the compressor section of a jet engine or the combustion chamber of a ramjet usually must be slowed to subsonic velocity. This process must be accomplished with the least possible waste of energy. At flight speeds just above the speed of sound only slight modifications to ordinary subsonic inlet design produce satisfactory performance. However, at supersonic flight speeds, the inlet design must slow the air with the weakest possible series or combination of shock waves to minimize energy losses and temperature rise. Figure 3.20 illustrates some of the various forms of supersonic inlets or "diffusers."

One of the least complicated types of inlet is the simple normal shock type diffuser. This type of inlet employs a single normal shock wave at the inlet with a subsequent internal subsonic compression. At low supersonic Mach numbers the strength of the normal shock wave is not too great and this type of inlet is quite practical. At higher supersonic Mach numbers, the single normal shock wave is very strong and causes a great reduction in the total pressure recovered by the inlet. In addition, it is necessary to consider that the wasted energy of the airstream will appear as an additional undesirable rise in temperature of the captured inlet airflow.

If the supersonic airstream can be captured, the shock wave formations will be swallowed and a gradual contraction will reduce the speed to just above sonic. Subsequent diverging flow section can then produce the normal shock wave which slows the airstream to subsonic. Further expansion continues to slow the air to lower subsonic speeds. This is the convergent-divergent type inlet shown in figure 3.20. If the initial contraction is too extreme for the inlet Mach number, the shock wave formation will not be swallowed and will move out in front of the inlet. The external location of the normal shock wave will produce subsonic flow immediately at the inlet. Since the airstream is suddenly slowed to subsonic through the strong normal shock a greater loss of airstream energy will occur.

Another form of diffuser employs an external oblique shock wave which slows the supersonic airstream before the normal shock occurs. Ideally, the supersonic airstream could be

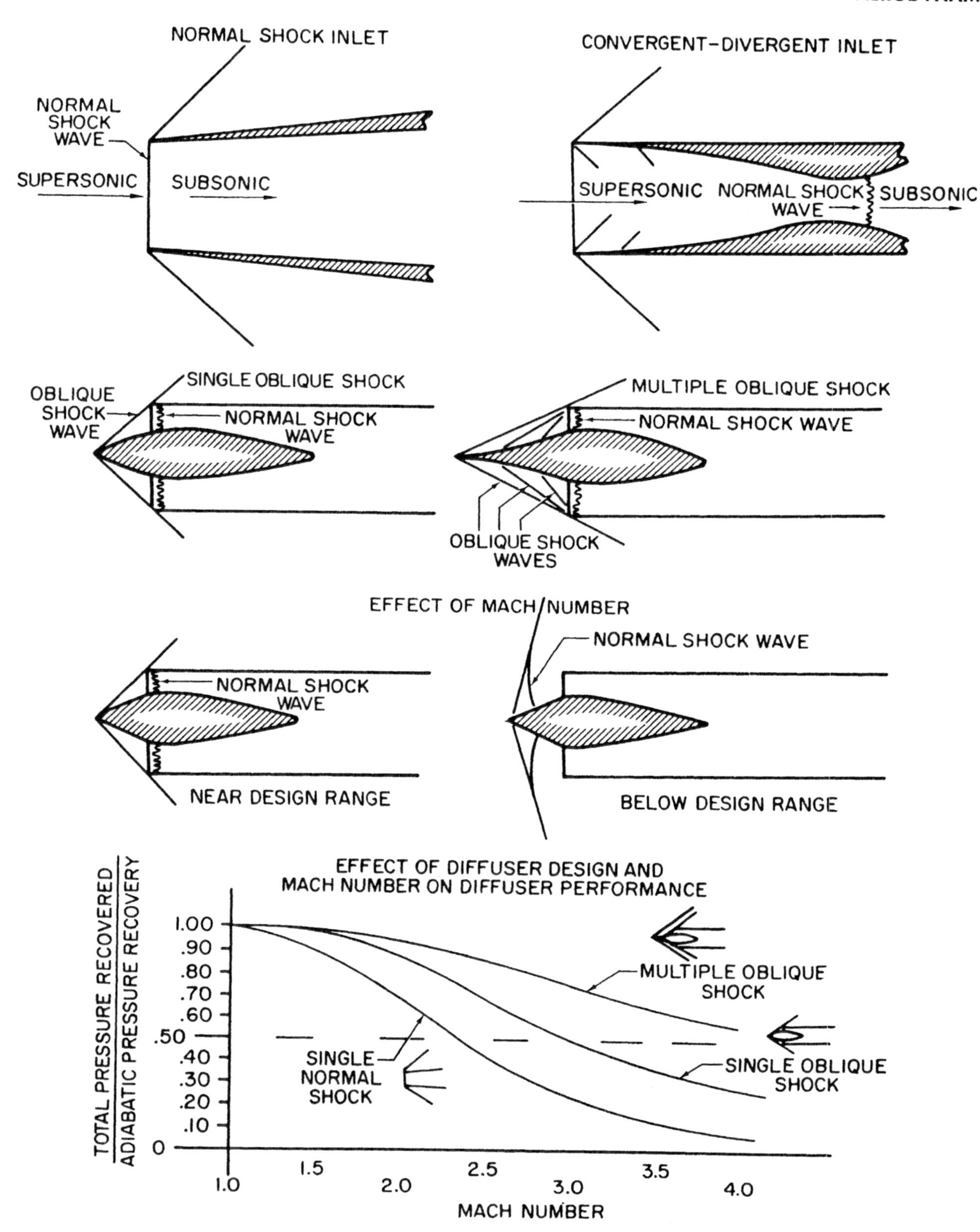

Figure 3.20. Various Types of Supersonic Inlets

slowed gradually through a series of very weak oblique shock waves to a speed just above sonic velocity. Then the subsequent normal shock to subsonic could be quite weak. Such a combination of the weakest possible waves would result in the least waste of energy and the highest pressure recovery. The efficiency of various types of diffusers is shown in figure 3.20 and illustrates this principle.

An obvious complication of the supersonic inlet is that the optimum shape is variable with inlet flow direction and Mach number. In other words, to derive highest efficiency and stability of operation, the geometry of the inlet would be different at each Mach number and angle of attack of flight. A typical supersonic military aircraft may experience large variations in angle of attack, sideslip angle, and flight Mach number during normal operation. These large variations in inlet flow conditions create certain important design considerations.

(1) The inlet should provide the highest practical efficiency. The ratio of recovered total pressure to airstream total pressure is an appropriate measure of this efficiency.

(2) The inlet should match the demands of the powerplant for airflow. The airflow captured by the inlet should match that necessary for engine operation.

(3) Operation of the inlet at flight conditions other than the design condition should not cause a noticeable loss of efficiency or excess drag. The operation of the inlet should be stable and not allow "buzz" conditions (an oscillation of shock location possible during off-design operation).

In order to develop a good, stable inlet design, the performance at the design condition may be compromised. A large variation of inlet flow conditions may require special geometric features for the inlet surfaces or a completely variable geometry inlet design.

SUPERSONIC CONFIGURATIONS. When all the various components of the supersonic airplane are developed, the most likely general configuration properties will be as follows:

(1) The *wing* will be of low aspect ratio, have noticeable taper, and have sweepback depending on the design speed range. The wing sections will be of low thickness ratio and require sharp leading edges.

(2) The *fuselage and nacelles* will be of high fineness ratio (long and slender). The supersonic pressure distribution may create significant lift and drag and require consideration of the stability contribution of these surfaces.

(3) The *tail surfaces* will be similar to the wing—low aspect ratio, tapered, swept and of thin section with sharp leading edge. The controls will be fully powered and irreversible with all movable surfaces the most likely configuration.

(4) In order to reduce interference drag in transonic and supersonic flight, the gross cross section of the aircraft may be "area ruled" to approach that of some optimum high speed shape.

One of the most important qualities of high speed configurations will be the low speed flight characteristics. The low aspect ratio swept wing planform has the characteristic of high induced drag at low flight speeds. Steep turns, excessively low airspeeds, and steep, power-off approaches can then produce extremely high rates of descent during landing. Sweepback and low aspect ratio can cause severe deterioration of handling qualities at speeds below those recommended for takeoff and landing. On the other hand, thin, swept wings at high wing loading will have relatively high landing speeds. Any excess of this basically high airspeed can create an impossible requirement of brakes, tires, and arrest ing gear. These characteristics require that the pilot account for the variation of optimum speeds with weight changes and adhere to the procedures and techniques outlined in the flight handbook.

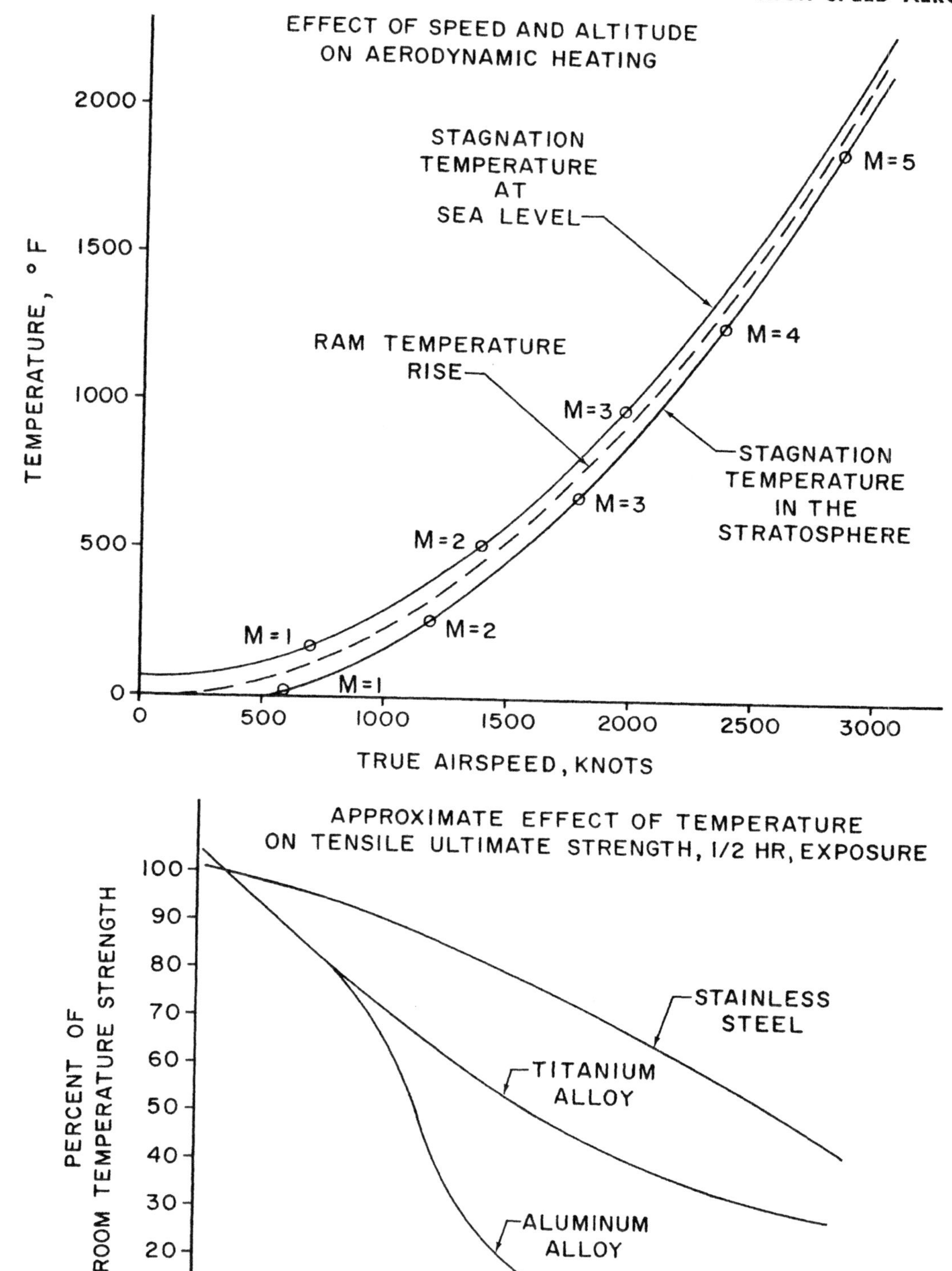

Figure 3.21. Aerodynamic Heating

AERODYNAMIC HEATING

When air flows over any aerodynamic surface certain reductions in velocity occur with corresponding increases in temperature. The greatest reduction in velocity and increase in temperature will occur at the various stagnation points on the aircraft. Of course, similar changes occur at other points on the aircraft but these temperatures can be related to the ram temperature rise at the stagnation point. While subsonic flight does not produce temperatures of any real concern, supersonic flight can produce temperatures high enough to be of major importance to the airframe and powerplant structure. The graph of figure 3.21 illustrates the variation of ram temperature rise with airspeed in the standard atmosphere. The ram temperature rise is independent of altitude and is a function of true airspeed. Actual temperatures would be the sum of the temperature *rise* and the *ambient* air temperature. Thus, low altitude flight at high Mach numbers will produce the highest temperatures.

In addition to the effect on the crew member environment, aerodynamic heating creates special problems for the airplane structure and the powerplant. The effect of temperature on the short time strength of three typical structural materials is shown in figure 3.21. Higher temperatures produce definite reductions in the strength of aluminum alloy and require the use of titanium alloys, stainless steels, etc., at very high temperatures. Continued exposure at elevated temperatures effects further reductions of strength and magnifies the problems of "creep" failure and structural stiffness.

The turbojet engine is adversely affected by high compressor inlet air temperatures. Since the thrust output of the turbojet is some function of the fuel flow, high compressor inlet air temperatures reduce the fuel flow that can be used within turbine operating temperature limits. The reduction in performance of the turbojet engines with high compressor inlet air temperatures requires that the inlet design produce the highest practical efficiency and minimize the temperature rise of the air delivered to the compressor face.

High flight speeds and compressible flow dictate airplane configurations which are much different from the ordinary subsonic airplane. To achieve safe and efficient operation, the pilot of the modern, high speed aircraft must understand and appreciate the advantages and disadvantages of the configuration. A knowledge of high speed aerodynamics will contribute greatly to this understanding.

Chapter 4
STABILITY AND CONTROL

An aircraft must have satisfactory handling qualities in addition to adequate performance. The aircraft must have adequate stability to maintain a uniform flight condition and recover from the various disturbing influences. It is necessary to provide sufficient stability to minimize the workload of the pilot. Also, the aircraft must have proper response to the controls so that it may achieve the inherent performance. There are certain conditions of flight which provide the most critical requirements of stability and control and these conditions must be understood and respected to accomplish safe and efficient operation of the aircraft.

DEFINITIONS

STATIC STABILITY

An aircraft is in a state of *equilibrium* when the sum of all forces and all moments is equal

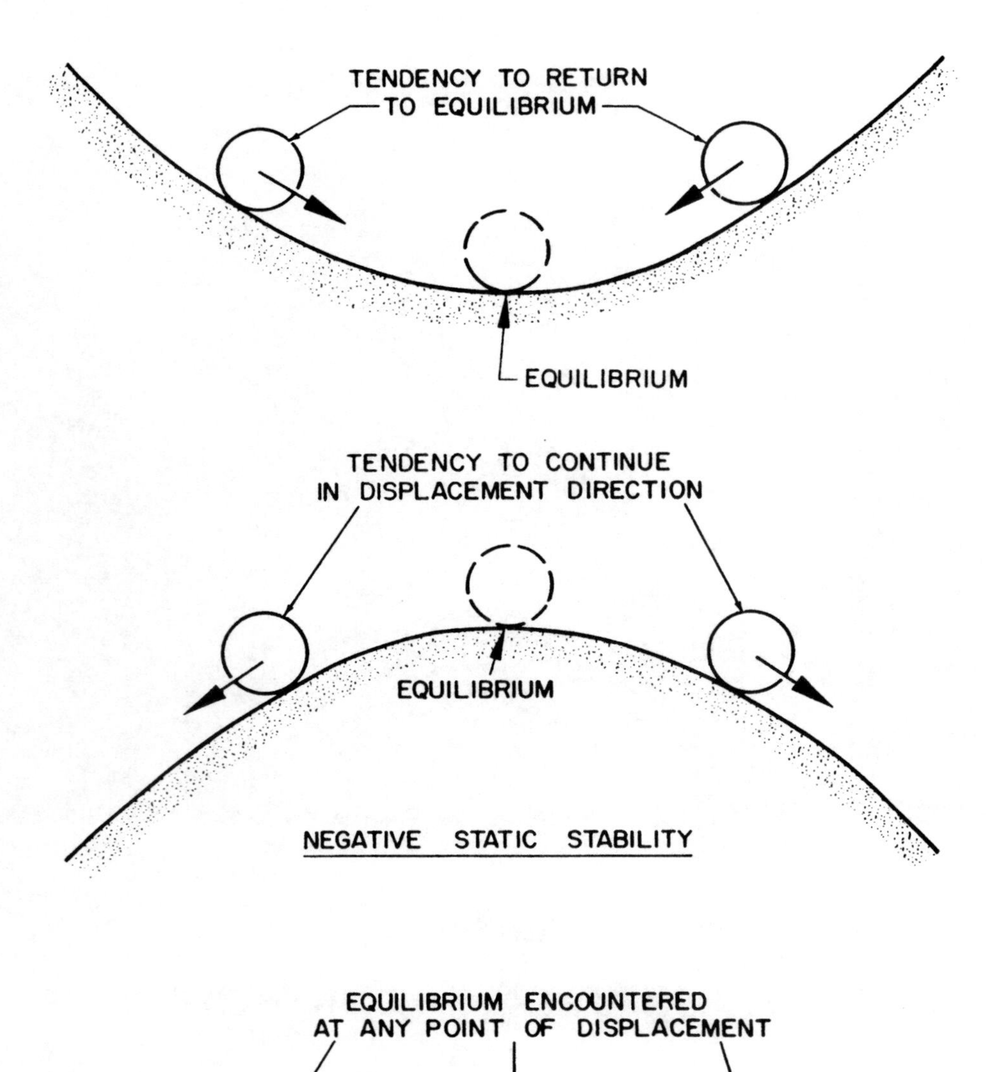

Figure 4.1. Static Stability

to zero. When an aircraft is in equilibrium, there are no accelerations and the aircraft continues in a steady condition of flight. If the equilibrium is disturbed by a gust or deflection of the controls, the aircraft will experience acceleration due to unbalance of moment or force.

The static stability of a system is defined by the initial tendency to return to equilibrium conditions following some disturbance from equilibrium. If an object is disturbed from equilibrium and has the tendency to return to equilibrium, *positive static stability* exists. If the object has a tendency to continue in the direction of disturbance, *negative static stability* or *static instability* exists. An intermediate condition could occur where an object displaced from equilibrium remains in equilibrium in the displaced position. If the object subject to a disturbance has neither the tendency to return nor the tendency to continue in the displacement direction, *neutral static stability* exists. These three categories of static stability are illustrated in figure 4.1. The ball in a trough illustrates the condition of positive static stability. If the ball is displaced from equilibrium at the bottom of the trough, the initial tendency of the ball is to return to the equilibrium condition. The ball may roll back and forth through the point of equilibrium but displacement to either side creates the initial tendency to return. The ball on a hill illustrates the condition of static instability. Displacement from equilibrium at the hilltop brings about the tendency for greater displacement. The ball on a flat, level surface illustrates the condition of neutral static stability. The ball encounters a new equilibrium at any point of displacement and has neither stable nor unstable tendencies.

The term "static" is applied to this form of stability since the resulting motion is not considered. Only the *tendency* to return to equilibrium conditions is considered in static stability. The static longitudinal stability of an aircraft is appreciated by displacing the aircraft from some trimmed angle of attack. If the aerodynamic pitching moments created by this displacement tend to return the aircraft to the equilibrium angle of attack the aircraft has positive static longitudinal stability.

DYNAMIC STABILITY

While *static* stability is concerned with the tendency of a displaced body to return to equilibrium, *dynamic* stability is defined by the resulting *motion with time*. If an object is disturbed from equilibrium, the time history of the resulting motion indicates the dynamic stability of the system. In general, the system will demonstrate positive dynamic stability if the amplitude of motion decreases with time. The various conditions of possible dynamic behavior are illustrated by the time history diagrams of figure 4.2.

The nonoscillatory modes shown in figure 4.2 depict the time histories possible without cyclic motion. If the system is given an initial disturbance and the motion simply subsides without oscillation, the mode is termed "subsidence" or "deadbeat return." Such a motion indicates positive static stability by the tendency to return to equilibrium and positive dynamic stability since the amplitude decreases with time. Chart B illustrates the mode of "divergence" by a noncyclic increase of amplitude with time. The initial tendency to continue in the displacement direction is evidence of static instability and the increasing amplitude is proof of dynamic instability. Chart C illustrates the mode of pure neutral stability. If the original disturbance creates a displacement which remains constant thereafter, the lack of tendency for motion and the constant amplitude indicate neutral static and neutral dynamic stability.

The oscillatory modes of figure 4.2 depict the time histories possible with cyclic motion. One feature common to each of these modes is that positive static stability is demonstrated in the cyclic motion by tendency to return to

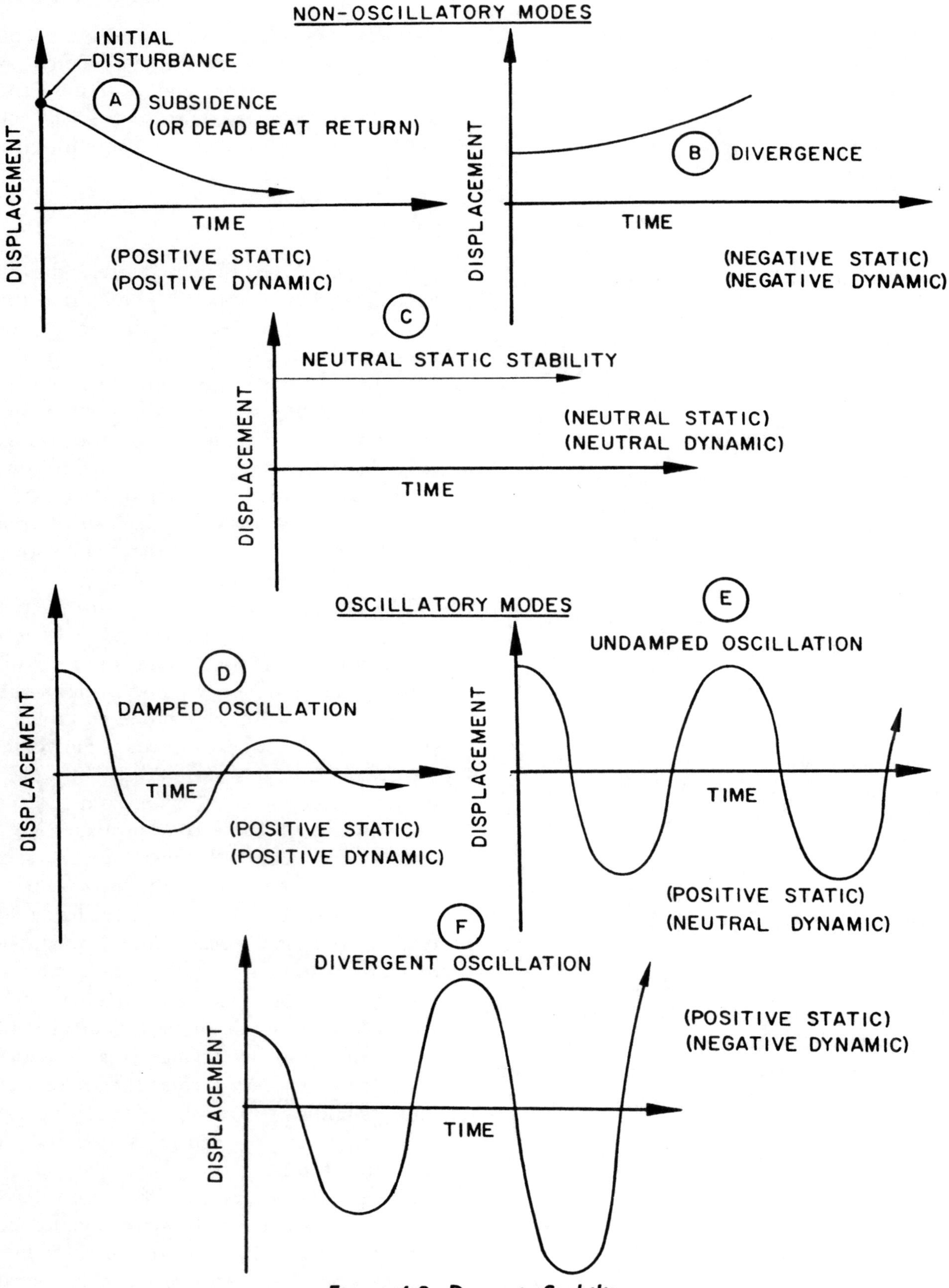

Figure 4.2. Dynamic Stability

equilibrium conditions. However, the dynamic behavior may be stable, neutral, or unstable. Chart D illustrates the mode of a damped oscillation where the amplitude decreases with time. The reduction of amplitude with time indicates there is resistance to motion and that energy is being dissipated. The dissipation of energy—or "damping"—is necessary to provide positive dynamic stability. If there is no damping in the system, the mode of chart E is the result, an undamped oscillation. Without damping, the oscillation continues with no reduction of amplitude with time. While such an oscillation indicates positive static stability, neutral dynamic stability exists. Positive damping is necessary to eliminate the continued oscillation. As an example, an automobile with worn shock absorbers (or "dampers") lacks sufficient dynamic stability and the continued oscillatory motion is neither pleasant nor conducive to safe operation. In the same sense, the aircraft must have sufficient damping to rapidly dissipate any oscillatory motion which would affect the operation of the aircraft. When natural aerodynamic damping cannot be obtained, a synthetic damping must be furnished to provide the necessary positive dynamic stability.

Chart F of figure 4.2 illustrates the mode of a divergent oscillation. This motion is statically stable since it tends to return to the equilibrium position. However, each subsequent return to equilibrium is with increasing velocity such that amplitude continues to increase with time. Thus, dynamic instability exists. The divergent oscillation occurs when energy is supplied to the motion rather than dissipated by positive damping. The most outstanding illustration of the divergent oscillation occurs with the short period pitching oscillation of an aircraft. If a pilot unknowingly supplies control functions which are near the natural frequency of the airplane in pitch, energy is added to the system, negative damping exists, and the "pilot induced oscillation" results.

In any system, the existence of static stability does not necessarily guarantee the existence of dynamic stability. However, the existence of dynamic stability implies the existence of static stability.

Any aircraft must demonstrate the required degrees of static and dynamic stability. If the aircraft were allowed to have static instability with a rapid rate of divergence, the aircraft would be very difficult—if not impossible—to fly. The degree of difficulty would compare closely with learning to ride a unicycle. In addition, positive dynamic stability is mandatory in certain areas to preclude objectionable continued oscillations of the aircraft.

TRIM AND CONTROLLABILITY

An aircraft is said to be trimmed if all moments in pitch, roll, and yaw are equal to zero. The establishment of equilibrium at various conditions of flight is the function of the controls and may be accomplished by pilot effort, trim tabs, or bias of a surface actuator.

The term "controllability" refers to the ability of the aircraft to respond to control surface displacement and achieve the desired condition of flight. Adequate controllability must be available to perform takeoff and landing and accomplish the various maneuvers in flight. An important contradiction exists between stability and controllability since adequate controllability does not necessarily exist with adequate stability. In fact, a high degree of stability tends to reduce the controllability of the aircraft. The general relationship between static stability and controllability is illustrated by figure 4.3.

Figure 4.3 illustrates various degrees of static stability by a ball placed on various surfaces. Positive static stability is shown by the ball in a trough; if the ball is displaced from equilibrium at the bottom of the trough, there is an initial tendency to return to equilibrium. If it is desired to "control" the ball

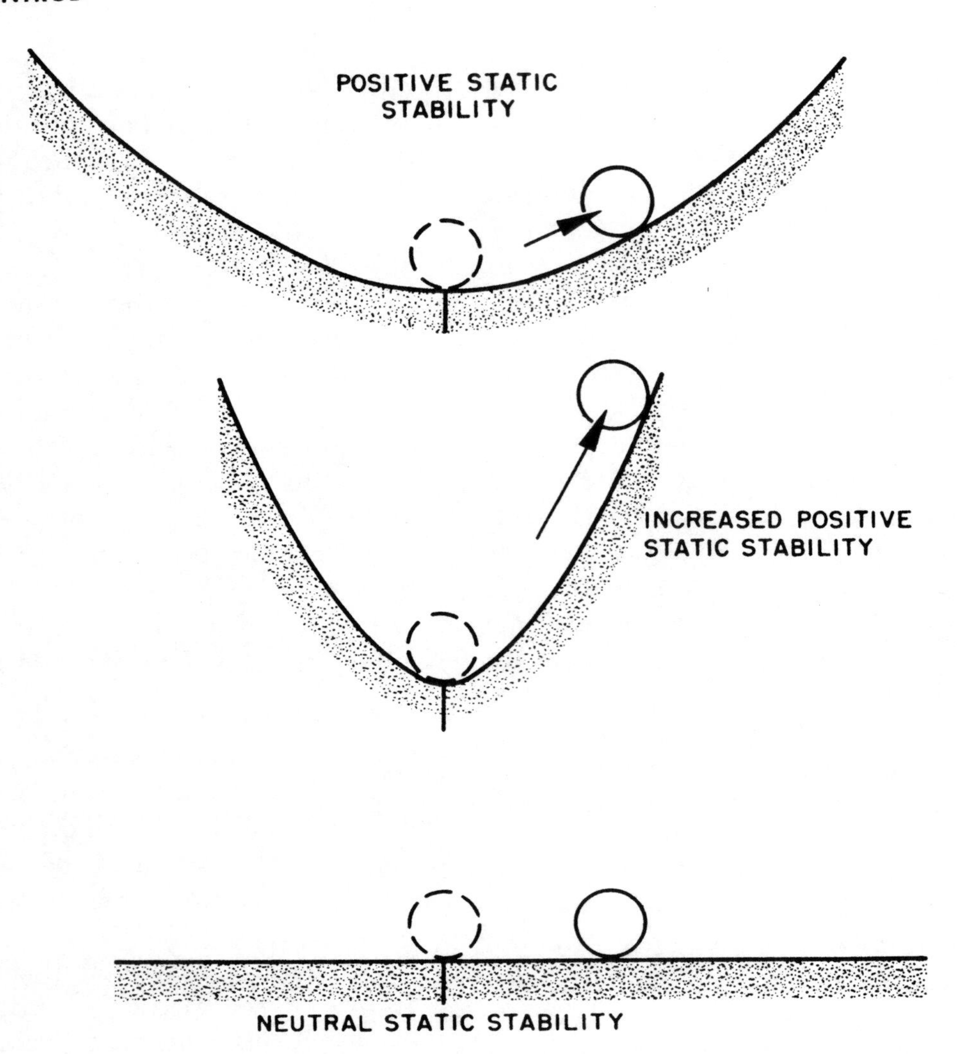

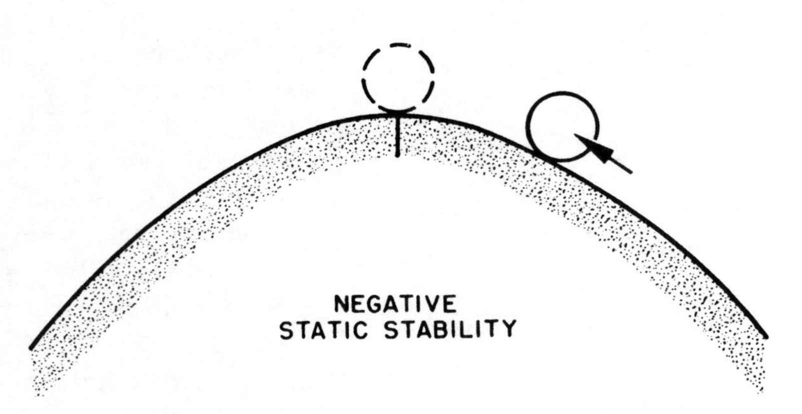

Figure 4.3. Stability and Controllability

and maintain it in the displaced position, a force must be supplied in the direction of displacement to balance the inherent tendency to return to equilibrium. This same stable tendency in an aircraft resists displacement from trim by pilot effort on the controls or atmospheric disturbances.

The effect of increased stability on controllability is illustrated by the ball in a steeper trough. A greater force is required to "control" the ball to the same lateral displacement when the stability is increased. In this manner, a large degree of stability tends to make the aircraft less controllable. It is necessary to achieve the proper balance between stability and controllability during the design of an aircraft because the *upper limits of stability are set by the lower limits of controllability*.

The effect of reduced stability on controllability is illustrated by the ball on a flat surface. When neutral static stability exists, the ball may be displaced from equilibrium and there is no stable tendency to return. A new point of equilibrium is obtained and no force is required to maintain the displacement. As the static stability approaches zero, controllability increases to infinity and the only resistance to displacement is a resistance to the motion of displacement—damping. For this reason, *the lower limits of stability may be set by the upper limits of controllability*. If the stability of the aircraft is too low, control deflections may create exaggerated displacements of the aircraft.

The effect of static instability on controllability is illustrated by the ball on a hill. If the ball is displaced from equilibrium at the top of the hill, the initial tendency is for the ball to continue in the displaced direction. In order to "control" the ball to some lateral displacement, a force must be applied *opposite* to the direction of displacement. This effect would be appreciated during flight of an unstable aircraft by an unstable "feel" of the aircraft. If the controls were deflected to increase the angle of attack, the aircraft would be trimmed at the higher angle of attack by a push force to keep the aircraft from continuing in the displacement direction. Such control force reversal would evidence the airplane instability; the pilot would be supplying the stability by his attempt to maintain the equilibrium. An unstable aircraft can be flown if the instability is slight with a low rate of divergence. Quick reactions coupled with effective controls can allow the pilot to cope with some degree of static instability. Since such flight would require constant attention by the pilot, slight instability can be tolerated only in airships, helicopters, and certain minor motions of the airplane. However, the airplane in high speed flight will react rapidly to any disturbances and any instability would create unsafe conditions. Thus, it is necessary to provide some positive static stability to the major aircraft degrees of freedom.

AIRPLANE REFERENCE AXES

In order to visualize the forces and moments on the aircraft, it is necessary to establish a set of mutually perpendicular reference axes originating at the center of gravity. Figure 4.4 illustrates a conventional right hand axis system. The longitudinal or X axis is located in a plane of symmetry and is given a positive direction pointing into the wind. A moment about this axis is a rolling moment, L, and the positive direction for a positive rolling moment utilizes the right hand rule. The vertical or Z axis also is in a plane of symmetry and is established positive downward. A moment about the vertical axis is a yawing moment, N, and a positive yawing moment would yaw the aircraft to the right (right hand rule). The lateral or Y axis is perpendicular to the plane of symmetry and is given a positive direction out the right side of the aircraft. A moment about the lateral axis is a pitching moment, M, and a positive pitching moment is in the nose-up direction.

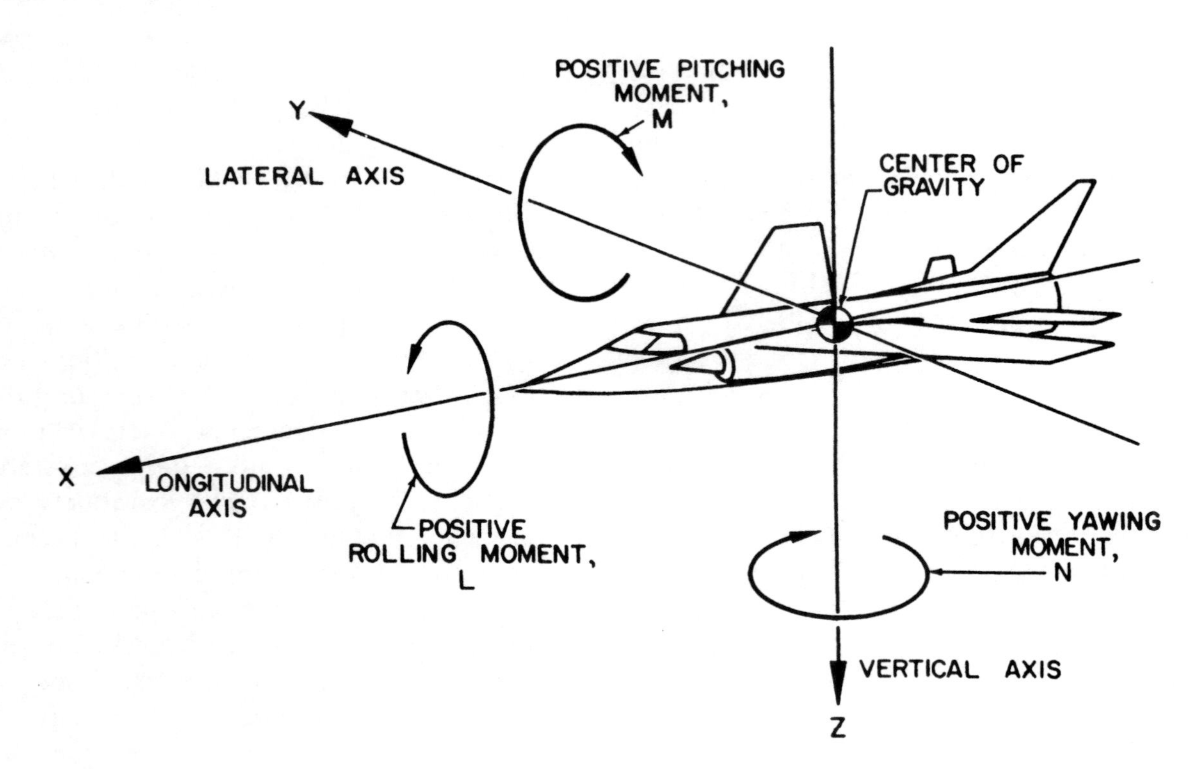

Figure 4.4. Airplane Reference Axes

LONGITUDINAL STABILITY AND CONTROL

STATIC LONGITUDINAL STABILITY

GENERAL CONSIDERATIONS. An aircraft will exhibit positive static longitudinal stability if it tends to return to the trim angle of attack when displaced by a gust or control movement. The aircraft which is unstable will continue to pitch in the disturbed direction until the displacement is resisted by opposing control forces. If the aircraft is neutrally stable, it tends to remain at any displacement to which it is disturbed. It is most necessary to provide an airplane with positive static longitudinal stability. The stable airplane is safe and easy to fly since the airplane seeks and tends to maintain a trimmed condition of flight. It also follows that control deflections and control "feel" are logical in direction and magnitude. Neutral static longitudinal stability usually defines the lower limit of airplane stability since it is the boundary between stability and instability. The airplane with neutral static stability may be excessively responsive to controls and the aircraft has no tendency to return to trim following a disturbance. The airplane with negative static longitudinal stability is inherently divergent from any intended trim condition. If it is at all possible to fly the aircraft, the aircraft cannot be trimmed and illogical control forces and deflections are required to provide equilibrium with a change of attitude and airspeed.

Since static longitudinal stability depends upon the relationship of angle of attack and pitching moments, it is necessary to study the pitching moment contribution of each component of the aircraft. In a manner similar to all other aerodynamic forces, the pitching moment about the lateral axis is studied in the coefficient form.

$$M=C_M qS(MAC)$$

or

$$C_M=\frac{M}{qS(MAC)}$$

where

M=pitching moment about the c.g., ft.-lbs., positive if in a nose-up direction
q=dynamic pressure, psf
S=wing area, sq. ft.
MAC=mean aerodynamic chord, ft.
C_M=pitching moment coefficient

The pitching moment coefficients contributed by all the various components of the aircraft are summed up and plotted versus lift coefficient. Study of this plot of C_M versus C_L will relate the static longitudinal stability of the airplane.

Graph *A* of figure 4.5 illustrates the variation of pitching moment coefficient, C_M, with lift coefficient, C_L, for an airplane with positive static longitudinal stability. Evidence of static stability is shown by the tendency to return to equilibrium—or "trim"—upon displacement. The airplane described by graph *A* is in trim or equilibrium when $C_M=0$ and, if the airplane is disturbed to some different C_L, the pitching moment change tends to return the aircraft to the point of trim. If the airplane were disturbed to some higher C_L (point *Y*), a negative or nose-down pitching moment is developed which tends to decrease angle of attack back to the trim point. If the airplane were disturbed to some lower C_L (point *X*), a positive or nose-up pitching moment is developed which tends to increase the angle of attack back to the trim point. Thus, positive static longitudinal stability is indicated by a negative slope of C_M versus C_L, i.e., positive stability is evidenced by a decrease in C_M with an increase in C_L.

The degree of static longitudinal stability is indicated by the slope of the curve of pitching moment coefficient with lift coefficient. Graph *B* of figure 4.5 provides comparison of the stable and unstable conditions. Positive stability is indicated by the curve with negative slope. Neutral static stability would be the result if the curve had zero slope. If neutral stability exists, the airplane could be disturbed to some higher or lower lift coefficient without change in pitching moment coefficient. Such a condition would indicate that the airplane would have no tendency to return to some original equilibrium and would not hold trim. An airplane which demonstrates a positive slope of the C_M versus C_L curve would be unstable. If the unstable airplane were subject to any disturbance from equilibrium at the trim point, the changes in pitching moment would only magnify the disturbance. When the unstable airplane is disturbed to some higher C_L, a positive change in C_M occurs which would illustrate a tendency for continued, greater displacement. When the unstable airplane is disturbed to some lower C_L, a negative change in C_M takes place which tends to create continued displacement.

Ordinarily, the static longitudinal stability of a conventional airplane configuration does not vary with lift coefficient. In other words, the slope of C_M versus C_L does not change with C_L. However, if the airplane has sweepback, large contribution of power effects to stability, or significant changes in downwash at the horizontal tail, noticeable changes in static stability can occur at high lift coefficients. This condition is illustrated by graph *C* of figure 4.5. The curve of C_M versus C_L of this illustration shows a good stable slope at low values of C_L. Increasing C_L effects a slight decrease in the negative slope hence a decrease in stability occurs. With continued increase in C_L, the slope becomes zero and neutral stability exists. Eventually, the slope becomes positive and the airplane becomes unstable or "pitch-up" results. Thus, at any lift coefficient, the static stability of the airplane is depicted by the slope of the curve of C_M versus C_L.

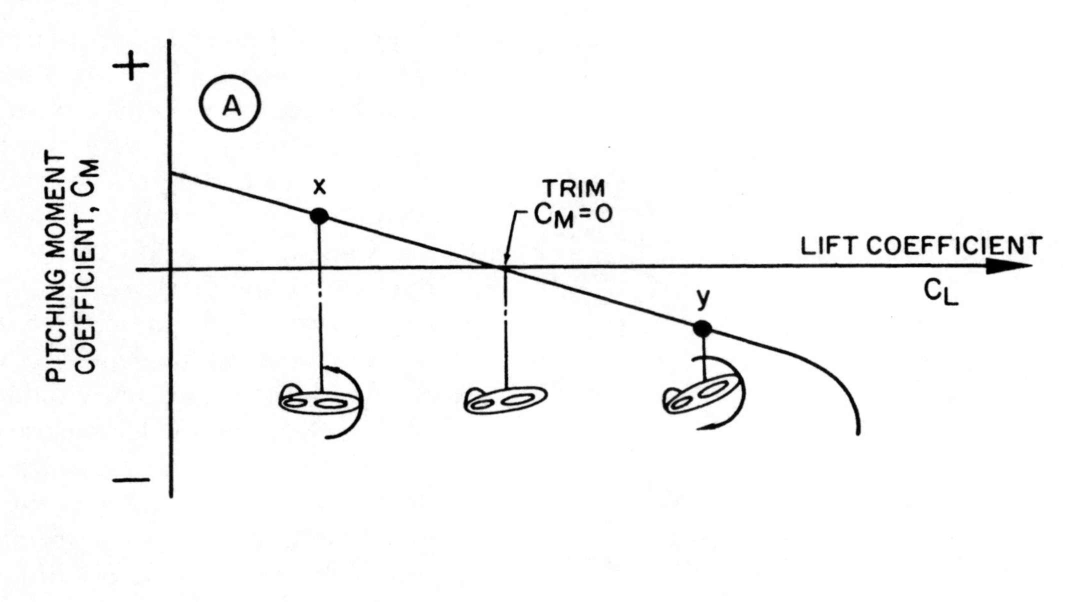

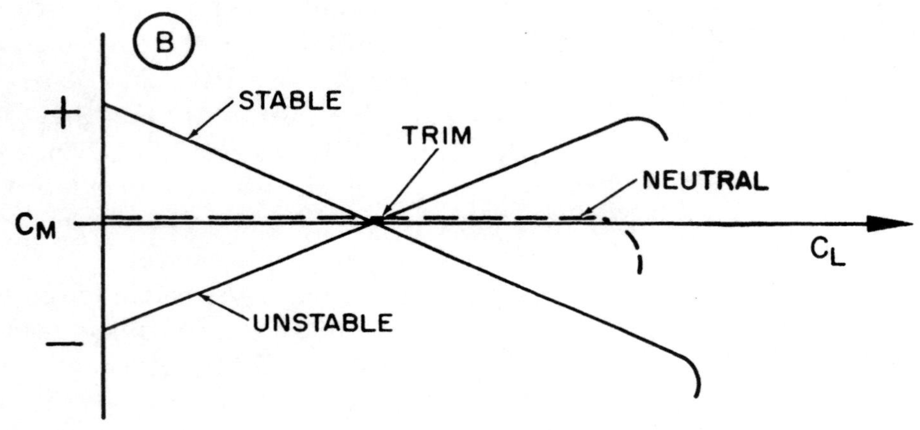

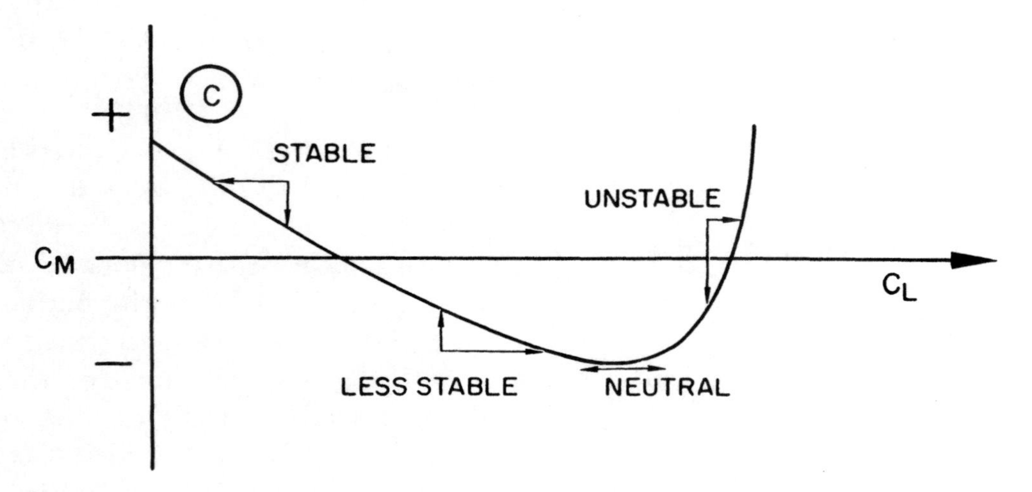

Figure 4.5. Airplane Static Longitudinal Stability

CONTRIBUTION OF THE COMPONENT SURFACES. The net pitching moment about the lateral axis is due to the contribution of each of the component surfaces acting in their appropriate flow fields. By study of the contribution of each component the effect of each component on the static stability may be appreciated. It is necessary to recall that the pitching moment coefficient is defined as:

$$C_M = \frac{M}{qS(MAC)}$$

Thus, any pitching moment coefficient—regardless of source—has the common denominator of dynamic pressure, q, wing area, S, and wing mean aerodynamic chord, MAC. This common denominator is applied to the pitching moments contributed by the fuselage and nacelles, horizontal tail, and power effects as well as pitching moments contributed by the wing.

WING. The contribution of the wing to stability depends primarily on the location of the aerodynamic center with respect to the airplane center of gravity. Generally, the aerodynamic center—or a.c.—is defined as the point on the wing mean aerodynamic chord where the wing pitching moment coefficient does not vary with lift coefficient. All changes in lift coefficient effectively take place at the wing aerodynamic center. Thus, if the wing experiences some change in lift coefficient, the pitching moment created will be a direct function of the relative location of the a.c. and c.g.

Since stability is evidenced by the development of restoring moments, the c.g. must be forward of the a.c. for the wing to contribute to positive static longitudinal stability. As shown in figure 4.6, a change in lift aft of the c.g. produces a stable restoring moment dependent upon the lever arm between the a.c. and c.g. In this case, the wing contribution would be stable and the curve of C_M versus C_L for the wing alone would have a negative slope. If the c.g. were located at the a.c., C_M would not vary with C_L since all changes in lift would take place at the c.g. In this case, the wing contribution to stability would be neutral. When the c.g. is located behind the a.c. the wing contribution is unstable and the curve of C_M versus C_L for the wing alone would have a positive slope.

Since the wing is the predominating aerodynamic surface of an airplane, any change in the wing contribution may produce a significant change in the airplane stability. This fact would be most apparent in the case of the flying wing or tailless airplane where the wing contribution determines the airplane stability. In order for the wing to achieve stability, the c.g. must be ahead of the a.c. Also, the wing must have a positive pitching moment about the aerodynamic center to achieve trim at positive lift coefficients. The first chart of figure 4.7 illustrates that the wing which is stable will trim at a negative lift coefficient if the $C_{M_{AC}}$ is negative. If the stable wing has a positive $C_{M_{AC}}$ it will then trim at a useful positive C_L. The only means available to achieve trim at a positive C_L with a wing which has a negative $C_{M_{AC}}$ is an unstable c.g. position aft of the a.c. As a result, the tailless aircraft cannot utilize high lift devices which incur any significant changes in $C_{M_{AC}}$.

While the trim lift coefficient may be altered by a change in c.g. position, the resulting change in stability is undesirable and is unsatisfactory as a primary means of control. The variation of trim C_L by deflection of control surfaces is usually more effective and is less inviting of disaster. The early attempts at manned flight led to this conclusion.

When the aircraft is operating in subsonic flight, the a.c. of the wing remains fixed at the 25 percent chord station. When the aircraft is flown in supersonic flight, the a.c. of the wing will approach the 50 percent chord station. Such a large variation in the location of the a.c. can produce large changes in the wing contribution and greatly alter the airplane longitudinal stability. The second chart

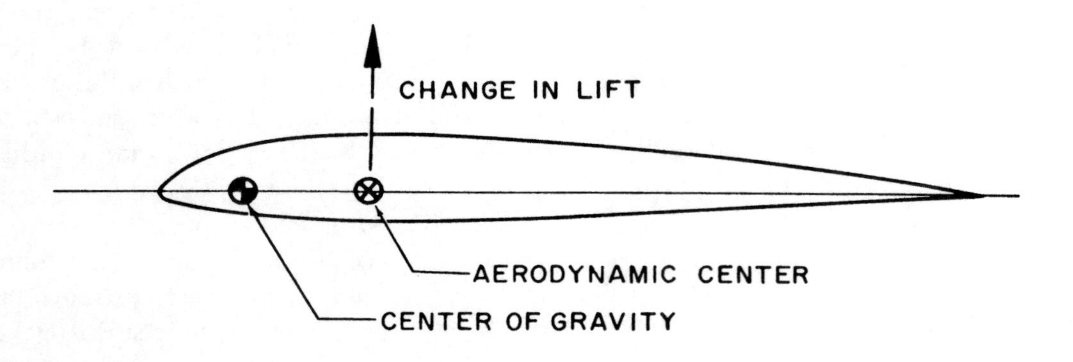

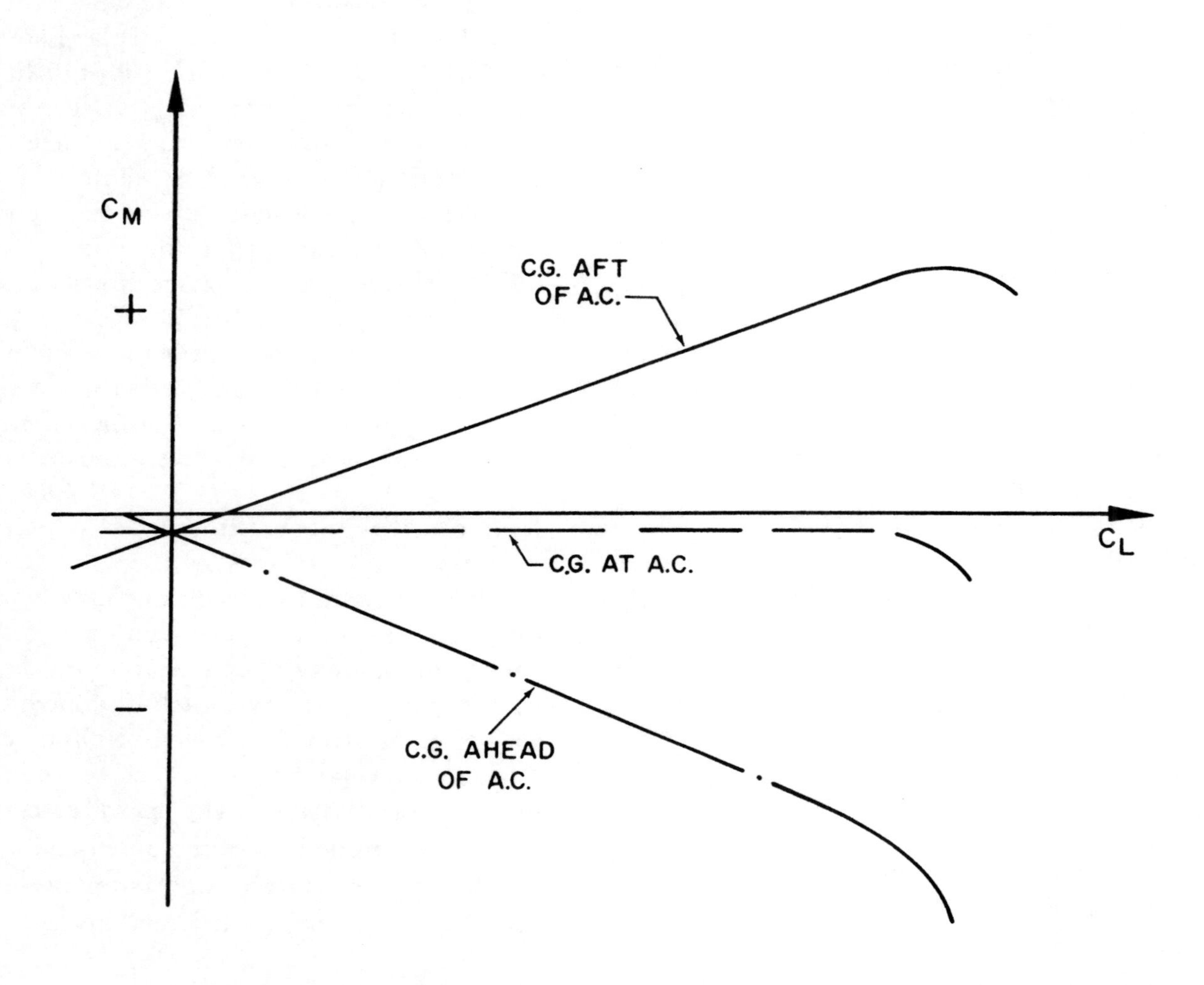

Figure 4.6. Wing Contribution

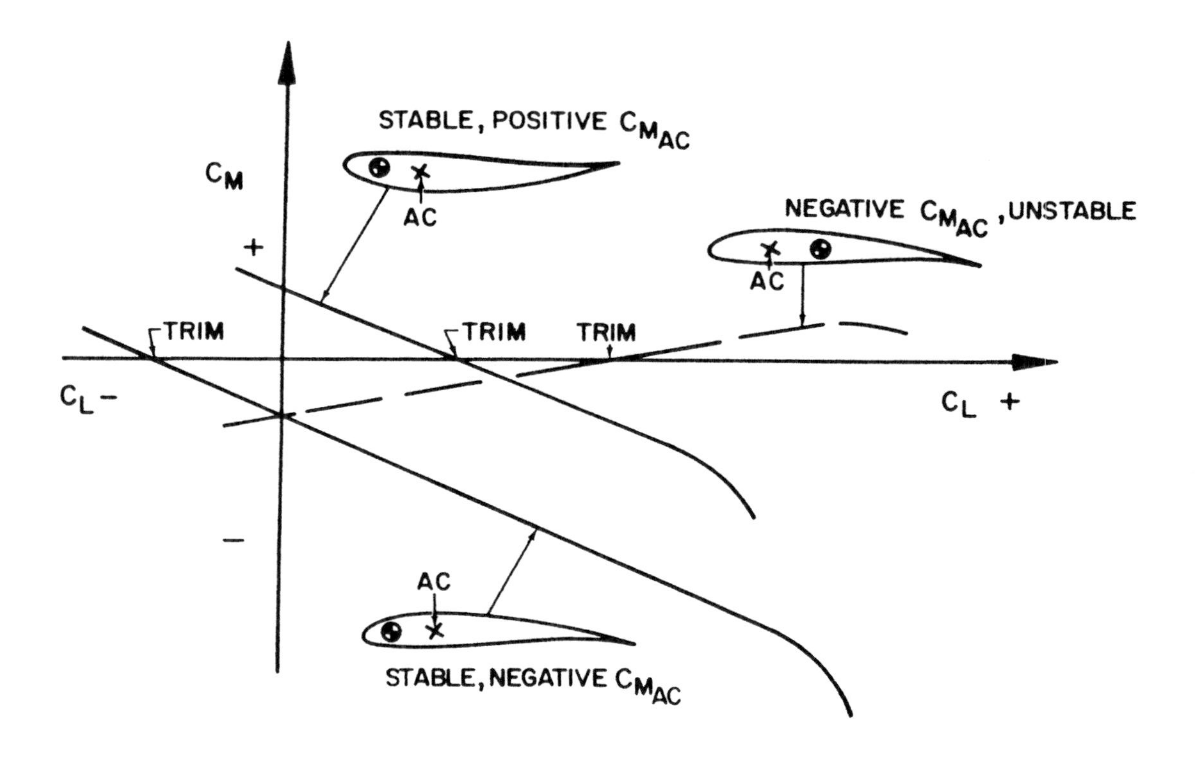

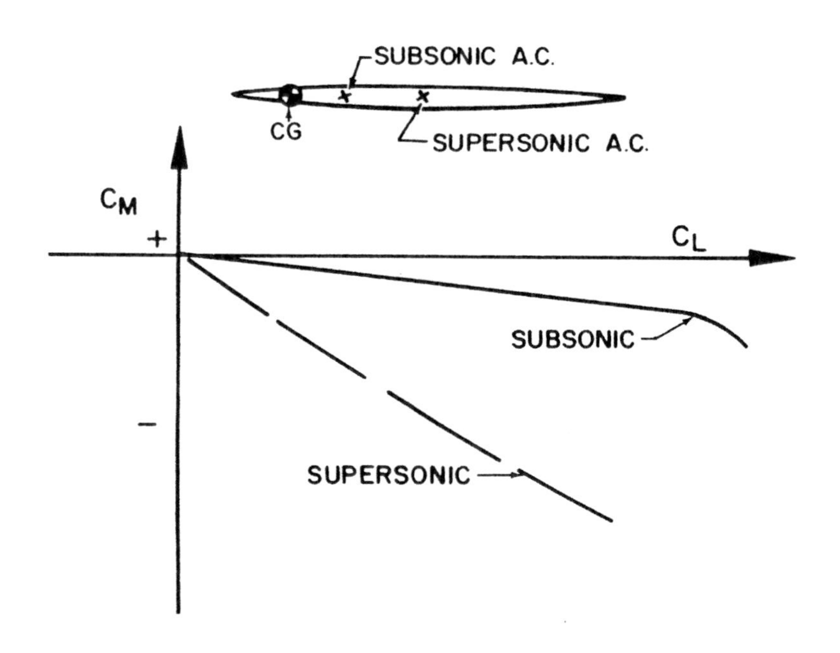

Figure 4.7. Effect of $C_{M_{AC}}$, C. G. Position and Mach Number

of figure 4.7 illustrates the change of wing contribution possible between subsonic and supersonic flight. The large increase in static stability in supersonic flight can incur high trim drag or require great control effectiveness to prevent reduction in maneuverability.

FUSELAGE AND NACELLES. In most cases, the contribution of the fuselage and nacelles is destabilizing. A symmetrical body of revolution in the flow field of a perfect fluid develops an unstable pitching moment when given an angle of attack. In fact, an increase in angle of attack produces an increase in the unstable pitching moment without the development of lift. Figure 4.8 illustrates the pressure distribution which creates this unstable moment on the body of revolution. In the actual case of real subsonic flow essentially the same effect is produced. An increase in angle of attack causes an increase in the unstable pitching moment but a negligible increase in lift.

An additional factor for consideration is the influence of the induced flow field of the wing. As illustrated in figure 4.8, the upwash ahead of the wing increases the destabilizing influence from the portions of the fuselage and nacelles ahead of the wing. The downwash behind the wing reduces the destabilizing influence from the portions of the fuselage and nacelles aft of the wing. Hence, the location of the fuselage and nacelles relative to the wing is important in determining the contribution to stability.

The body of revolution in supersonic flow can develop lift of a magnitude which cannot be neglected. When the body of revolution in supersonic flow is given an angle of attack, a pressure distribution typical of figure 4.8 is the result. Since the center of pressure is well forward, the body contributes a destabilizing influence. As is usual with supersonic configurations, the fuselage and nacelles may be quite large in comparison with the wing area and the contribution to stability may be large. Interaction between the wing and fuselage and nacelles deserves consideration in several instances. Body upwash and variation of local Mach number can influence the wing lift while lift carryover and downwash can effect the fuselage and nacelles forces and moments.

HORIZONTAL TAIL. The horizontal tail usually provides the greatest stabilizing influence of all the components of the airplane. To appreciate the contribution of the horizontal tail to stability, inspect figure 4.9. If the airplane is given a change in angle of attack, a change in tail lift will occur at the aerodynamic center of the tail. An increase in lift at the horizontal tail produces a negative moment about the airplane c.g. and tends to return the airplane to the trim condition. While the contribution of the horizontal tail to stability is large, the magnitude of the contribution is dependent upon the change in tail lift and the lever arm of the surface. It is obvious that the horizontal tail will produce a stabilizing effect only when the surface is aft of the c.g. For this reason it would be inappropriate to refer to the forward surface of a canard (tail-first) configuration as a horizontal "stabilizer." In a logical sense, the horizontal "stabilizer" must be aft of the c.g. and—generally speaking—the farther aft, the greater the contribution to stability.

Many factors influence the change in tail lift which occurs with a change in airplane angle of attack. The area of the horizontal tail has the obvious effect that a large surface would generate a large change in lift. In a similar manner, the change in tail lift would depend on the slope of the lift curve for the horizontal tail. Thus, aspect ratio, taper, sweepback, and Mach number would determine the sensitivity of the surface to changes in angle of attack. It should be appreciated that the flow at the horizontal tail is not of the same flow direction or dynamic pressure as the free stream. Due to the wing wake, fuselage boundary layer, and power effects, the q at the horizontal tail may be greatly different from the q of the free stream. In most in-

BODY OF REVOLUTION IN PERFECT FLUID

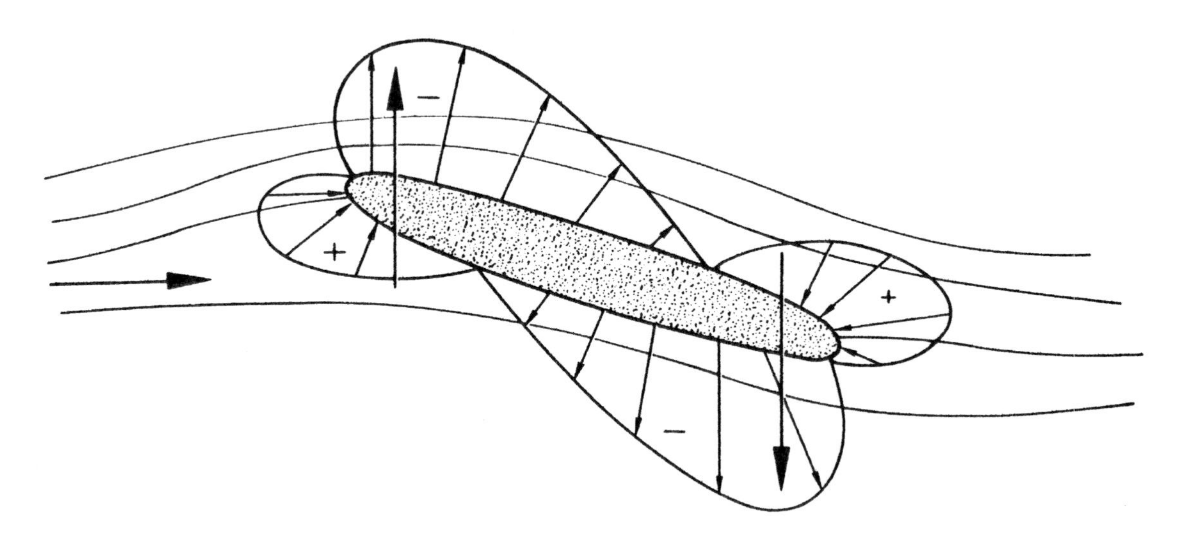

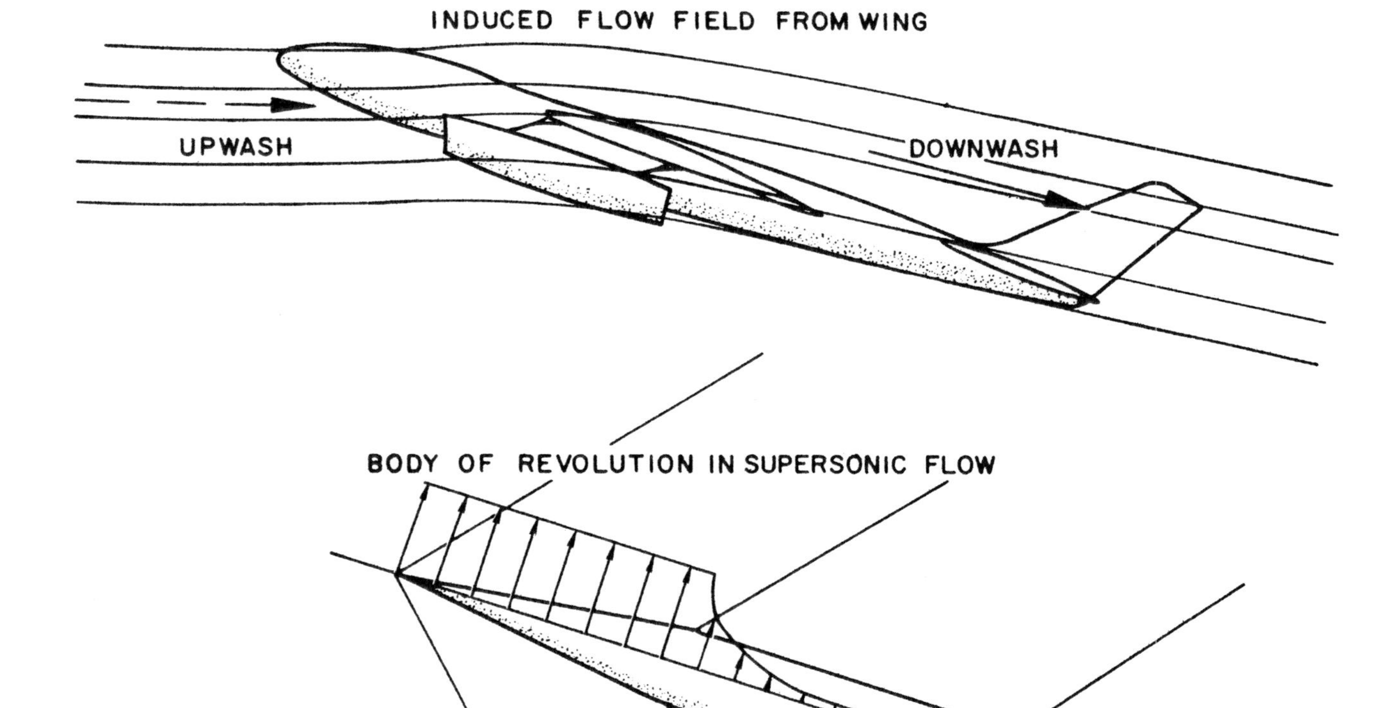

Figure 4.8. Body or Nacelle Contribution

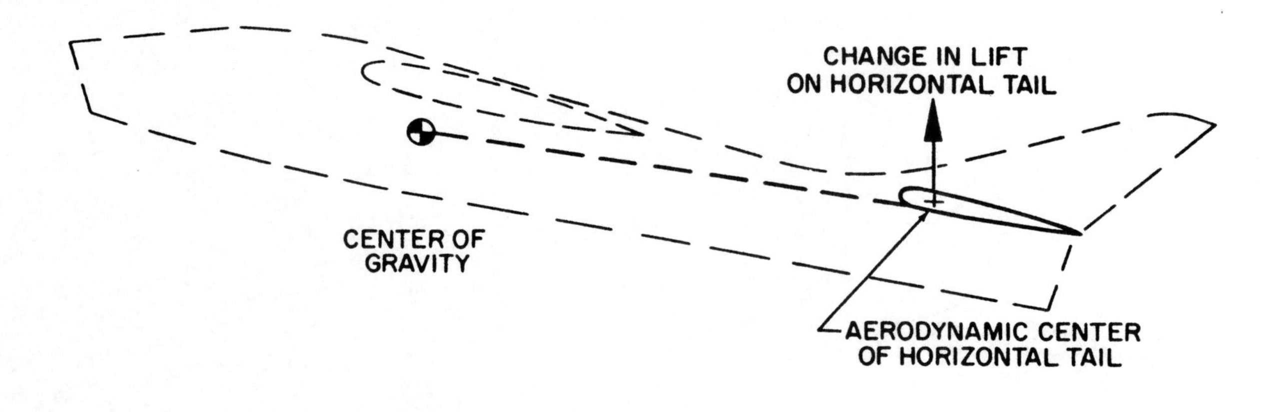

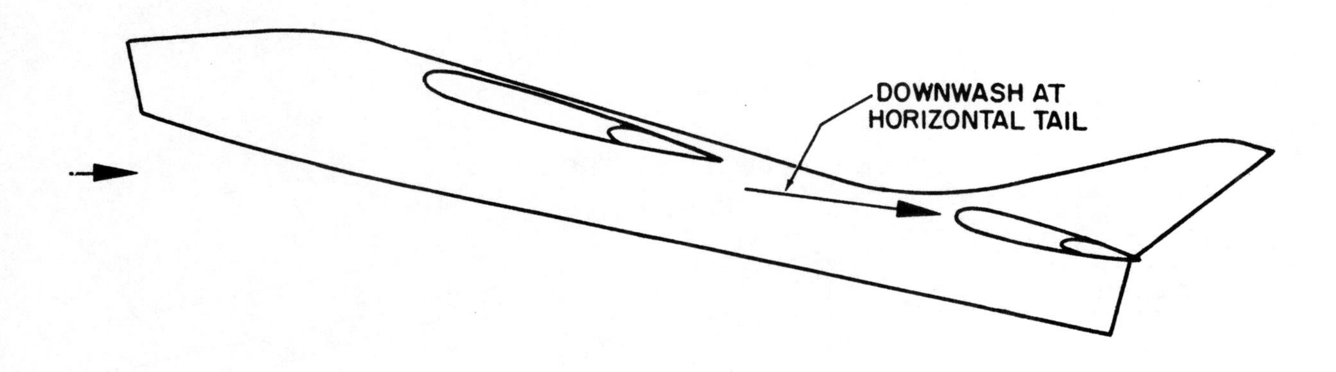

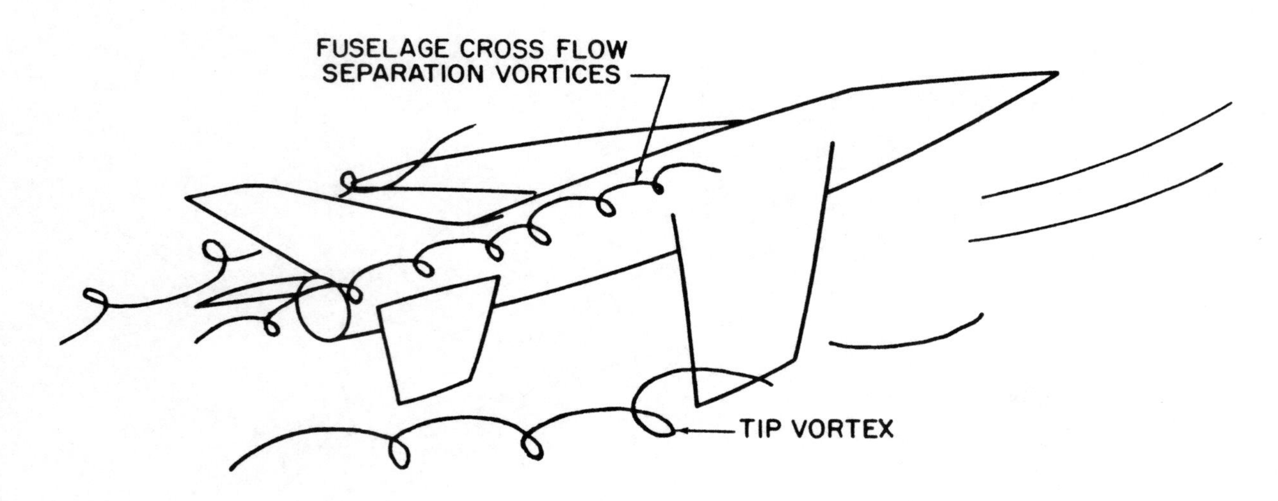

Figure 4.9. Contribution of Tail and Downwash Effects

stances, the q at the tail is usually less and this reduces the efficiency of the tail.

When the airplane is given a change in angle of attack, the horizontal tail does not experience the same change in angle of attack as the wing. Because of the increase in downwash behind the wing, the horizontal tail will experience a smaller change in angle of attack, e.g., if a 10° change in wing angle of attack causes a 4° increase in downwash at the horizontal tail, the horizontal tail experiences only a 6° change in angle of attack. In this manner, the downwash at the horizontal tail reduces the contribution to stability. Any factor which alters the rate of change of downwash at the horizontal tail will directly affect the tail contribution and airplane stability.

Power effects can alter the downwash at the horizontal tail and affect the tail contribution. Also, the downwash at the tail is affected by the lift distribution on the wing and the flow condition on the fuselage. The low aspect ratio airplane requires large angles of attack to achieve high lift coefficients and this positions the fuselage at high angles of attack. The change in the wing downwash can be accompanied by crossflow separation vortices on the fuselage. It is possible that the net effect obviates or destabilizes the contribution of the horizontal tail and produces airplane instability.

POWER-OFF STABILITY. When the intrinsic stability of a configuration is of interest, power effects are neglected and the stability is considered by a buildup of the contributing components. Figure 4.10 illustrates a typical buildup of the components of a conventional airplane configuration. If the c.g. is arbitrarily set at 30 percent *MAC*, the contribution of the wing alone is destabilizing as indicated by the positive slope of C_M versus C_L. The combination of the wing and fuselage increases the instability. The contribution of the tail alone is highly stabilizing from the large negative slope of the curve. The contribution of the tail must be sufficiently stabilizing so that the complete configuration will exhibit positive static stability at the anticipated c.g. locations. In addition, the tail and wing incidence must be set to provide a trim lift coefficient near the design condition.

When the configuration of the airplane is fixed, a variation of c.g. position can cause large changes in the static stability. In the conventional airplane configuration, the large changes in stability with c.g. variation are primarily due to the large changes in the wing contribution. If the incidence of all surfaces remains fixed, the effect of c.g. position on static longitudinal stability is typified by the second chart of figure 4.10. As the c.g. is gradually moved aft, the airplane static stability decreases, then becomes neutral then unstable. The c.g. position which produces zero slope and neutral static stability is referred to as the "neutral point." The neutral point may be imagined as the effective aerodynamic center of the entire airplane configuration, i.e., with the c.g. at this position, all changes in net lift effectively occur at this point and no change in pitching moment results. The neutral point defines the most aft c.g. position without static instability.

POWER EFFECTS. The effects of power may cause significant changes in trim lift coefficient and static longitudinal stability. Since the contribution to stability is evaluated by the change in moment coefficients, power effects will be most significant when the airplane operates at high power and low airspeeds such as the power approach or waveoff condition.

The effects of power are considered in two main categories. First, there are the direct effects resulting from the forces created by the propulsion unit. Next, there are the indirect effects of the slipstream and other associated flow which alter the forces and moments of the aerodynamic surfaces. The direct effects of power are illustrated in figure 4.11. The vertical location of the thrust line defines one of the direct contributions to stability. If the

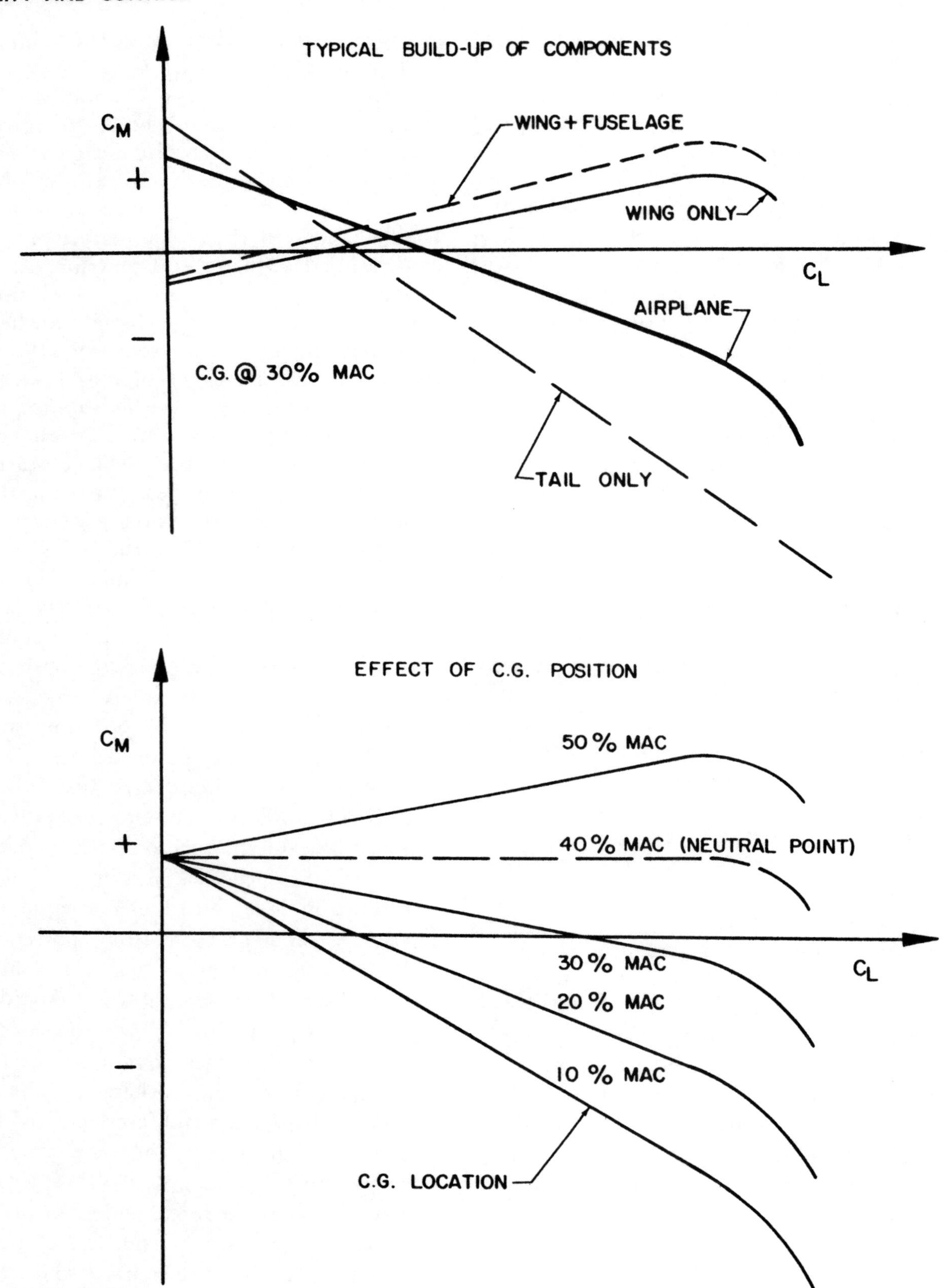

Figure 4.10. Stability Build-up and Effect of C.G. Position

thrust line is below the c.g., thrust produces a positive or noseup moment and the effect is destabilizing. On the other hand, if the thrust line is located above the c.g., a negative moment is created and the effect is stabilizing.

A propeller or inlet duct located ahead of the c.g. contributes a destabilizing effect. As shown in figure 4.11, a rotating propeller inclined to the windstream causes a deflection of the airflow. The momentum change of the slipstream creates a normal force at the plane of the propeller similar to a wing creating lift by deflecting an airstream. As this normal force will increase with an increase in airplane angle of attack, the effect will be destabilizing when the propeller is ahead of the c.g. The magnitude of the unstable contribution depends on the distance from the c.g. to the propeller and is largest at high power and low dynamic pressure. The normal force created

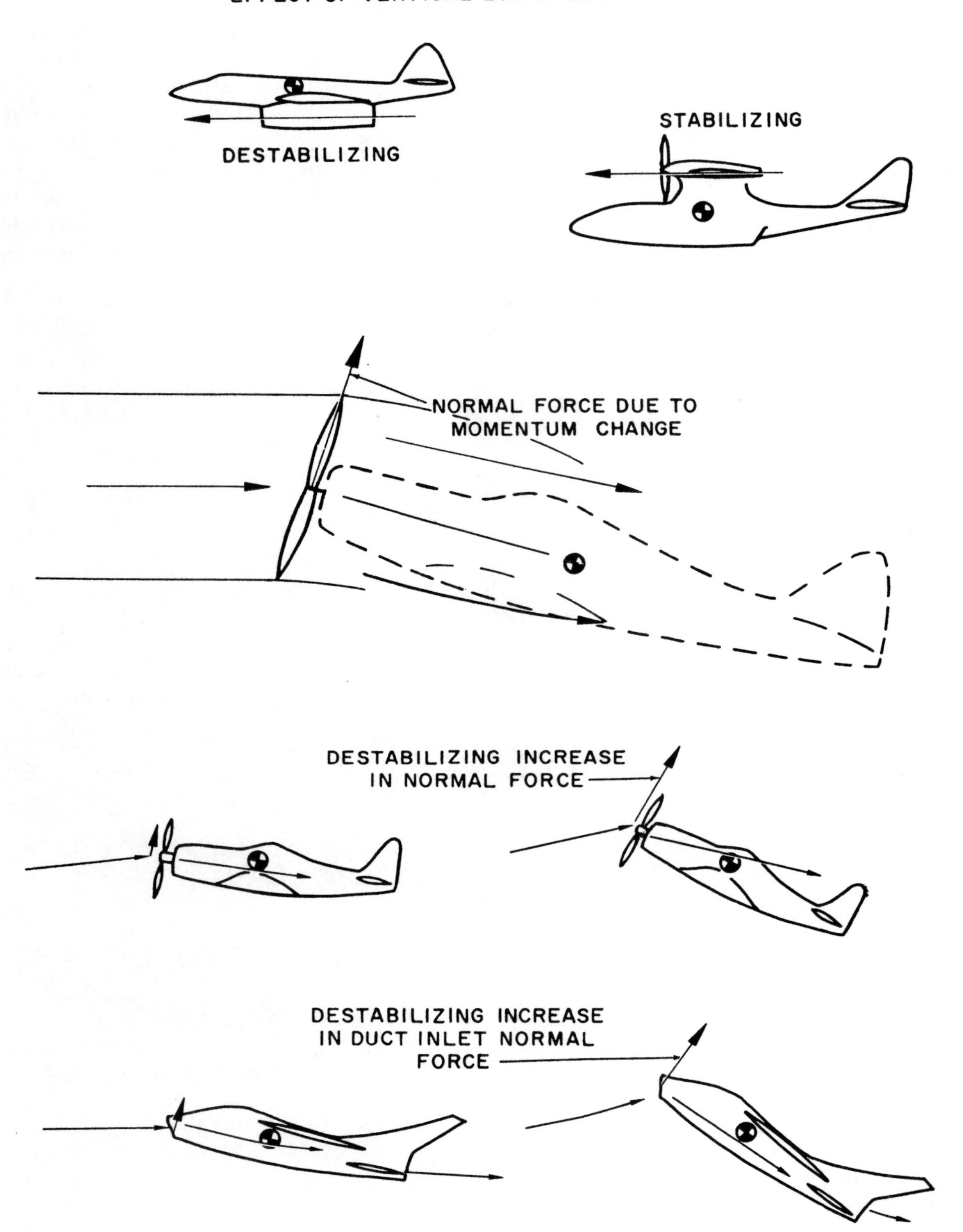

Figure 4.11. Direct Power Effects

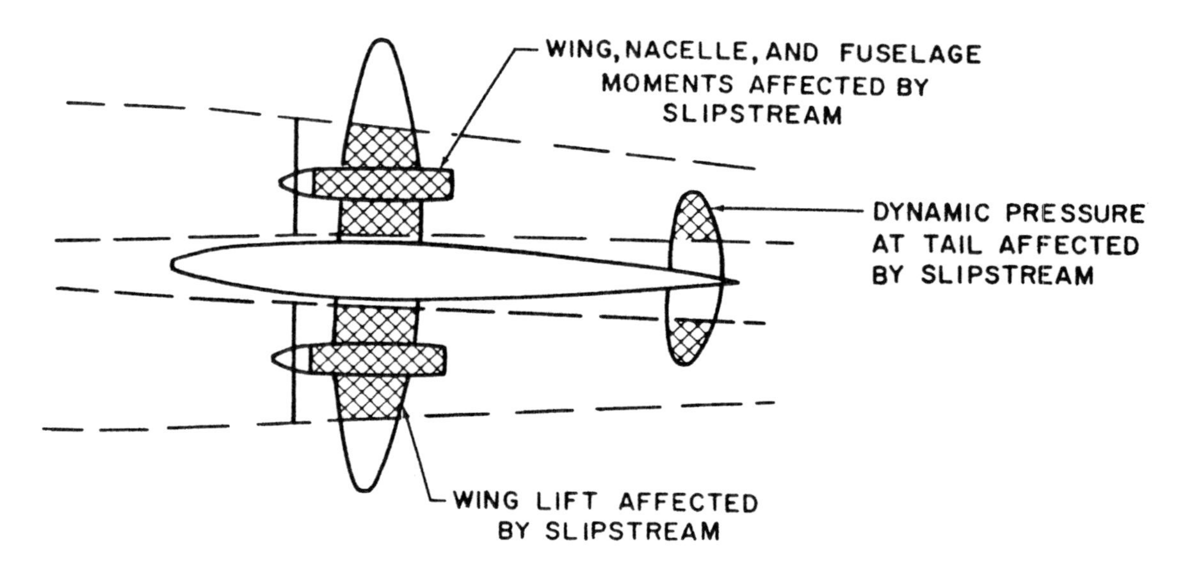

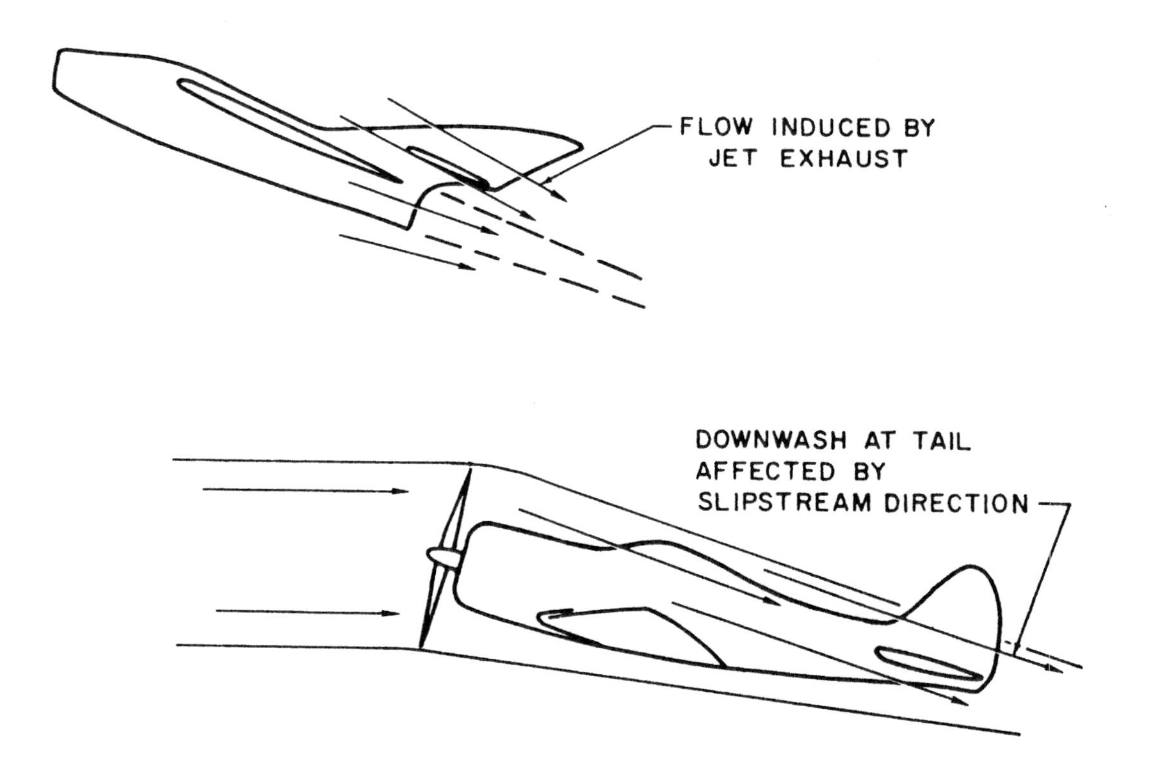

Figure 4.12. Indirect Power Effects

at the inlet of a jet engine contributes a similar destabilizing effect when the inlet is ahead of the c.g. As with the propeller, the magnitude of the stability contribution is largest at high thrust and low flight speed.

The indirect effects of power are of greatest concern in the propeller powered airplane rather than the jet powered airplane. As shown in figure 4.12, the propeller powered airplane creates slipstream velocities on the various surfaces which are different from the flow field typical of power-off flight. Since the various wing, nacelle, and fuselage surfaces are partly or wholly immersed in this slipstream, the contribution of these components to stability can be quite different from the power-off flight condition. Ordinarily, the change of fuselage and nacelle contribution with power is relatively small. The added lift on the portion of the wing immersed in the slipstream requires that the airplane operate at a lower angle of attack to produce the same effective lift coefficient. Generally, this reduction in angle of attack to effect the same C_L reduces the tail contribution to stability. However, the increase in dynamic pressure at the tail tends to increase the effectiveness of the tail and may be a stabilizing effect. The magnitude of this contribution due to the slipstream velocity on the tail will depend on the c.g. position and trim lift coefficient.

The deflection of the slipstream by the normal force at the propeller tends to increase the downwash at the horizontal tail and reduce the contribution to stability. Essentially the same destabilizing effect is produced by the flow induced at the exhaust of the jet powerplant. Ordinarily, the induced flow at the horizontal tail of a jet airplane is slight and is destabilizing when the jet passes underneath the horizontal tail. The magnitude of the indirect power effects on stability tends to be greatest at high C_L, high power, and low flight speeds.

The combined direct and indirect power effects contribute to a general reduction of static stability at high power, high C_L, and low q. It is generally true that any airplane will experience the lowest level of static longitudinal stability under these conditions. Because of the greater magnitude of both direct and indirect power effects, the propeller powered airplane usually experiences a greater effect than the jet powered airplane.

An additional effect on stability can be from the extension of high lift devices. The high lift devices tend to increase downwash at the tail and reduce the dynamic pressure at the tail, both of which are destabilizing. However, the high lift devices may prevent an unstable contribution of the wing at high C_L. While the effect of high lift devices depends on the airplane configuration, the usual effect is destabilizing. Hence, the airplane may experience the most critical forward neutral point during the power approach or waveoff. During these conditions of flight the static stability is usually the weakest and particular attention must be given to precise control of the airplane. The power-on neutral point may set the most aft limit of c.g. position.

CONTROL FORCE STABILITY. The static longitudinal stability of an airplane is defined by the tendency to return to equilibrium upon displacement. In other words, the stable airplane will resist displacement from the trim or equilibrium. The control forces of the airplane should reflect the stability of the airplane and provide suitable reference for precise control of the airplane.

The effect of elevator deflection on pitching moments is illustrated by the first graph of figure 4.13. If the elevators of the airplane are fixed at zero deflection, the resulting line of C_M versus C_L for 0° depicts the static stability and trim lift coefficient. If the elevators are fixed at a deflection of 10° up, the airplane static stability is unchanged but the trim lift coefficient is increased. A change in elevator or stabilizer position does not alter the tail contribution to stability but the change in pitching moment will alter the lift coefficient

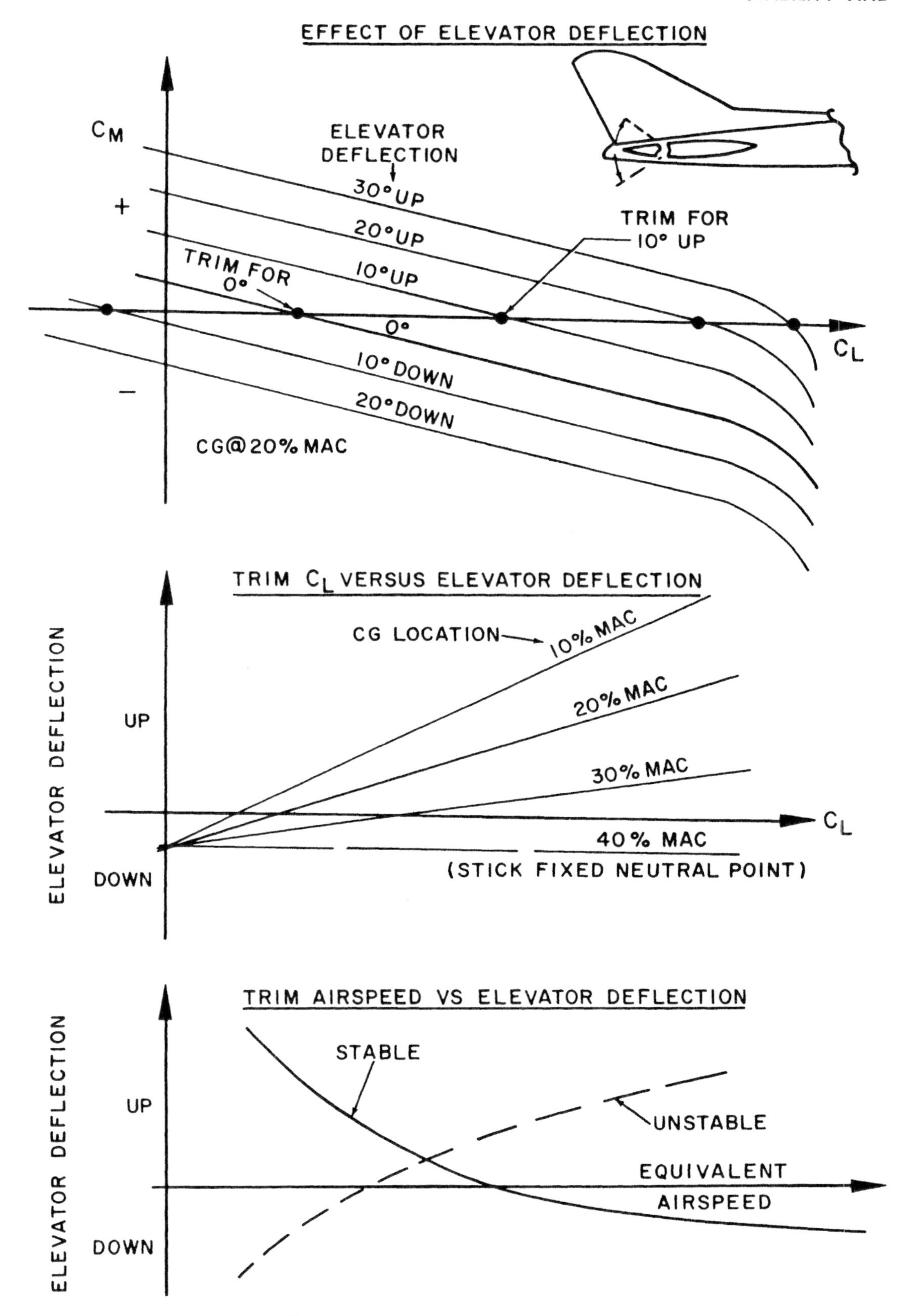

Figure 4.13. Longitudinal Control

at which equilibrium will occur. As the elevator is fixed in various positions, equilibrium (or trim) will occur at various lift coefficients and the trim CL can be correlated with elevator deflection as in the second graph of figure 4.13.

When the c.g. position of the airplane is fixed, each elevator position corresponds to a particular trim lift coefficient. As the c.g. is moved aft the slope of this line decreases and the decrease in stability is evident by a given control displacement causing a greater change in trim lift coefficient. This is evidence that decreasing stability causes increased controllability and, of course, increasing stability decreases controllability. If the c.g. is moved aft until the line of trim C_L versus elevator deflection has zero slope, neutral static stability is obtained and the "stick-fixed" neutral point is determined.

Since each value of lift coefficient corresponds to a particular value of dynamic pressure required to support an airplane in level flight, trim airspeed can be correlated with elevator deflection as in the third graph of figure 4.13. If the c.g. location is ahead of the stick-fixed neutral point and control position is directly related to surface deflection, the airplane will give evidence of *stick position stability*. In other words, the airplane will require the stick to be moved aft to increase the angle of attack and trim at a lower airspeed and to be moved forward to decrease the angle of attack and trim at a higher airspeed. To be sure, it is desirable to have an airplane demonstrate this feature. If the airplane were to have stick position instability, the airplane would require the stick to be moved aft to trim at a higher airspeed or to be moved forward to trim at a lower airspeed.

There may be slight differences in the static longitudinal stability if the elevators are allowed to float free. If the elevators are allowed to float free as in "hands-off" flight, the elevators may have a tendency to "float" or streamline when the horizontal tail is given a change in angle of attack. If the horizontal tail is subject to an increase in angle of attack and the elevators tend to float up, the change in lift on the tail is less than if the elevators remain fixed and the tail contribution to stability is reduced. Thus, the "stick-free" stability of an airplane is usually less than the stick-fixed stability. A typical reduction of stability by free elevators is shown in figure 4.14(A) where the airplane stick-free demonstrates a reduction of the slope of C_M versus C_L. While aerodynamic balance may be provided to reduce control forces, proper balance of the surfaces will reduce floating and prevent great differences between stick-fixed and stick-free stability. The greatest floating tendency occurs when the surface is at a high angle of attack hence the greatest difference between stick-fixed and stick-free stability occurs when the airplane is at high angle of attack.

If the controls are fully powered and actuated by an irreversible mechanism, the surfaces are not free to float and there is no difference between the stick-fixed and stick-free static stability.

The control forces in a conventional airplane are made up of two components. First, the basic stick-free stability of the airplane contributes an increment of force which is independent of airspeed. Next, there is an increment of force dependent on the trim tab setting which varies with the dynamic pressure or the square of equivalent airspeed. Figure 4.14(B) indicates the variation of stick force with airspeed and illustrates the effect of tab setting on stick force. In order to trim the airplane at point (1) a certain amount of up elevator is required and zero stick force is obtained with the use of the tab. To trim the airplane for higher speeds corresponding to points (2) and (3) less and less nose-up tab is required. Note that when the airplane is properly trimmed, a push force is required to increase airspeed and a pull force is required to decrease airspeed. In this manner, the airplane would indicate positive stick force stability with a stable "feel" for air-

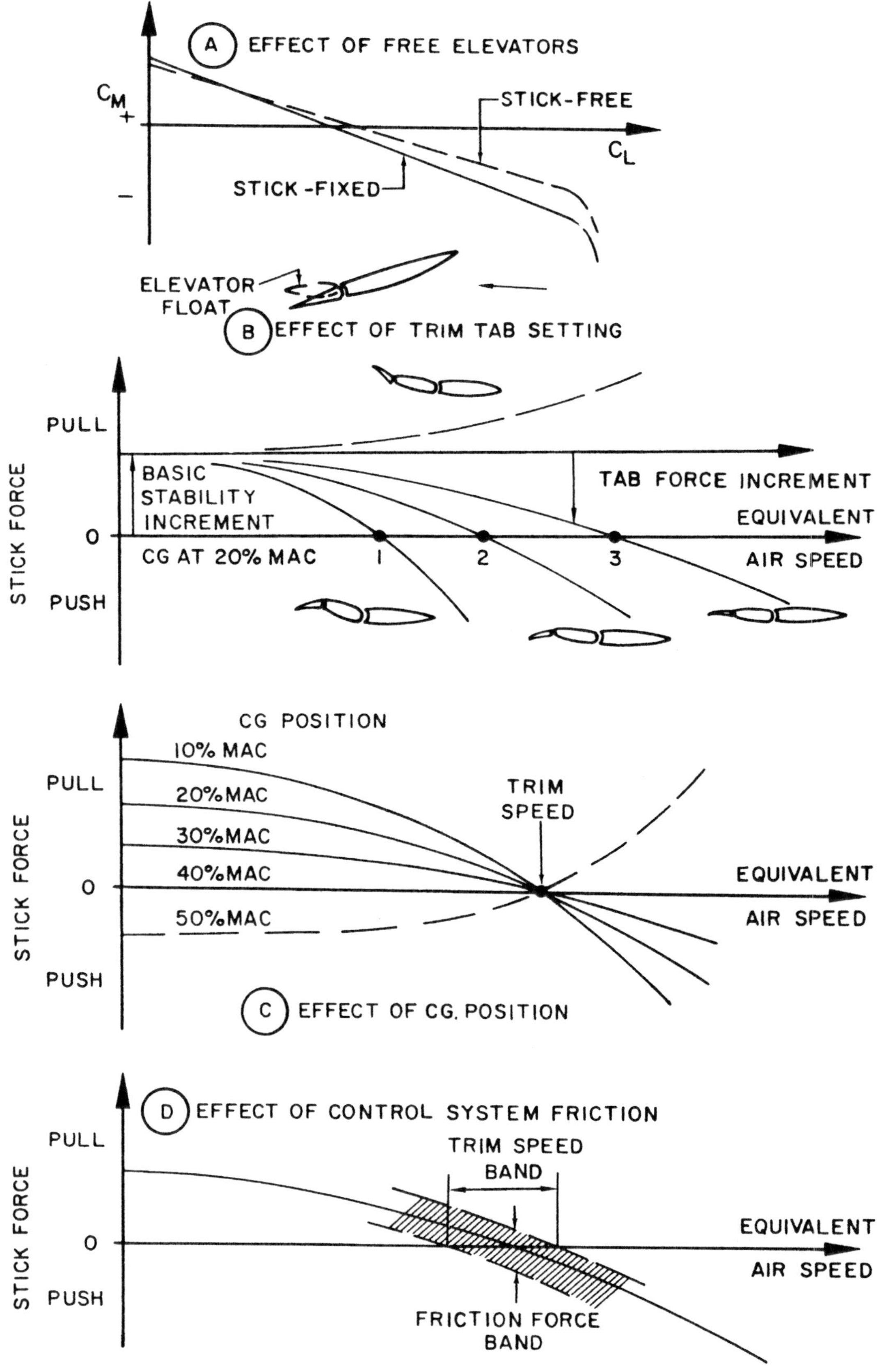

Figure 4.14. Control Force Stability

speed. If the airplane were given a large nose down tab setting the pull force would increase with airspeed. This fact points out the possibility of "feel" as not being a true indication of airplane static stability.

If the c.g. of the airplane were varied while maintaining trim at a constant airspeed, the effect of c.g. position on stick force stability could be appreciated. As illustrated in figure 4.14(C), moving the c.g. aft decreases the slope of the line of stick force through the trim speed. Thus, decreasing stick force stability is evident in that smaller stick forces are necessary to displace the airplane from the trim speed. When the stick force gradient (or slope) becomes zero, the c.g. is at the stick-free neutral point and neutral stability exists. If the c.g. is aft of the stick-free neutral point, stick force instability will exist, e.g. the airplane will require a push force at a lower speed or a pull force at a higher speed. It should be noted that the stick force gradient is low at low airspeeds and when the airplane is at low speeds, high power, and a c.g. position near the aft limit, the "feel" for airspeed will be weak.

Control system friction can create very undesirable effects on control forces. Figure 4.14(D) illustrates that the control force versus airspeed is a band rather than a line. A wide friction force band can completely mask the stick force stability when the stick force stability is low. Modern flight control systems require precise maintenance to minimize the friction force band and preserve proper feel to the airplane.

MANEUVERING STABILITY. When an airplane is subject to a normal acceleration, the flight path is curved and the airplane is subject to a pitching velocity. Because of the pitching velocity in maneuvering flight, the longitudinal stability of the airplane is slightly greater than in steady flight conditions. When an airplane is subject to a pitching velocity at a given lift coefficient, the airplane develops a pitching moment resisting the pitch motion which adds to the restoring moment from the basic static stability. The principal source of this additional pitching moment is illustrated in figure 4.15.

During a pull-up the airplane is subject to an angular rotation about the lateral axis and the horizontal tail will experience a component of wind due to the pitching velocity. The vector addition of this component velocity to the flight velocity provides a change in angle of attack for the tail and the change in lift on the tail creates a pitching moment resisting the pitching motion. Since the pitching moment opposes the pitching motion but is due to the pitching motion, the effect is a damping in pitch. Of course, the other components of the airplane may develop resisting moments and contribute to pitch damping but the horizontal tail is usually the largest contribution. The added pitching moment from pitch damping will effect a higher stability in maneuvers than is apparent in steady flight. From this consideration, the neutral point for maneuvering flight will be aft of the neutral point for unaccelerated flight and in most cases will not be a critical item. If the airplane demonstrates static stability in unaccelerated flight, it will most surely demonstrate stability in maneuvering flight.

The most direct appreciation of the maneuvering stability of an airplane is obtained from a plot of stick force versus load factor such as shown in figure 4.15. The airplane with positive maneuvering stability should demonstrate a steady increase in stick force with increase in load factor or "*G*". The maneuvering stick force gradient—or stick force per *G*—must be positive but should be of the proper magnitude. The stick force gradient must not be excessively high or the airplane will be difficult and tiring to maneuver. Also, the stick force gradient must not be too low or the airplane may be overstressed inadvertently when light control forces exist. A maneuvering stick force gradient of 3 to 8 lbs. per *G* is satisfactory for most fighter and

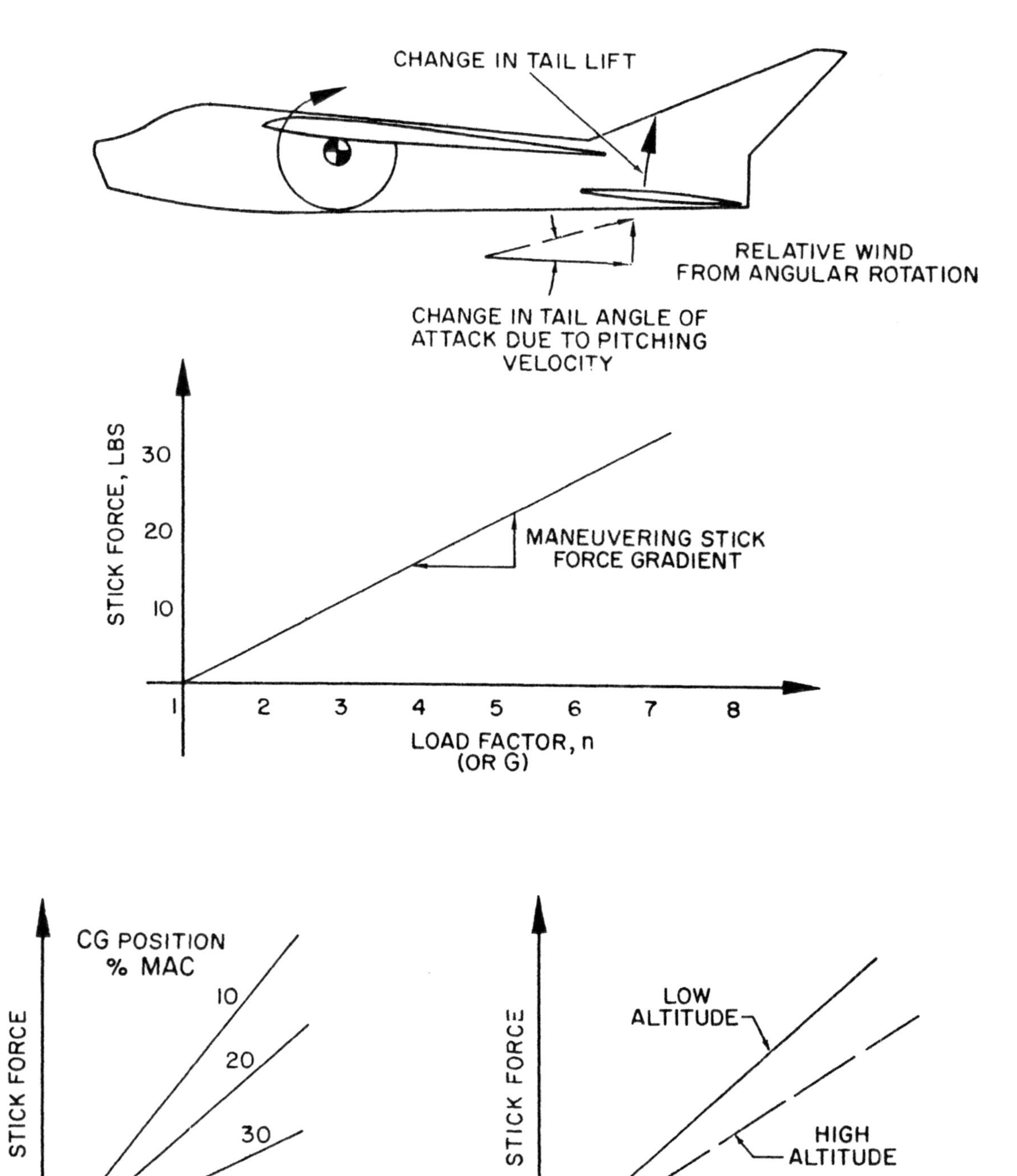

Figure 4.15. Maneuvering Stability

attack airplanes. A large patrol or transport type airplane would ordinarily show a much higher maneuvering stick force gradient because of the lower limit load factor.

When the airplane has high static stability, the maneuvering stability will be high and a high stick force gradient will result. A possibility exists that the forward c.g. limit could be set to prevent an excessively high maneuvering stick force gradient. As the c.g. is moved aft, the stick force gradient decreases with decreasing maneuvering stability and the lower limit of stick force gradient may be reached.

The pitch damping of the airplane is obviously related to air density. At high altitudes, the high true airspeed reduces the change in tail angle of attack for a given pitching velocity and reduces the pitch damping. Thus, a decrease in maneuvering stick force stability can be expected with increased altitude.

TAILORING CONTROL FORCES. The control forces should reflect the stability of the airplane but, at the same time, should be of a tolerable magnitude. The design of the surfaces and control system may employ an infinite variety of techniques to provide satisfactory control forces.

Aerodynamic balance must be thought of in two different senses. First, the control surface must be balanced to reduce hinge moments due to changes in angle of attack. This is necessary to reduce the floating tendency of the surface which reduces the stick-free stability. Next, aerodynamic balance can reduce the hinge moments due to deflection of the control surface. Generally, it is difficult to obtain a high degree of deflection balance without incurring a large overbalance of the surface for changes in angle of attack.

Some of the types of aerodynamic balance are illustrated in figure 4.16. The simple horn type balance employs a concentrated balance area located ahead of the hinge line. The balance area may extend completely to the leading edge (unshielded) or part way to the leading edge (shielded). Aerodynamic balance can be achieved by the provision of a hinge line aft of the control surface leading edge. The resulting overhang of surface area ahead of the hinge line will provide a degree of balance depending on the amount of overhang. Another variation of aerodynamic balance is an internal balance surface ahead of the hinge line which is contained within the surface. A flexible seal is usually incorporated to increase the effectiveness of the balance area. Even the bevelling of the trailing edge of the control surface is effective also as a balancing technique. The choice of the type of aerodynamic balance will depend on many factors such as required degree of balance, simplicity, drag, etc.

Many devices can be added to a control system to modify or tailor the stick force stability to desired levels. If a spring is added to the control system as shown in figure 4.16, it will tend to center the stick and provide a force increment depending on stick displacement. When the control system has a fixed gearing between stick position and surface deflection, the centering spring will provide a contribution to stick force stability according to stick position. The contribution to stick force stability will be largest at low flight speeds where relatively large control deflections are required. The contribution will be smallest at high airspeed because of the smaller control deflections required. Thus, the stick centering bungee will increase the airspeed and maneuvering stick force stability but the contribution decreases at high airspeeds. A variation of this device would be a spring stiffness which would be controlled to vary with dynamic pressure, q. In this case, the contribution of the spring to stick force stability would not diminish with speed.

A "downspring" added to a control system is a means of increasing airspeed stick force stability without a change in airplane static

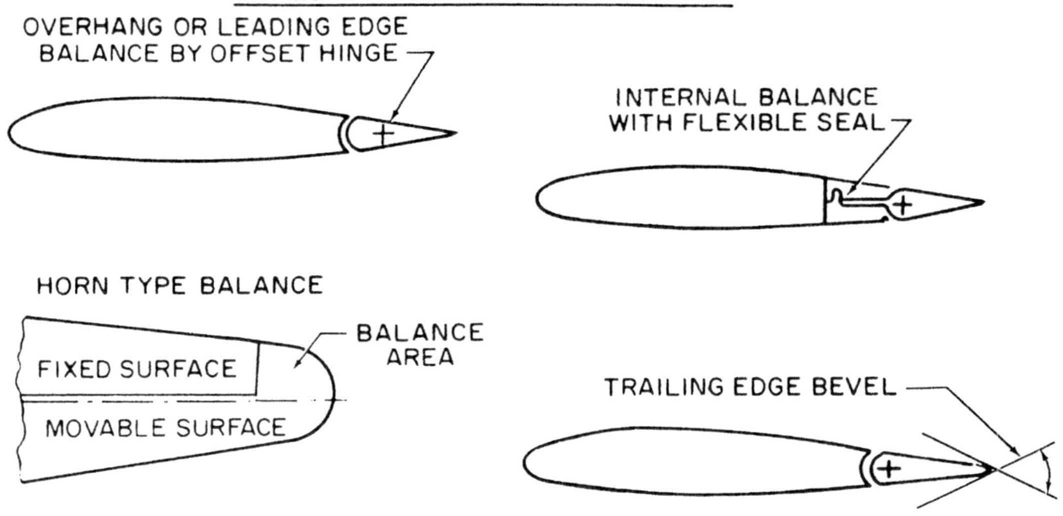

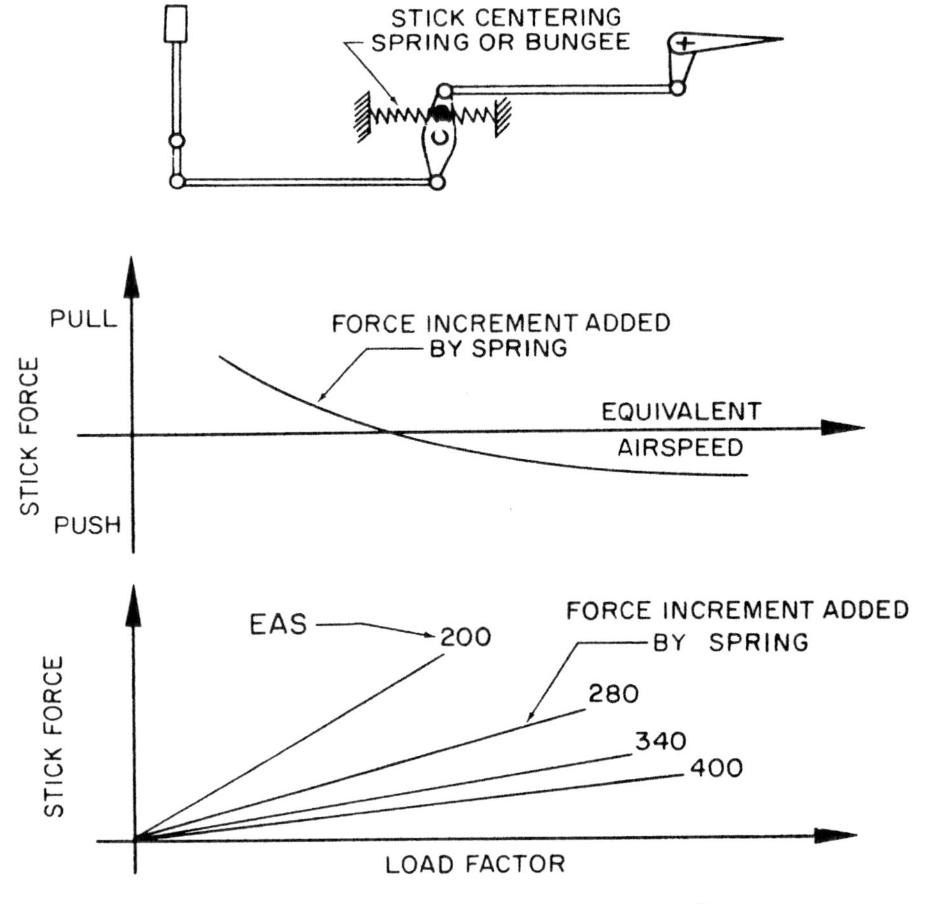

Figure 4.16. Tailoring Control Forces

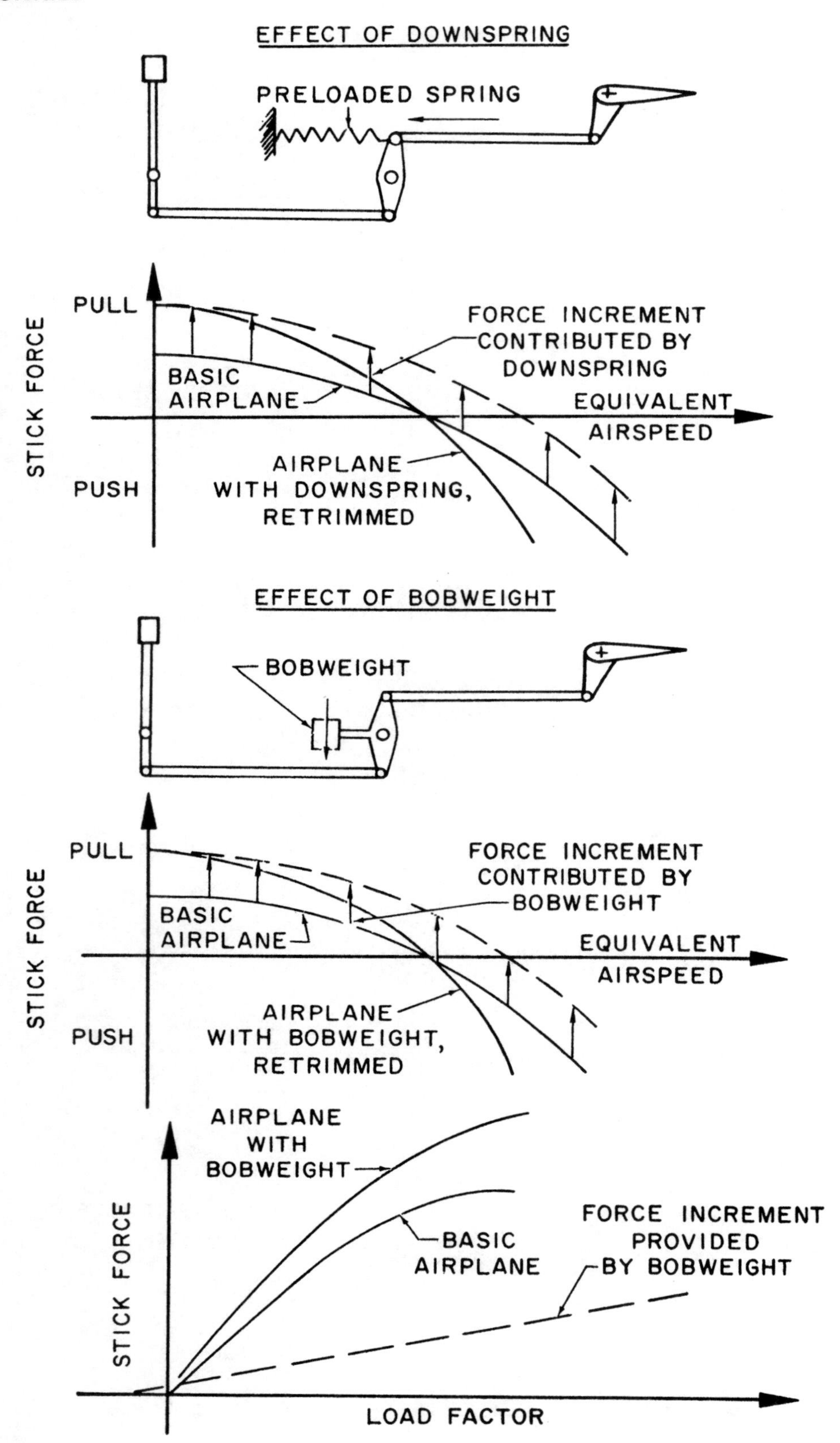

Figure 4.17. Tailoring Control Forces

stability. As shown in figure 4.17, a downspring consists of a long preloaded spring attached to the control system which tends to rotate the elevators down. The effect of the downspring is to contribute an increment of pull force independent of control deflection or airspeed. When the downspring is added to the control system of an airplane and the airplane is retrimmed for the original speed, the airspeed stick force gradient is increased and there is a stronger feel for airspeed. The downspring would provide an "ersatz" improvement to an airplane deficient in airspeed stick force stability. Since the force increment from the downspring is unaffected by stick position or normal acceleration, the maneuvering stick force stability would be unchanged.

The bobweight is an effective device for improving stick force stability. As shown in figure 4.17, the bobweight consists of an eccentric mass attached to the control system which—in unaccelerated flight—contributes an increment of pull force identical to the downspring. In fact, a bobweight added to the control system of an airplane produces an effect identical to the downspring. The bobweight will increase the airspeed stick force gradient and increase the feel for airspeed.

A bobweight will have an effect on the maneuvering stick force gradient since the bobweight mass is subjected to the same acceleration as the airplane. Thus, the bobweight will provide an increment of stick force in direct proportion to the maneuvering acceleration of the airplane. Because of the linear contribution of the bobweight, the bobweight can be applied to increase the maneuvering stick force stability if the basic airplane has too low a value or develops a decreasing gradient at high lift coefficients.

The example of the bobweight is useful to point out the effect of the control system distributed masses. All carrier aircraft must have the control system mass balanced to prevent undesirable control forces from the longitudinal accelerations during catapult launching.

Various control surface tab devices can be utilized to modify control forces. Since the deflection of a tab is so powerful in creating hinge moments on a control surface, the possible application of tab devices is almost without limit. The basic trim tab arrangement is shown in figure 4.18 where a variable linkage connects the tab and the control surface. Extension or contraction of this linkage will deflect the tab relative to the control surface and create a certain change in hinge moment coefficient. The use of the trim tab will allow the pilot to reduce the hinge moment to zero and trim the control forces to zero for a given flight condition. Of course, the trim tab should have adequate effectiveness so that control forces can be trimmed out throughout the flight speed range.

The *lagging tab* arrangement shown in figure 4.18 employs a linkage between the fixed surface and the tab surface. The geometry is such that upward deflection of the control surface displaces the tab down relative to the control surface. Such relative displacement of the tab will aid in deflection of the control surface and thus reduce the hinge moments due to deflection. An obvious advantage of this device is the reduction of deflection hinge moments without a change in aerodynamic balance.

The *leading tab* arrangement shown in figure 4.18 also employs a linkage between the fixed surface and the tab surface. However, the geometry of the linkage is such that upward deflection of the control surface displaces the tab up relative to the control surface. This relationship serves to increase the control surface hinge moments due to deflection of the surface.

The *servo tab* shown in figure 4.18 utilizes a horn which has no direct connection to the control surface and is free to pivot about the hinge axis. However, a linkage connects this free horn to the tab surface. Thus, the control system simply deflects the tab and the resulting hinge moments deflect the control surface.

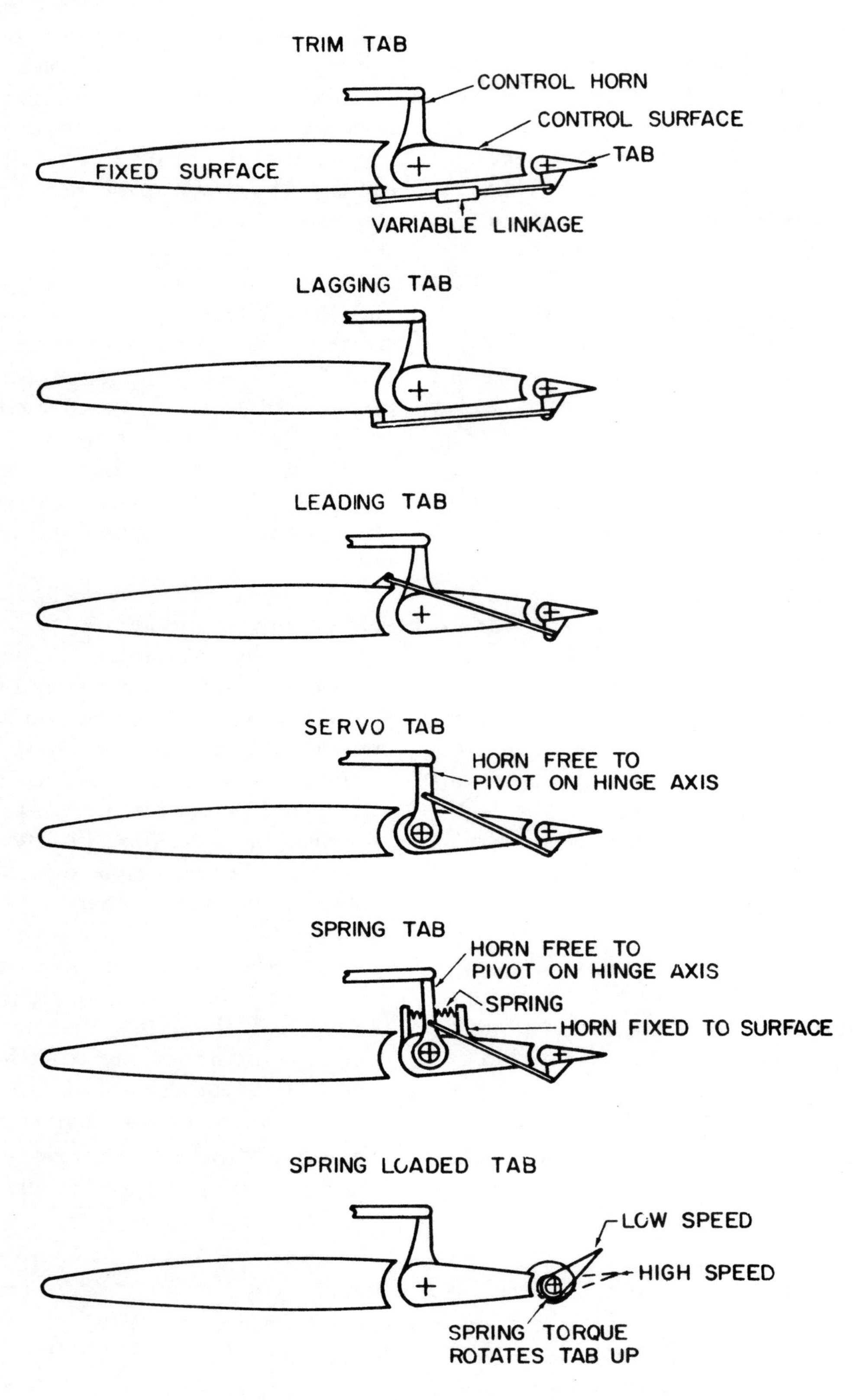

Figure 4.18. Various Tab Devices

Since the only control forces are those of the tab, this device makes possible the deflection of large surfaces with relatively small control forces.

A variation of the basic servo tab layout is the *spring tab* arrangement of figure 4.18. When the control horn is connected to the control surface by springs, the function of the tab is to provide a given portion of the required control forces. The spring tab arrangement can then function as a boost to reduce control forces. The servo tab and spring tab are usually applied to large or high speed subsonic airplanes to provide tolerable stick forces.

The *spring loaded tab* of figure 4.18 consists of a free tab preloaded with a spring which furnishes a constant moment about the tab hinge line. When the airplane is at zero airspeed, the tab is rotated up to the limit of deflection. As airspeed is increased, the aerodynamic hinge moment on the tab will finally equal the spring torque and the tab will begin to streamline. The effect of this arrangement is to provide a constant hinge moment to the control system and contribute a constant push force requirement at speeds above the preload speed. Thus, the spring loaded tab can improve the stick force gradient in a manner similar to the downspring. Generally, the spring loaded tab may be more desirable because of greater effectiveness and the lack of undesirable control forces during ground operation.

The various tab devices have almost unlimited possibilities for tailoring control forces. However, these devices must receive proper care and maintenance in order to function properly. In addition, much care must be taken to ensure that no slop or play exists in the joints and fittings, otherwise destructive flutter may occur.

LONGITUDINAL CONTROL

To be satisfactory, an airplane must have adequate controllability as well as adequate stability. An airplane with high static longitudinal stability will exhibit great resistance to displacement from equilibrium. Hence, the most critical conditions of controllability will occur when the airplane has high stability, i.e., the lower limits of controllability will set the upper limits of stability.

There are three principal conditions of flight which provide the critical requirements of longitudinal control power. Any one or combination of these conditions can determine the longitudinal control power and set a limit to forward c.g. position.

MANEUVERING CONTROL REQUIREMENT. The airplane should have sufficient longitudinal control power to attain the maximum usable lift coefficient or limit load factor during maneuvers. As shown in figure 4.19, forward movement of the c.g. increases the longitudinal stability of an airplane and requires larger control deflections to produce changes in trim lift coefficient. For the example shown, the maximum effective deflection of the elevator is not capable of trimming the airplane at $C_{L_{max}}$ for c.g. positions ahead of 18 percent *MAC*.

This particular control requirement can be most critical for an airplane in supersonic flight. Supersonic flight is usually accompanied by large increases in static longitudinal stability and a reduction in the effectiveness of control surfaces. In order to cope with these trends, powerful all-movable surfaces must be used to attain limit load factor or maximum usable C_L in supersonic flight. This requirement is so important that once satisfied, the supersonic configuration usually has sufficient longitudinal control power for all other conditions of flight.

TAKEOFF CONTROL REQUIREMENT. At takeoff, the airplane must have sufficient control power to assume the takeoff attitude prior to reaching takeoff speed. Generally, for airplanes with tricycle landing gears, it is desirable to have at least sufficient control power to attain the takeoff attitude at 80

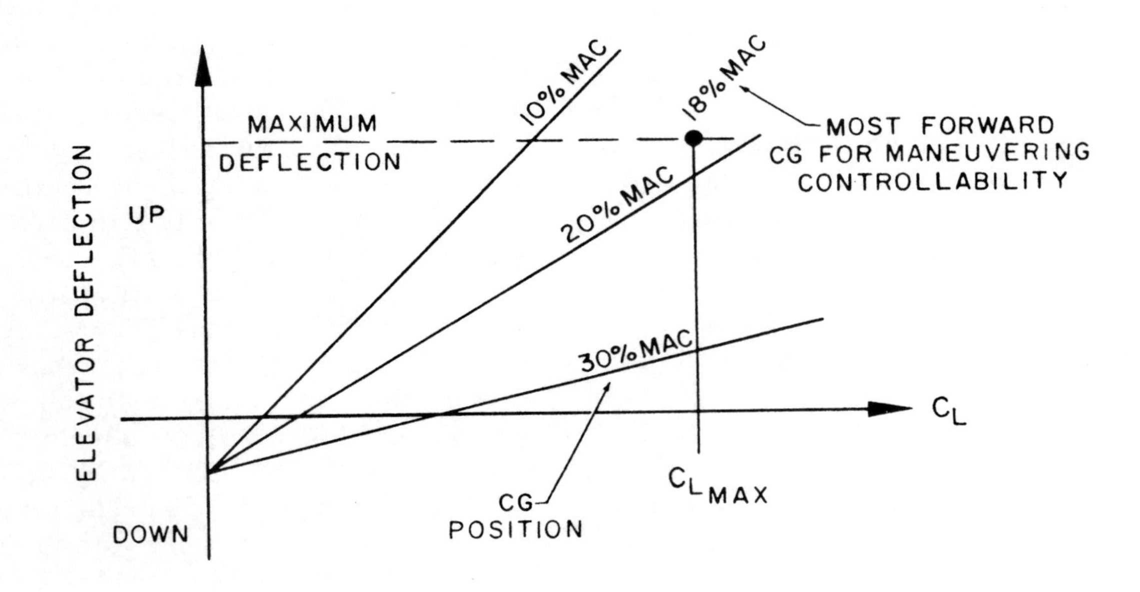

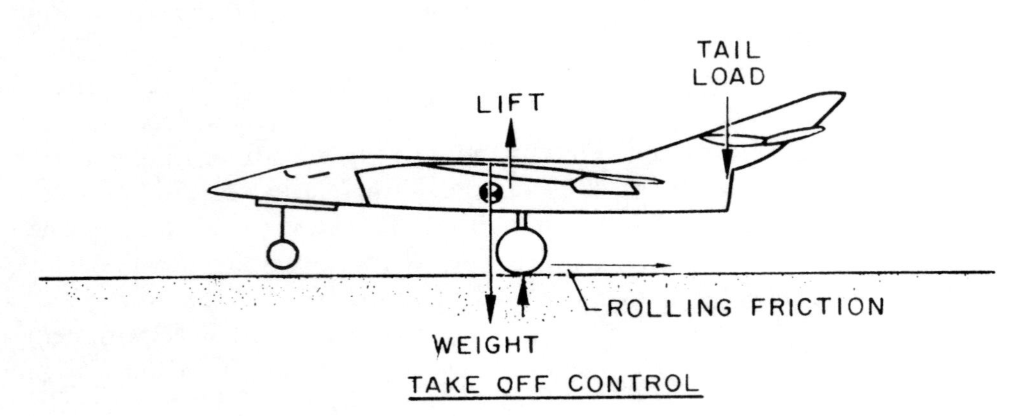

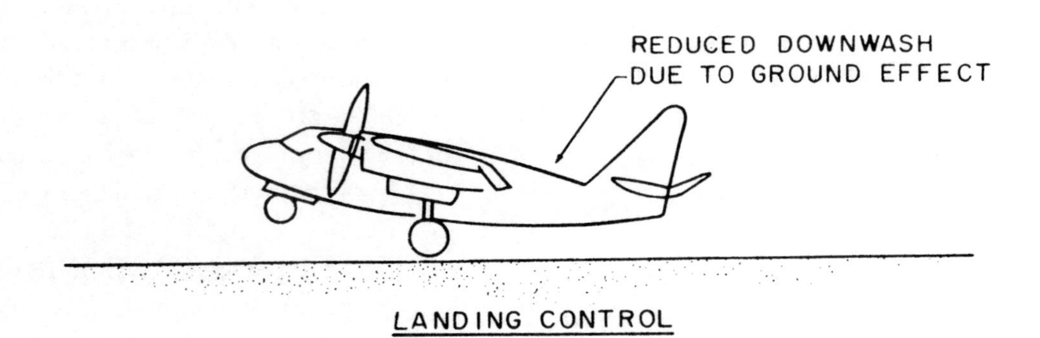

Figure 4.19. Longitudinal Control Requirements

percent of the stall speed for propeller airplanes or 90 percent of the stall speed for jet airplanes. This feat must be accomplished on a smooth runway at all normal service takeoff loading conditions.

Figure 4.19 illustrates the principal forces acting on an airplane during takeoff roll. When the airplane is in the three point attitude at some speed less than the stall speed, the wing lift will be less than the weight of the airplane. As the elevators must be capable of rotating to the takeoff attitude, the critical condition will be with zero load on the nose wheel and the net of lift and weight supported on the main gear. Rolling friction resulting from the normal force on the main gear creates an adverse nose down moment. Also, the center of gravity ahead of the main gear contributes a nose down moment and this consideration could decide the most aft location of the main landing gear during design. The wing may contribute a large nose down moment when flaps are deflected but this effect may be countered by a slight increase in downwash at the tail. To balance these nose down moments, the horizontal tail should be capable of producing sufficient nose up moment to attain the takeoff attitude at the specified speeds.

The propeller airplane at takeoff power may induce considerable slipstream velocity at the horizontal tail which can provide an increase in the efficiency of the surface. The jet airplane does not experience a similar magnitude of this effect since the induced velocities from the jet are relatively small compared to the slipstream velocities from a propeller.

LANDING CONTROL REQUIREMENT. At landing, the airplane must have sufficient control power to ensure adequate control at specified landing speeds. Adequate landing control is usually assured if the elevators are capable of holding the airplane just off the runway at 105 percent of the stall speed. Of course, the most critical requirement will exist when the c.g. is in the most forward position, flaps are fully extended, and power is set at idle. This configuration will provide the most stable condition which is most demanding of controllability. The full deflection of flaps usually provides the greatest wing diving moment and idle power will produce the most critical (least) dynamic pressure at the horizontal tail.

The landing control requirement has one particular difference from the maneuvering control requirement of free flight. As the airplane approaches the ground surface, there will be a change in the three-dimensional flow of the airplane due to ground effect. A wing in proximity to the ground plane will experience a decrease in tip vortices and downwash at a given lift coefficient. The decrease in downwash at the tail tends to increase the static stability and produce a nosedown moment from the reduction in download on the tail. Thus, the airplane just off the runway surface will require additional control deflection to trim at a given lift coefficient and the landing control requirement may be critical in the design of longitudinal control power.

As an example of ground effect, a typical propeller powered airplane may require as much as 15° more up elevator to trim at $C_{L_{max}}$ in ground effect than in free flight away from the ground plane. Because of this effect, many airplanes have sufficient control power to achieve full stall out of ground effect but do not have the ability to achieve full stall when in close proximity to the ground.

In some cases the effectiveness of the control surface is adversely affected by the use of trim tabs. If trim tabs are used to excess in trimming stick forces, the effectiveness of the elevator may be reduced to hinder landing or takeoff control.

Each of the three principal conditions requiring adequate longitudinal control are critical for high static stability. If the forward c.g. limit is exceeded, the airplane may encounter a deficiency of controllability in any of these conditions. Thus, the forward c.g.

4
NAVY

limit is set by the minimum permissible controllability while the aft c.g. limit is set by the minimum permissible stability.

LONGITUDINAL DYNAMIC STABILITY. All previous considerations of longitudinal stability have been concerned with the initial tendency of the airplane to return to equilibrium when subjected to a disturbance. The considerations of longitudinal dynamic stability are concerned with time history response of the airplane to these disturbances, i.e., the variation of displacement amplitude with time following a disturbance. From previous definition, dynamic stability will exist when the amplitude of motion decreases with time and dynamic instability will exist if the amplitude increases with time.

Of course, the airplane must demonstrate positive dynamic stability for the major longitudinal motions. In addition, the airplane must demonstrate a certain degree of longitudinal stability by reducing the amplitude of motion at a certain rate. The required degree of dynamic stability is usually specified by the time necessary for the amplitude to reduce to one-half the original value—the time to damp to half-amplitude.

The airplane in free flight has six degrees of freedom: rotation in roll, pitch, and yaw and translation in the horizontal, vertical, and lateral directions. In the case of longitudinal dynamic stability, the degrees of freedom can be limited to pitch rotation, vertical and horizontal translation. Since the airplane is usually symmetrical from port to starboard, there will be no necessity for consideration of coupling between longitudinal and lateral-directional motions. Thus, the principal variables in the longitudinal motion of an airplane will be:

(1) The pitch attitude of the airplane.

(2) The angle of attack (which will differ from the pitch attitude by the inclination of the flight path).

(3) The flight velocity.

(4) The displacement or deflection of the elevator when the stick-free condition is considered.

The longitudinal dynamic stability of an airplane generally consists of three basic modes (or manners) of oscillation. While the longitudinal motion of the airplane may consist of a combination of these modes, the characteristics of each mode are sufficiently distinct that each oscillatory tendency may be studied separately.

The *first mode* of dynamic longitudinal stability consists of a very long period oscillation referred to as the *phugoid*. The phugoid or long period oscillation involves noticeable variations in pitch attitude, altitude, and airspeed but nearly constant angle of attack. Such an oscillation of the airplane could be considered as a gradual interchange of potential and kinetic energy about some equilibrium airspeed and altitude. Figure 4.20 illustrates the characteristic motion of the phugoid.

The period of oscillation in the phugoid is quite large, typical values being from 20 to 100 seconds. Since the pitching rate is quite low and only negligible changes in angle of attack take place, damping of the phugoid is weak and possibly negative. However, such weak or negative damping does not necessarily have any great consequence. Since the period of oscillation is so great, the pilot is easily able to counteract the oscillatory tendency by very slight and unnoticed control movements. In most cases, the necessary corrections are so slight that the pilot may be completely unaware of the oscillatory tendency.

Due to the nature of the phugoid, it is not necessary to make any specific aerodynamic provisions to contend with the oscillation. The inherent long period of the oscillation allows study to be directed to more important oscillatory tendencies. Similarly, the differences between the stick-fixed and stick-free phugoid are not of great importance.

The *second mode* of longitudinal dynamic stability is a relatively short period motion that

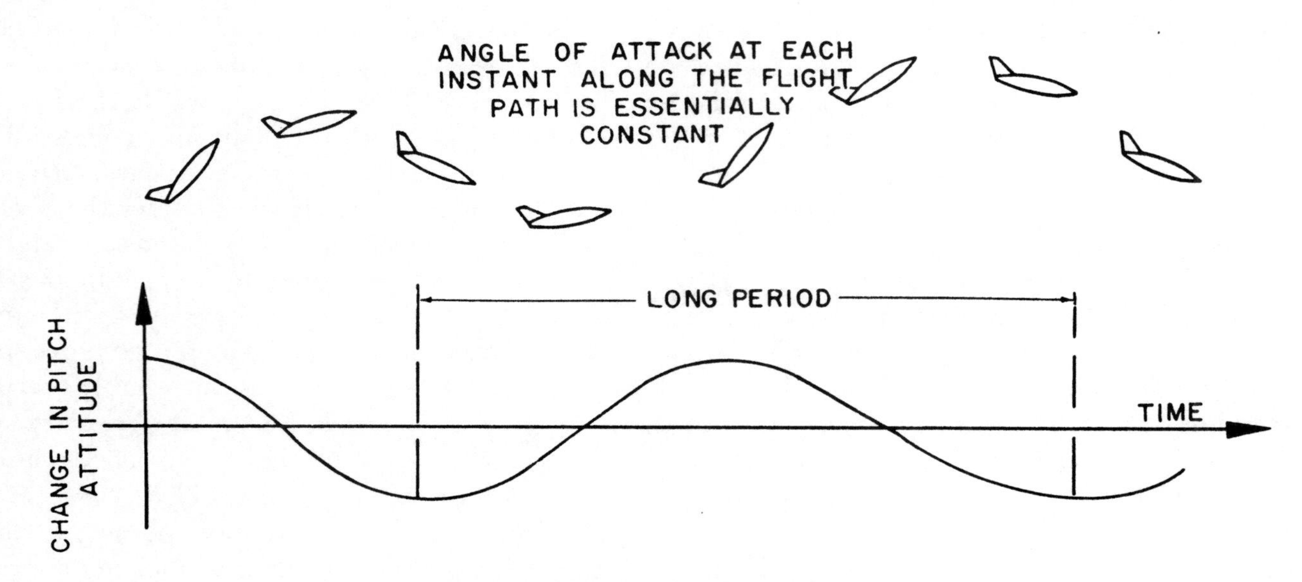

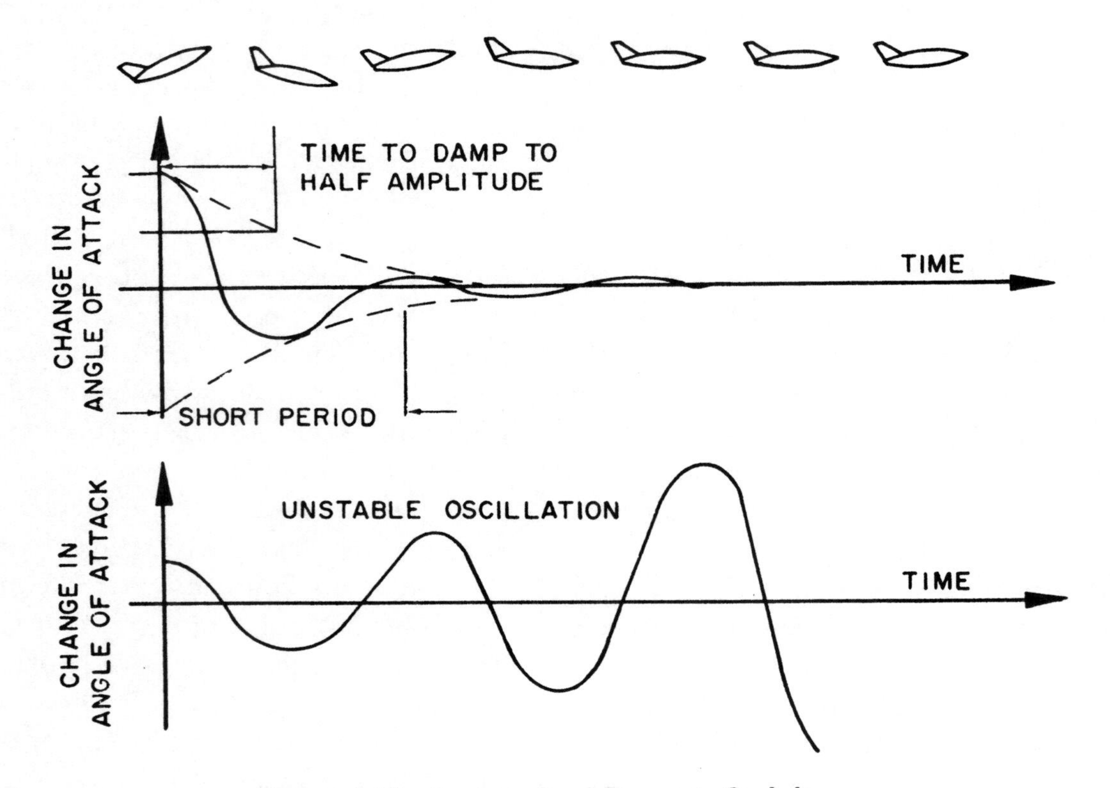

Figure 4.20. Longitudinal Dynamic Stability

can be assumed to take place with negligible changes in velocity. The second mode consists of a pitching oscillation during which the airplane is being restored to equilibrium by the static stability and the amplitude of oscillation decreased by pitch damping. The typical motion is of relatively high frequency with a period of oscillation on the order of 0.5 to 5 seconds.

For the conventional subsonic airplane, the second mode stick-fixed is characterized by heavy damping with a time to damp to half amplitude of approximately 0.5 seconds. Usually, if the airplane has static stability stick-fixed, the pitch damping contributed by the horizontal tail will assume sufficient dynamic stability for the short period oscillation. However, the second mode *stick-free* has the possibility of weak damping or unstable oscillations. This is the case where static stability does not automatically imply adequate dynamic stability. The second mode stick-free is essentially a coupling of motion between the airplane short period pitching motion and elevator in rotation about the hinge line. Extreme care must be taken in the design of the control surfaces to ensure dynamic stability for this mode. The elevators must be statically balanced about the hinge line and aerodynamic balance must be within certain limits. Control system friction must be minimized as it contributes to the oscillatory tendency. If instability were to exist in the second mode, "porpoising" of the airplane would result with possibility of structural damage. An oscillation at high dynamic pressures with large changes in angle of attack could produce severe flight loads.

The second mode has relatively short periods that correspond closely with the normal pilot response lag time, e.g., 1 or 2 seconds or less. There is the possibility that an attempt to forceably damp an oscillation may actually reinforce the oscillation and produce instability. This is particularly true in the case of powered controls where a small input energy into the control system is greatly magnified. In addition, response lag of the controls may add to the problem of attempting to forceably damp the oscillation. In this case, should an oscillation appear, the best rule is to release the controls as the airplane stick-free will demonstrate the necessary damping. Even an attempt to fix the controls when the airplane is oscillating may result in a small unstable input into the control system which can reinforce the oscillation to produce failing flight loads. Because of the very short period of the oscillation, the amplitude of an unstable oscillation can reach dangerous proportions in an extremely short period of time.

The *third mode* occurs in the elevator free case and is usually a very short period oscillation. The motion is essentially one of the elevator flapping about the hinge line and, in most cases, the oscillation has very heavy damping. A typical flapping mode may have a period of 0.3 to 1.5 seconds and a time to damp to half-amplitude of approximately 0.1 second.

Of all the modes of longitudinal dynamic stability, the second mode or porpoising oscillation is of greatest importance. The porpoising oscillation has the possibility of damaging flight loads and can be adversely affected by pilot response lag. It should be remembered that when stick-free the airplane will demonstrate the necessary damping.

The problems of dynamic stability are acute under certain conditions of flight. Low static stability generally increases the period (decreases frequency) of the short period oscillations and increases the time to damp to half-amplitude. High altitude—and consequently low density—reduces the aerodynamic damping. Also, high Mach numbers of supersonic flight produce a decay of aerodynamic damping.

MODERN CONTROL SYSTEMS

In order to accomplish the stability and control objectives, various configurations of control systems are necessary. Generally, the type of flight control system is decided by the size and flight speed range of the airplane.

The *conventional control system* consists of direct mechanical linkages from the controls to the control surfaces. For the subsonic airplane, the principal means of producing proper control forces utilize aerodynamic balance and various tab, spring, and bobweight devices. Balance and tab devices are capable of reducing control forces and will allow the use of the conventional control system on large airplanes to relatively high subsonic speeds.

When the airplane with a conventional control system is operated at transonic speeds, the great changes in the character of flow can produce great aberrations in control surface hinge moments and the contribution of tab devices. Shock wave formation and separation of flow at transonic speeds will limit the use of the conventional control system to subsonic speeds.

The *power-boosted control system* employs a mechanical actuator in parallel with the mechanical linkages of a conventional control system. The principle of operation is to provide a fixed, percentage of the required control forces thus reducing control forces at high speeds. The power-boosted control system requires a hydraulic actuator with a control valve which supplies boost force in fixed proportion to control force. Thus, the pilot is given an advantage by the boost ratio to assist in deflecting the control surface, e.g., with a boost ratio of 14, the actuator provides 14 lbs. of force for each 1 lb. of stick force.

The power-boosted control system has the obvious advantage of reducing control forces at high speeds. However, at transonic speeds, the changes in control forces due to shock waves and separation still take place but to a lesser degree. The "feedback" of hinge moments is reduced but the aberrations in stick forces may still exist.

The *power-operated, irreversible control system* consists of mechanical actuators controlled by the pilot. The control surface is deflected by the actuator and none of the hinge moments are fed back through the controls. In such a control system, the control position decides the deflection of the control surfaces regardless of the airloads and hinge moments. Since the power-operated control system has zero feedback, control feel must be synthesized otherwise an infinite boost would exist.

The advantages of the power-operated control system are most apparent in transonic and supersonic flight. In transonic flight, none of the erratic hinge moments are fed back to the pilot. Thus, no unusual or erratic control forces will be encountered in transonic flight. Supersonic flight generally requires the use of an all-movable horizontal surface to achieve the necessary control effectiveness. Such control surfaces must then be actuated and positively positioned by an irreversible device.

The most important item of an artificial feel system is the stick-centering spring or bungee. The bungee develops a stick force in proportion to stick displacement and thus provides feel for airspeed and maneuvers. A bobweight may be included in the feel system to develop a steady positive maneuvering stick force gradient which is independent of airspeed for ordinary maneuvers.

The gearing between the stick position and control surface deflection is not necessarily a linear relationship. The majority of powered control systems will employ a nonlinear gearing such that relatively greater stick deflection per surface deflection will occur at the neutral stick position. This sort of gearing is to advantage for airplanes which operate at flight conditions of high dynamic pressure. Since the airplane at high q is very sensitive to small deflections of the control surface, the nonlinear gearing provides higher stick force stability with less sensitive control movements than the system with a linear gearing. Figure 4.21 illustrates a typical linear and nonlinear control system gearing.

The second chart of figure 4.21 illustrates the typical control system stick force variation

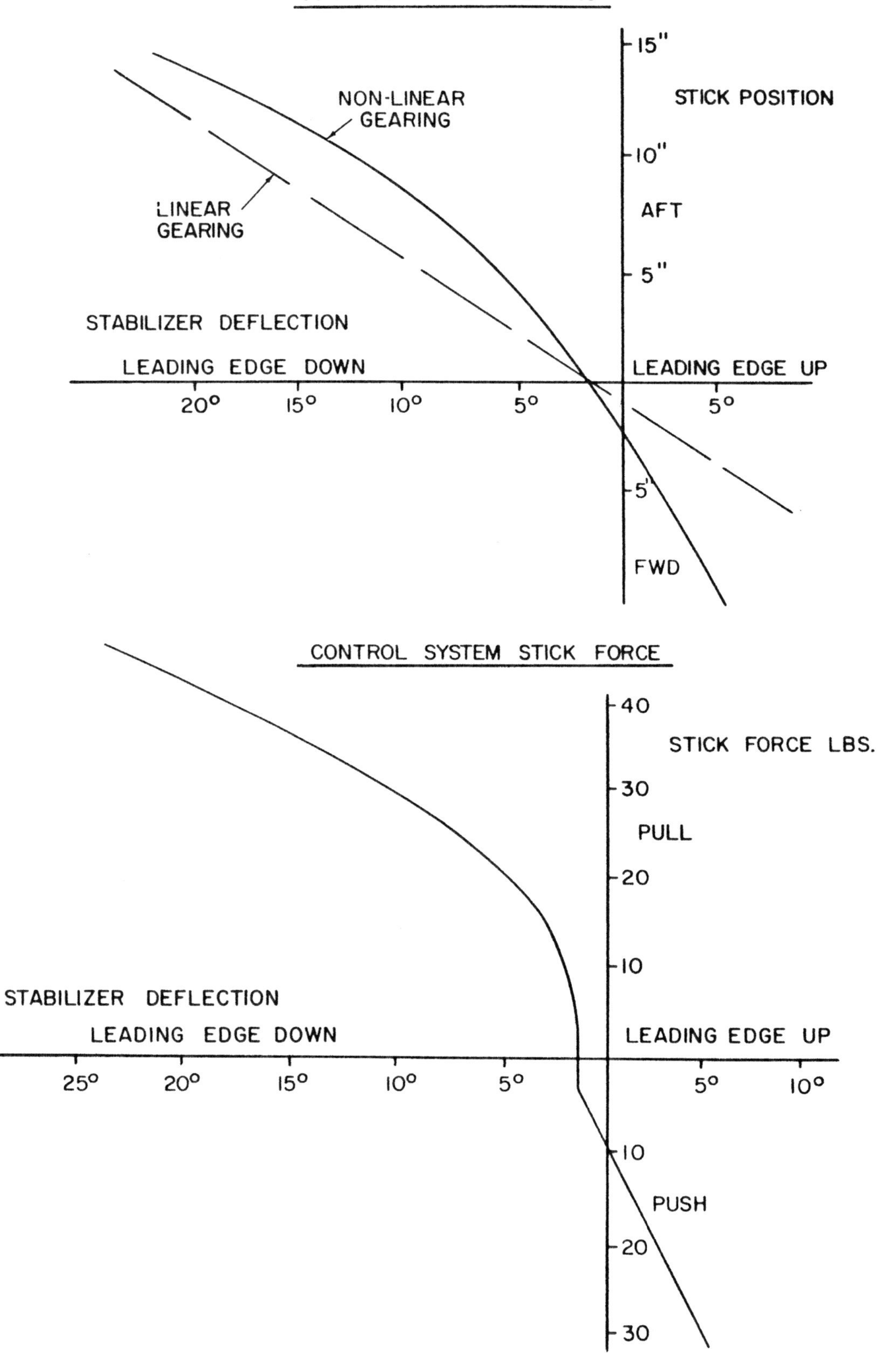

Figure 4.21. Longitudinal Control System

with control surface deflection. While it is desirable to have a strong centering of the stick near the neutral position, the amount of force required to create an initial displacement must be reasonable. If the control system "break-out" forces are too high, precise control of the airplane at high speeds is difficult. As the solid friction of the control system contributes to the break-out forces, proper maintenance of the control system is essential. Any increase in control system friction can create unusual and undesirable control forces.

The trim of the powered control system is essentially any device to produce zero control force for a given control surface deflection. One system may trim off bungee force at a given stick position while another system may trim by returning the stick to neutral position.

Flight at high supersonic Mach numbers might require a great variety of devices in the longitudinal control system. The deterioration of pitch damping with Mach number may require that dynamic stability be obtained synthetically by pitch dampers in the control system. The response of the airplane to longitudinal control may be adversely affected by flight at high dynamic pressures. In such conditions of flight stick forces must be adequate to prevent an induced oscillation. Stick forces must relate the transients of flight as well as the steady state conditions. Such a contribution to control system forces may be provided by a pitching acceleration bobweight and a control system viscous damper.

DIRECTIONAL STABILITY AND CONTROL

DIRECTIONAL STABILITY

The directional stability of an airplane is essentially the "weathercock" stability and involves moments about the vertical axis and their relationship with yaw or sideslip angle. An airplane which has static directional stability would tend to return to an equilibrium when subjected to some disturbance from equilibrium. Evidence of static directional stability would be the development of yawing moments which tend to restore the airplane to equilibrium.

DEFINITIONS. The axis system of an airplane will define a positive yawing moment, N, as a moment about the vertical axis which tends to rotate the nose to the right. As in other aerodynamic considerations, it is convenient to consider yawing moments in the coefficient form so that static stability can be evaluated independent of weight, altitude, speed, etc. The yawing moment, N, is defined in the coefficient form by the following equation:

$$N = C_n q S b$$

or

$$C_n = \frac{N}{qSb}$$

where

N = yawing moment, ft.-lbs; positive to the right
q = dynamic pressure, psf
S = wing area, sq. ft.
b = wing span, ft.
C_n = yawing moment coefficient, positive to the right

The yawing moment coefficient, C_n, is based on the wing dimensions S and b as the wing is the characteristic surface of the airplane.

The *yaw angle* of an airplane relates the displacement of the airplane centerline from some reference azimuth and is assigned the shorthand notation ψ (psi). A positive yaw angle occurs when the nose of the airplane is displaced to the right of the azimuth direction. The definition of sideslip angle involves a significant difference. *Sideslip angle* relates the displacement of the airplane centerline from the relative wind rather than some reference azimuth. Sideslip angle is provided the shorthand notation β (beta) and is positive when the relative wind is displaced to the right of the airplane centerline. Figure 4.22 illustrates the definitions of sideslip and yaw angles.

The sideslip angle, β, is essentially the directional angle of attack of the airplane and is the primary reference in lateral stability as well as directional stability considerations. The yaw angle, ψ, is a primary reference for wind tunnel tests and time history motion of an airplane. From the definitions there is no direct relationship between β and ψ for an airplane in free flight, e.g., an airplane flown through a 360° turn has yawed 360° but sideslip may have been zero throughout the entire turn. Since the airplane has no directional sense, static directional stability of the airplane is appreciated by response to sideslip.

The *static directional stability* of an airplane can be illustrated by a graph of yawing moment coefficient, C_n, versus sideslip angle, β, such as shown in figure 4.22. When the airplane is subject to a positive sideslip angle, static directional stability will be evident if a positive yawing moment coefficient results. Thus, when the relative wind comes from the right ($+\beta$), a yawing moment to the right ($+C_n$) should be created which tends to weathercock the airplane and return the nose into the wind. Static directional stability will exist when the curve of C_n versus β has a positive slope and the degree of stability will be a function of the slope of this curve. If the curve has zero slope, there is no tendency to return to equilibrium and neutral static directional stability exists. When the curve of C_n versus β has a negative slope, the yawing moments developed by sideslip tend to diverge rather than restore and static directional instability exists.

The final chart of figure 4.22 illustrates the fact that the instantaneous slope of the curve of C_n versus β will describe the static directional stability of the airplane. At small angles of sideslip a strong positive slope depicts strong directional stability. Large angles of sideslip produce zero slope and neutral stability. At very high sideslip the negative slope of the curve indicates directional instability. This decay of directional stability with increased sideslip is not an unusual condition. However, directional instability should not occur at the angles of sideslip of ordinary flight conditions.

Static directional stability must be in evidence for all the critical conditions of flight. Generally, good directional stability is a fundamental quality directly affecting the pilots' impression of an airplane.

CONTRIBUTION OF THE AIRPLANE COMPONENTS. The static directional stability of the airplane is a result of contribution of each of the various airplane components. While the contribution of each component is somewhat dependent upon and related to other components, it is necessary to study each component separately.

The *vertical tail* is the primary source of directional stability for the airplane. As shown in figure 4.23, when the airplane is in a sideslip the vertical tail will experience a change in angle of attack. The change in lift—or side force—on the vertical tail creates a yawing moment about the center of gravity which tends to yaw the airplane into the relative wind. The magnitude of the vertical tail contribution to static directional stability then depends on the change in tail lift and the tail moment arm. Obviously, the tail moment arm is a powerful factor but essentially dictated by the major configuration properties of the airplane.

When the location of the vertical tail is set, the contribution of the surface to directional stability depends on its ability to produce changes in lift—or side force—with changes in sideslip. The surface area of the vertical tail is a powerful factor with the contribution of the vertical tail being a direct function of the area. When all other possibilities are exhausted, the required directional stability may be obtained by increases in tail area. However, increased surface area has the obvious disadvantage of increased drag.

The lift curve slope of the vertical tail relates how sensitive the surface is to changes in angle of attack. While it is desirable to have a high lift curve slope for the vertical surface, a high aspect ratio surface is not necessarily practical or desirable. The stall

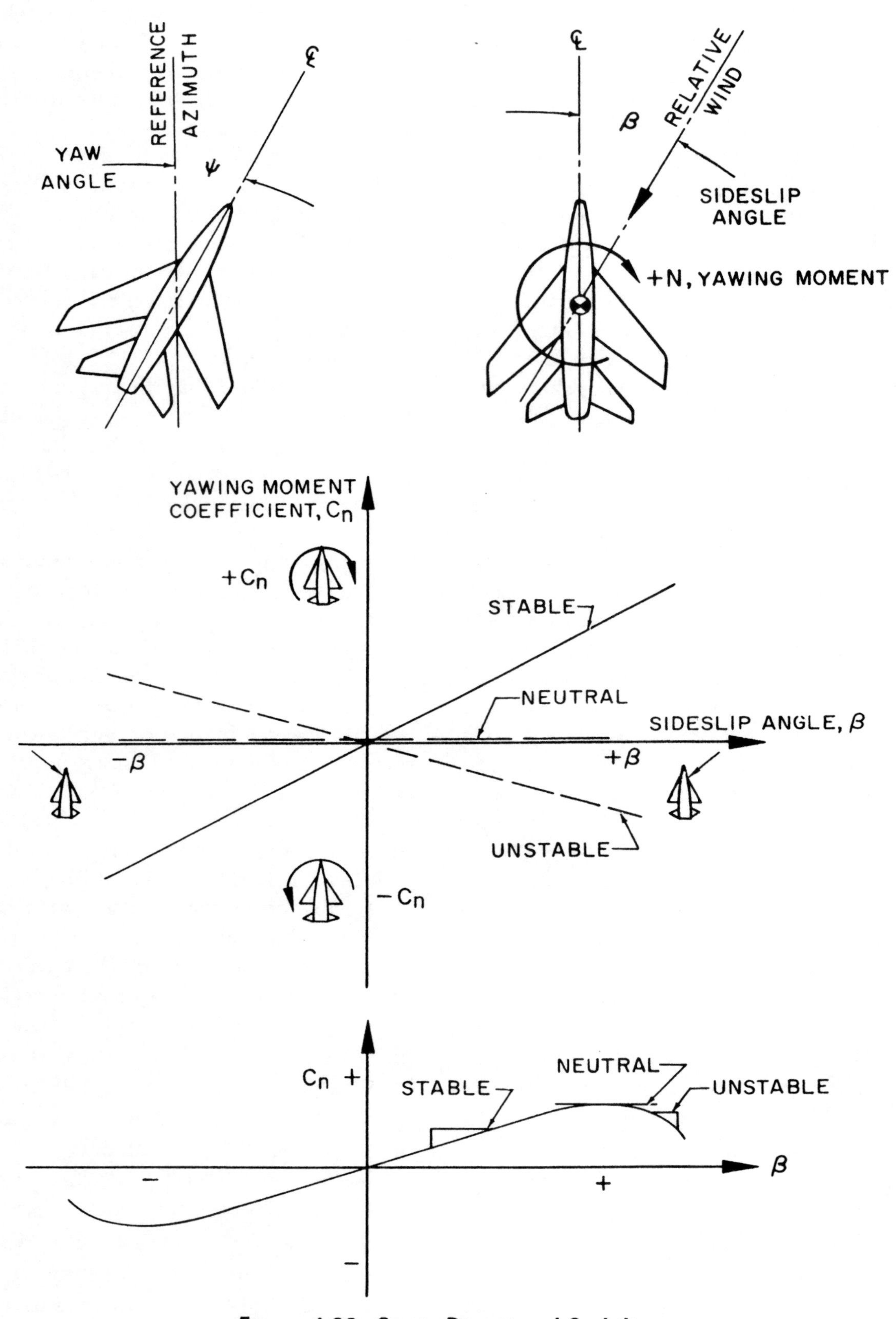

Figure 4.22. Static Directional Stability

angle of the surface must be sufficiently great to prevent stall and subsequent loss of effectiveness at ordinary sideslip angles. The high Mach numbers of supersonic flight produces a decrease in lift curve slope with the consequent reduction in tail contribution to stability. In order to have sufficient directional stability at high Mach numbers, the typical supersonic configuration will exhibit relatively large vertical tail surfaces.

The flow field in which the vertical tail operates is affected by the other components of the airplane as well as power effects. The dynamic pressure at the vertical tail could depend on the slipstream of a propeller or the boundary layer of the fuselage. Also, the local flow direction at the vertical tail is influenced by the wing wake, fuselage crossflow, induced flow of the horizontal tail, or the direction of slipstream from a propeller. Each of these factors must be considered as possibly affecting the contribution of the vertical tail to directional stability.

The contribution of the *wing* to static directional stability is usually small. The swept wing provides a stable contribution depending on the amount of sweepback but the contribution is relatively weak when compared with other components.

The contribution of the *fuselage and nacelles* is of primary importance since these components furnish the greatest destabilizing influence. The contribution of the fuselage and nacelles is similar to the longitudinal case with the exception that there is no large influence of the induced flow field of the wing. The subsonic center of pressure of the fuselage will be located at or forward of the quarter-length point and, since the airplane c.g. is usually considerably aft of this point, the fuselage contribution will be destabilizing. However, at large angles of sideslip the large destabilizing contribution of the fuselage diminishes which is some relief to the problem of maintaining directional stability at large displacements. The supersonic pressure distribution on the body provides a relatively greater aerodynamic force and, generally, a continued destabilizing influence.

Figure 4.23 illustrates a typical buildup of the directional stability of an airplane by separating the contribution of the fuselage and tail. As shown by the graph of C_n versus β, the contribution of the fuselage is destabilizing but the instability decreases at large sideslip angles. The contribution of the vertical tail alone is highly stabilizing up to the point where the surface begins to stall. The contribution of the vertical tail must be large enough so that the complete airplane (wing-fuselage-tail combination) exhibits the required degree of stability.

The dorsal fin has a powerful effect on preserving the directional stability at large angles of sideslip which would produce stall of the vertical tail. The addition of a dorsal fin to the airplane will allay the decay of directional stability at high sideslip in two ways. The least obvious but most important effect is a large increase in the fuselage stability at large sideslip angles. In addition, the effective aspect ratio of the vertical tail is reduced which increases the stall angle for the surface. By this twofold effect, the addition of the dorsal fin is a very useful device.

Power effects on static directional stability are similar to the power effects on static longitudinal stability. The direct effects are confined to the normal force at the propeller plane or the jet inlet and, of course, are destabilizing when the propeller or inlet is located ahead of the c.g. The indirect effects of power induced velocities and flow direction changes at the vertical tail are quite significant for the propeller driven airplane and can produce large directional trim changes. As in the lontitudinal case, the indirect effects are negligible for the jet powered airplane.

The contribution of the direct and indirect power effects to static directional stability is greatest for the propeller powered airplane and usually slight for the jet powered airplane. In either case, the general effect of power is

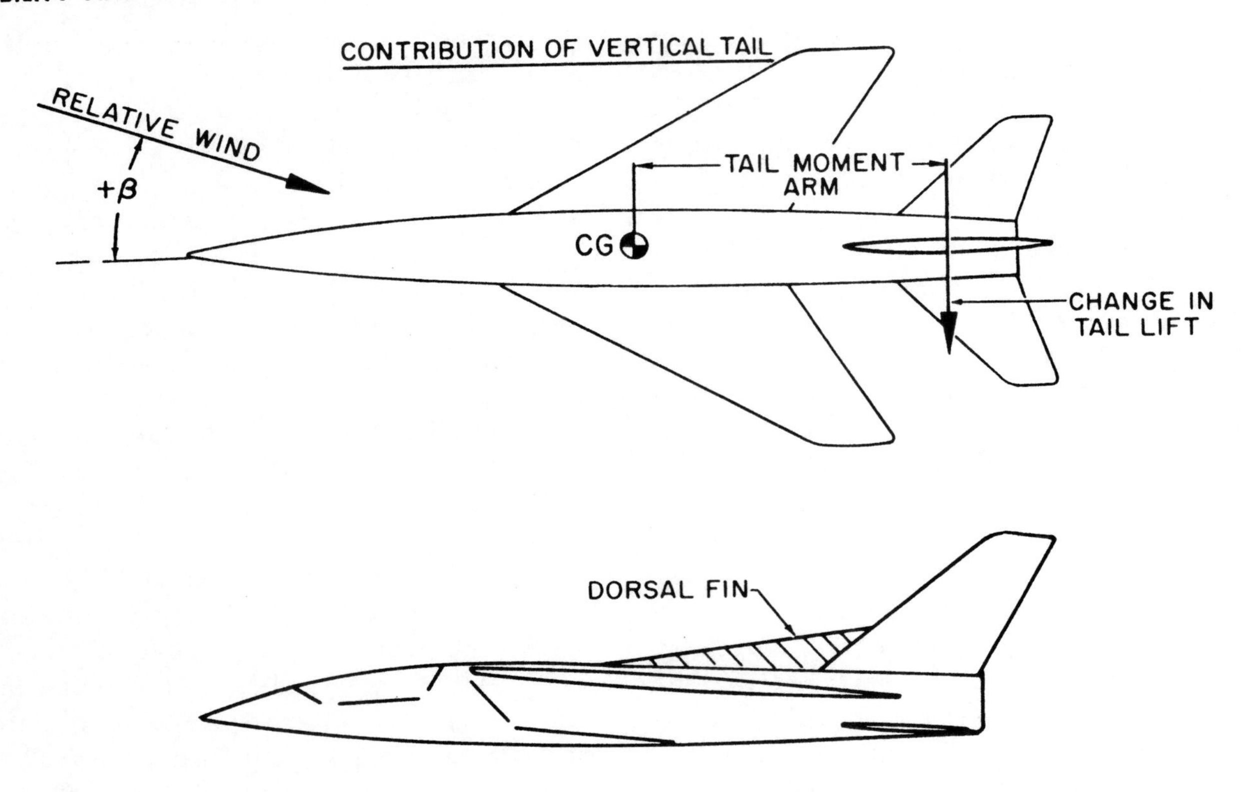

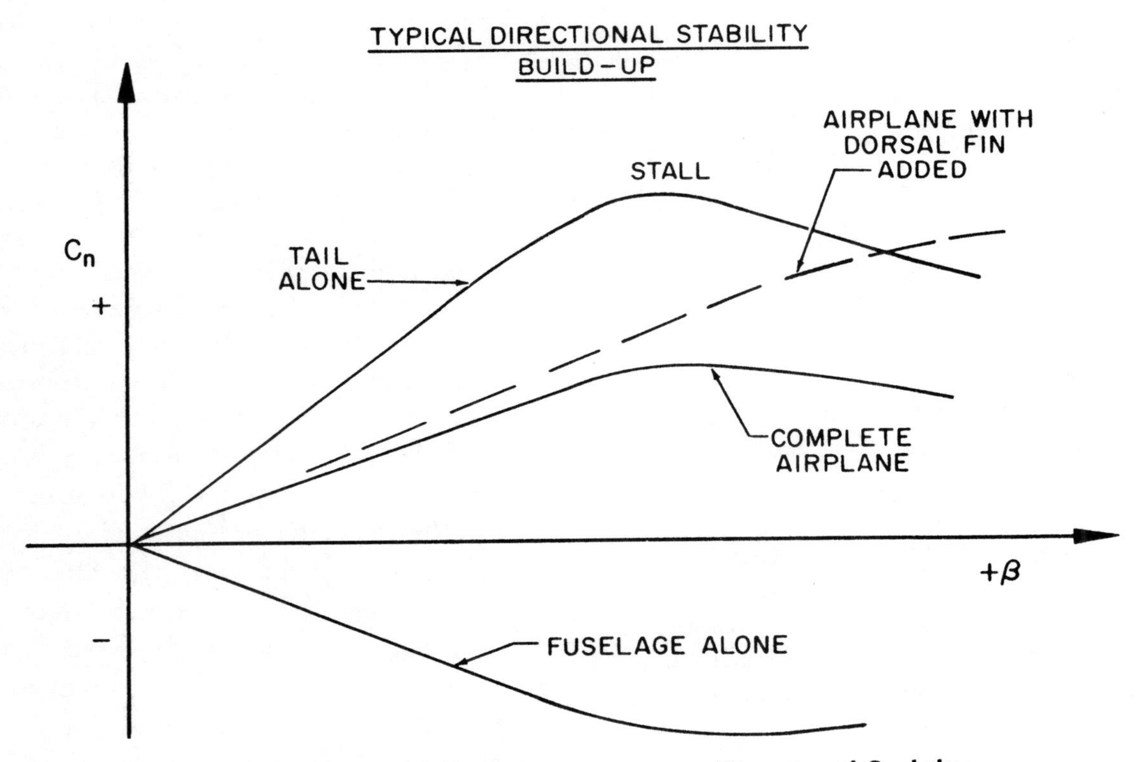

Figure 4.23. Contribution of Components to Directional Stability

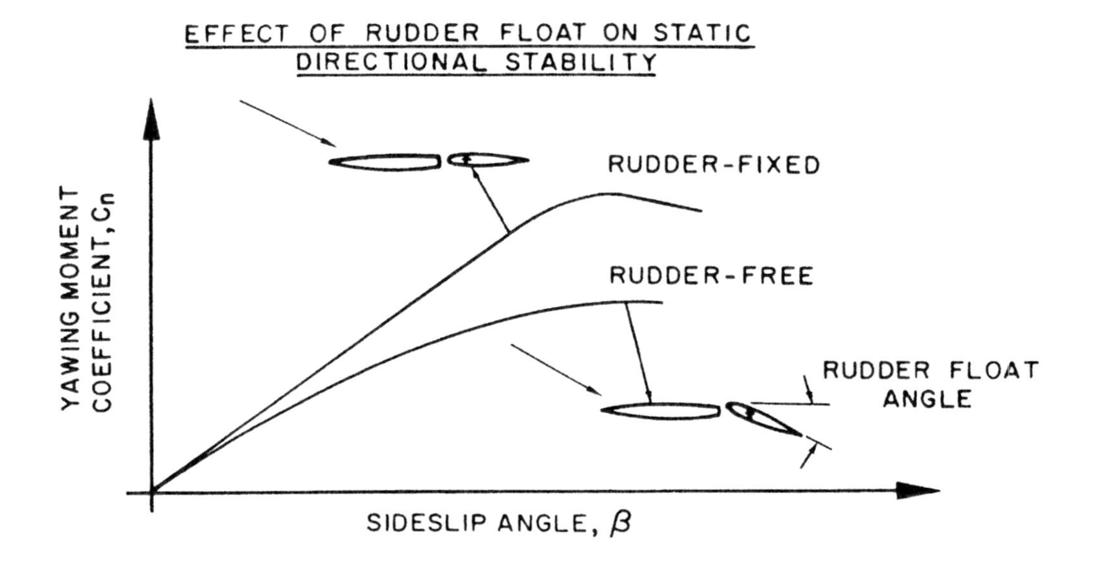

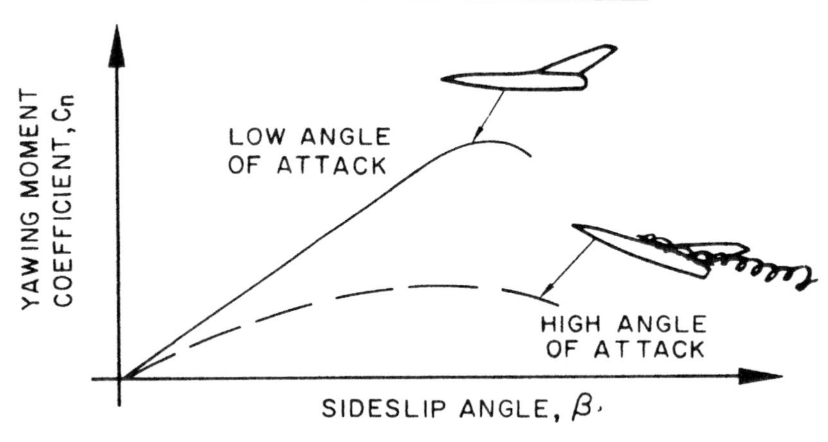

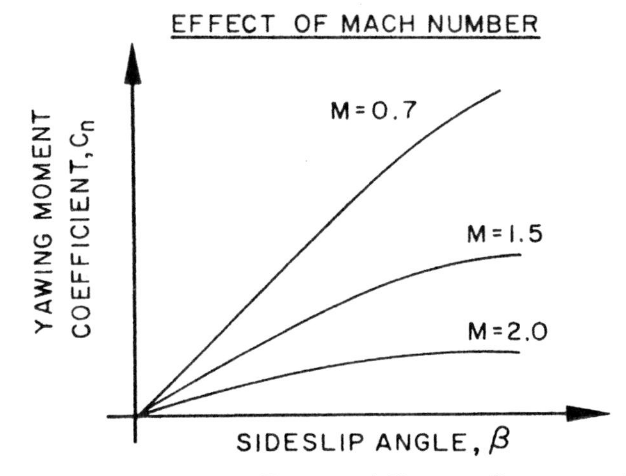

Figure 4.24. Factors Affecting Directional Stability

destabilizing and the greatest contribution will occur at high power and low dynamic pressure as during a waveoff.

As in the case of longitudinal static stability, freeing the controls will reduce the effectiveness of the tail and alter the stability. While the rudder must be balanced to reduce control pedal forces, the rudder will tend to float or streamline and reduce the contribution of the vertical tail to static directional stability. The floating tendency is greatest at large angles of sideslip where large angles of attack for the vertical tail tend to decrease aerodynamic balance. Figure 4.24 illustrates the difference between rudder-fixed and rudder-free static directional stability.

CRITICAL CONDITIONS. The most critical conditions of static directional stability are usually the combination of several separate effects. The combination which produces the most critical condition is much dependent upon the type and mission of the airplane. In addition, there exists a coupling of lateral and directional effects such that the required degree of static directional stability may be determined by some of these coupled conditions.

Center of gravity position has a relatively negligible effect on static directional stability. The usual range of c.g. position on any airplane is set by the linits of *longitudinal* stability and control. Within this limiting range of c.g. position, no significant changes take place in the contribution of the vertical tail, fuselage, nacelles, etc. Hence, the static directional stability is essentially unaffected by the variation of c.g. position within the longitudinal limits.

When the airplane is at a *high angle of attack* a decrease in static directional stability can be anticipated. As shown by the second chart of figure 4.24, a high angle of attack reduces the stable slope of the curve of C_n versus β. The decrease in static directional stability is due in great part to the reduction in the contribution of the vertical tail. At high angles of attack, the effectiveness of the vertical tail is reduced because of increase in the fuselage boundary layer at the vertical tail location. The decay of directional stability with angle of attack is most significant for the low aspect ratio airplane with sweepback since this configuration requires such high angles of attack to achieve high lift coefficients. Such decay in directional stability can have a profound effect on the response of the airplane to adverse yaw and spin characteristics.

High Mach numbers of supersonic flight reduce the contribution of the vertical tail to directional stability because of the reduction of lift curve slope with Mach number. The third chart of figure 4.24 illustrates the typical decay of directional stability with Mach number. To produce the required directional stability at high Mach numbers, a very large vertical tail area may be necessary. Ventral fins may be added as an additional contribution to directional stability but landing clearance requirements may limit their size or require the fins to be retractable.

Hence, the most critical demands of static directional stability will occur from some combination of the following effects:

(1) high angle of sideslip
(2) high power at low airspeed
(3) high angle of attack
(4) high Mach number

The propeller powered airplane may have such considerable power effects that the critical conditions may occur at low speed while the effect of high Mach numbers may produce the critical conditions for the typical supersonic airplane. In addition, the coupling of lateral and directional effects may require prescribed degrees of directional stability.

DIRECTIONAL CONTROL

In addition to directional stability, the airplane must have adequate directional control to coordinate turns, balance power effects, create sideslip, balance unsymmetrical power, etc. The principal source of directional control is the rudder and the rudder must be capable of producing sufficient yawing moment for the critical conditions of flight.

The effect of rudder deflection is to produce a yawing moment coefficient according to control deflection and produce equilibrium at some angle of sideslip. For small deflections of the rudder, there is no change in stability but a change in equilibrium. Figure 4.25 shows the effect of rudder deflection on yawing moment coefficient curves with the change in equilibrium sideslip angle.

If the airplane exhibits static directional stability with rudder fixed, each angle of sideslip requires a particular deflection of the rudder to achieve equilibrium. Rudder-free directional stability will exist when the float angle of the rudder is less than the rudder deflection required for equilibrium. However at high angles of sideslip, the floating tendency of the rudder increases. This is illustrated by the second chart of figure 4.25 where the line of rudder float angle shows a sharp increase at large values of sideslip. If the floating angle of the rudder catches up with the required rudder angle, the rudder pedal force will decrease to zero and rudder lock will occur. Sideslip angles beyond this point produce a floating angle greater than the required rudder deflection and the rudder tends to float to the limit of deflection.

Rudder lock is accompanied by a reversal of pedal force and rudder-free instability will exist. The dorsal fin is a useful addition in this case since it will improve the directional stability at high angles of sideslip. The resulting increase in stability requires larger deflections of the rudder to achieve equilibrium at high sideslip and the tendency for rudder lock is reduced.

Rudder-free directional stability is appreciated by the pilot as the rudder pedal force to maintain a given sideslip. If the rudder pedal force gradient is too low near zero sideslip, it will be difficult to maintain zero sideslip during various maneuvers. The airplane should have a stable rudder pedal feel through the available range of sideslip.

DIRECTIONAL CONTROL REQUIREMENTS. The control power of the rudder must be adequate to contend with the many unsymmetrical conditions of flight. Generally, there are five conditions of flight which provide the most critical requirements of directional control power. The type and mission of the airplane will decide which of these conditions is most important.

ADVERSE YAW. When an airplane is rolled into a turn yawing moments are produced which require rudder deflection to maintain zero sideslip, i.e., coordinate the turn. The usual source of adverse yawing moment is illustrated in figure 4.26. When the airplane shown is subject to a roll to the left, the downgoing port wing will experience a new relative wind and an increase in angle of attack. The inclination of the lift vector produces a component force forward on the downgoing wing. The upgoing starboard wing has its lift inclined with a component force aft. The resulting yawing moment due to rolling motion is in a direction opposite to the roll and is hence "adverse yaw." The yaw due to roll is primarily a function of the wing lift coefficient and is greatest at high C_L.

In addition to the yaw due to rolling motion there will be a yawing moment contribution due to control surface deflection. Conventional ailerons usually contribute an adverse yaw while spoilers may contribute a favorable or "proverse" yaw. The high wing airplane with a large vertical tail may encounter an influence from inboard ailerons. Such a configuration may induce flow directions at the vertical tail to cause proverse yaw.

Since adverse yaw will be greatest at high C_L and full deflection of the ailerons, coordinating steep turns at low speed may produce a critical requirement for rudder control power.

SPIN RECOVERY. In the majority of airplanes, the rudder is the principal control for spin recovery. Powerful control of sideslip at

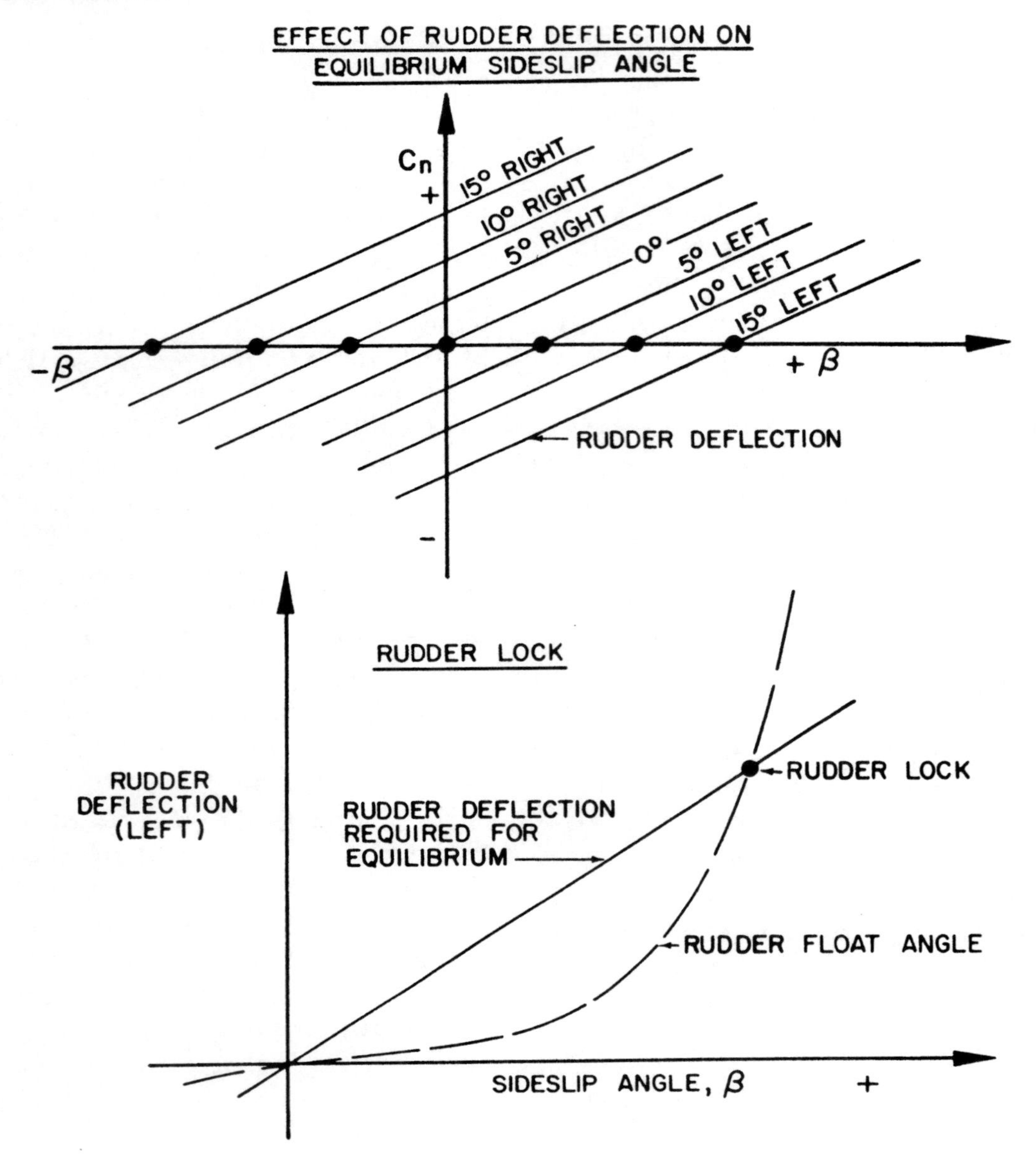

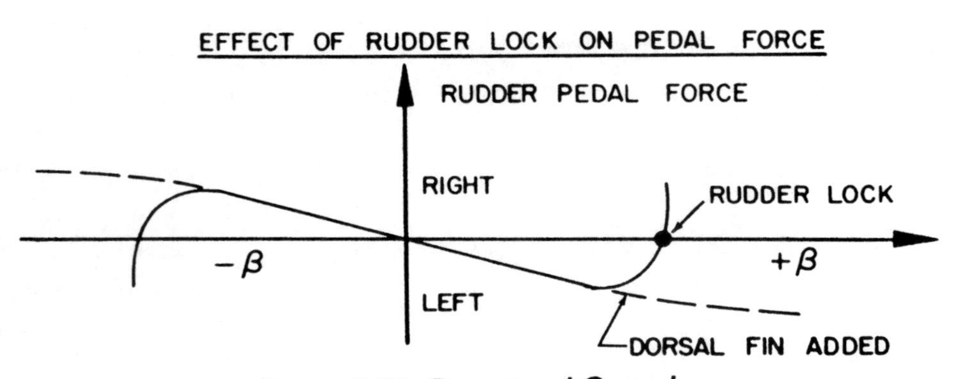

Figure 4.25. Directional Control

ADVERSE YAW DUE TO ROLL

AIRPLANE IN ROLL TO LEFT

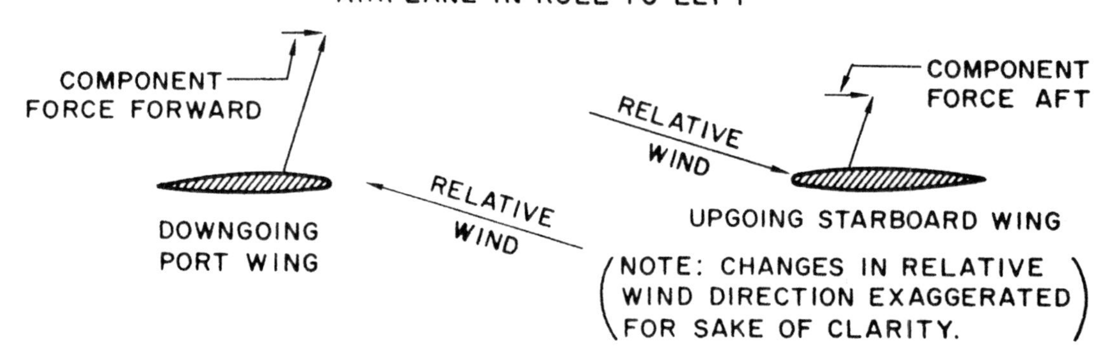

SLIPSTREAM SWIRL ON THE PROPELLER POWERED AIRPLANE

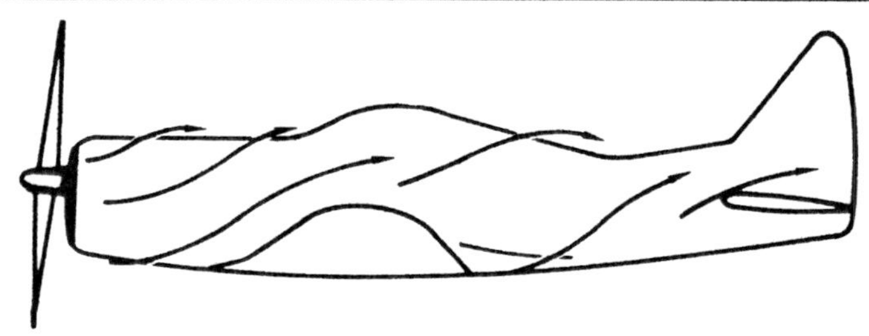

YAWING MOMENT DUE TO ASYMMETRICAL THRUST

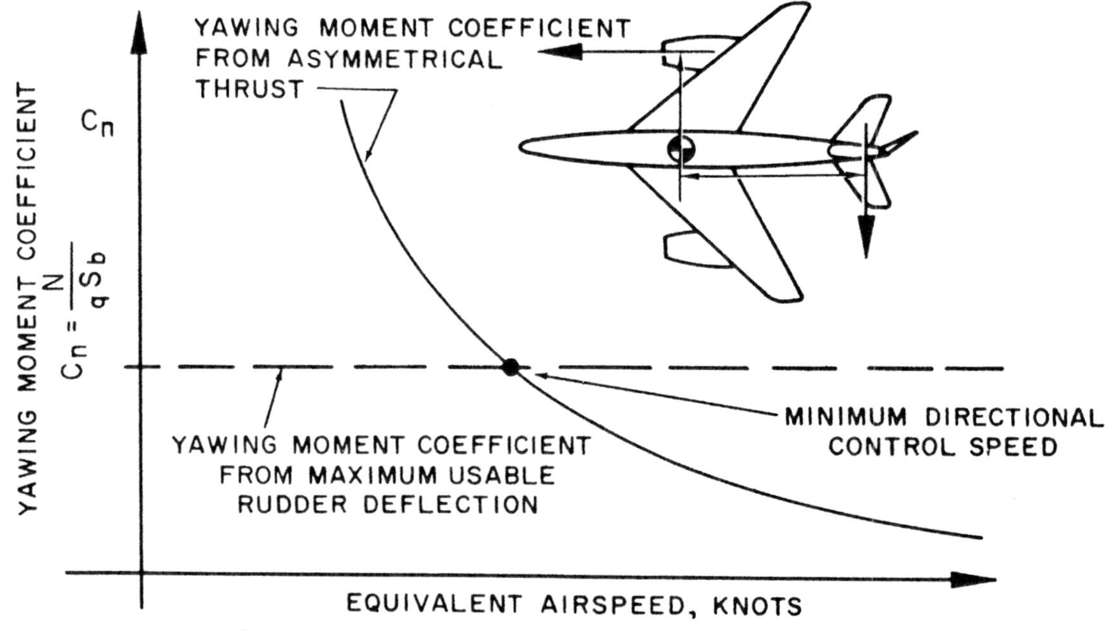

Figure 4.26. Requirements for Directional Control

high angles of attack is required to effect recovery during a spin. Since the effectiveness of the vertical tail is reduced at large angles of attack, the directional control power necessary for spin recovery may produce a critical requirement of rudder power.

SLIPSTREAM ROTATION. A critical directional control requirement may exist when the propeller powered airplane is at high power and low airspeed. As shown in figure 4.26, the single rotation propeller induces a slipstream swirl which causes a change in flow direction at the vertical tail. The rudder must furnish sufficient control power to balance this condition and achieve zero sideslip.

CROSSWIND TAKEOFF AND LANDING. Since the airplane must make a true path down the runway, a crosswind during takeoff or landing will require that the airplane be controlled in a sideslip. The rudder must have sufficient control power to create the required sideslip for the expected crosswinds.

ASYMMETRICAL POWER. The design of a multiengine airplane must account for the possibility of an engine failure at low airspeed. The unbalance of thrust from a condition of unsymmetrical power produces a yawing moment dependent upon the thrust unbalance and the lever arm of the force. The deflection of the rudder will create a side force on the tail and contribute a yawing moment to balance the yawing moment due to the unbalance of thrust. Since the yawing moment coefficient from the unbalance of thrust will be greatest at low speed, the critical requirement will be at a low speed with the one critical engine out and the remaining engines at maximum power. Figure 4.26 compares the yawing moment coefficient for maximum rudder deflection with the yawing moment coefficient for the unbalance of thrust. The intersection of the two lines determines the minimum speed for directional control, i.e., the lowest speed at which the rudder control moment can equal the moment of unbalanced thrust. It is usually specified that the minimum directional control speed be no greater than 1.2 times the stall speed of the airplane in the lightest practical takeoff configuration. This will provide adequate directional control for the remaining conditions of flight.

Once defined, the minimum directional control speed is not a function of weight, altitude, etc., but is simply the equivalent airspeed (or dynamic pressure) to produce a required yawing moment with the maximum rudder deflection. If the airplane is operated in the critical unbalance of power below the minimum control speed, the airplane will yaw uncontrollably into the inoperative engine. In order to regain directional control below the minimum speed certain alternatives exist: reduce power on the operating engines or sacrifice altitude for airspeed. Neither alternative is satisfactory if the airplane is in a marginal condition of powered flight so due respect must be given to the minimum control speed.

Due to the side force on the vertical tail, a slight bank is necessary to prevent turning flight at zero sideslip. The inoperative engine will be raised and the inclined wing lift will provide a component of force to balance the side force on the tail.

In each of the critical conditions of required directional control, high directional stability is desirable as it will reduce the displacement of the aircraft from any disturbing influence. Of course, directional control must be sufficient to attain zero sideslip. The critical control requirement for the multiengine airplane is the condition of asymmetrical power since spinning is not common to this type of airplane. The single engine propeller airplane may have either the spin recovery or the slipstream rotation as a critical design condition. The single engine jet airplane may have a variety of critical items but the spin recovery requirement usually predominates.

LATERAL STABILITY AND CONTROL

LATERAL STABILITY

The static lateral stability of an airplane involves consideration of rolling moments due to sideslip. If an airplane has favorable rolling moment due to sideslip, a lateral displacement from wing level flight produces sideslip and the sideslip creates rolling moments tending to return the airplane to wing level flight. By this action, static lateral stability will be evident. Of course, a sideslip will produce yawing moments depending on the nature of the static directional stability but the considrations of static lateral stability will involve only the relationship of rolling moments and sideslip.

DEFINITIONS. The axis system of an airplane defines a positive rolling, L, as a moment about the longitudinal axis which tends to rotate the right wing down. As in other aerodynamic considerations, it is convenient to consider rolling moments in the coefficient form so that lateral stability can be evaluated independent of weight, altitude, speeds, etc. The rolling moment, L, is defined in the coefficient form by the following equation:

$$L = C_l q S b$$

or

$$C_l = \frac{L}{qSb}$$

where

L = rolling moment, ft.-lbs., positive to the right
q = dynamic pressure, psf.
S = wing area, sq. ft.
b = wingspan, ft.
C_l = rolling moment coefficient, positive to the right

The angle of sideslip, β, has been defined previously as the angle between the airplane centerline and the relative wind and is positive when the relative wind is to the right of the centerline.

The *static lateral stability* of an airplane can be illustrated by a graph of rolling moment coefficient, C_l, versus sideslip angle, β, such as shown in figure 4.27. When the airplane is subject to a positive sideslip angle, lateral stability will be evident if a negative rolling moment coefficient results. Thus, when the relative wind comes from the right ($+\beta$), a rolling moment to the left ($-C_l$) should be created which tends to roll the airplane to the left. Lateral stability will exist when the curve of C_l versus β has a negative slope and the degree of stability will be a function of the slope of this curve. If the slope of the curve is zero, neutral lateral stability exists; if the slope is positive lateral instability is present.

It is desirable to have lateral stability or favorable roll due to sideslip. However, the required magnitude of lateral stability is determined by many factors. Excessive roll due to sideslip complicates crosswind takeoff and landing and may lead to undesirable oscillatory coupling with the directional motion of the airplane. In addition, a high lateral stability may combine with adverse yaw to hinder rolling performance. Generally, favorable handling qualities are obtained with a relatively light—or weak positive—lateral stability.

CONTRIBUTION OF THE AIRPLANE COMPONENTS. In order to appreciate the development of lateral stability in an airplane, each of the contribution components must be inspected. Of course, there will be interference between the components which will alter the contribution to stability of each component on the airplane.

The principal surface contributing to the lateral stability of an airplane is the *wing*. The effect of the geometric dihedral of a wing is a powerful contribution to lateral stability. As shown in figure 4.28, a wing with dihedral will develop stable rolling moments with sideslip. If the relative wind comes from the side, the wing into the wind is subject to an increase in angle of attack and develops an increase in lift. The wing away from the wind is subject to a decrease in angle of attack and develops a decrease in lift. The changes in lift effect a rolling moment tending to raise the windward wing hence dihedral contributes a stable roll due to sideslip.

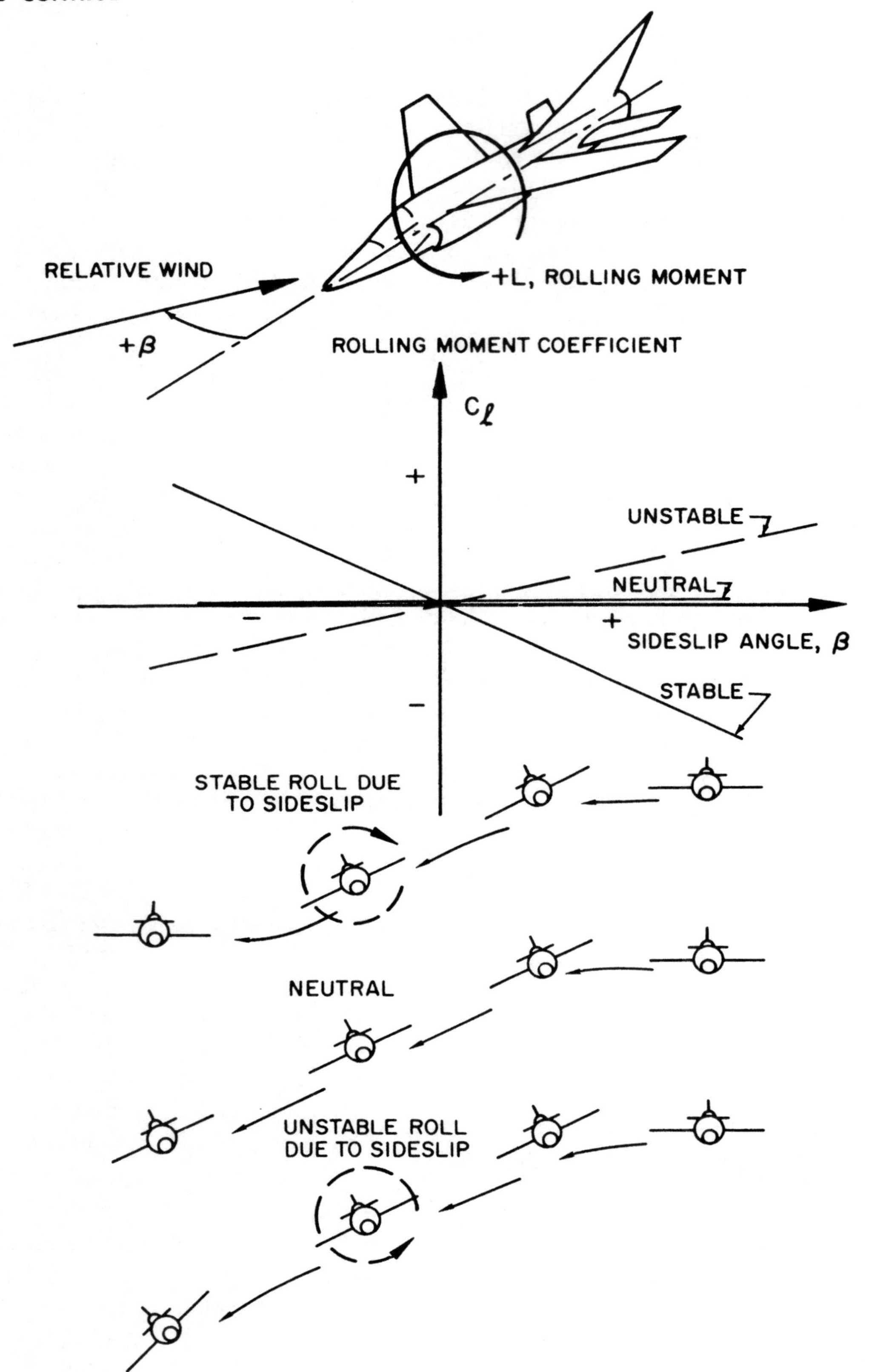

Figure 4.27. Static Lateral Stability

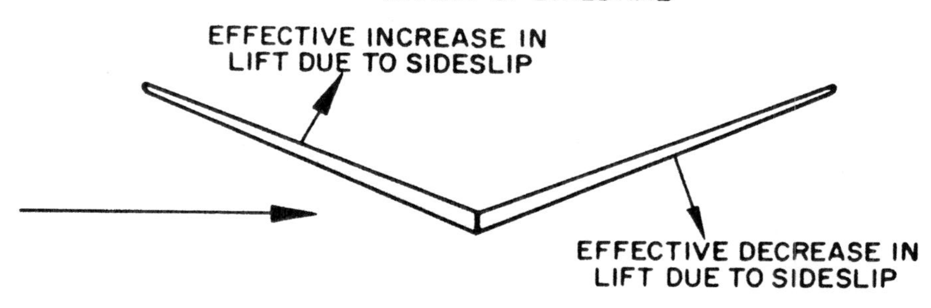

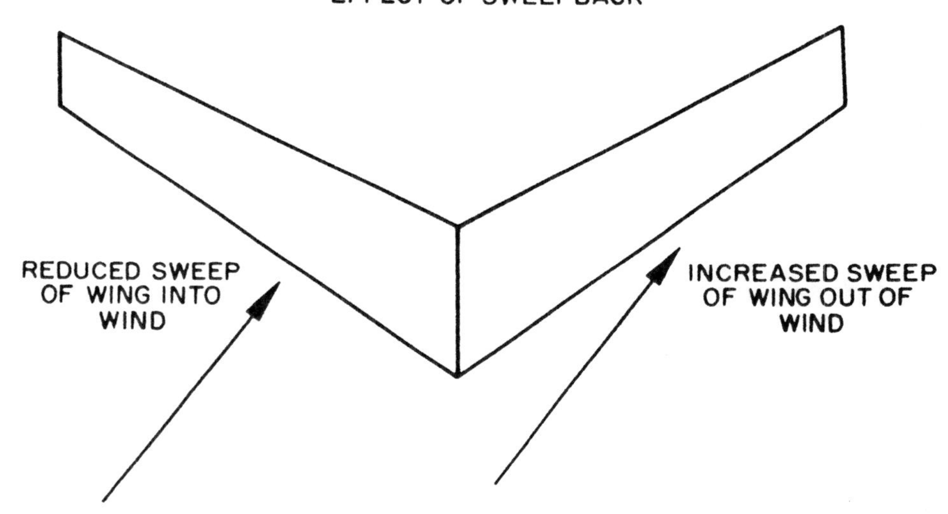

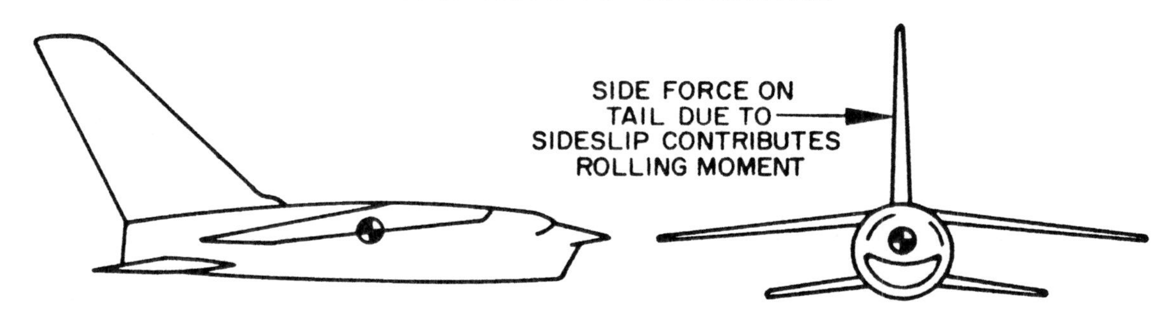

Figure 4.28. Contribution of Components to Lateral Stability

Since wing dihedral is so powerful in producing lateral stability it is taken as a common denominator of the lateral stability contribution of all other components. Generally, the contribution of wing position, flaps, power, etc., is expressed as an equivalent amount of "effective dihedral" or "dihedral effect."

The contribution of the *fuselage* alone is usually quite small depending on the location of the resultant aerodynamic side force on the fuselage. However, the effect of the wing-fuselage-tail combination is significant since the vertical placement of the wing on the fuselage can greatly affect the stability of the combination. A wing located at the mid wing position will generally exhibit a dihedral effect no different from that of the wing alone. A low wing location on the fuselage may contribute an effect equivalent to 3° or 4° of negative dihedral while a high wing location may contribute a positive dihedral of 2° or 3°. The magnitude of dihedral effect contributed by vertical position of the wing is large and may necessitate a noticeable dihedral angle for the low wing configuration.

The contribution of *sweepback* to dihedral effect is important because of the nature of the contribution. As shown in figure 4.28, the swept wing in a sideslip has the wing into wind operating with an effective decrease in sweepback while the wing out of the wind is operating with an effective increase in sweepback. If the wing is at a positive lift coefficient, the wing into the wind has less sweep and an increase in lift and the wing out of the wind has more sweep and a decrease in lift. In this manner the swept back wing would contribute a positive dihedral effect and the swept forward wing would contribute a negative dihedral effect.

The unusual nature of the contribution of sweepback to dihedral effect is that the contribution is proportional to the wing lift coefficient as well as the angle of sweepback. It should be clear that the swept wing at zero lift will provide no roll due to sideslip since there is no wing lift to change. Thus, the dihedral effect due to sweepback is zero at zero lift and increases directly with wing lift coefficient. When the demands of high speed flight require a large amount of sweepback, the resulting configuration may have an excessively high dihedral effect at low speeds (high C_L) while the dihedral effect may be satisfactory in normal flight (low or medium C_L).

The *vertical tail* of modern configurations can provide a significant—and, at times, undesirable—contribution to the effective dihedral. If the vertical tail is large, the side force produced by sideslip may produce a noticeable rolling moment as well as the important yawing moment contribution. Such an effect is usually small for the conventional airplane configuration but the modern high speed airplane configuration induces this effect to a great magnitude. It is difficult then to obtain a large vertical tail contribution to directional stability without incurring an additional contribution to dihedral effect.

The amount of effective dihedral necessary to produce satisfactory flying qualities varies greatly with the type and purpose of the airplane. Generally, the effective dihedral should not be too great since high roll due to sideslip can create certain problems. Excessive dihedral effect can lead to "Dutch roll," difficult rudder coordination in rolling maneuvers, or place extreme demands for lateral control power during crosswind takeoff and landing. Of course, the effective dihedral should not be negative during the predominating conditions of flight, e.g., cruise, high speed, etc. If the airplane demonstrates satisfactory dihedral effect for these conditions of flight, certain exceptions can be considered when the airplane is in the takeoff and landing configuration. Since the effects of flaps and power are destablizing and reduce the dihedral effect, a certain amount of negative dihedral effect may be possible due to these sources.

The deflection of flaps causes the inboard sections of the wing to become relatively more effective and these sections have a small spanwise moment arm. Therefore, the changes in wing lift due to sideslip occur closer inboard and the dihedral effect is reduced. The effect of power on dihedral effect is negligible for the jet airplane but considerable for the propeller driven airplane. The propeller slipstream at high power and low airspeed makes the inboard wing sections much more effective and reduces the dihedral effect. The reduction in dihedral effect is most critical when the flap and power effects are combined, e.g., the propeller driven airplane in the power approach or waveoff.

With certain exceptions during the conditions of landing and takeoff, the dihedral effect or lateral stability should be positive but light. The problems created by excessive dihedral effect are considerable and difficult to contend with. Lateral stability will be evident to a pilot by stick forces and displacements required to maintain sideslip. Positive stick force stability will be evident by stick forces required in the direction of the controlled sideslip.

LATERAL DYNAMIC EFFECTS

Previous discussion has separated the lateral and directional response of the airplane to sideslip. This separation is convenient for detailed study of each the airplane static lateral stability and the airplane static directional stability. However, when the airplane in free flight is placed in a sideslip, the lateral and directional response will be coupled, i.e., simultaneously the airplane produces rolling moment due to sideslip and yawing moment due to sideslip. Thus, the lateral dynamic motion of the airplane in free flight must consider the coupling or interaction of the lateral and directional effects.

The principal effects which determine the lateral dynamic characteristics of an airplane are:

(1) Rolling moment due to sideslip or dihedral effect (lateral stability).

(2) Yawing moment due to sideslip or static directional stability.

(3) Yawing moment due to rolling velocity or the adverse (or proverse) yaw.

(4) Rolling moment due to yawing velocity—a cross effect similar to (3). If the aircraft has a yawing motion to the right, the left wing will move forward faster and momentarily develop more lift than the right and cause a rolling moment to the right.

(5) Aerodynamic side force due to sideslip.

(6) Rolling moment due to rolling velocity or damping in roll.

(7) Yawing moment due yawing velocity or damping in yaw.

(8) The moments of inertia of the airplane about the roll and yaw axes.

The complex interaction of these effects produces three possible types of motion of the airplane: (*a*) a directional divergence, (*b*) a spiral divergence, and (*c*) an oscillatory mode termed Dutch roll.

Directional divergence is a condition which cannot be tolerated. If the reaction to a small initial sideslip is such as to create moments which tend to increase the sideslip, directional divergence will exist. The sideslip would increase until the airplane is broadside to the wind or structural failure occurs. Of course, increasing the static directional stability reduces the tendency for directional divergence.

Spiral divergence will exist when the static directional stability is very large when compared with the dihedral effect. The character of spiral divergence is by no means violent. The airplane, when disturbed from the equilibrium of level flight, begins a slow spiral which gradually increases to a spiral dive. When a small sideslip is introduced, the strong directional stability tends to restore the nose into the wind while the relatively weak dihedral effect lags in restoring the airplane laterally. In the usual case, the rate of divergence in the spiral motion is so gradual that the pilot can control the tendency without difficulty.

Dutch roll is a coupled lateral-directional oscillation which is usually dynamically stable but is objectionable because of the oscillatory nature. The damping of this oscillatory mode may be weak or strong depending on the properties of the airplane. The response of the airplane to a disturbance from equilibrium is a combined rolling-yawing oscillation in which the rolling motion is phased to precede the yawing motion. Such a motion is quite undesirable because of the great havoc it would create with a bomb, rocket, or gun platform.

Generally, Dutch roll will occur when the dihedral effect is large when compared to static directional stability. Unfortunately, Dutch roll will exist for relative magnitudes of dihedral effect and static directional stability between the limiting conditions for directional divergence and spiral divergence. When the dihedral effect is large in comparison with static directional stability, the Dutch roll motion has weak damping and is objectionable. When the static directional stability is strong in comparison with the dihedral effect, the Dutch roll motion has such heavy damping that it is not objectionable. However, these qualities tend toward spiral divergence.

The choice is then the least of three evils. Directional divergence cannot be tolerated, Dutch roll is objectionable, and spiral divergence is tolerable if the rate of divergence is low. For this reason the dihedral effect should be no more than that required for satisfactory lateral stability. If the static directional stability is made adequate to prevent objectionable Dutch roll, this will automatically be sufficient to prevent directional divergence. Since the more important handling qualities are a result of high static directional stability and minimum necessary dihedral effect, most airplanes demonstrate a mild spiral tendency. As previously mentioned, a weak spiral tendency is of little concern to the pilot and certainly preferable to Dutch roll.

The contribution of sweepback to the lateral dynamics of an airplane is significant. Since the dihedral effect from sweepback is a function of lift coefficient, the dynamic characteristics may vary throughout the flight speed range. When the swept wing airplane is at low C_L, the dihedral effect is small and the spiral tendency may be apparent. When the swept wing airplane is at high C_L, the dihedral effect is increased and the Dutch Roll oscillatory tendency is increased.

An additional oscillatory mode is possible in the lateral dynamic effects with the rudder free and the mode is termed a "snaking" oscillation. This yawing oscillation is greatly affected by the aerodynamic balance of the rudder and requires careful consideration in design to prevent light or unstable damping of the oscillation.

CONTROL IN ROLL

The lateral control of an airplane is accomplished by producing differential lift on the wings. The rolling moment created by the differential lift can be used to accelerate the airplane to some rolling motion or control the airplane in a sideslip by opposing dihedral effect. The differential lift for control in roll is usually obtained by some type of ailerons or spoilers.

ROLLING MOTION OF AN AIRPLANE. When an airplane is given a rolling motion in flight, the wing tips move in a helical path through the air. As shown in figure 4.29, a rolling velocity to the right gives the right wing tip a downward velocity component and the left wing tip an upward velocity component. By inspection of the motion of the left wing tip, the velocity of the tip due to roll combines with the airplane flight path velocity to define the resultant motion. The resulting angle between the flight path vector and the resultant path of the tip is the helix angle of roll. From the trigonometry of small angles, the helix angle of roll can be defined as:

$$\text{Roll helix angle} = \frac{pb}{2V} \text{ (radians)}$$

where

p=rate of roll, radians per second

b=wing span, ft.

V=airplane flight velocity, ft. per sec.

and, one radian=57.3 degrees

Generally, the maximum values of $\frac{pb}{2V}$ obtained by control in roll are approximately 0.1 to 0.07. The helix angle of roll, $\frac{pb}{2V}$, is actually a common denominator of rolling performance.

The deflection of the lateral control surfaces creates the differential lift and the rolling moment to accelerate the airplane in roll. The roll rate increases until an equal and opposite moment is created by the resistance to rolling motion or "damping in roll." The second illustration of figure 4.29 defines the source of the damping in roll. When the airplane is given a rolling velocity to the right, the downgoing wing experiences an increase in angle of attack due to the helix angle of roll. Of course, the upgoing wing experiences a decrease in angle of attack. In flight at angles of attack less than that for maximum lift, the downgoing wing experiences an increase in lift and the upgoing wing experiences a decrease in lift and a rolling moment is developed which opposes the rolling motion. Thus, the steady state rolling motion occurs when the damping moment equals the control moment.

The response of the airplane to aileron deflection is shown by the time history diagram of figure 4.29. When the airplane is restrained so that pure rolling motion is obtained, the initial response to an aileron deflection is a steady increase in roll rate. As the roll rate increases so does the damping moment and the roll acceleration decreases. Finally, the damping moment approaches the control moment and a steady state roll rate is achieved. If the airplane is unrestrained and sideslip is allowed, the affect of the directional stability and dihedral effect can be appreciated. The conventional airplane will develop adverse yawing moments due to aileron deflection and rolling motion. Adverse yaw tends to produce yawing displacements and sideslip but this is resisted by the directional stability of the airplane. If adverse yaw produces sideslip, dihedral effect creates a rolling moment opposing the roll and tends to reduce the roll rate. The typical transient motions (A) and (B) of the time history diagram of figure 4.29 show that high directional stability with low dihedral effect is the preferable combination. Such a combination provides an airplane which has no extreme requirement of coordinating aileron and rudder in order to achieve satisfactory rolling performance. While the coupled motion of the airplane in roll is important, further discussion of lateral control will be directed to pure uncoupled rolling performance.

ROLLING PERFORMANCE. The required rolling performance of an airplane is generally specified as certain necessary values of the roll helix angle, $\frac{pb}{2V}$. However, in certain conditions of flight, it may be more appropriate to specify minimum times for the airplane to accelerate through a given angle of roll. Usually, the maximum value of $\frac{pb}{2V}$ should be on the order of 0.10. Of course, fighters and attack airplanes have a more specific requirement for high rolling performance and 0.09 may be considered a minimum necessary $\frac{pb}{2V}$. Patrol, transport, and bomber airplanes have less requirement for high rolling performance and a $\frac{pb}{2V}$ of 0.07 may be adequate for these types.

The ailerons or spoilers must be powerful enough to provide the required $\frac{pb}{2V}$. While the size and effectiveness of the lateral control devices is important, consideration must be

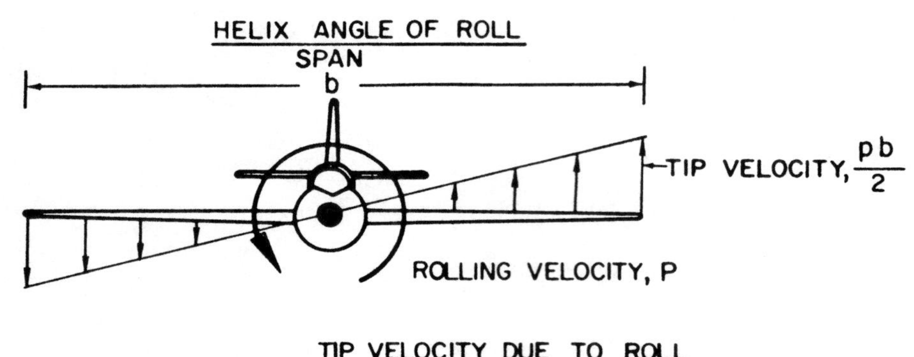

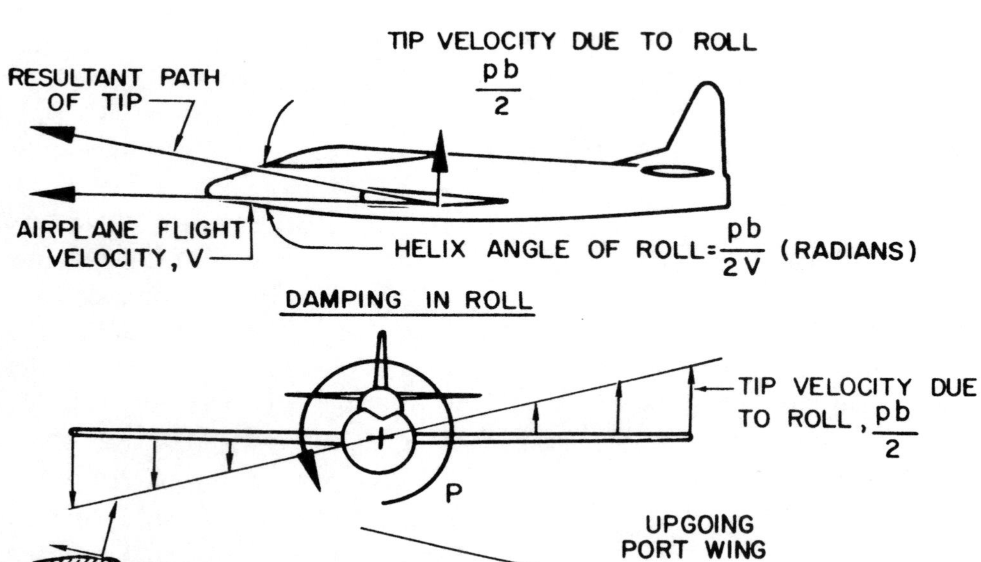

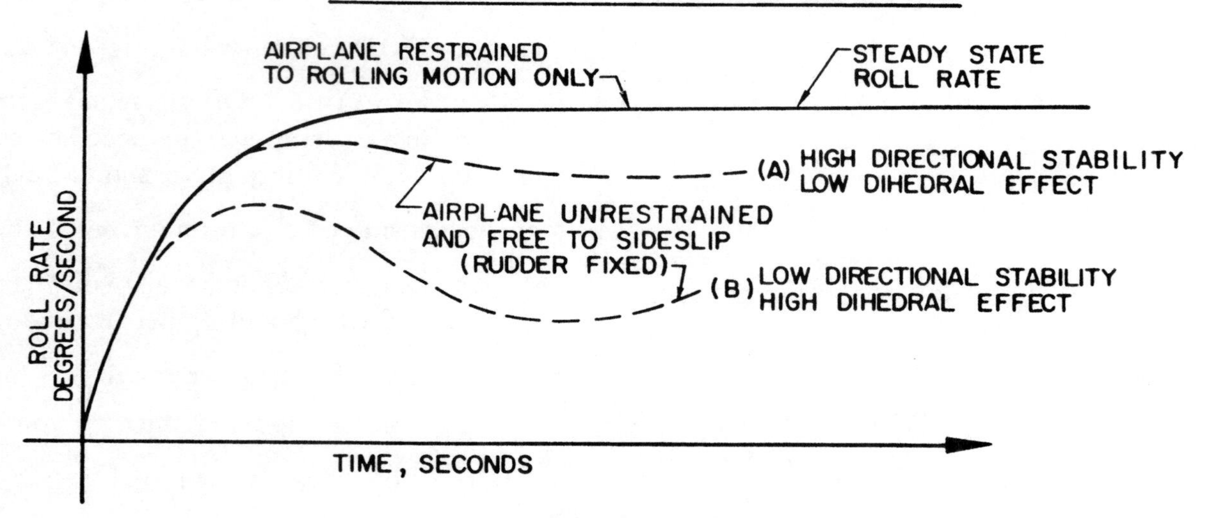

Figure 4.29. Rolling Performance

given to the airplane size. For geometrically similar airplanes, a certain deflection of the ailerons will produce a fixed value of $\frac{pb}{2V}$ independent of the airplane size. However, the roll rate of the geometrically similar airplanes at a given speed will vary inversely with the span, b.

If

$$\frac{pb}{2V}=\text{constant}$$

$$p=(\text{constant})\left(\frac{2V}{b}\right)$$

Thus, the smaller airplane will have an advantage in roll rate or in time to accelerate through a prescribed angle of roll. For example, a one-half scale airplane will develop twice the rate of roll of the full scale airplane. This relationship points to the favor of the small, short span airplane for achieving high roll performance.

An important variable affecting the rate of roll is the true airspeed or flight velocity, V. If a certain deflection of the ailerons creates a specific value of $\frac{pb}{2V}$, the rate of roll varies directly with the true airspeed. Thus, if the roll helix angle is held constant, the rate of roll at a particular true airspeed will not be affected by altitude. The linear variation of roll rate with airspeed points out the fact that high roll rates will require high airspeeds. The low roll rates at low airspeeds are simply a consequence of the low flight speed and this condition may provide a critical lateral control requirement for satisfactory handling qualities.

Figure 4.30 illustrates the typical rolling performance of a low speed airplane. When the ailerons are at full deflection, the maximum roll helix angle is obtained. The rate of roll increases linearly with speed until the control forces increase to limit of pilot effort and full control deflection cannot be maintained. Past the critical speed, with some limited amount of force applied by the pilot (usually the limit of lateral force is assumed to be 30 lbs.), the ailerons cannot be held at full deflection, $\frac{pb}{2V}$ drops, and rate of roll decreases. In this example, the rolling performance at high speeds is limited by the ability of the pilot to maintain full deflection of the controls. In an effort to reduce the aileron hinge moments and control forces, extensive application is made of aerodynamic balance and various tab devices. However, 100 percent aerodynamic balance is not always feasible or practical but a sufficient value of $\frac{pb}{2V}$ must be maintained at high speeds.

Rather than developing an extensive weight lifting program mandatory for all Naval Aviators, mechanical assistance in lateral control can be provided. If a power boost is provided for the lateral control system, the rolling performance of the airplane may be extended to higher speeds since pilot effort will not be a limiting factor. The effect of a power boost is denoted by the dashed line extensions of figure 4.30. A full powered, irreversible lateral control system is common for high speed airplanes. In the power operated system there is no immediate limit to the deflection of the control surfaces and none of the aberrations in hinge moments due to compressibility are fed back to the pilot. Control forces are provided by the stick centering lateral bungee or spring.

A problem particular to the high speed is due to the interaction of aerodynamic forces and the elastic deflections of the wing in torsion. The deflection of ailerons creates twisting moments on the wing which can cause significant torsional deflections of the wing. At the low dynamic pressures of low flight speeds, the twisting moments and twisting deflections are too small to be of importance. However, at high dynamic pressures, the deflection of an aileron creates significant

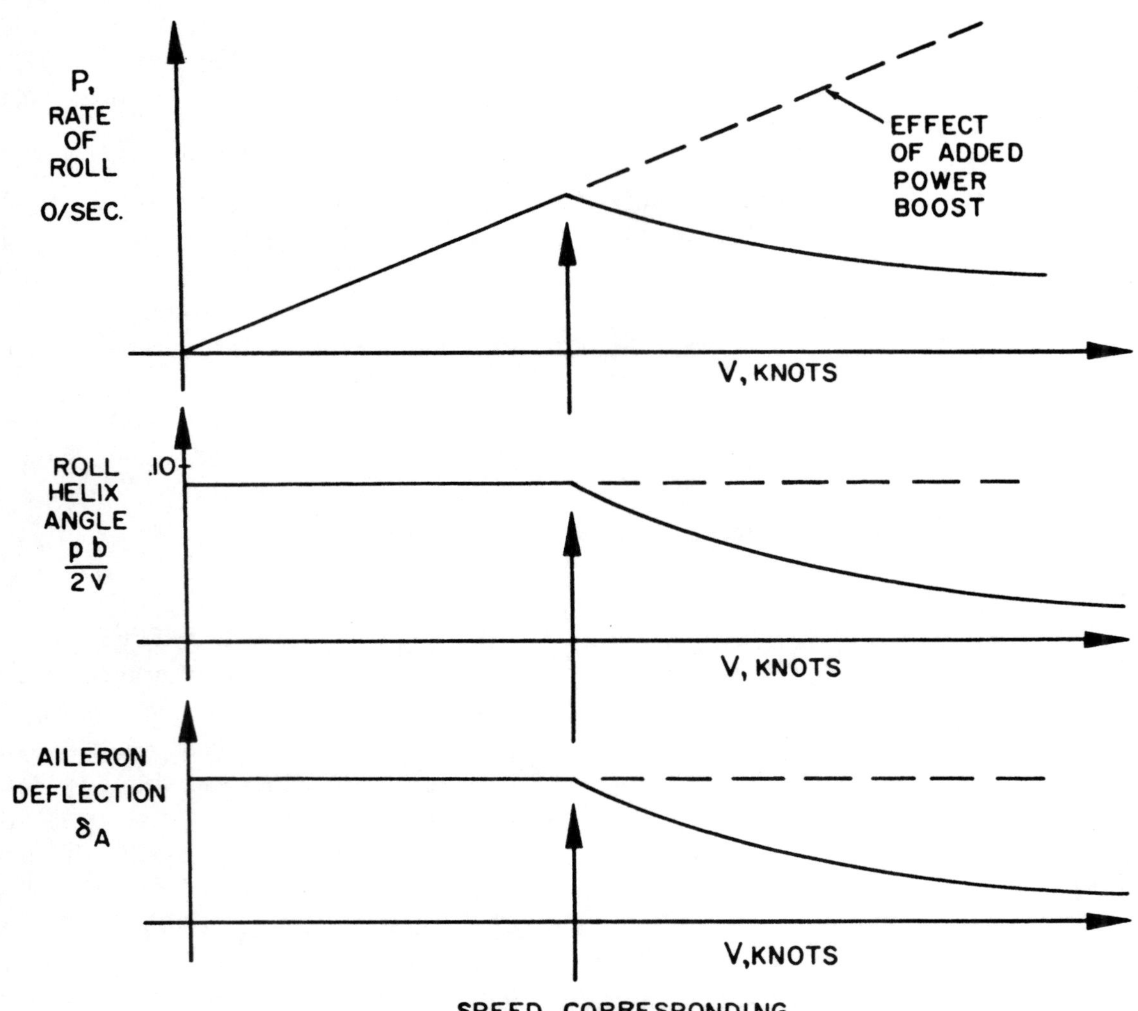

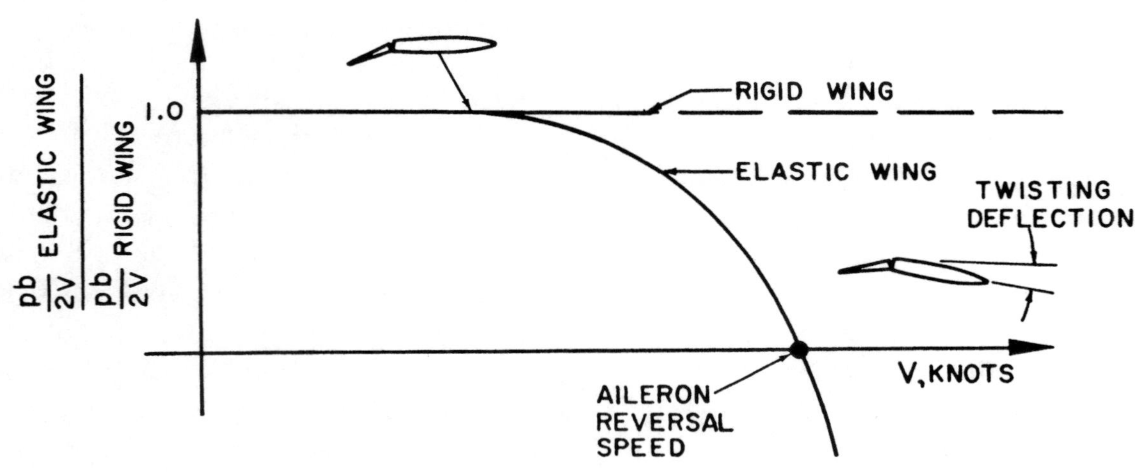

Figure 4.30. Control in Roll

twisting deflections which reduce the effectiveness of the aileron, e.g., downward deflection of an aileron creates a nose down twist of the wing which reduces the rolling moment due to aileron deflection. At very high speeds, the torsional deflection of the wing may be so great than a rolling moment is created opposite to the direction controlled and "aileron reversal" occurs. Prior to the speed for aileron reversal, a serious loss of roll helix angle may be encountered. The effect of this aeroelastic phenomenon on rolling performance is illustrated in figure 4.30.

To counter the undesirable interaction between aerodynamic forces and wing torsional deflections, the trailing edge ailerons may be moved inboard to reduce the portion of the span subjected to twisting moments. Of course, the short span, highly tapered wing planform is favorable for providing relatively high stiffness. In addition, various configurations of spoilers may be capable of producing the required rolling performance without the development of large twisting moments.

CRITICAL REQUIREMENTS. The critical conditions for requiring adequate lateral control power may occur at either high speed or low speed depending on the airplane configuration and intended use. In transonic and supersonic flight, compressibility effects tend to reduce the effectiveness of lateral control devices to produce required roll helix angles. These effects are most significant when combined with a loss of control effectiveness due to aeroelastic effects. Airplanes designed for high speed flight must maintain sufficient lateral control effectiveness at the design dive speed and this is usually the predominating requirement.

During landing and takeoff, the airplane must have adequate lateral control power to contend with the ordinary conditions of flight. The lateral controls must be capable of achieving required roll helix angles and acceleration through prescribed roll displacements. Also, the airplane must be capable of being controlled in a sideslip to accomplish crosswind takeoff and landing. The lateral control during crosswind takeoff and landing is a particular problem when the dihedral effect is high. Since the sweepback contributes a large dihedral effect at high lift coefficients, the problem is most important for the airplane with considerable sweepback. The limiting crosswind components must be given due respect especially when the airplane is at low gross weight. At low gross weight the specified takeoff and landing speeds will be low and the controlled angle of sideslip will be largest for a given crosswind velocity.

MISCELLANEOUS STABILITY PROBLEMS

There are several general problems of flying which involve certain principles of stability as well as specific areas of longitudinal, directional and lateral stability. Various conditions of flight will exist in which certain problems of stability (or instability) are unavoidable for some reason or another. Many of the following items deserve consideration because of the possible unsafe condition of flight and the contribution to an aircraft accident.

LANDING GEAR CONFIGURATIONS

There are three general configurations for the aircraft landing gear: the tricycle, bicycle, and "conventional" tail wheel arrangement. At low rolling speeds where the airplane aerodynamic forces are negligible, the "control-fixed" static stability of each of these configurations is determined by the side force characteristics of the tires and is not a significant problem.

The instability which allows ground loops in an aircraft with a conventional tail wheel landing gear is quite basic and can be appreciated from the illustration of figure 4.31. Centrifugal force produced by a turn must be balanced and the aircraft placed in equilibrium. The greatest side force is produced at the main wheels but to achieve equilibrium with the

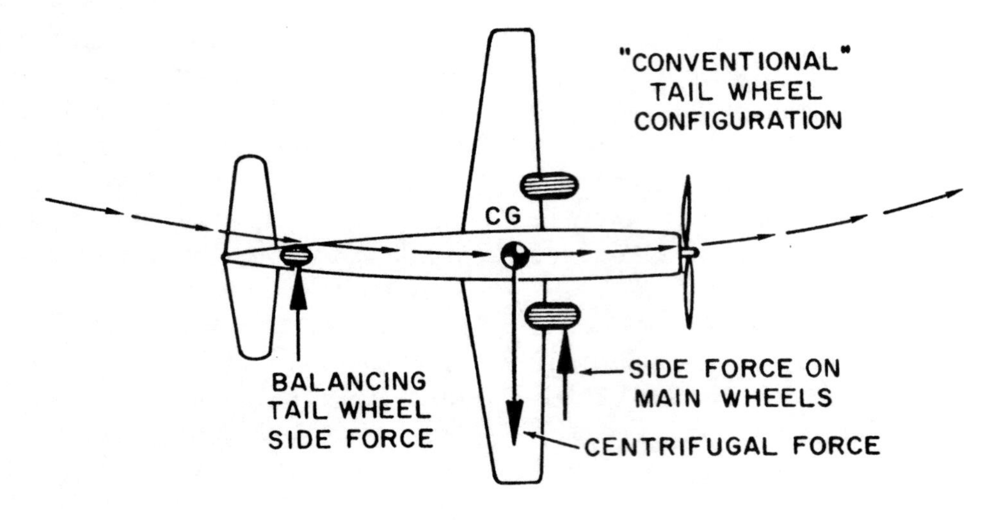

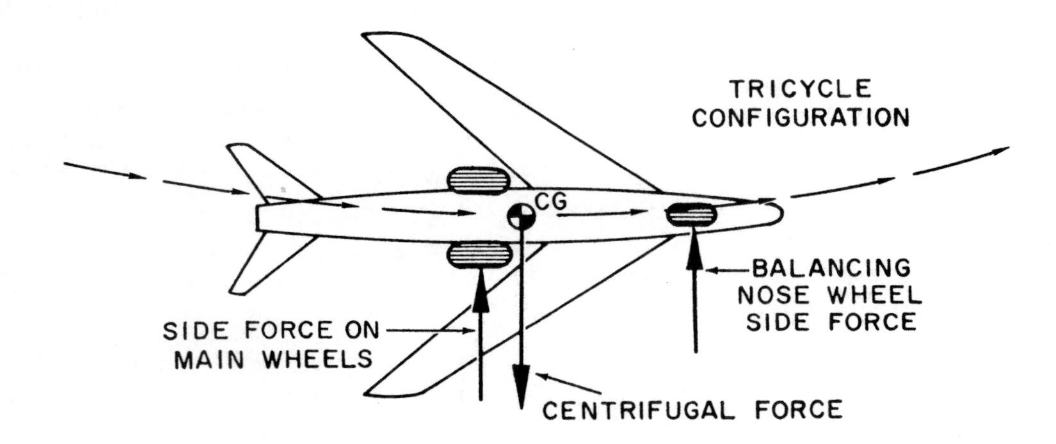

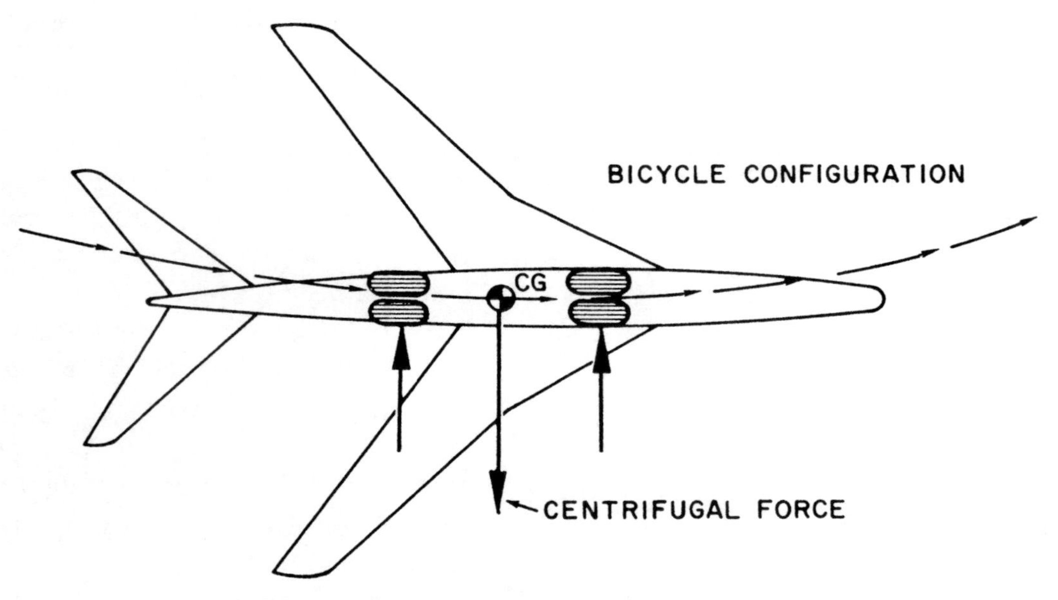

Figure 4.31. Landing Gear Configurations

center of gravity aft of the main wheels a balancing load on the tail wheel must be produced toward the center of turn. When the tail wheel is free to swivel, the equilibrium of the turn requires a control force opposite to the direction of turn—i.e., control force instability. The inherent stability problem exists because the center of gravity is aft of the point where the main side forces are developed. This condition is analogous to the case of static longitudinal stability with the center of gravity aft of the neutral point.

The conventional tail wheel configuration has this basic instability or ground loop tendency which must be stabilized by the pilot. At high rolling speeds where aerodynamic forces are significant, the aerodynamic directional stability of the airplane resists the ground looping tendency. The most likely times for a ground loop exist when rolling speeds are not high enough to provide a contribution of the aerodynamic forces. When the tail wheel is free to swivel or when the normal force on the tail wheel is small, lack of pilot attention can allow the ground loop to take place.

The tricycle landing gear configuration has an inherent stability due to the relative position of the main wheels and the center of gravity. Centrifugal force produced by a turn is balanced by the side force on the main wheels and a side force on the nose wheel in the direction of turn. Note that the freeing the nose wheel to swivel produces moments which bring the aircraft out of the turn. Thus, the tricycle configuration has a basic stability which is given evidence by control displacement and a wheel side force in the direction of turn. Because of the contrast in stability, the tricycle configuration is much less difficult to maneuver than the tail wheel configuration and does not provide an inherent ground loop tendency. However, a steerable nose wheel is usually necessary to provide satisfactory maneuvering capabilities.

The bicycle configuration of landing gear has stability characteristics more like the automobile. If directional control is accomplished with the front wheels operated by power controls, no stability problem exists at low speeds. A problem can exist when the airplane is at high speeds because of a distribution of normal force being different from the ordinary static weight distribution. If the airplane is held onto the runway at speeds well above the normal takeoff and landing speeds, the front wheels carry a greater than ordinary amount of normal force and a tendency for instability exists. However, at these same high speeds the rudder is quite powerful and the condition is usually well within control.

The basically stable nature of the tricycle and bicycle landing gear configurations is best appreciated by the ease of control and ground maneuvering of the airplane. Operation of a conventional tail wheel configuration after considerable experience with tricycle configurations requires careful consideration of the stability that must be furnished by the pilot during ground maneuvering.

SPINS AND PROBLEMS OF SPIN RECOVERY

The motion of an airplane in a spin can involve many complex aerodynamic and inertia forces and moments. However, there are certain fundamental relationships regarding spins and spin recoveries with which all aviators should be familiar. The spin differs from a spiral dive in that the spin always involves flight at high angle of attack while the spiral dive involves a spiral motion of the airplane at relatively low angle of attack.

The stall characteristics and stability of the airplane at high lift coefficients are important in the initial tendencies of the airplane. As previously mentioned, it is desirable to have the wing initiate stall at the root first rather than tip first. Such a stall pattern prevents the undesirable rolling moments at high lift coefficients, provides suitable stall

warning, and preserves lateral control effectiveness at high angles of attack. Also, the airplane must maintain positive static longitudinal stability at high lift coefficients and should demonstrate satisfactory stall recovery characteristics.

In order to visualize the principal effects of an airplane entering a spin, suppose the airplane is subjected to the rolling and yawing velocities shown in figure 4.32. The yawing velocity to the right tends to produce higher local velocities on the left wing than on the right wing. The rolling velocity tends to increase the angle of attack for the downgoing right wing (α_r) and decrease the angle of attack for the upgoing left wing (α_l). At airplane angles of attack below the stall this relationship produces roll due to yaw, damping in roll, etc., and some related motion of the airplane in unstalled flight. However, at angles of attack above the stall, important changes take place in the aerodynamic characteristics.

Figure 4.32 illustrates the aerodynamic characteristics typical of a conventional airplane configuration, i.e., moderate or high aspect ratio and little—if any—sweepback. If this airplane is provided a rolling displacement when at some angle of attack above the stall, the upgoing wing experiences a decrease in angle of attack with a corresponding increase in C_L and decrease in C_D. In other words, the upgoing wing becomes less stalled. Similarly, the downgoing wing experiences an increase in angle of attack with a corresponding decrease in C_L and increase in C_D. Essentially, the downgoing wing becomes more stalled. Thus, the rolling motion is aided rather than resisted and a yawing moment is produced in the direction of roll. At angles of attack below stall the rolling motion is resisted by damping in roll and adverse yaw is usually present. At angles of attack above the stall, the damping in roll is negative and a rolling motion produces a rolling moment in the direction of the roll. This negative damping in roll is generally referred to as "autorotation."

When the conventional airplane is stalled and some rolling-yawing displacement takes place, the resulting autorotation rolling moments and yawing moments start the airplane into a self-sustaining rolling-yawing motion. The autorotation rolling and yawing tendencies of the airplane at high angles of attack are the principal prospin moments of the conventional airplane configuration and these tendencies accelerate the airplane into the spin until some limiting condition exists. The stabilized spin is not necessarily a simple steady vertical spiral but may involve some coupled unsteady oscillatory motion.

An important characteristic of the more conventional airplane configuration is that the spin shows a predominating contribution of the autorotation tendency. Generally, the conventional configuration has a spin motion which is primarily rolling with moderate yaw. High directional stability is favorable since it will limit or minimize the yaw displacement of the spinning airplane.

The fundamental requirement of the spin is that the airplane be placed at an excessive angle of attack to produce the autorotation rolling and yawing tendencies. Generally speaking, the conventional airplane must be stalled before a spin can take place. This relationship establishes a fundamental principle of recovery—the airplane must be unstalled by decreasing the wing angle of attack. The most effective procedure for the conventional configuration is to use opposite rudder to stop the sideslip, then lower the angle of attack with the elevators. With sufficient rudder power this procedure will produce a positive recovery with a minimum loss of altitude. Care should be taken during pullout from the ensuing dive to prevent excessive angle of attack and entry into another spin.

It should be appreciated that a spin is always a possible corollary of a stall and the self-sustaining motion of a spin will take place at

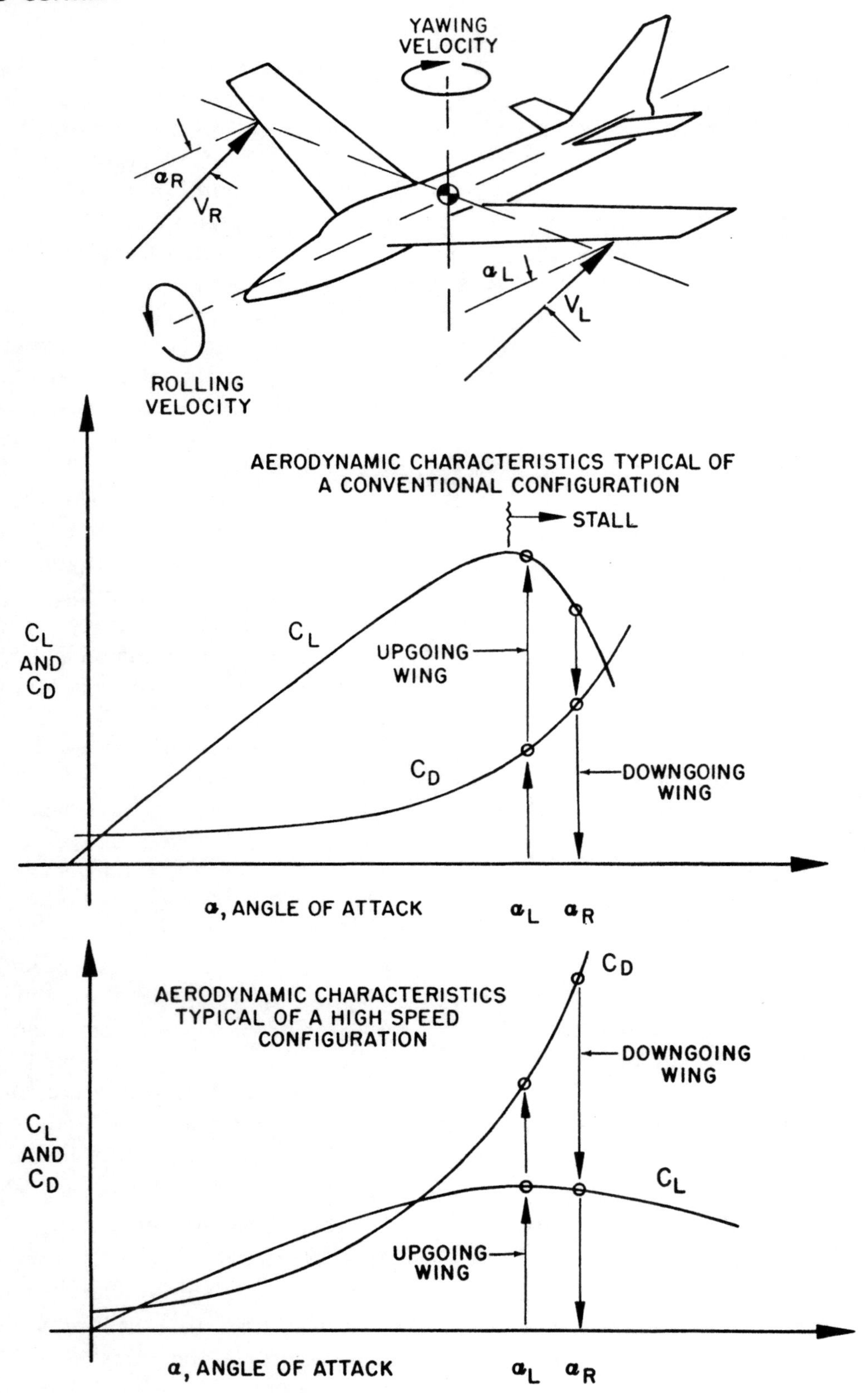

Figure 4.32. Spin Characteristics

excessive angles of attack. Of course, a low speed airplane could be designed to be spinproof by making it stallproof. By limiting the amount of control deflection, the airplane may not have the longitudinal control power to trim to maximum lift angle of attack. Such a provision may be possible for certain light planes and commercial aircraft but would create an unrealistic and impractical limitation on the utility of a military airplane.

The modern high speed airplane configuration is typified by low aspect ratio, swept wing planforms with relatively large yaw and pitch inertia. The aerodynamic characteristics of such a configuration are shown in figure 4.32. The lift curve (C_L versus α) is quite shallow at high angles of attack and maximum lift is not clearly defined. When this type of airplane is provided a rolling motion at high angles of attack, relatively small changes in C_L take place. When this effect is combined with the relatively short span of this type airplane, it is apparent that the wing autorotation contribution will be quite weak and will not be a predominating pro-spin moment. The relatively large changes in drag coefficient with rolling motion imply a predominance of yaw for the spin of the high speed airplane configuration.

Actually, various other factors contribute to the predominating yaw tendency for the spin of the modern airplane configuration. The static directional stability deteriorates at high angles of attack and may be so weak that extemely large yaw displacements result. In certain instances, very high angles of attack may bring such a decay in directional stability that a "slice" or extreme yaw displacement takes place before a true spin is apparent. At these high angles of attack, the adverse yaw due to roll and aileron deflection can be very strong and create large yaw displacements of the airplane prior to realizing a stall.

The aircraft with the relatively large, long fuselage can exhibit a significant moment contribution from the fuselage alone. The cross flow pattern on the fuselage at high angles of attack is capable of producing pro-spin moments of considerable magnitude which contribute to the self-sustaining nature of the spin. Also, the large distributed mass of the fuselage in rolling-yawing rotation contributes to inertia moments which flatten the spin and place the aircraft at extreme angles of attack.

The spin recovery of the modern high speed airplane involves principles which are similar to those of the spin recovery of the conventional airplane. However, the nature of the spin for the modern configuration may involve specific differences in technique necessary to reduce the sideslip and angle of attack. The use of opposite rudder to control the sideslip and effect recovery will depend on the effectiveness of the rudder when the airplane is in the spin. At high positive angles of attack and high sideslip the rudder effectiveness may be reduced and additional anti-spin moments must be provided for rapid recovery. The deflection of ailerons into the spin reduces the autorotation rolling moment and can produce adverse yaw to aid the rudder yawing moment in effecting recovery.

There may be many other specific differences in the technique necessary to effect spin recovery. The effectiveness of the rudder during recovery may be altered by the position of elevators or horizontal tail. Generally, full aft stick may be necessary during the *initial* phase of recovery to increase the effectiveness of the rudder. The use of power during the spin recovery of a propeller powered airplane may or may not aid recovery depending on the specific airplane and the particular nature of the slipstream effects. The use of power during the spin recovery of a jet powered airplane induces no significant or helpful flow but does offer the possibility of a severe compressor stall and adverse gyroscopic moments. Since the airplane is at high angle of attack and sideslip, the flow at the inlet may be very poor and the stall limits considerably reduced. These items serve to point out possible differences in technique required for various configurations. The spin recovery specific for

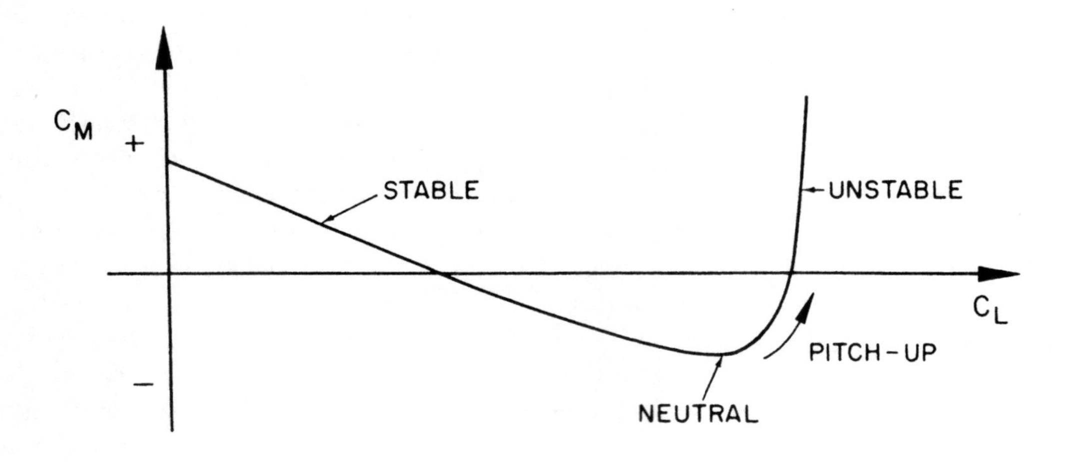

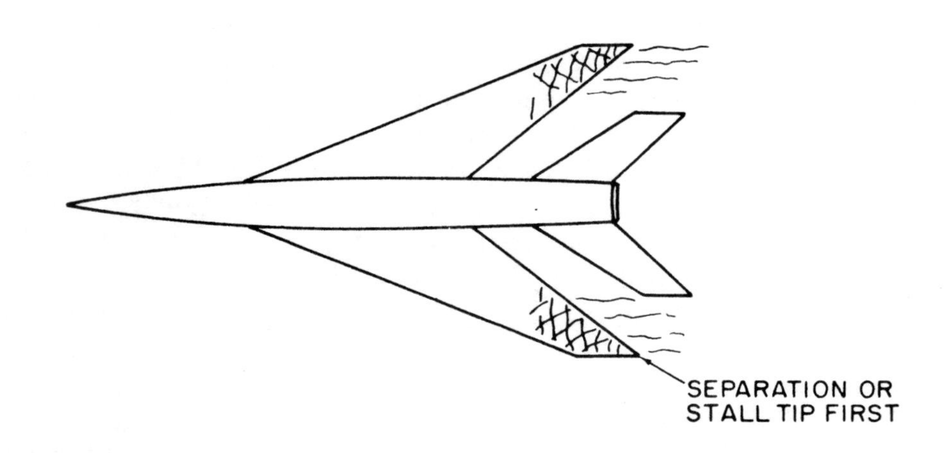

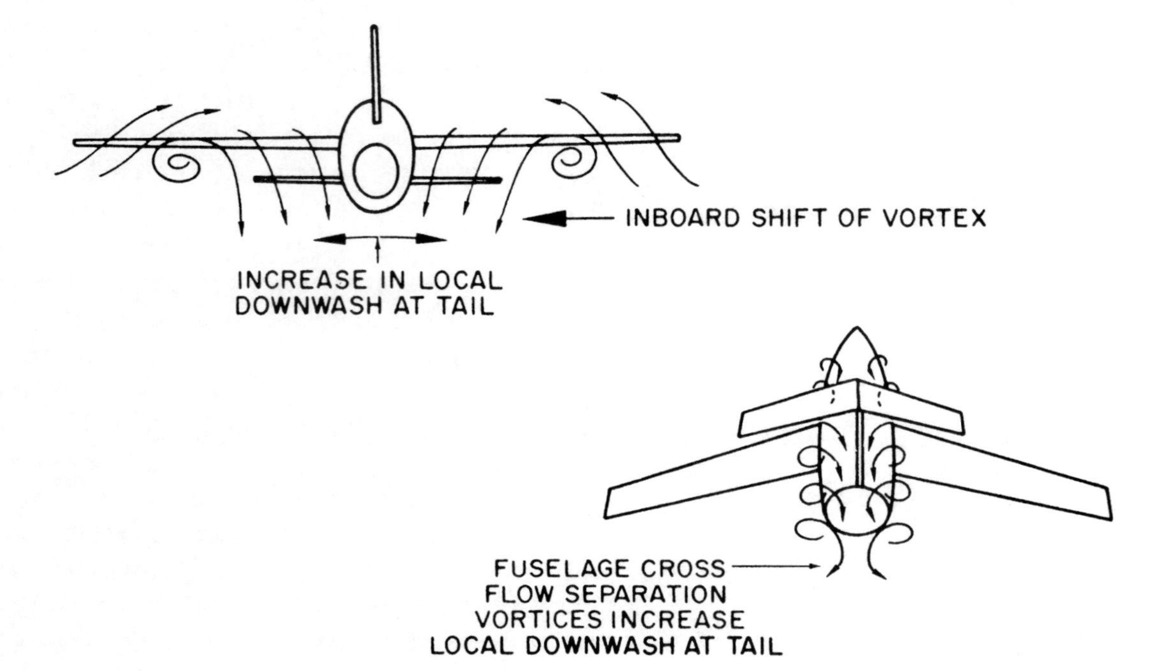

Figure 4.33. Pitch-up

each airplane is outlined in the pilot's handbook and it is *imperative* that the specific technique be followed for successful recovery.

PITCH-UP

The term of "pitch-up" generally applies to the static longitudinal instability encountered by certain configurations at high angle of attack. The condition of pitch-up is illustrated by the graph of C_M versus C_L in figure 4.33. Positive static longitudinal stability is evident at low values of C_L by the negative slope of the curve. At higher values of C_L the curve changes to a positive slope and large positive pitching moments are developed. This sort of instability implies that an increase in angle of attack produces nose up moments which tend to bring about further increases in angle of attack hence the term "pitch-up" is applied.

There are several items which may contribute to a pitch-up tendency. Sweepback of the wing planform can contribute unstable moments when separation or stall occurs at the tips first. The combination of sweepback and taper alters the lift distribution to produce high local lift coefficients and low energy boundary layer near the tip. Thus, the tip stall is an inherent tendency of such a planform. In addition, if high local lift coefficients exist near the tip, the tendency will be to incur the shock induced separation first in these areas. Generally, the wing will contribute to pitch-up only when there is large sweepback.

Of course, the wing is not the only item contributing to the longitudinal stability of the airplane. Another item important as a source of pitch-up is the downwash at the horizontal tail. The contribution of the tail to stability depends on the change in tail lift when the airplane is given a change in angle of attack. Since the downwash at the tail reduces the change in angle of attack at the tail, any increase in downwash at the tail is destabilizing. For certain low aspect ratio airplane configurations, an increase in airplane angle of attack may physically locate the horizontal tail in the wing flow field where higher relative downwash exists. Thus, a decrease in stability would take place.

Certain changes in the flow field behind the wing at high angles of attack can produce large changes in the tail contribution to stability. If the wing tips stall first, the vortices shift inboard and increase the local downwash at the tail for a given airplane C_L. Also, the fuselage at high angle of attack can produce strong cross flow separation vortices which increase the local downwash for a horizontal tail placed above the fuselage. Either one or a combination of these downwash influences may provide a large unstable contribution of the horizontal tail.

The pitch-up instability is usually confined to the high angle of attack range and may be a consequence of a configuration that otherwise has very desirable flying qualities. In such a case it would be necessary to provide some automatic control function to prevent entry into the pitch-up range or to provide synthetic stability for the condition. Since the pitch-up is usually a strong instability with a high rate of divergence, most pilots would not be capable of contending with the condition. At high q, pitch-up would be of great danger in that structural failure could easily result. At low q, failing flight loads may not result but the strong instability may preclude a successful recovery from the ensuing motion of the airplane.

EFFECTS OF HIGH MACH NUMBER

Certain stability problems are particular to supersonic flight. While most of the problem areas have been treated in particular in previous discussion, it is worthwhile to review the effects of supersonic flight on the various items of stability.

The static longitudinal stability of an airplane increases during the transition from subsonic to supersonic flight. Usually the principal source of the change in stability is due to the shift of the wing aerodynamic center with

Mach number. As a corollary of this increase in stability is a decrease in controllability and an increase in trim drag.

The static directional stability of an airplane decreases with Mach number in supersonic flight. The influence of the fuselage and the decrease in vertical tail lift curve slope bring about this condition.

The dynamic stability of the airplane generally deteriorates with Mach number in supersonic flight. Since a large part of the damping depends on the tail surfaces, the decrease in lift curve slope with Mach number will account in part for the decrease in damping. Of course, all principal motions of the aircraft must have satisfactory damping and if the damping is not available aerodynamically it must be provided synthetically to obtain satisfactory flying qualities. For many high speed configurations the pitch and yaw dampers, flight stabilization systems, etc., are basic necessities rather than luxuries.

Generally, flight at high Mach number will take place at high altitude hence the effect of high altitude must be separated for study. All of the basic aerodynamic damping is due to moments created by pitching, rolling, or yawing motion of the aircraft. These moments are derived from the changes in angles of attack on the tail surfaces with angular rotation (see fig. 4.15). The very high true airspeeds common to high altitude flight reduce the angle of attack changes and reduce the aerodynamic damping. In fact, the aerodynamic damping is proportional to $\sqrt{\sigma}$, similar to the proportion of true airspeed to equivalent airspeed. Thus, at the altitude of 40,000 ft., the aerodynamic damping would be reduced to one-half the sea level value and at the altitude of 100,000 ft. the aerodynamic damping would be reduced to one-tenth the sea level value.

High dynamic pressures (high q) can be common to flight at high Mach number and adverse aeroelastic effects may be encountered. If the aircraft surfaces encounter significant deflection when subject to load, the tendency may be to lower the contribution to static stability and reduce the damping contribution. Thus, the problem of adequate stability of the various airplane motions is aggravated.

PILOT INDUCED OSCILLATIONS

The pilot may purposely induce various motions to the airplane by the action of the controls. In addition, certain undesirable motions may occur due to inadvertent action on the controls. The most important condition exists with the short period longitudinal motion of the airplane where pilot-control system response lag can produce an unstable oscillation. The coupling possible in the pilot-control system-airplane combination is most certainly capable of producing damaging flight loads and loss of control of the airplane.

When the normal human response lag and control system lag are coupled with the airplane motion, inadvertent control reactions by the pilot may furnish a negative damping to the oscillatory motion and dynamic instability exists. Since the short period motion is of relatively high frequency, the amplitude of the pitching oscillation can reach dangerous proportions in an unbelievably short time. When the pilot induced oscillation is encountered, the most effective solution is an immediate release of the controls. Any attempt to forcibly damp the oscillation simply continues the excitation and amplifies the oscillation. Freeing the controls removes the unstable (but inadvertent) excitation and allows the airplane to recover by virtue of its inherent dynamic stability.

The pilot induced oscillation is most likely under certain conditions. Most obvious is the case of the pilot unfamiliar with the "feel" of the airplane and likely to overcontrol or have excessive response lag. High speed flight at low altitude (high q) is most likely to provide low stick-force gradients and periods of oscillation which coincide with the pilot-control system response lag. Also, the high q flight condition provides the aerodynamic capability for failing flight loads during the oscillation.

If a pilot induced oscillation is encountered the pilot must rely on the inherent dynamic stability of the airplane and immediately release the controls. If the unstable excitation is continued, dangerous oscillation amplitudes will develop in a very short time.

ROLL COUPLING

The appearance of "inertia coupling" problems in modern airplanes was the natural result of the progressive change in aerodynamic and inertia characteristics to meet the demands of high speed flight. Inertia coupling problems were unexpected only when dynamic stability analyses did not adequately account for the rapid changes in aerodynamic and inertia characteristics of airplane configurations. The The term of "intertia coupling" is somewhat misleading because the complete problem is one of aerodynamic as well as inertia coupling.

"Coupling" results when some disturbance about one airplane axis causes a disturbance about another axis. An example of uncoupled motion is the disturbance provided an airplane when subjected to an elevator deflection. The resulting motion is restricted to pitching motion without disturbance in yaw or roll. An example of coupled motion could be the disturbance provided an airplane when subjected to rudder deflection. The ensuing motion can be some combination of yawing and rolling motion. Hence, the rolling motion is coupled with the yawing motion to define the resulting motion. This sort of interaction results from aerodynamic characteristics and is termed "aerodynamic coupling."

A separate type of coupling results from the inertia characteristics of the airplane configuration. The inertia characteristics of the complete airplane can be divided into the roll, yaw, and pitch inertia and each inertia is a measure of the resistance to rolling, yawing, or pitching acceleration of the airplane. The long,slender, high-density fuselage with short, thin wings produces a roll inertia which is quite small in comparison to the pitch and yaw inertia. These characteristics are typical of the modern airplane configuration. The more conventional low speed airplane may have a wingspan greater than the fuselage length. This type of configuration produces a relatively large roll inertia. A comparison of these configurations is shown in figure 4.34.

Inertia coupling can be illustrated by considering the mass of the airplane to be concentrated in two elements, one representing the mass ahead of the c.g. and one representing the mass behind the c.g. There are two principal axis systems to consider: (1) the aerodynamic, or wind axis is through the c.g. in the relative wind direction, and (2) the inertia axis is through the c.g. in the direction of the two element masses. This axis system is illustrated in figure 4.34.

If the airplane shown in figure 4.34 were in some flight condition where the inertia axis and the aerodynamic axis are alined, no inertia coupling would result from rolling motion. However, if the inertia axis is inclined to the aerodynamic axis, rotation about the aerodynamic axis will create centrifugal forces and cause a pitching moment. In this case, a rolling motion of the aircraft induces a pitching moment through the action of inertia forces. This is "inertia coupling" and is illustrated by part *B* of figure 4.34.

When the airplane is rotated about the inertia axis no inertia coupling will exist but aerodynamic coupling will be present. Part *C* of figure 4.34 shows the airplane after rolling 90° about the inertia axis. The inclination which was initially the angle of attack (α) is now the angle of sideslip ($-\beta$). Also the original zero sideslip has now become zero angle of attack. The sideslip induced by this 90° displacement will affect the roll rate

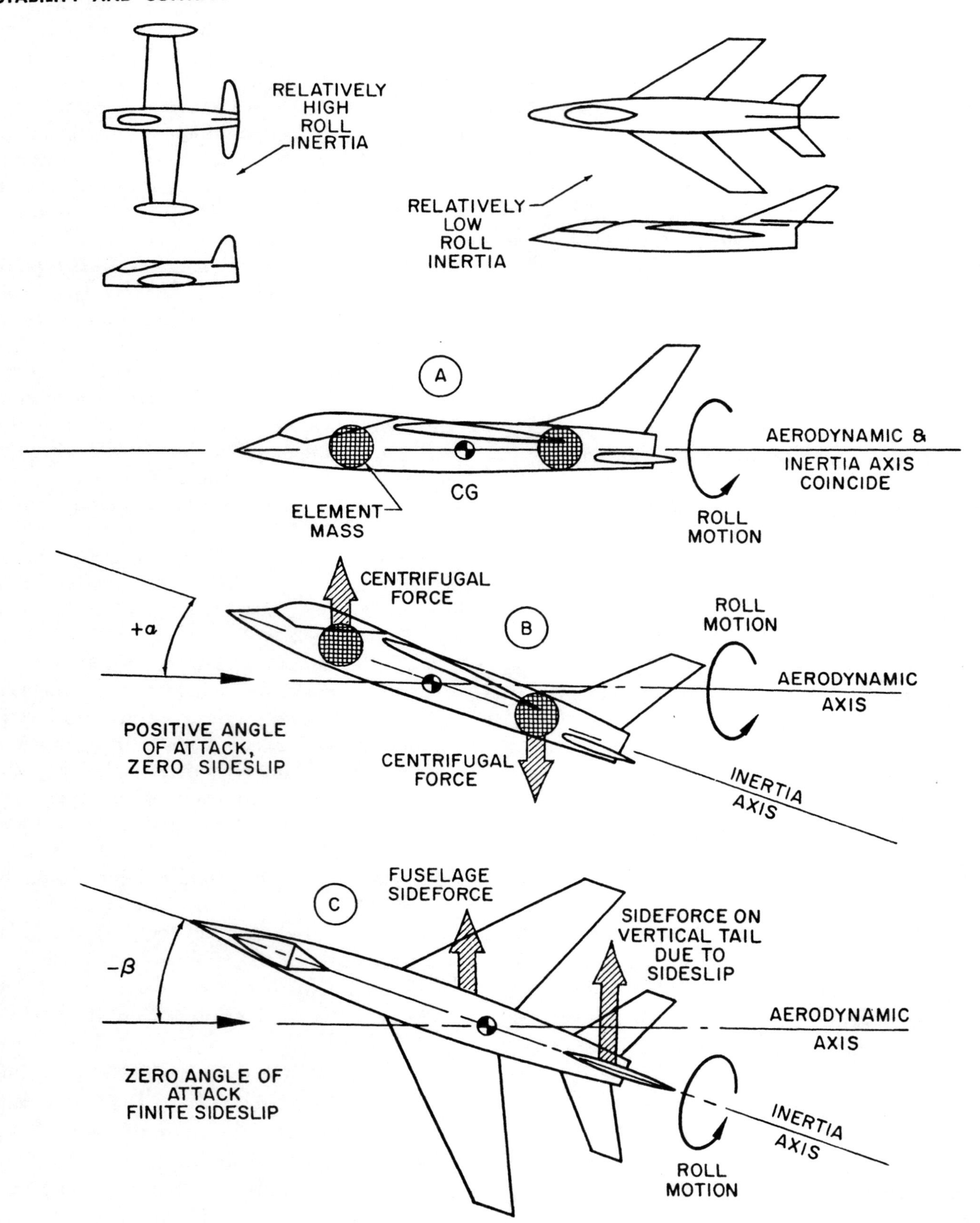

Figure 4.34. Roll Coupling

depending on the nature of the dihedral effect of the airplane.

It should be noted that initial inclination of the inertia axis above the aerodynamic axis will cause the inertia couple to provide adverse yaw with rolling motion. If the inertia axis were initially inclined below the aerodynamic axis (as may happen at high q or negative load factors), the roll induced inertia couple would provide proverse yaw. Thus, roll coupling may present a problem at both positive and negative inclination of the inertia axis depending on the exact aerodynamic and inertia characteristics of the configuration.

As a result of the aerodynamic and inertia coupling, rolling motion can induce a great variety of longitudinal, directional, and lateral forces and moments. The actual motion of the airplane is a result of a complex combination of the aerodynamic and inertia coupling. Actually, all airplanes exhibit aerodynamic and inertia coupling but of varying degrees. The roll coupling causes no problem when the moments resulting from the inertia couple are easily counteracted by the aerodynamic restoring moments. The very short span, high speed modern aircraft has the capability for the high roll rates which cause large magnitudes of the inertia couple. The low aspect ratio planform and flight at high Mach number allow large inclination of the inertia axis with respect to the aerodynamic axis and also add to the magnitude of the inertia couple. In addition, the aerodynamic restoring moments deteriorate as a result of high Mach number and angle of attack and can create the most serious roll coupling conditions.

Since the roll coupling induces pitching and yawing motion, the longitudinal and directional stability is important in determining the overall characteristics of the coupled motion. A stable airplane, when disturbed in pitch and yaw, will return to equilibrium after a series of oscillations. For each flight condition, the airplane will have a coupled pitch-yaw frequency between the uncoupled and separate pitch frequency and yaw frequency. Generally, the greater the static longitudinal and directional stability, the higher will be the coupled pitch-yaw frequency. When the airplane is subject to rolling motion, the inertia couple disturbs the airplane in pitch and yaw with each roll revolution and provides a disturbing forcing function. If the airplane is rolled at a rate equal to the coupled pitch-yaw frequency, the oscillatory motion will either diverge or stabilize at some maximum amplitude depending on the airplane characteristics.

The longitudinal stability of the typical high speed configuration is much greater than the directional stability and results in a pitch frequency higher than the yaw frequency. Increasing the directional stability by increasing the vertical tail area, addition of ventral fins, or use of stabilization systems will increase the coupled pitch-yaw frequency and raise the roll rate at which a possible divergent condition could exist. Increasing directional stability by the addition of ventral fins rather than by addition to the vertical tail has an advantage of not contributing to the positive dihedral effect at low or negative angles of attack. High dihedral effect makes higher roll rates more easily attainable in roll motion where proverse yaw occurs.

Since the uncoupled yawing frequency is lower than the pitching frequency, a divergent condition would first reach critical proportions in yaw, closely followed by pitch. Of course, whether the airplane motion becomes divergent directionally or longitudinally is of academic interest only.

There is one additional type of coupling problem that is referred to as "autorotative rolling." A rolling airplane which has a high positive dihedral effect may reach a large proverse sideslip as a result of the inertia couple and the rolling moment due to sideslip may exceed that available from lateral control. In such a case it would not be possible to stop the airplane from rolling although lateral control was held full against the roll direction. The

210 NF
34881
NAVY

design features which result in a large positive dihedral effect are high sweepback, high wing position, or large, high vertical tail. When the inertia axis is inclined below the aerodynamic axis at low or negative angles of attack, the roll induced inertia couple results in proverse yaw.

Depending on the flight condition where the roll coupling problem exists, four basic types of airplane behavior are possible:

(1) *Coupled motion stable but unacceptable.* In this case the motion is stable but proves unacceptable because of poor damping of the motion. Poor damping would make it difficult to track a target or the initial amplitudes of the motion may be great enough to cause structural failure of loss of control.

(2) *Coupled motion stable and acceptable.* The behavior of the airplane is stable and adequately damped to allow acceptable target tracking. The amplitudes of motion are too slight to result in structural failure or loss of control.

(3) *Coupled motion divergent and unacceptable.* The rate of divergence is too rapid for the pilot to recognize the condition and recover prior to structural failure or complete loss of control.

(4) *Coupled motion divergent but acceptable.* For such a condition the rate of divergence is quite slow and considerable roll displacement is necessary to produce a critical amplitude. The condition can be recognized easily in time to take corrective action.

There are available various means to cope with the problem of roll coupling. The following items can be applied to control the problem of roll coupling:

(*a*) Increase directional stability.

(*b*) Reduce dihedral effect.

(*c*) Minimize the inclination of the inertia axis at normal flight conditions.

(*d*) Reduce undesirable aerodynamic coupling.

(*e*) Limit roll rate, roll duration, and angle of attack or load factor for performing rolling maneuvers.

The first four items can be effected only during design or by design changes. Some roll performance restriction is inevitable since all of the desirable characteristics are difficult to obtain without serious compromise elsewhere in the airplane design. The typical high speed airplane will have some sort of roll performance limitation provided by flight restrictions or automatic control devices to prevent reaching some critical condition from which recovery is impossible. Any roll restriction provided an airplane must be regarded as a principal flight operating limitation since the more severe motions can cause complete loss of control and structural failure.

HELICOPTER STABILITY AND CONTROL

In discussing many of the problems of stability and control that occur in high speed airplanes, one might be prone to believe that the slow flying helicopter does not have any such problems. Unfortunately, this is not the case. Flying qualities that would be considered totally unsatisfactory by fixed-wing standards are normal for helicopters. Helicopter pilots are living evidence that an unstable aircraft can be controlled. Also, they are evidence that control without stability requires constant attention and results in considerable pilot fatigue.

"Inertia coupling" problems are relatively new to fixed-wing aircraft but a similar effect in the helicopter rotor has resulted in some of its most important characteristics. This aerodynamic-dynamic coupling effect is so important that it must be considered in discussing both stability and control. The helicopter derives both longitudinal and lateral control by tilting the main rotor and thus producing a pitching or rolling moment as indicated in figure 4.35. The magnitude of the rotor thrust the angle of tilt, and the height of the rotor hub above the c.g. determine the control moment produced. It should be noted that low control effectiveness would result when the rotor thrust is low. Some helicopters

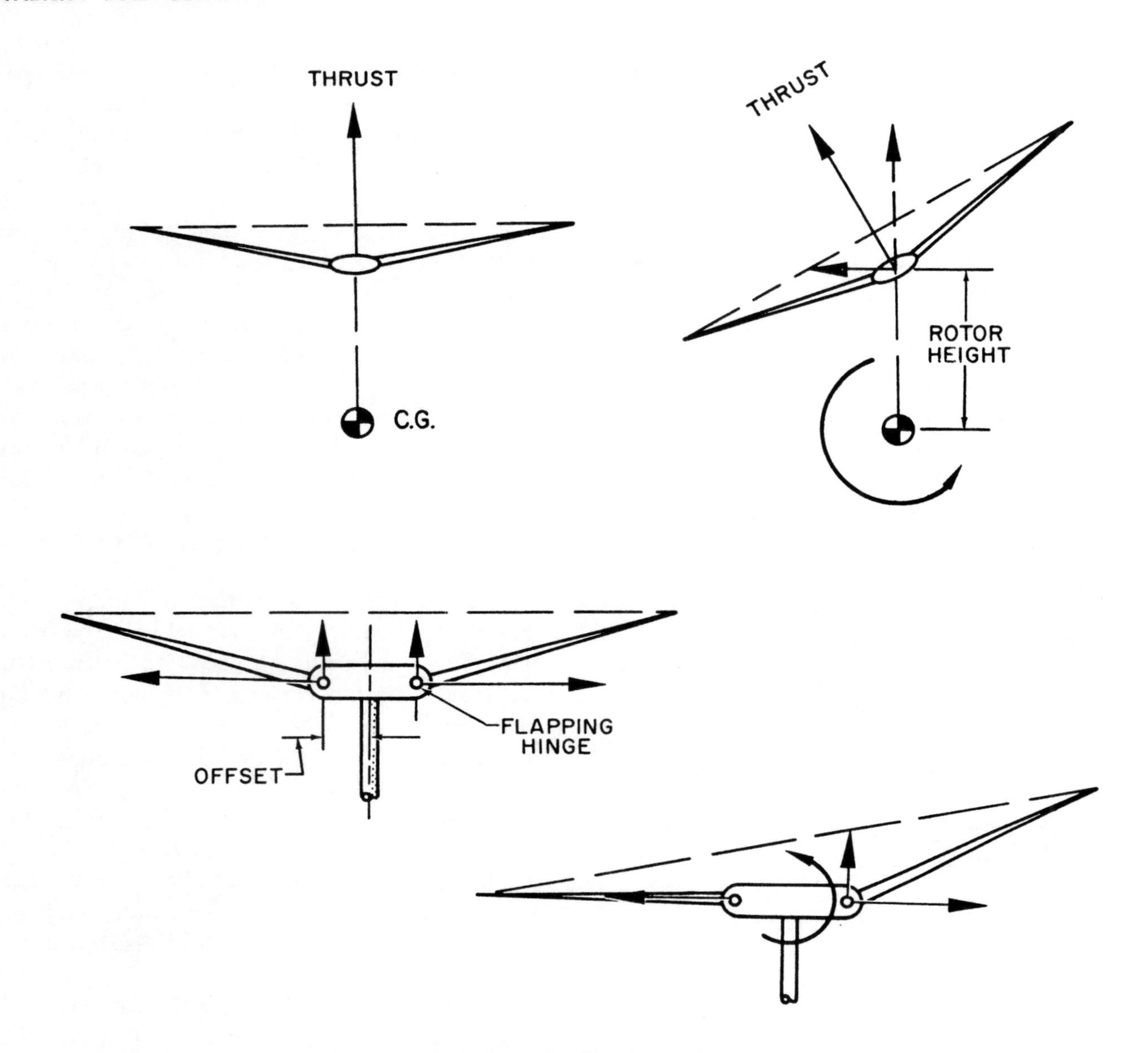

ROTOR GYROSCOPIC ACTION

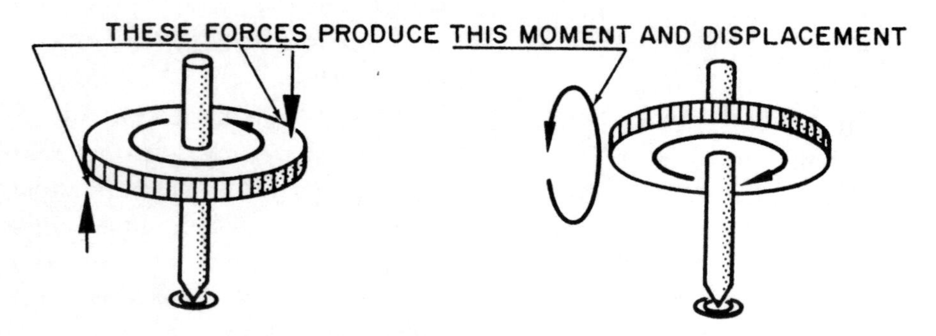

Figure 4.35. Rotor Forces and Moments

employ an offset flapping hinge to increase the control effectiveness by creating a centrifugal force couple when the rotor is tilted. This is shown in figure 4.35.

The rotor is tilted by taking advantage of the gyroscopic effect of the rotor system. This effect causes a rotating mass which is disturbed about one axis to respond about another axis, as shown in figure 4.35. A forward tilt to the rotor is obtained by decreasing the pitch of the blade when at the starboard position and increasing the pitch of the blade when at the port position. The lateral dissymmetry of lift which results causes the rotor to tilt forward by the gyroscopic effect.

A differential blade pitch change like this is called a cyclic pitch change since each blade goes through a complete cycle of varying pitch angles as it completes one revolution of rotation about the hub. A cyclic pitch change is accomplished by the pilot by the use of the cyclic stick. The control arrangement is such that the rotor tilts in the same direction that the cyclic stick is deflected.

A variation in rotor thrust is accomplished by increasing the pitch of the blades simultaneously or collectively. This type of control action is called "collective pitch" and is accomplished by the use of the collective pitch stick. In operation, the cyclic stick is analogous to the control stick of an airplane, and the collective stick is analogous to the throttle of an airplane.

There are several possibilities for longitudinal control of a tandem-rotor helicopter. A pitching moment can be produced by tilting both rotors by a cyclic pitch change in each rotor, by a differential collective pitch change that increases the thrust on one rotor and decreases it on the other, or by some combination of these methods. The two basic methods are illustrated in figure 4.36. Obviously, a change in fuselage attitude must accompany the differential collective method of longitudinal control.

Adequate pitch and lateral control effectiveness are easy to obtain in the typical helicopter and usually present no problems. The more usual problem is an excess of control effectiveness which results in an overly sensitive helicopter. The helicopter control specifications attempt to assure satisfactory control characteristics by requiring adequate margins of control travel and effectiveness without objectionable sensitivity.

Directional control in a single rotor helicopter is obtained by a tail rotor (antitorque rotor) since a conventional aerodynamic surface would not be effective at low speeds or hovering. The directional control requirements of the tail rotor on a typical shaft-driven helicopter are quite demanding since it must counteract the engine torque being supplied to the main rotor as well as provide directional control. Being a rotor in every respect, the tail rotor requires some of the engine power to generate its control forces. Unfortunately, the maximum demands of the tail rotor occur at conditions when engine power is also in great demand. The most critical condition is while hovering at maximum gross weight. The tail rotor effectiveness is determined by the rotor characteristics and the distance the tail rotor is behind the c.g. The control specifications require the helicopter to be able to turn in the most critical direction at some specified rate while hovering at maximum gross weight in a specified wind condition. Also, it is required that the helicopter have sufficient directional control to fly sideways up to 30 knots, an important requirement for plane guard duties.

The directional control requirements are easily met by a tip-driven helicopter since the directional control does not have to counter the engine torque.

Directional control of a tandem-rotor helicopter is accomplished by differential cyclic control of the main rotors. For a pedal turn to the starboard, the forward rotor is tilted to the starboard and the rear rotor is tilted to port, creating a turning moment as shown in

TANDEM ROTOR LONGITUDINAL CONTROL

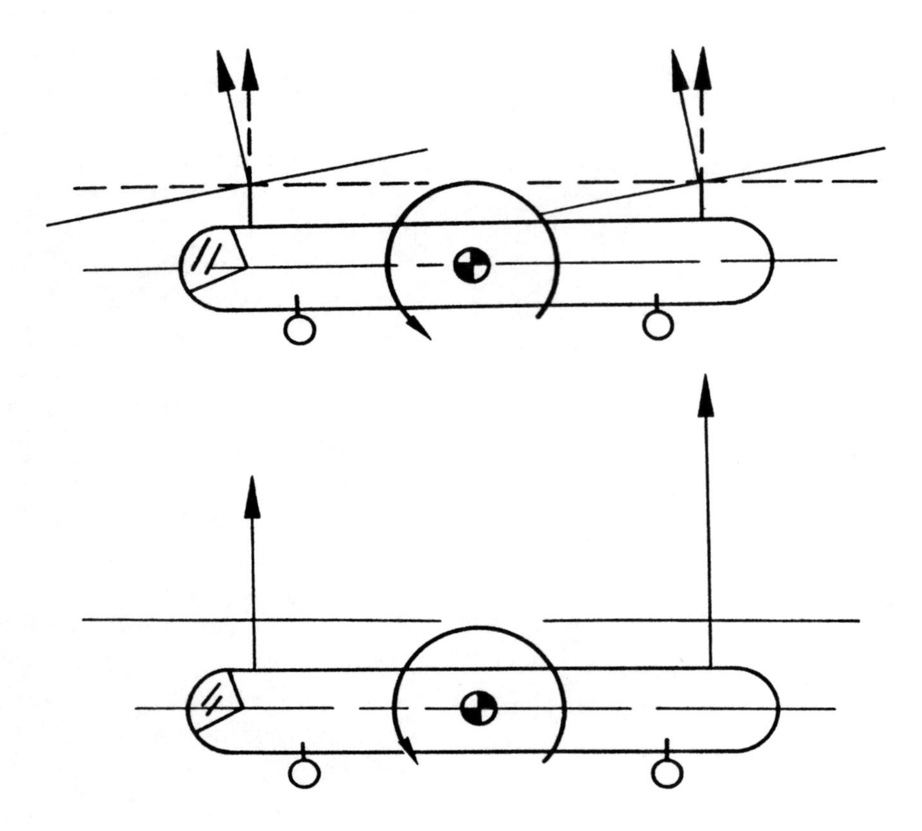

TANDEM ROTOR DIRECTIONAL CONTROL

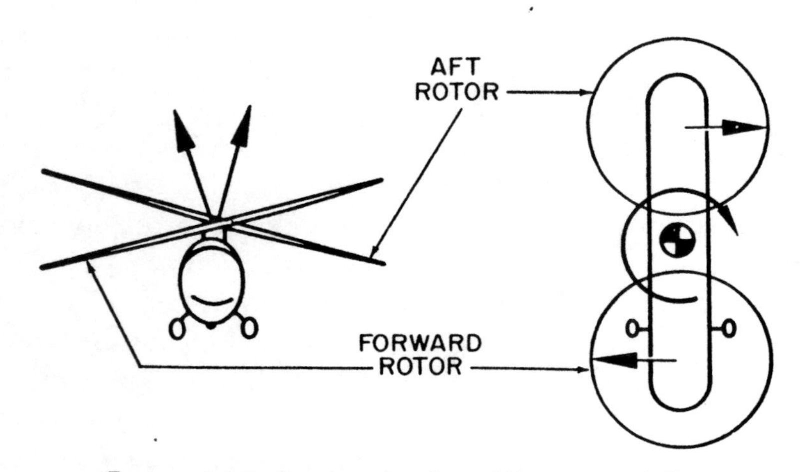

Figure 4.36. Longitudinal and Directional Control

figure 4.36. The directional control requirements are easily met in a tandem-rotor helicopter because the engine torque from one rotor is opposed by the torque of the other rotor thereby eliminating one directional moment. Of course, some net unbalance of torque may have to be overcome if the engine torque on the two rotors is different.

When a tandem-rotor helicopter is rotated rapidly about one of the rotors rather than about the c.g., the other rotor picks up "translational lift" as a result of the velocity due to rotation and an increase in rotor thrust results. This causes pitch-up or pitch-down depending on which rotor the helicopter is being rotated about. Rotation about the forward rotor, which is more common, results in pitch-down.

The overall stability of a helicopter results from the individual stability contributions of the various components just as in the case of the fixed-wing airplane. The stability contributions can be divided as follows:

(1) Rotor
(2) Fuselage
(3) Stabilizers
(4) Mechanical devices

The destabilizing contribution of the fuselage and the stabilizing contribution of a stabilizing surface are similar in effect to an airplane and will not be discussed here. The principal stability characteristics that make the helicopter different from an airplane are those of the rotor.

Two types of stability are important in the rotor: (1) angle of attack stability and (2) velocity stability. In hovering flight the relative wind velocity, angle of attack, and lift on each blade of the rotor is the same. If the rotor is displaced through some angle, no changes in forces result. Therefore, the rotor has neutral angle of attack stability when hovering. However, in forward flight, an increase in rotor angle of attack increases the lift on the advancing blade more than on the retreating blade since the relative wind velocities are greater on the advancing blade. This lateral dissymmetry of lift causes the rotor to tilt back due to the gyroscopic effect of the rotor, further increasing the rotor angle of attack. Thus, the rotor is unstable with changes in angle of attack at forward flight speeds. Since the magnitude of the unstable moment is affected by the magnitude of the rotor thrust as well as the tilt of the thrust force, a greater instability exists for increases in angle of attack than for decreases in angle of attack. In addition, the instability is greater for increases in angle of attack when the rotor thrust also increases.

If the rotor angle of attack is held constant and the rotor is given a translational velocity, a dissymmetry of lift results since the velocity of the advancing blade is increased while the velocity of the retreating blade is decreased. This dissymmetry of lift causes the rotor to tilt in a direction to oppose the change in velocity due to the gyroscopic effect of the rotor. Hence, the rotor has velocity stability.

A hovering helicopter exhibits some degree of apparent stability by virtue of its velocity stability although it has neutral angle of attack stability. This type of hovering stability is analogous to the apparent lateral-directional stability an airplane exhibits due to dihedral effect. Additional hovering stability can be obtained by the use of mechanical stabilizers such as the Bell stabilizer bar, by the use of offset flapping hinges, or by synthetic or artificial stabilization devices.

The total static stability of a helicopter is determined by combining the stability contributions of all the components. The usual result for a typical helicopter is instability with angle of attack and a variable velocity stability which becomes neutral or unstable at high speeds. Of course, the helicopter could be made stable with angle of attack by providing a large enough horizontal stabilizer. Unfortunately, adverse effects at low speed or hovering and large trim moments upon entering autorotation will limit the stabilizer size to a relatively small surface. Usually the horizontal stabilizer is used only to give the fuselage the desired moment characteristics.

The angle of attack stability of a tandem-rotor helicopter is adversely affected by the downwash from the forward rotor reducing the angle of attack and thrust of the rear rotor. This reduction of thrust behind the c.g. causes the helicopter to pitch up to a higher angle of attack, thereby adding to the angle of attack instability.

As in the airplane, several oscillatory modes of motion are characteristic of the dynamic stability of a helicopter. The phugoid is the most troublesome for the helicopter. The phugoid mode is unstable in the majority of helicopters which operate without the assistance of artificial stabilization devices. The dynamic instability of the helicopter is given evidence by the flying qualities specification for helicopters. These specifications essentially limit the rate of divergence of the dynamic oscillations for the ordinary helicopter. Although this dynamic instability can be controlled, it requires constant attention by the pilot and results in pilot fatigue. The elimination of the dynamic instability would contribute greatly to improving the flying qualities of the helicopter.

This dynamic instability characteristic is particularly important if the helicopter is expected to be used for instrument flight in all-weather operations. In fact, a seriously divergent phugoid mode would make instrument flight impractical. For this reason, the flying qualities specification requires that helicopters with an instrument capability exhibit varying degrees of stability or instability depending on the period of the oscillation. Long period oscillations (over 20 seconds) must not double in amplitude in less than 15 seconds whereas short period oscillations (under 10 seconds) must damp to half amplitude in two cycles.

The only immediate solution for the dynamic instability is an attitude stabilization system which is essentially an autopilot. Other solutions to the dynamic instability problem involve mechanical, aerodynamic, or electronic control feedback of pitch attitude, pitch velocity, normal acceleration, or angle of attack. The improvement of the helicopter's stability is mandatory to fully utilize its unique capability. As more of the helicopter problems are analyzed and studied, the flying qualities of helicopters will improve and be comparable to the fixed wing aircraft.

Chapter 5

OPERATING STRENGTH LIMITATIONS

The weight of the structural components of an aircraft is an extremely important factor in the development of an efficient aircraft configuration. In no other field of mechanical design is there such necessary importance assigned to structural weight. The efficient aircraft and powerplant structure is the zenith of highly refined minimum weight design. In order to obtain the required service life from his aircraft, the Naval Aviator must understand, appreciate, and observe the operating strength limitations. Failure to do so will incur excessive maintenance costs and a high incidence of failure during the service life of an aircraft.

GENERAL DEFINITIONS AND STRUCTURAL REQUIREMENTS

There are strength requirements which are common to all aircraft. In general, these requirements can be separated into three particular areas. These are detailed in the following discussion.

STATIC STRENGTH

The static strength requirement is the consideration given to the effect of simple static loads with none of the ramifications of the repetition or cyclic variation of loads. An important reference point in the static strength requirement is the "limit load" condition. When the aircraft is at the design configuration, there will be some maximum of load which would be anticipated from the mission requirement of the airplane. For example, a fighter or attack type aircraft, at the design configuration, may encounter a very peak load factor of 7.5 in the accomplishment of its mission. Of course, such an aircraft may be subject to load factors of 3, 4, 5, 6, 1, etc., but no more than 7.5 should be required to accomplish the mission. Thus, the limit load condition is the maximum of loads anticipated in normal operation of the aircraft. Various types of aircraft will have different limit load factors according to the primary mission of the aircraft. Typical values are tabulated below:

Type of aircraft:	*Positive limit load factor*
Fighter or attack	7.5
Trainer	7.5
Transport, patrol, antisubmarine	3.0 or 2.5

Of course, these examples are quite general and it is important to note that there may be variations according to specific mission requirements.

Since the limit load is the maximum of the normally anticipated loads, the aircraft structure must withstand this load with no ill effects. Specifically, the primary structure of the aircraft should experience no objectionable permanent deformation when subjected to the limit load. In fact, the components must withstand this load with a positive margin. This requirement implies that the aircraft should withstand successfully the limit load and then return to the original unstressed shape when the load is removed. Obviously, if the aircraft is subjected to some load which is in excess of the limit load, the overstress may incur an objectionable permanent deformation of the primary structure and require replacement of the damaged parts.

Many different flight and ground load conditions must be considered to define the most critical conditions for the structural components. In addition to positive lift flight, negative lift flight must be considered. Also, the effect of flap and landing gear configuration, gross weight, flight Mach number, symmetry of loading, c.g. positions, etc., must be studied to account for all possible sources of critical loads. To verify the capability of the structure, ground static tests are conducted and flight demonstrations are required.

To provide for the rare instances of flight when a load greater than the limit is required to prevent a disaster, an "ultimate factor of safety" is provided. Experience has shown that an ultimate factor of safety of 1.5 is sufficient for piloted aircraft. Thus, the aircraft must be capable of withstanding a load which is 1.5 times the design limit load. The primary structure of the aircraft must withstand the "ultimate load" (1.5 times limit) without failure. Of course, permanent deformation may be expected with this "overstress" but no actual failure of the major load-carrying components should take place at ultimate load. Ground static tests are necessary to verify this capability of the structure.

An appreciation of the static strength requirements may be obtained by inspection of the basic properties of a typical aircraft metal. Figure 5.1 illustrates the typical static strength properties of a metal sample by a plot of applied stress versus resulting strain. At low values

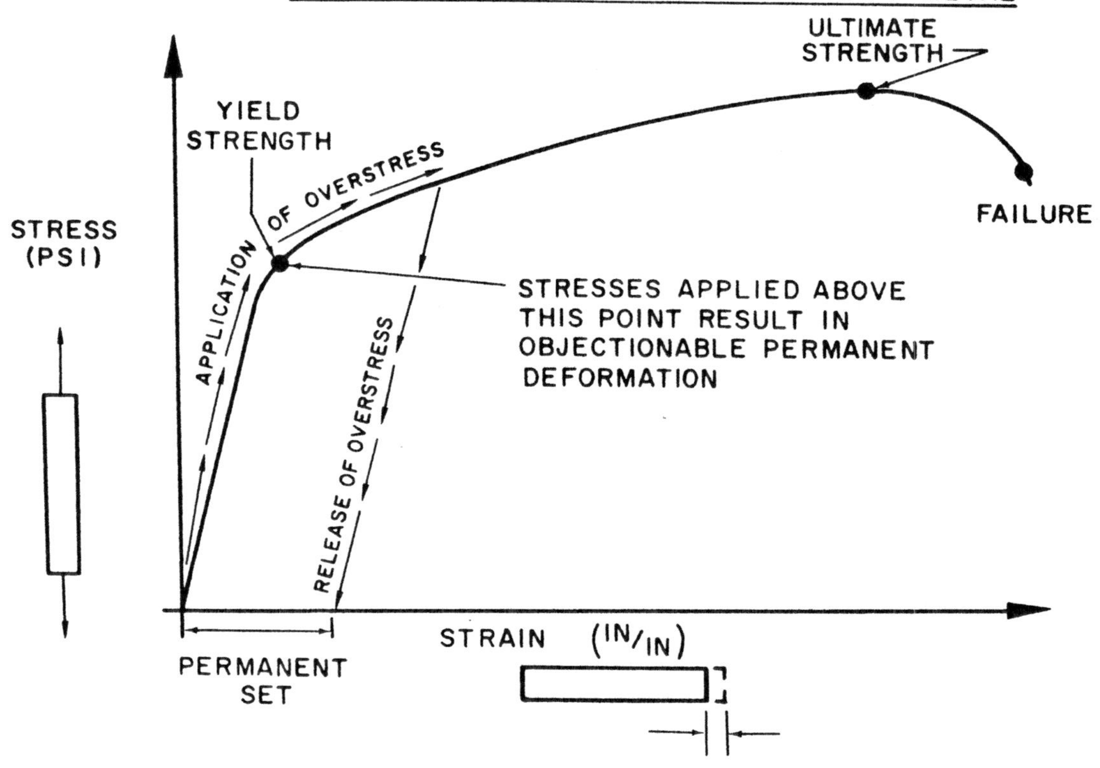

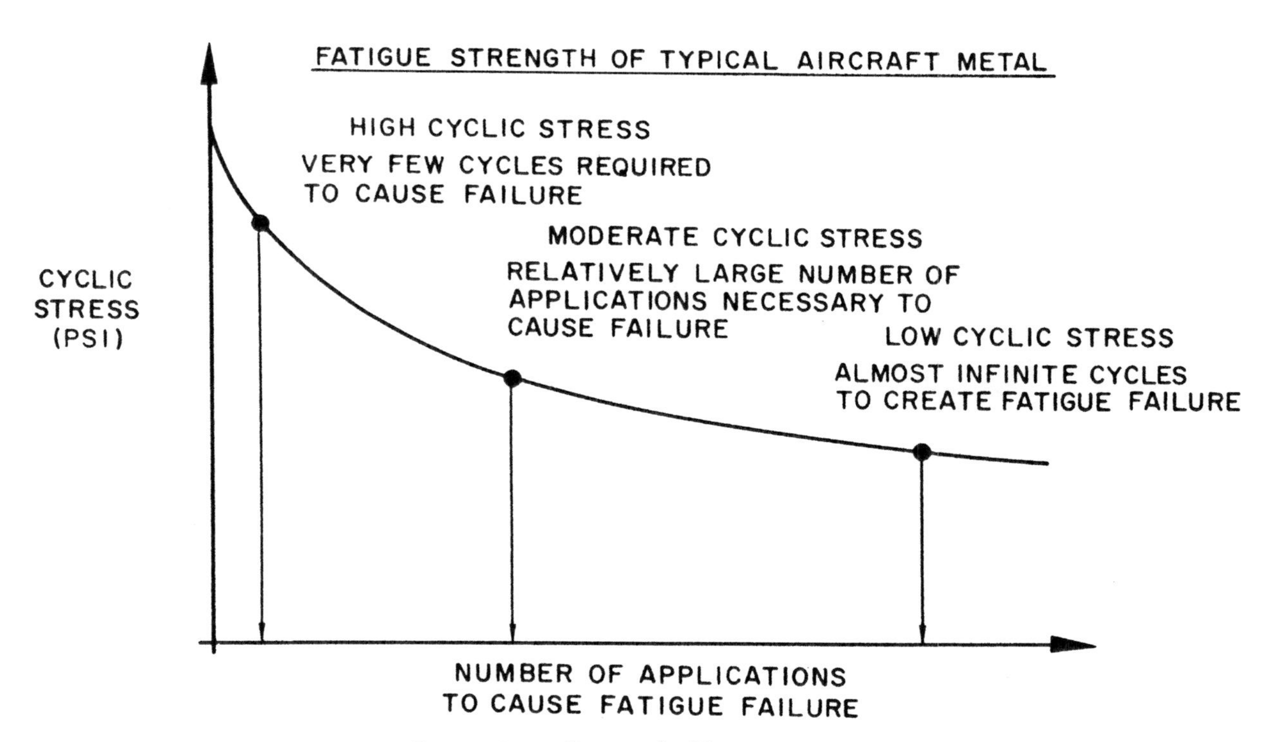

Figure 5.1. Strength Characteristics

of stress the plot of stress and strain is essentially a straight line, i.e., the material in this range is *elastic*. A stress applied in this range incurs no permanent deformation and the material returns to the original unstressed shape when the stress is released. At higher values of stress the plot of stress versus strain develops a distinct curvature in the strain direction and the material incurs disproportionate strains.

High levels of stress applied to the part and then released produce a permanent deformation. Upon release of some high stress, the metal snaps back—but not all the way. The stress defining the limit of tolerable permanent strain is the "yield stress" and stresses applied above this point produce objectionable permanent deformation. The very highest stress the material can withstand is the "ultimate stress." Noticeable permanent deformation usually occurs in this range, but the material does have the capability for withstanding one application of the ultimate stress.

The relationship between the stress-strain diagram and operating strength limits should be obvious. If the aircraft is subjected to a load greater than the limit, the yield stress may be exceeded and objectionable permanent deformation may result. If the aircraft is subject to a load greater than the ultimate, failure is imminent.

SERVICE LIFE

The various components of the aircraft and powerplant structure must be capable of operating without failure or excessive deformation throughout the intended service life. The repetition of various service loads can produce fatigue damage in the structure and special attention must be given to prevent fatigue failure within the service life. Also, the sustaining of various service loads can produce creep damage and special attention must be given to prevent excessive deformation or creep failure within the service life. This is a particular feature of components which are subjected to operation at high temperatures.

FATIGUE CONSIDERATIONS. The fatigue strength requirement is the consideration given the cumulative effect of repeated or cyclic loads during service. While there is a vague relationship with the static strength, repeated cyclic loads produce a completely separate effect. If a cyclic, tensile stress is applied to a metal sample, the part is subject to a "fatigue" type loading. After a period of time, the cyclic stressing will produce a minute crack at some critical location in the sample. With continued application of the varying stress, the crack will enlarge and propagate into the cross section. When the crack has progressed sufficiently, the remaining cross section is incapable of withstanding the imposed stress and a sudden, final rupture occurs. In this fashion, a metal can be failed at stresses much lower than the static ultimate strength.

Of course, the time necessary to produce fatigue failure is related to the magnitude of the cyclic stress. This relationship is typified by the graph of figure 5.1. The fatigue strength of a material can be demonstrated by a plot of cyclic stress versus cycles of stress required to produce fatigue failure. As might be expected, a very high stress level requires relatively few cycles to produce fatigue failure. Moderate stress levels require a fairly large number of cycles to produce failure and a very low stress may require nearly an infinite number of cycles to produce failure. The very certain implication is that the aircraft must be capable of withstanding the gamut of service loads without producing fatigue failure of the primary structure.

For each mission type of aircraft there is a probable spectrum of loads which the aircraft will encounter. That is, various loads will be encountered with a frequency particular to the mission profile. The fighter or attack type of aircraft usually experiences a predominance of maneuver loads while the transport or patrol type usually encounters a predominance of gust loads. Since fatigue damage

is *cumulative* during cyclic stressing, the useful service life of the aircraft must be anticipated to predict the gross effect of service loads. Then, the primary structure is required to sustain the typical load spectrum through the anticipated service life without the occurrence of fatigue failure. To prove this capability of the structure, various major components must be subjected to an accelerated fatigue test to verify the resistance to repeated loads.

The design of a highly stressed or long life structure emphasizes the problems of fatigue. Great care must be taken during design and manufacture to minimize stress concentrations which enhance fatigue. When the aircraft enters service operation, care must be taken in the maintenance of components to insure proper adjustment, torquing, inspection, etc., as proper maintenance is a necessity for achieving full service life. Also, the structure must not be subjected to a load spectrum more severe than was considered in design or fatigue failures may occur within the anticipated service life. With this additional factor in mind, any pilot should have all the more respect for the operating strength limits—recurring overstress causes a high rate of fatigue damage.

There are many examples of the detrimental effect of repeated overstress on service life. One major automobile manufacturer advertised his product as "guaranteed to provide 100,000 miles of normal driving without mechanical failure." The little old lady from Pasadena—the original owner of ALL used cars—will probably best the guaranteed mileage by many times. On the other hand, the hot-rod artist and freeway Grand Prix contender do not qualify for the guarantee since their manner of operation could not be considered *normal*. The typical modern automobile may be capable of 60,000 to 100,000 miles of normal operation before an overhaul is necessary. However, this same automobile may encounter catastrophic failures in a few hundred miles if operated continually at maximum torque in low drive range. Obviously, there are similar relationships for aircraft and powerplant structures.

CREEP CONSIDERATIONS. By definition, creep is the structural deformation which occurs as a function of time. If a part is subjected to a constant stress of sufficient magnitude, the part will continue to develop plastic strain and deform with time. Eventually, failure can occur from the accumulation of creep damage. Creep conditions are most critical at high stress and high temperature since both factors increase the rate of creep damage. Of course, any structure subject to creep conditions should not encounter excessive deformation or failure within the anticipated service life.

The high operating temperatures of gas turbine components furnish a critical environment for creep conditions. The normal operating temperatures and stresses of gas turbine components create considerable problems in design for service life. Thus, operating limitations deserve very serious respect since excessive engine speed or excessive turbine temperatures will cause a large increase in the rate of creep damage and lead to premature failure of components. Gas turbines require high operating temperatures to achieve high performance and efficiency and short periods of excessive temperatures can incur highly damaging creep rates.

Airplane structures can be subject to high temperatures due to aerodynamic heating at high Mach numbers. Thus, very high speed airplanes can be subject to operating limitations due to creep conditions.

AEROELASTIC EFFECTS

The requirement for structural stiffness and rigidity is the consideration given to the interaction of aerodynamic forces and deflections of the structure. The aircraft and its components must have sufficient stiffness to prevent or minimize aeroelastic influences in the normal flight range. Aileron reversal, divergence, flutter, and vibration should not occur in the range of flight speeds which will be normal operation for the aircraft.

It is important to distinguish between strength and stiffness. Strength is simply the resistance to load while stiffness is the resistance to deflection or deformation. While strength and stiffness are related, it is necessary to appreciate that adequate structural strength does not automatically provide adequate stiffness. Thus, special consideration is necessary to provide the structural components with specific stiffness characteristics to prevent undesirable aeroelastic effects during normal operation.

An obvious solution to the apparent problems of static strength, fatigue strength, stiffness and rigidity would be to build the airplane like a product of an anvil works, capable of withstanding all conceivable loads. However, high performance airplane configurations cannot be developed with inefficient, lowly stressed structures. The effect of additional weight is best illustrated by preliminary design studies of a very long range, high altitude bomber. In the preliminary phases of design, each additional pound of *any* weight would necessitate a 25-pound increase in gross weight to maintain the same performance. An increase in the weight of any item produced a chain reaction—more fuel, larger tanks, bigger engines, more fuel, heavier landing gear, more fuel, etc. In the competitive sense of design, no additional structural weight can be tolerated to provide more strength than is specified as necessary for the design mission requirement.

AIRCRAFT LOADS AND OPERATING LIMITATIONS

FLIGHT LOADS—MANEUVERS AND GUSTS

The loads imposed on an aircraft in flight are the result of maneuvers and gusts. The maneuver loads may predominate in the design of fighter airplanes while gust loads may predominate in the design of the large multiengine aircraft. The maneuver loads an airplane may encounter depend in great part on the mission type of the airplane. However, the maximum maneuvering capability is of interest because of the relationship with strength limits.

The flight load factor is defined as the proportion between airplane lift and weight, where

$n = L/W$

n = load factor

L = lift, lbs.

W = weight, lbs.

MANEUVERING LOAD FACTORS. The maximum lift attainable at any airspeed occurs when the airplane is at $C_{L_{max}}$. With the use of the basic lift equation, this maximum lift is expressed as:

$$L_{max} = C_{L_{max}} \tfrac{1}{2} \rho V^2 S$$

Since maximum lift must be equal to the weight at the stall speed,

$$W = C_{L_{max}} \tfrac{1}{2} \rho V_s^2 S$$

If the effects of compressibility and viscosity on $C_{L_{max}}$ are neglected for simplification, the maximum load factor attainable is determined by the following relationship.

$$n_{max} = \frac{L_{max}}{W} = \frac{C_{L_{max}} \frac{1}{2} \rho V^2 S}{C_{L_{max}} \frac{1}{2} \rho V_s^2 S} = \left(\frac{V}{V_s}\right)^2$$

Thus, if the airplane is flying at twice the stall speed and the angle of attack is increased to obtain maximum lift, a maximum load factor of four will result. At three times the stall speed, nine "*g*'s" would result; four times the stall speed, sixteen *g*'s result; five times the stall speed, twenty-five *g*'s result; etc. Therefore, any airplane which has high speed performance may have the capability of high maneuvering load factors. The airplane which is capable of flight speeds that are many times the stall speed will require due consideration of the operating strength limits.

The structural design of the aircraft must consider the possibility of negative load factors from maneuvers. Since the pilot cannot comfortably tolerate large prolonged negative "g", the aircraft need not be designed for negative load factors as great as the positive load factors.

The effect of airplane gross weight during maneuvers must be appreciated because of the particular relation to flight operating strength limitations. During flight, the pilot appreciates the degree of a maneuver from the inertia forces produced by various load factors; the airplane structure senses the degree of a maneuver principally by the airloads involved. Thus, the pilot recognizes *load factor* while the structure recognizes only *load*. To better understand this relationship, consider an example airplane whose basic configuration gross weight is 20,000 lbs. At this basic configuration assume a limit load factor for symmetrical flight of 5.6 and an ultimate load factor of 8.4. If the airplane is operated at any other configuration, the load factor limits will be altered. The following data illustrate this fact by tabulating the load factors required to produce identical airloads at various gross weights.

Gross weight, lbs.	Limit load factor	Ultimate load factor
20,000 (basic)	5.60	8.40
30,000 (max. takeoff)	3.73	5.60
13,333 (min. fuel)	8.40	12.60

As illustrated, at high gross weights above the basic configuration weight, the limit and ultimate load factors may be seriously reduced. For the airplane shown, a 5-*g* maneuver immediately after a high gross weight takeoff could be very near the "disaster regime," especially if turbulence is associated with the maneuver. In the same sense, this airplane at very low operating weights below that of the basic configuration would experience greatly increased limit and ultimate load factors. Operation in this region of high load factors at low gross weight may create the impression that the airplane has great excess strength capability. This effect must be understood and intelligently appreciated since it is not uncommon to have a modern airplane configuration with more than 50 percent of its gross weight as fuel.

GUST LOAD FACTORS. Gusts are associated with the vertical and horizontal velocity gradients in the atmosphere. A horizontal gust produces a change in dynamic pressure on the airplane but causes relatively small and unimportant changes in flight load factor. The more important gusts are the vertical gusts which cause changes in angle of attack. This process is illustrated in figure 5.2. The vectorial addition of the gust velocity to the airplane velocity causes the change in angle of attack and change in lift. The change in angle of attack at some flight condition causes a change in the flight load factor. The increment change in load factor due to the vertical gust can be determined from the following equation:

$$\Delta n = 0.115 \frac{m\sqrt{\sigma}}{(W/S)} V_e (KU)$$

where

Δn = change in load factor due to gust
m = lift curve slope, unit of C_L per degree of α
σ = altitude density ratio
W/S = wing loading, psf
V_e = equivalent airspeed, knots
KU = equivalent sharp edged gust velocity ft. per sec.

As an example, consider the case of an airplane with a lift curve slope $m=0.08$ and wing loading, $(W/S)=60$ psf. If this airplane were flying at sea level at 350 knots and encountered an effective gust of 30 ft. per sec., the gust would produce a load factor increment of *1.61*. This increment would be added to the flight load factor of the airplane prior to the gust,

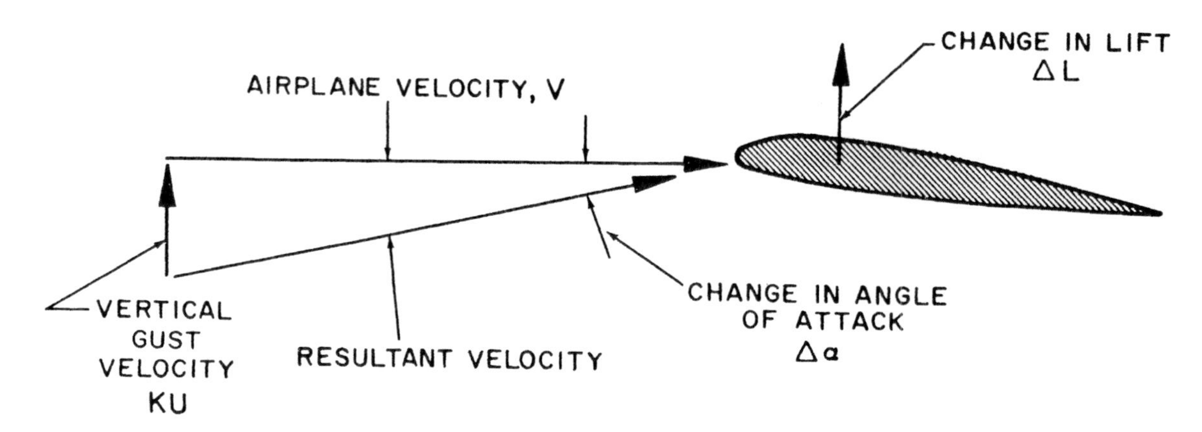

Figure 5.2. Effect of Vertical Gust

e.g., if in level flight before encountering the gust, a final load factor of $1.0+1.61=2.61$ would result. As a general requirement all airplanes must be capable of withstanding an approximate effective ± 30 ft. per sec. gust when at maximum level flight speed for normal rated power. Such a gust intensity has relatively low frequency of occurrence in ordinary flying operations.

The equation for gust load increment provides a basis for appreciating many of the variables of flight. The gust load increment varies directly with the equivalent sharp edged gust velocity, KU, since this factor effects the change in angle of attack. The highest reasonable gust velocity that may be anticipated is an actual vertical velocity, U, of 50 ft. per sec. This value is tempered by the fact that the airplane does not effectively encounter the full effect because of the response of the airplane and the gradient of the gust. A gust factor, K (usually on the order of 0.6), reduces the actual gust to the equivalent sharp edged gust velocity, KU.

The properties of the airplane exert a powerful influence on the gust increment. The lift curve slope, m, relates the sensitivity of the airplane to changes in angle of attack. An aircraft with a straight, high aspect ratio wing would have a high lift curve slope and would be quite sensitive to gusts. On the other hand, the low aspect ratio, swept wing airplane has a low lift curve slope and is comparatively less sensitive to turbulence. The apparent effect of wing loading, W/S, is at times misleading and is best understood by considering a particular airplane encountering a fixed gust condition at various gross weights. If the airplane encounters the gust at lower than ordinary gross weight, the accelerations due to the gust condition are higher. This is explained by the fact that essentially the same lift change acts on the lighter mass. The high accelerations and inertia forces magnify the impression of the magnitude of turbulence. If this same airplane encounters the gust condition at higher than ordinary gross weight, the accelerations due to the gust condition are lower, i.e., the same lift change acts on the greater mass. Since the pilot primarily senses the degree of turbulence by the resulting accelerations and inertia forces, this effect can produce a very misleading impression.

The effect of airspeed and altitude on the gust load factor is important from the standpoint of flying operations. The effect of altitude is related by the term $\sqrt{\sigma}$, which would related that an airplane flying at a given *EAS* at 40,000 ft. (σ=0.25) would experience a gust load factor increment only one-half as great as at sea level. This effect results because the *true* airspeed is twice as great and only one-half the change in angle of attack occurs for a given gust velocity. The effect of airspeed is illustrated by the linear variation of gust increment with equivalent airspeed. Such a variation emphasizes the effect of gusts at high flight speeds and the probability of structural damage at excessive speeds in turbulence.

The operation of any aircraft is subject to specific operating strength limitations. A single large overstress may cause structural failure or damage severe enough to require costly overhaul. Less severe overstress repeated for sufficient time will cause fatigue cracking and require replacement of parts to prevent subsequent failure. A combat airplane need not be operated in a manner like the "little old lady from Pasadena" driving to church on Sunday but each aircraft type has strength capability only specific to the mission requirement. Operating limitations must be given due regard.

THE *V–n* OR *V–g* DIAGRAM

The operating flight strength limitations of an airplane are presented in the form of a *V–n* or *V–g* diagram. This chart usually is included in the aircraft flight handbook in the section dealing with operating limitations. A typical *V–n* diagram is shown in figure 5.3. The *V–n* diagram presented in figure 5.3 is intended to present the most important general features of such a diagram and does not necessarily represent the characteristics of any particular airplane. Each airplane type has its own particular *V–n* diagram with specific *V*'s and *n*'s.

The flight operating strength of an airplane is presented on a graph whose horizontal scale is airspeed (*V*) and vertical scale is load factor (*n*). The presentation of the airplane strength is contingent on four factors being known: (1) the aircraft gross weight, (2) the configuration of the aircraft (clean, external stores, flaps and landing gear position, etc.), (3) symmetry of loading (since a rolling pullout at high speed can reduce the structural limits to approximately two-thirds of the symmetrical load limits) and (4) the applicable altitude. A change in any one of these four factors can cause important changes in operating limits.

For the airplane shown, the positive limit load factor is 7.5 and the positive ultimate load factor is 11.25 (7.5×1.5). For negative lift flight conditions the negative limit load factor is 3.0 and the negative ultimate load factor is 4.5 (3.0×1.5). The limit airspeed is stated as 575 knots while the wing level stall speed is apparently 100 knots.

Figure 5.4 provides supplementary information to illustrate the significance of the *V–n* diagram of figure 5.3. The lines of maximum lift capability are the first points of importance on the *V–n* diagram. The subject aircraft is capable of developing no more than one positive "*g*" at 100 knots, the wing level stall speed of the airplane. Since the maximum load factor varies with the square of the airspeed,

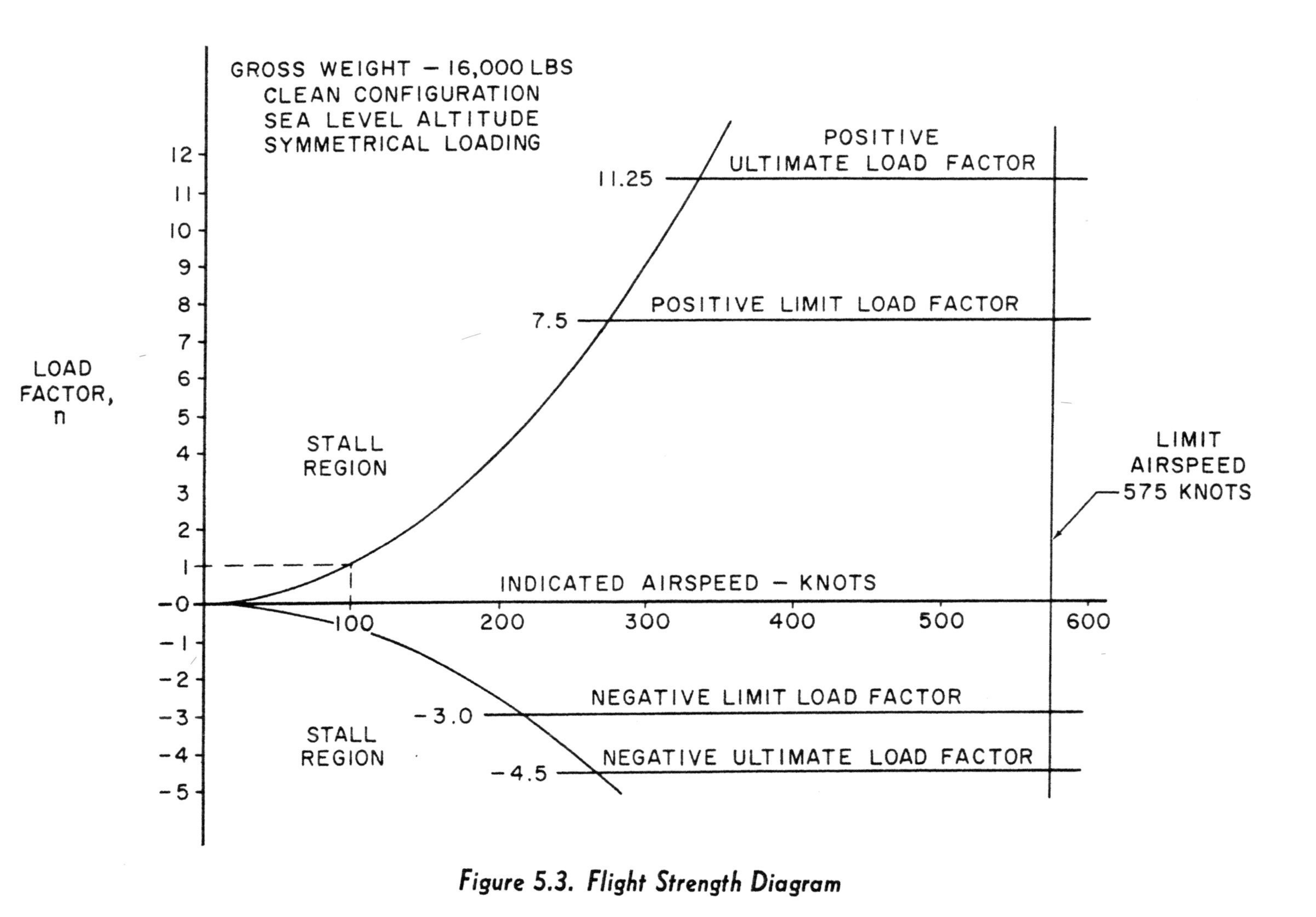

Figure 5.3. Flight Strength Diagram

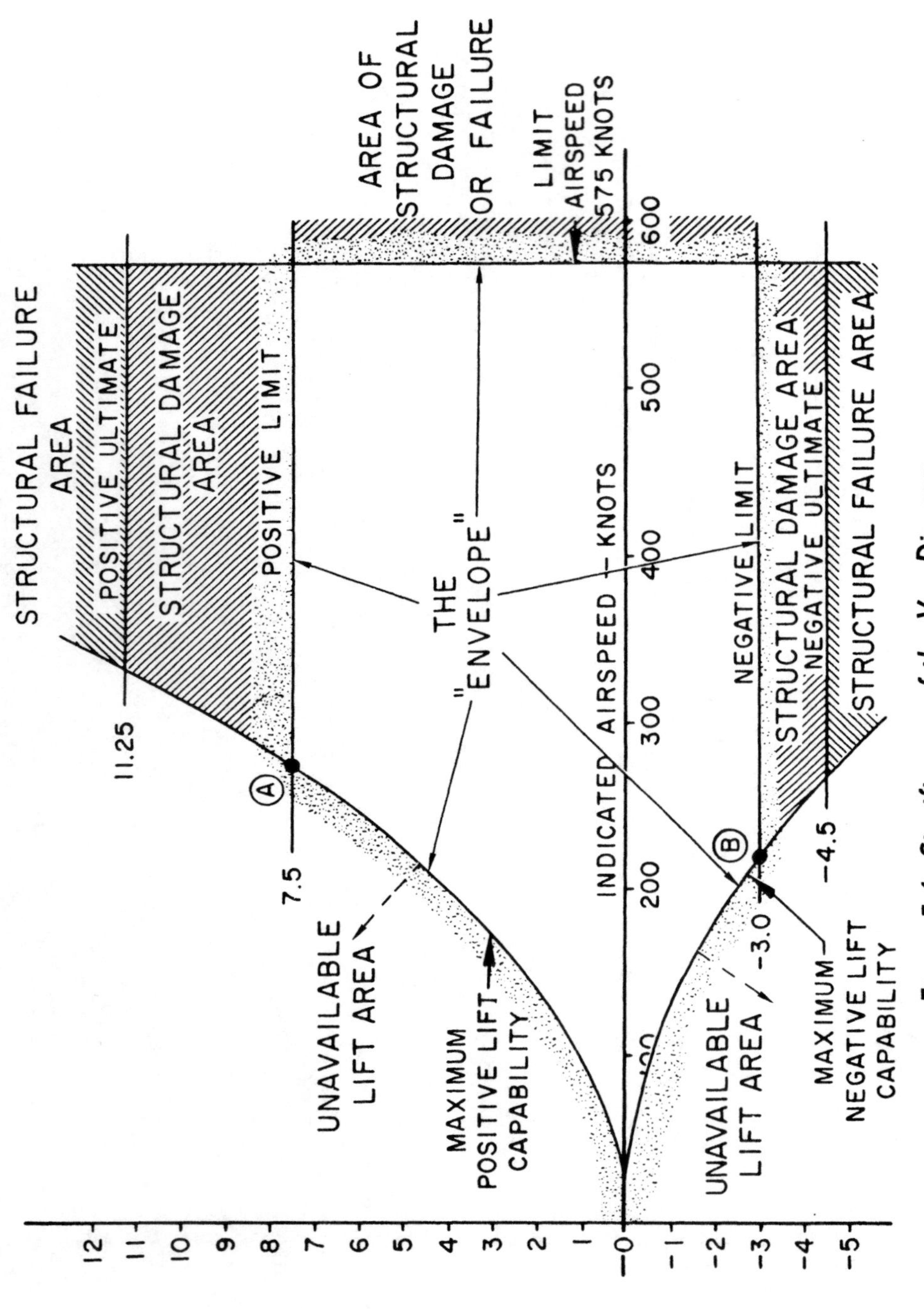

Figure 5.4. Significance of the V–n Diagram

the maximum positive lift capability of this airplane is 4 "g" at 200 knots, 9 g at 300 knots, 16 g at 400 knots, etc. Any load factor above this line is unavailable aerodynamically, i.e., the subject airplane cannot fly above the line of maximum lift capability. Essentially the same situation exists for negative lift flight with the exception that the speed necessary to produce a given negative load factor is higher than that to produce the same positive load factor. Generally, the negative $C_{L_{max}}$ is less than the positive $C_{L_{max}}$ and the airplane may lack sufficient control power to maneuver in this direction.

If the subject airplane is flown at a positive load factor greater than the positive limit load factor of 7.5, structural damage will be possible. When the airplane is operated in this region, objectionable permanent deformation of the primary structure may take place and a high rate of fatigue damage is incurred. Operation above the limit load factor must be avoided in normal operation. If conditions of extreme emergency require load factors above the limit to prevent an immediate disaster, the airplane should be capable of withstanding the ultimate load factor without failure. The same situation exists in negative lift flight with the exception that the limit and ultimate load factors are of smaller magnitude and the negative limit load factor may not be the same value at all airspeeds. At speeds above the maximum level flight airspeed the negative limit load factor may be of smaller magnitude.

The limit airspeed (or redline speed) is a design reference point for the airplane—the subject airplane is limited to 575 knots. If flight is attempted beyond the limit airspeed structural damage *or* structural failure may result from a variety of phenomena. The airplane in flight above the limit airspeed may encounter:

(*a*) critical gust
(*b*) destructive flutter
(*c*) aileron reversal
(*d*) wing or surface divergence
(*e*) critical compressibility effects such as stability and control problems, damaging buffet, etc.

The occurrence of any one of these items could cause structural damage or failure of the primary structure. A reasonable accounting of these items is required during the design of an airplane to prevent such occurrences in the required operating regions. The limit airspeed of an airplane may be any value between terminal dive speed and 1.2 times the maximum level flight speed depending on the aircraft type and mission requirement. Whatever the resulting limit airspeed happens to be, it deserves due respect.

Thus, the airplane in flight is limited to a regime of airspeeds and *g*'s which do not exceed the limit (or redline) speed, do not exceed the limit load factor, and cannot exceed the maximum lift capability. The airplane must be operated within this "envelope" to prevent structural damage and ensure that the anticipated service life of the airplane is obtained. The pilot must appreciate the *V–n* diagram as describing the allowable combination of airspeeds and load factors for safe operation. Any maneuver, gust, or gust plus maneuver outside the structural envelope can cause structural damage and effectively shorten the service life of the airplane.

There are two points of great importance on the *V–n* diagram of figure 5.4. Point B is the intersection of the negative limit load factor and line of maximum negative lift capability. Any airspeed greater than point B provides a negative lift capability sufficient to damage the airplane; any airspeed less than point B does not provide negative lift capability sufficient to damage the airplane from excessive flight loads. Point A is the intersection of the positive limit load factor and the line of maximum positive lift capability. The airspeed at this point is the minimum airspeed at which the limit load can be developed aerodynamically. Any airspeed greater than point A provides a positive lift capability sufficient to damage the airplane; any airspeed less than point A does *not* provide positive lift capability sufficient to

NAVY

cause damage from excessive flight loads. The usual term given to the speed at point A is the "maneuver speed," since consideration of subsonic aerodynamics would predict minimum usable turn radius to occur at this condition. The maneuver speed is a valuable reference point since an airplane operating below this point cannot produce a damaging positive flight load. Any combination of maneuver and gust cannot create damage due to excess airload when the airplane is below the maneuver speed.

The maneuver speed can be computed from the following equation:

$$V_P = V_s \sqrt{n \text{ limit}}$$

where

V_P = maneuver speed

V_s = stall speed

n limit = limit load factor

Of course, the stall speed and limit load factor must be appropriate for the airplane gross weight. One notable fact is that this speed, once properly computed, remains a constant value if no significant change takes place in the spanwise weight distribution. The maneuver speed of the subject aircraft of figure 5.4. would be

$$V_P = 100\sqrt{7.5}$$
$$= 274 \text{ knots}$$

EFFECT OF HIGH SPEED FLIGHT

Many different factors may be of structural importance in high speed flight. Any one or combination of these factors may be encountered if the airplane is operated beyond the limit (or redline) airspeed.

At speeds beyond the limit speed the airplane may encounter a *critical gust*. This is especially true of a high aspect ratio airplane with a low limit load factor. Of course, this is also an important consideration for an airplane with a high limit load factor if the gust should be superimposed on a maneuver. Since the gust load factor increment varies directly with airspeed and gust intensity, high airspeeds must be avoided in turbulent conditions.

When it is impossible to avoid turbulent conditions and the airplane must be subject to gusts, the flight condition must be properly controlled to minimize the effect of turbulence. If possible, the airplane airspeed and power should be adjusted prior to entry into turbulence to provide a stabilized attitude. Obviously, penetration of turbulence should not be accomplished at an excess airspeed because of possible structural damage. On the other hand, an excessively low speed should not be chosen to penetrate turbulence for the gusts may cause stalling of the aircraft and difficulty of control. To select a proper penetration airspeed the speed should not be excessively high or low—the two extremes must be tempered. The "maneuver" speed is an important reference point since it is the highest speed that can be taken to alleviate stall due to gust and the lowest speed at which limit load factor can be developed aerodynamically. The optimum penetration speed occurs at or very near the maneuver speed.

Aileron reversal is a phenomenon particular to high speed flight. When in flight at very high dynamic pressures, the wing torsional deflections which occur with aileron deflection are considerable and cause noticeable change in aileron effectiveness. The deflection of an aileron on a rigid wing creates a change in lift and produces a rolling moment. In addition the deflection of the control surface creates a twisting moment on the wing. When the actual elastic wing is subject to this condition at high dynamic pressures, the twisting moment produces measurable twisting deformations which affect the rolling performance of the aircraft. Figure 5.5 illustrates this process and the effect of airspeed on aileron effectiveness. At some high dynamic pressure, the

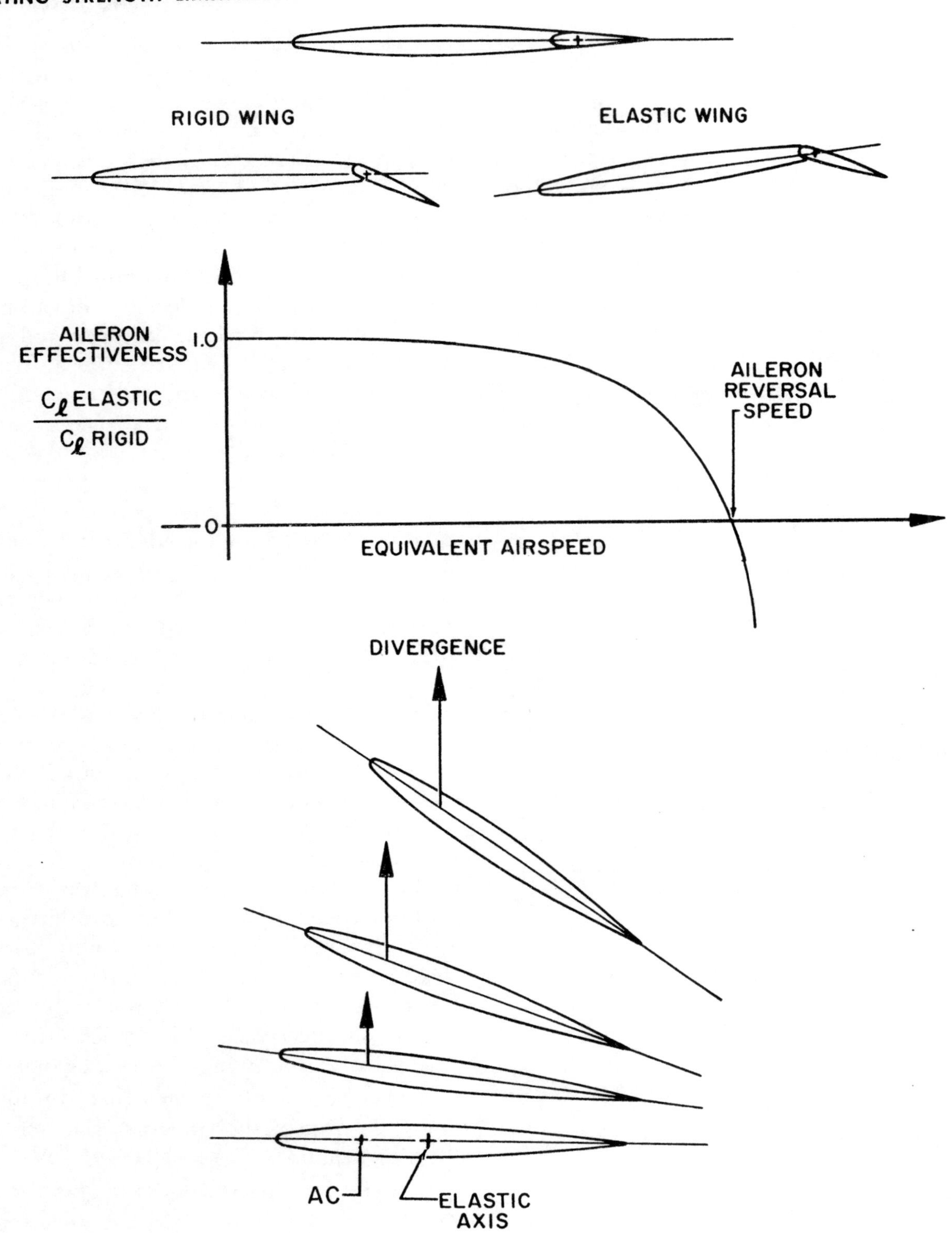

Figure 5.5. Aeroelastic Effects (Sheet 1 of 2)

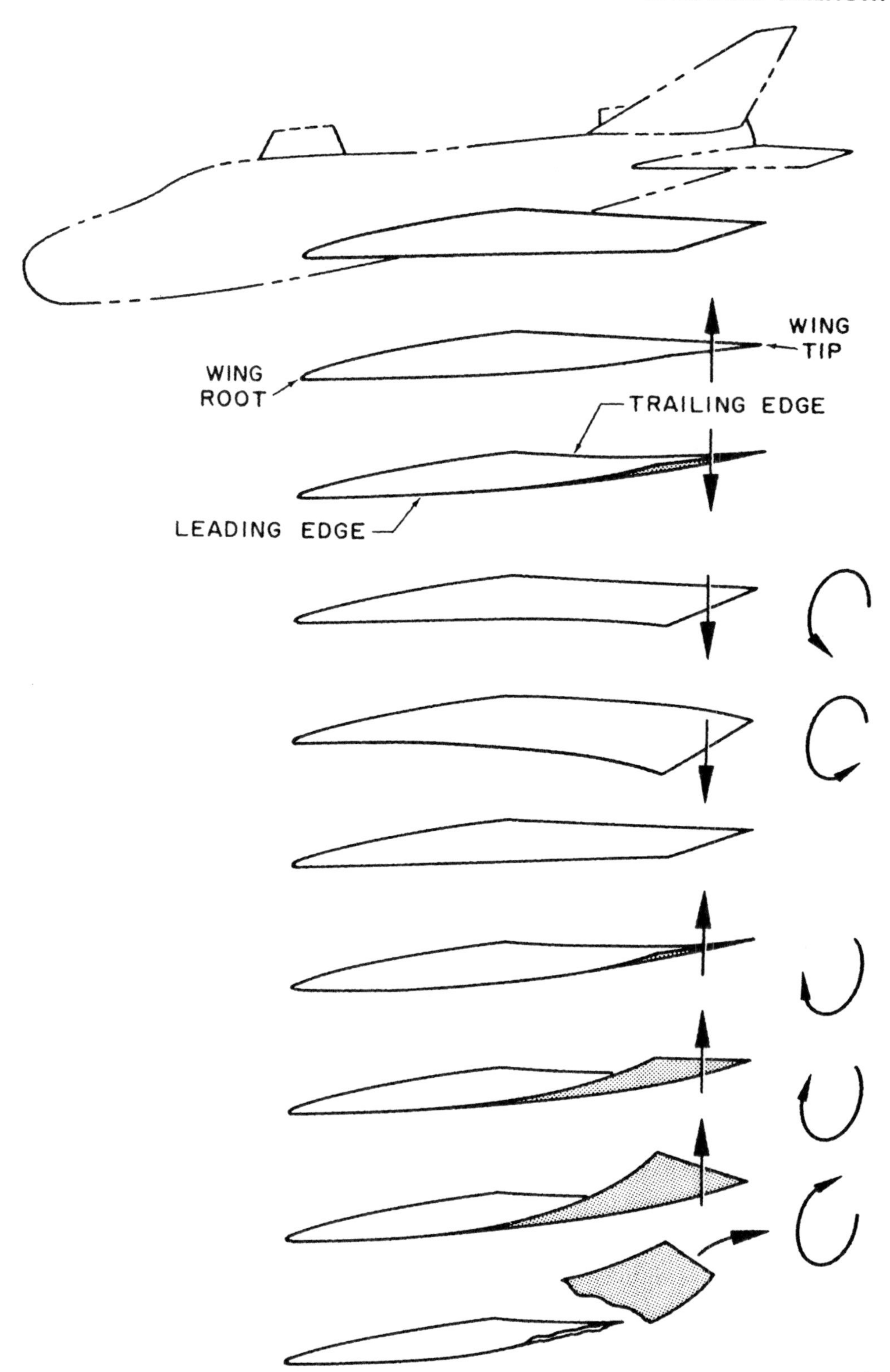

Figure 5.5. Aeroelastic Effects (Sheet 2 of 2)

twisting deformation will be great enough to nullify the effect on aileron deflection and the aileron effectiveness will be zero. Since speeds above this point create rolling moments opposite to the direction controlled, this point is termed the "aileron reversal speed." Operation beyond the reversal speed would create an obvious control difficulty. Also, the extremely large twisting moments which produce loss of aileron effectiveness create large twisting moments capable of structural damage.

In order to prevent loss of aileron effectiveness at high airspeeds, the wing must have high torsional stiffness. This may be a feature difficult to accomplish in a wing of very thin section and may favor the use of inboard ailerons to reduce the twisted span length and effectively increase torsional stiffness. The use of spoilers for lateral control minimizes the twisting moments and alleviates the reversal problem.

Divergence is another phenomenon common to flight at high dynamic pressures. Like aileron reversal, it is an effect due to the interaction of aerodynamic forces and elastic deflections of the structure. However, it differs from aileron reversal in that it is a violent instability which produces immediate failure. Figure 5.5 illustrates the process of instability. If the surface is above the divergence speed, any disturbance precipitates this sequence. Any change in lift takes place at the aerodynamic center of the section. The change in lift ahead of the elastic axis produces a twisting moment and a consequent twisting deflection. The change in angle of attack creates greater lift at the a.c., greater twisting deflection, more lift, etc., until failure occurs.

At low flight speeds where the dynamic pressure is low, the relationship between aerodynamic force buildup and torsional deflection is stable. However, the change in lift per angle of attack is proportional to V^2 but the structural torsional stiffness of the wing remains constant. This relationship implies that at some high speed, the aerodynamic force buildup may overpower the resisting torsional stiffness and "divergence" will occur. The divergence speed of the surfaces must be sufficiently high that the airplane does not encounter this phenomenon within the normal operating envelope. Sweepback, short span, and high taper help raise the divergence speed.

Flutter involves aerodynamic forces, inertia forces and the elastic properties of a surface. The distribution of mass and stiffness in a structure determine certain natural frequencies and modes of vibration. If the structure is subject to a forcing frequency near these natural frequencies, a resonant condition can result with an unstable oscillation. The aircraft is subject to many aerodynamic excitations while in operation and the aerodynamic forces at various speeds have characteristic properties for rate of change of force and moment. The aerodynamic forces may interact with the structure in a fashion which may excite or negatively damp the natural modes of the structure and allow flutter. Flutter must not occur within the normal flight operating envelope and the natural modes must be damped if possible or designed to occur beyond the limit speed. A typical flutter mode is illustrated in figure 5.5.

Since the problem is one of high speed flight, it is generally desirable to have very high natural frequencies and flutter speeds well above the normal operating speeds. Any change of stiffness or mass distribution will alter the modes and frequencies and thus allow a change in the flutter speeds. If the aircraft is not properly maintained and excessive play and flexibility exist, flutter could occur at flight speeds below the limit airspeed.

Compressibility problems may define the limit airspeed for an airplane in terms of Mach number. The supersonic airplane may experience a great decay of stability at some high Mach number or encounter critical structural or engine inlet temperatures due to aerodynamic heating. The transonic airplane at an excessive speed may encounter a variety of stability, control, or buffet problems associated with transonic flight. Since the equivalent airspeed for a given Mach number decreases with altitude, the magnitude of compressibility effects at high altitude may be negligible for the transonic airplane. In this sense, the airplane may not be able to fly at high enough dynamic pressures within a certain range of Mach numbers to create any significant stability or control problem.

The transonic airplane which is buffet limited requires due consideration of the effect of load factor on the onset of buffet. Since critical Mach number decreases with lift coefficient, the limit Mach number will decrease with load factor. If the airplane is subject to prolonged or repeated buffet for which it was not designed, structural fatigue will be the certain result.

The limit airspeed for each type aircraft is set sufficiently high that full intended application of the aircraft should be possible. Each of the factors mentioned about the effect of excess airspeed should provide due respect for the limit airspeed.

LANDING AND GROUND LOADS

The most critical loads on the landing gear occur at high gross weight and high rate of descent at touchdown. Since the landing gear has requirements of static strength and fatigue strength similar to any other component, overstress must be avoided to prevent failure and derive the anticipated service life from the components.

The most significant function of the landing gear is to absorb the vertical energy of the aircraft at touchdown. An aircraft at a given weight and rate of descent at touchdown has a certain kinetic energy which must be dissipated in the shock absorbers of the landing gear. If the energy were not absorbed at touchdown, the aircraft would bounce along similar to an automobile with faulty shock absorbers. As the strut deflects on touchdown, oil is forced through an orifice at high velocity and the energy of the aircraft is absorbed. To have an efficient strut the orifice size must be controlled with a tapered pin to absorb the energy with the most uniform force on the strut.

The vertical landing loads resulting at touchdown can be simplified to an extent by assuming the action of the strut to produce a uniformly accelerated motion of the aircraft. The landing load factor for touchdown at a constant rate of descent can be expressed by the following equation:

$$n = F/W$$

$$n = \frac{(ROD)^2}{2gS}$$

where

n = landing load factor—the ratio of the load in the strut, F, to the weight, W

ROD = rate of descent, ft. per sec.

g = acceleration due to gravity

= 32 ft. per sec.^{2}

S = effective stroke of the strut, ft.

As an example, assume that an aircraft touches down at a constant rate of descent of 18 ft. per sec. and the effective stroke of the strut is 18 inches (1.5 ft.). The landing load factor for the condition would be 3.37; the average force would be 3.37 times the weight of the aircraft. (NOTE: there is no specific correlation between the landing load factor and the indication of a cockpit mounted flight accelerometer. The response of the instrument, its mounting, and the onset of landing loads usually prevent direct correlation.)

This simplified equation points out two important facts. The effective stroke of the strut should be large to minimize the loads since a greater distance of travel reduces the force necessary to do the work of arresting the vertical descent of the aircraft. This should emphasize the necessity of proper maintenance of the struts. An additional fact illustrated is that the landing load factor varies as the square of the touchdown rate of descent. Therefore, a 20 percent higher rate of descent increases the landing load factor 44 percent. This fact should emphasize the need for proper landing technique to prevent a hard landing and overstress of the landing gear components and associated structure.

The effect of landing gross weight is twofold. A higher gross weight at some landing load factor produces a higher force in the landing gear. The higher gross weight requires a higher approach speed and, if the same glide path is used, a higher rate of descent results. In addition to the principal vertical loads on the landing gear, there are varied side loads, wheel spin up and spring back loads, etc., all of which tend to be more critical at high gross weight, high touchdown ground speed, and high rate of descent.

The function of the landing gear as a shock absorbing device has an important application when a forced landing must be accomplished on an unprepared surface. If the terrain is rough and the landing gear is not extended, initial contact will be made with relatively solid structure and whatever energy is absorbed will be accompanied by high vertical accelerations. These high vertical accelerations encountered with a gear-up landing on an unprepared surface are the source of a very incapacitating type injury—vertical compression fracture of the vertebrae. Unless some peculiarity of the configuration makes it inadvisable, it is generally recommended that the landing gear be down for forced landing on an unprepared surface. (Note: for those prone to forget, it is also recommended that the gear be down for landing on prepared surfaces.)

EFFECT OF OVERSTRESS ON SERVICE LIFE

Accumulated periods of overstress can create a very detrimental effect on the useful service life of any structural component. This fact is certain and irreversible. Thus, the operation of the airplane, powerplant, and various systems must be limited to design values to prevent failure or excessive maintenance costs early in the anticipated service life. The operating limitations presented in the handbook must be adhered to in a very strict fashion.

In many cases of modern aircraft structures it is very difficult to appreciate the effect of a moderate overstress. This feature is due in great part to the inherent strength of the materials used in modern aircraft construction. As a general airframe static strength requirement, the primary structure must not experience objectionable permanent deformation at limit load or failure at 150 percent of limit load (ultimate load is 1.5 times limit load). To satisfy each part of the requirement, limit load must not exceed the yield stress and ultimate load must not exceed the ultimate stress capability of the parts.

Many of the high strength materials used in aircraft construction have stress-strain diagrams typical of figure 5.6. One feature of these materials is that the yield point is at some stress much *greater* than two-thirds of the ultimate stress. Thus, the critical design condition is the ultimate load. If 150 percent of limit load corresponds to ultimate stress of the material, 100 percent of limit load corresponds to a stress much lower than the yield stress. Because of the inherent properties of the high strength material and the ultimate factor of safety of 1.5, the limit load condition is rarely the critical design point and usually possesses a large positive margin of static strength. This fact alone implies that the structure must be grossly overstressed to produce damage *easily visible* to the naked eye. This lack of immediate visible damage with "overstress" makes it quite difficult to recognize or appreciate the long range effect.

A reference point provided on the stress strain diagram of figure 5.6 is a stress termed

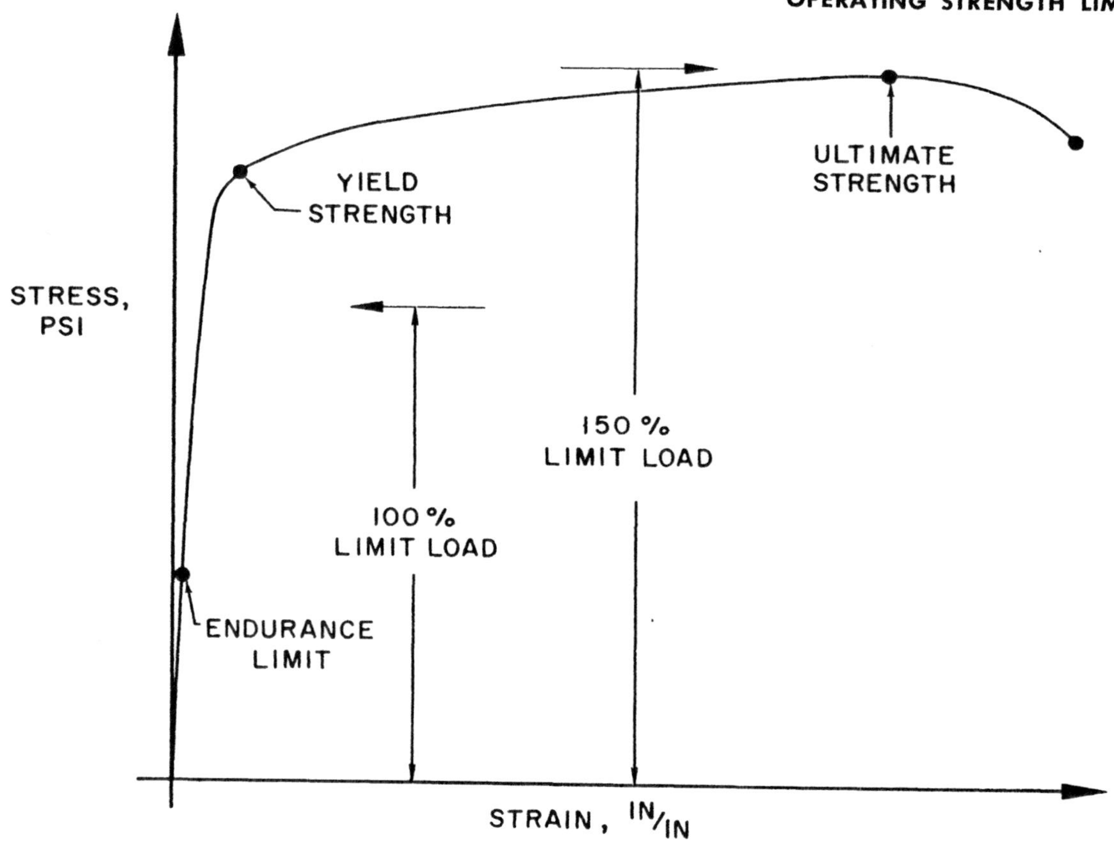

Figure 5.6. Typical Stress Strain Diagram for a High Strength Aluminum Alloy

the "endurance limit." If the operating cyclic stresses never exceed this "endurance limit" an infinite (or in some cases "near infinite") number of cycles can be withstood without fatigue failure. No significant fatigue damage accrues from stresses below the endurance limit but the value of this endurance limit is approximately 30 to 50 percent of the yield strength for the light alloys used in aircraft construction. The rate of fatigue damage caused by stresses only *slightly* above the endurance limit is insignificant. Even stresses near the limit load do not cause a significant accumulation of fatigue damage if the frequency of application is reasonable and within the intended mission requirement. However, stresses above the limit load—and especially stresses well above the limit load—create a very rapid rate of fatigue damage.

A puzzling situation then exists. "Overstress" is difficult to recognize because of the inherent high yield strength and low ductility of typical aircraft metals. These same overstresses cause high rate of fatigue damage and create premature failure of parts in service. The effect of accumulated overstress is the formation and propagation of fatigue cracks. While it is sure that fatigue crack always will be formed before final failure of a part, accumulated overstress is most severe and fatigue provoking at the inevitable stress concentrations. Hence, disassembly and detailed inspection is both costly and time-consuming. To prevent in-service failures of a basically sound structure, the part must be properly maintained and operated within the design "envelope." Examples of in-service fatigue failures are shown in figure 5.7.

The operation of any aircraft and powerplant must be conducted within the operating limitations prescribed in the flight handbook. No hearsay or rumors can be substituted for the

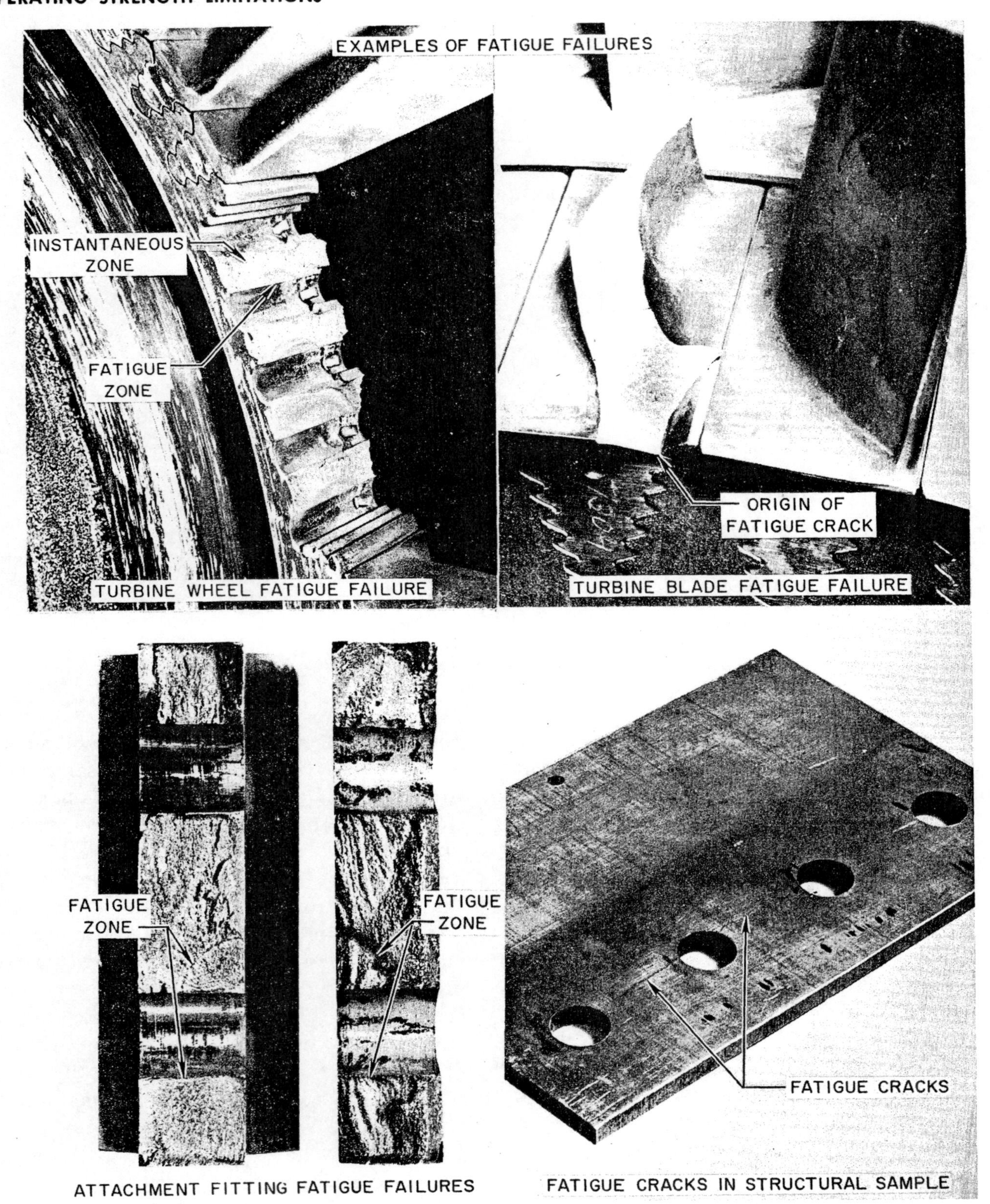

Figure 5.7. Examples of Fatigue Failures

accepted data presented in the aircraft handbook. All of the various static strength, service life, and aeroelastic effects must be given proper respect. An airplane can be overstressed with the possibility that no immediate damage is apparent. A powerplant may be operated past the specified time, speed, or temperature limits without immediate apparent damage. In each case, the cumulative effect will tell at some later time when in-service failures occur and maintenance costs increase.

Chapter 6

APPLICATION OF AERODYNAMICS TO SPECIFIC PROBLEMS OF FLYING

While the previous chapters have presented the detailed parts of the general field of aerodynamics, there remain various problems of flying which require the application of principles from many parts of aerodynamics. The application of aerodynamics to these various problems of flying will assist the Naval Aviator in understanding these problems and developing good flying techniques.

PRIMARY CONTROL OF AIRSPEED AND ALTITUDE

For the conditions of steady flight, the airplane must be in equilibrium Equilibrium will be achieved when there is no unbalance of force or moment acting on the airplane. If it is assumed that the airplane is trimmed so that no unbalance of pitching, yawing, or rolling moments exists, the principal concern is for the forces acting on the airplane, i.e., lift, thrust, weight, and drag.

ANGLE OF ATTACK VERSUS AIRSPEED. In order to achieve equilibrium in the vertical direction, the net lift must equal the airplane weight. This is a contingency of steady, level flight or steady climbing and descending flight when the flight path inclination is slight. A refinement of the basic lift equation defines the relationship of speed, weight, lift coefficient, etc., for the condition of lift equal to weight.

$$V=17.2\sqrt{\frac{W/S}{C_L\sigma}}$$

or

$$V_E=17.2\sqrt{\frac{W/S}{C_L}}$$

where

V = velocity, knots (TAS)
V_E = equivalent airspeed, knots (EAS)
W = gross weight, lbs.
S = wing surface area, sq. ft.
W/S = wing loading, psf
σ = altitude density ratio
C_L = lift coefficient

From this relationship it is appreciated that a given configuration of airplane with a specific wing loading, W/S, will achieve lift equal to weight at particular combinations of velocity, V, and lift coefficient, C_L. In steady flight, each equivalent airspeed demands a particular value of C_L and each value of C_L demands a particular equivalent airspeed to provide lift equal to weight. Figure 6.1 illustrates a typical lift curve for an airplane and shows the relationship between C_L and α, angle of attack. For this relationship, some specific value of α will create a certain value of C_L for any given aerodynamic configuration.

For the conditions of steady flight with a given airplane, each angle of attack corresponds to a specific airspeed. Each angle of attack produces a specific value of C_L and each value of C_L requires a specific value of equivalent airspeed to provide lift equal to weight. *Hence, angle of attack is the primary control of airspeed in steady flight.* If an airplane is established in steady, level flight at a particular airspeed, any increase in angle of attack will result in some reduced airspeed common to the increased C_L. A decrease in angle of attack will result in some increased airspeed common to the decreased C_L. As a result of the change in airspeed, the airplane may climb or descend if there is no change in power setting but the change in airspeed was provided by the change in angle of attack. The state of the airplane during the change in speed will be some transient condition between the original and final steady state conditions.

Primary control of airspeed in steady flight by angle of attack is an important principle. With some configurations of airplanes, low speed flight will bring about a low level of longitudinal stick force stability and possibility of low airplane static longitudinal stability. In such a case, the "feel" for airspeed will be light and may not furnish a ready reference for easy control of the airplane. In addition, the high angles of attack common to low speed flight are likely to provide large position errors to the airspeed indicating system. Thus, proper control of airspeed will be enhanced by good "attitude" flying or—when the visual reference field is poor—an angle of attack indicator.

RATE OF CLIMB AND DESCENT. In order for an airplane to achieve equilibrium at constant altitude, lift must be equal to weight and thrust must be equal to drag. Steady, *level* flight requires equilibrium in both the vertical and horizontal directions. For the case of climbing or descending flight conditions, a component of weight is inclined along the flight path direction and equilibrium is achieved when thrust is not equal to the drag. When the airplane is in a steady climb or descent, the rate of climb is related by the following expression:

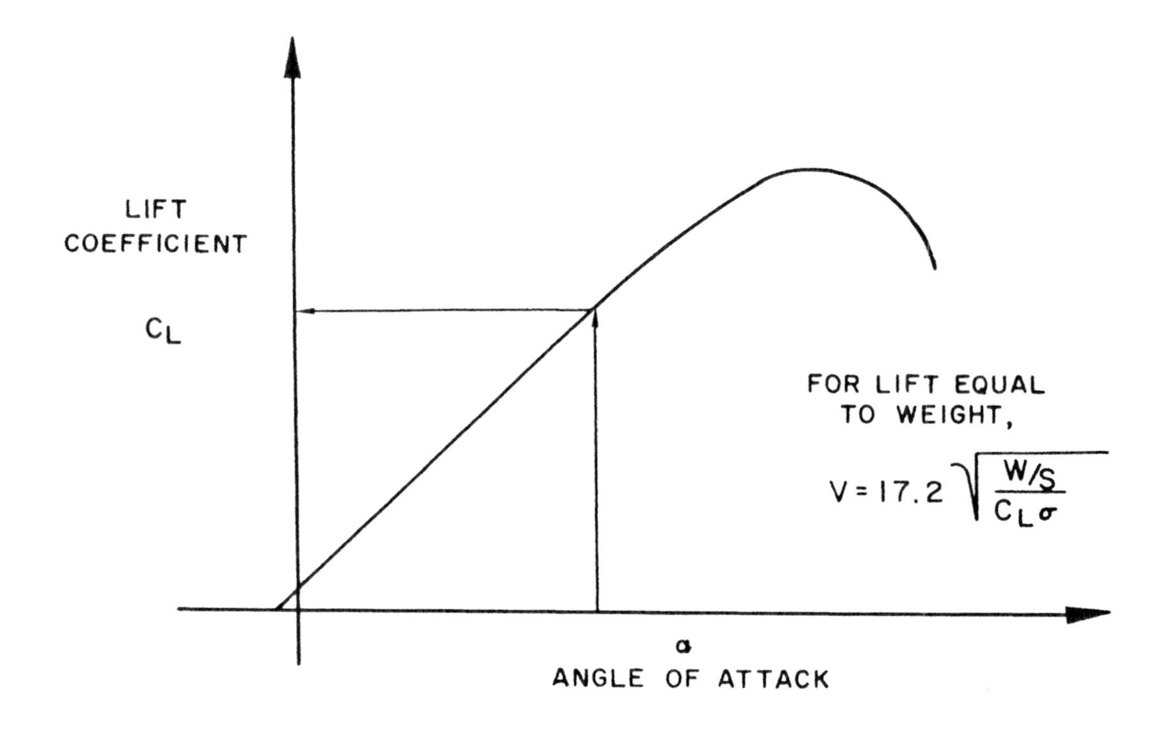

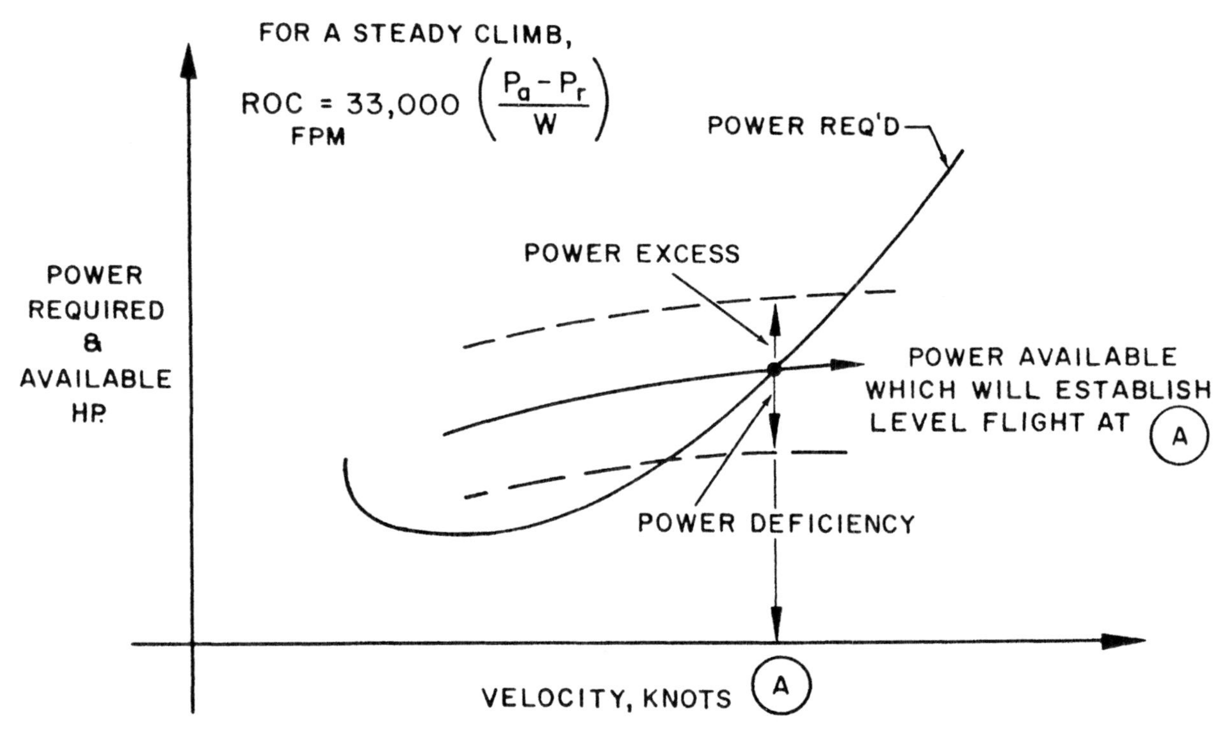

Figure 6.1. Primary Control of Airspeed and Altitude

$$RC_{fpm}=33{,}000\left(\frac{Pa-Pr}{W}\right)$$

where

RC=rate of climb, ft. per min.
Pa=propulsive power available, h.p.
Pr=power required for level flight, h.p.
W=gross weight, lbs.

From this relationship it is appreciated that the rate of climb in steady flight is a direct function of the difference between power available and power required. If a given airplane configuration is in lift-equal-to-weight flight at some specific airspeed and altitude, there is a specific power required to maintain these conditions. If the power available from the powerplant is adjusted to equal the power required, the rate of climb is zero ($Pa-Pr=0$). This is illustrated in figure 6.1 where the power available is set equal to the power required at velocity (A). If the airplane were in steady level flight at velocity (A), an increase in power available would create an excess of power which will cause a rate of climb. Of course, if the speed were allowed to increase by a decreased angle of attack, the increased power setting could simply maintain altitude at some higher airspeed. However, if the original aerodynamic conditions are maintained, speed is maintained at (A) and an increased power available results in a rate of climb. Also, a decrease in power available at point (A) will produce a deficiency in power and result in a negative rate of climb (or a rate of descent). For this reason, it is apparent that *power setting is the primary control of altitude in steady flight*. There is the direct correlation between the excess power ($Pa-Pr$), and the airplane rate of climb, RC.

FLYING TECHNIQUE. Since the conditions of steady flight predominate during a majority of all flying, the fundamentals of flying technique are the principles of steady flight:

(1) Angle of attack is the primary control of airspeed.

(2) Power setting is the primary control of altitude, i.e., rate of climb/descent.

With the exception of the transient conditions of flight which occur during maneuvers and acrobatics, the conditions of steady flight will be applicable during such steady flight conditions as cruise, climb, descent, takeoff, approach, landing, etc. A clear understanding of these two principles will develop good, safe flying techniques applicable to any sort of airplane.

The primary control of airspeed during steady flight conditions is the angle of attack. However, changes in airspeed will necessitate changes in power setting to maintain altitude because of the variation of power required with velocity. The primary control of altitude (rate of climb/descent) is the power setting. If an airplane is being flown at a particular airspeed in level flight, an increase or decrease in power setting will result in a rate of climb or descent at this airspeed. While the angle of attack must be maintained to hold airspeed in steady flight, a change in power setting will necessitate a change in *attitude* to accommodate the new flight path direction. These principles form the basis for "attitude" flying technique, i.e., "attitude plus power equals performance," and provide a background for good instrument flying technique as well as good flying technique for all ordinary flying conditions.

One of the most important phases of flight is the landing approach and it is during this phase of flight that the principles of steady flight are so applicable. If, during the landing approach, it is realized that the airplane is below the desired glide path, an increase in nose up attitude will not insure that the airplane will climb to the desired glide path. In fact, an increase in nose-up attitude may produce a greater rate of descent and cause the airplane to sink more below the desired glide path. At a given airspeed, only an increase in power setting can cause a rate of climb (or lower rate of descent) and an increase in nose up attitude without the appropriate power change only controls the airplane to a lower speed.

REGION OF REVERSED COMMAND

The variation of power or thrust required with velocity defines the power settings necessary to maintain steady level flight at various airspeeds. To simplify the situation, a generality could be assumed that the airplane configuration and altitude define a variation of *power setting* required (jet thrust required or prop power required) versus velocity. This general variation of required power setting versus velocity is illustrated by the first graph of figure 6.2. This curve illustrates the fact that at low speeds near the stall or minimum control speed the power setting required for steady level flight is quite high. However, at low speeds, an increase in speed reduces the required power setting until some minimum value is reached at the conditions for maximum endurance. Increased speed beyond the conditions for maximum endurance will then increase the power setting required for steady level flight.

REGIONS OF NORMAL AND REVERSED COMMAND. This typical variation of required power setting with speed allows a sort of terminology to be assigned to specific regimes of velocity. Speeds greater than the speed for maximum endurance require increasingly greater power settings to achieve steady, level flight. Since the normal command of flight assumes a higher power setting will achieve a greater speed, the regime of flight speeds greater than the speed for minimum required power setting is termed the "region of normal command." Obviously, parasite drag or parasite power predominates in this regime to produce the increased power setting required with increased velocity. Of course, the major items of airplane flight performance take place in the region of normal command.

Flight speeds below the speed for maximum endurance produce required power settings which increase with a *decrease* in speed. Since the increase in required power setting with decreased velocity is contrary to the normal command of flight, the regime of flight speeds between the speed for minimum required power setting and the stall speed (or minimum control speed) is termed the "region of reversed command." In this regime of flight, a decrease in airspeed must be accompanied by an increased power setting in order to maintain steady flight. Obviously, induced drag or induced power required predominates in this regime to produce the increased power setting required with decreased velocity. One fact should be made clear about the region of reversed command: flight in the "reversed" region of command does *not* imply that a decreased power setting will bring about a higher airspeed or an increased power setting will produce a lower airspeed. To be sure, the primary control of airspeed is not the power setting. Flight in the region of reversed command only implies that a higher airspeed will *require* a lower power setting and a lower airspeed will *require* a higher power setting to hold altitude.

Because of the variation of required power setting throughout the range of flight speeds, it is possible that one particular power setting may be capable of achieving steady, level flight at two different airspeeds. As shown on the first curve of figure 6.2, one given power setting would meet the power requirements and allow steady, level flight at both points *1* and 2. At speeds lower than point 2, a deficiency of power would exist and a rate of descent would be incurred. Similarly, at speeds greater than point *1*, a deficiency of power would exist and the airplane would descend. The speed range between points *1* and 2 would provide an excess of power and climbing flight would be produced.

FEATURES OF FLIGHT IN THE NORMAL AND REVERSED REGIONS OF COMMAND. The majority of all airplane flight is conducted in the region of normal command,

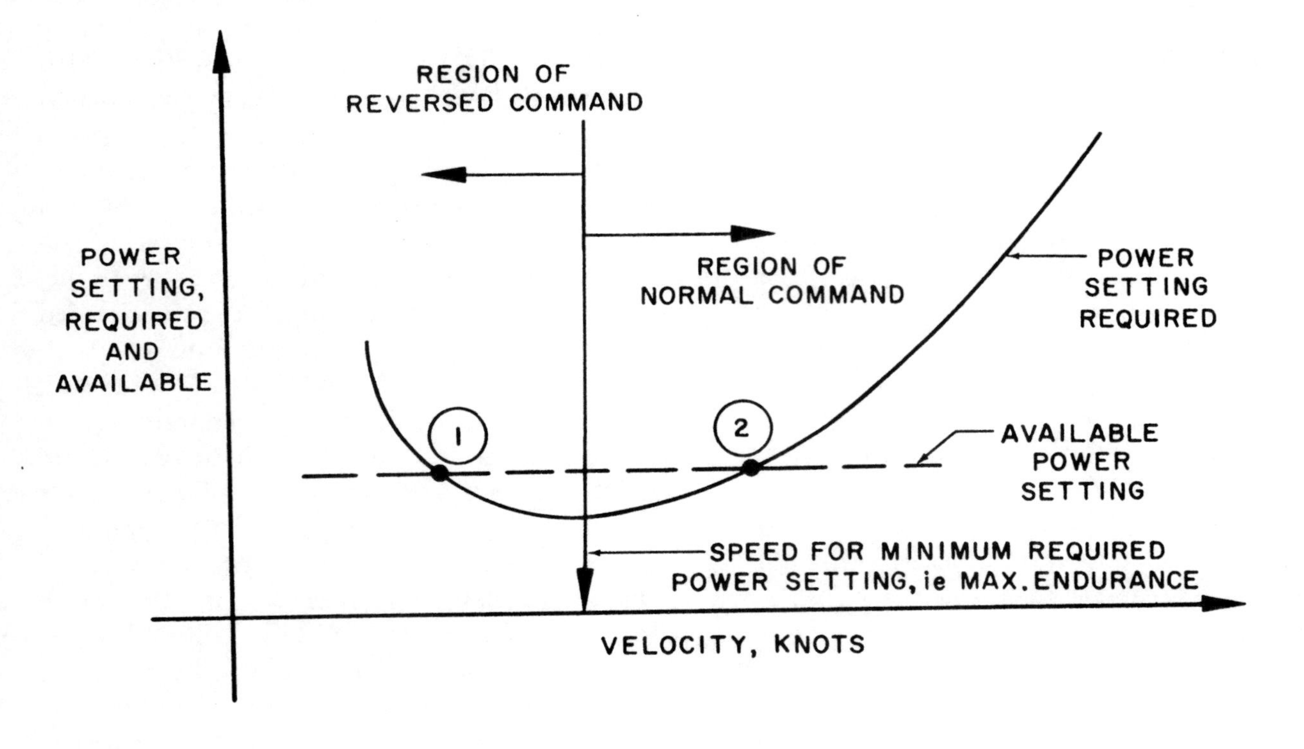

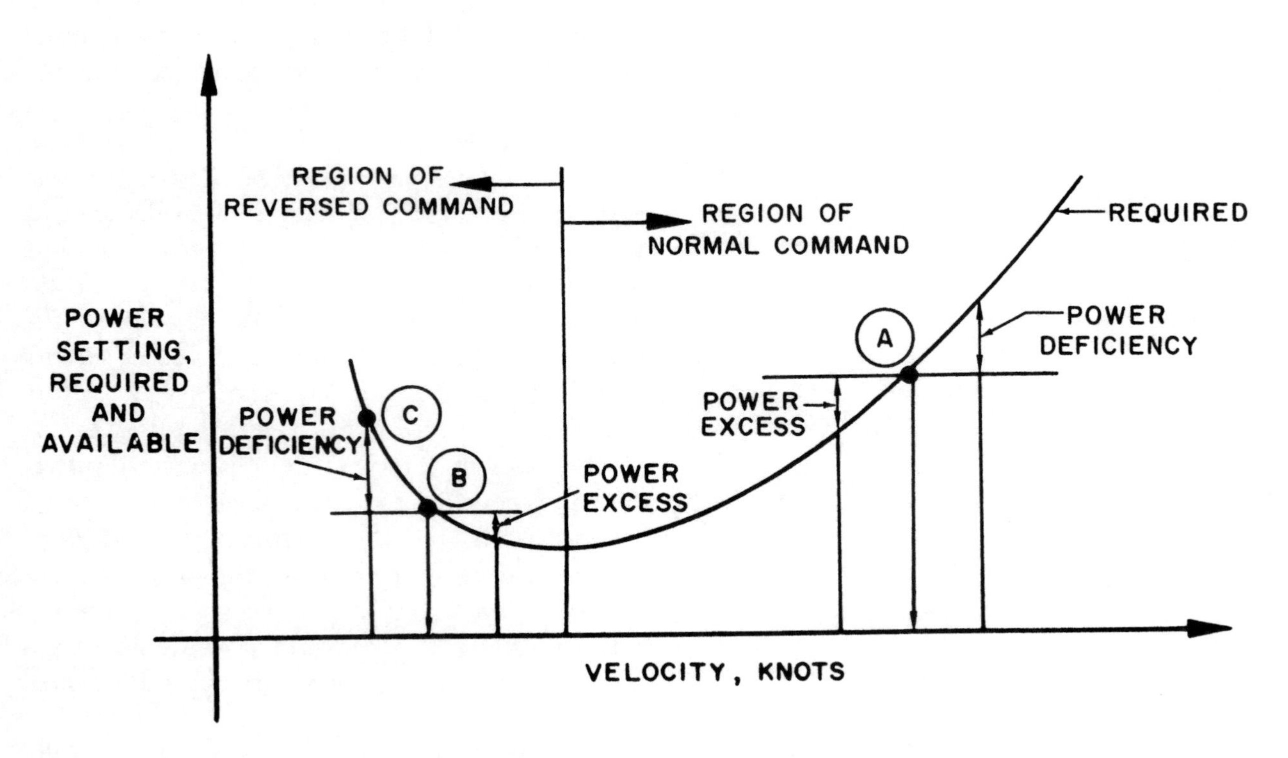

Figure 6.2. Region of Reversed Command

e.g., cruise, climb, maneuvers, etc. The region of reversed command is encountered primarily in the low speed phases of flight during takeoff and landing. Because of the extensive low speed flight during carrier operations, the Naval Aviator will be more familiar with the region of reversed command than the ordinary pilot.

The characteristics of flight in the region of normal command are illustrated at point *A* on the second curve of figure 6.2. If the airplane is established in steady, level flight at point *A*, lift is equal to weight and the power available is set equal to the power required. When the airplane is disturbed to some airspeed slightly greater than point *A*, a power deficiency exists and, when the airplane is disturbed to some airspeed slightly lower than point *A*, a power excess exists. This relationship provides a tendency for the airplane to return to the equilibrium of point *A* and resume the original flight condition following a disturbance. Also, the static longitudinal stability of the airplane tends to return the airplane to the original trimmed C_L and velocity corresponding to this C_L. The phugoid usually has most satisfactory qualities at low values of C_L so the high speed of the region of normal command provides little tendency of the airplane's airspeed to vary or wander about.

With all factors considered, flight in the region of normal command is characterized by a relatively strong tendency of the airplane to maintain the trim speed quite naturally. However, flight in the region of normal command can lead to some unusual and erroneous impressions regarding proper flying technique. For example, if the airplane is established at point *A* in steady level flight, a controlled increase in airspeed without a change in power setting will create a deficiency of power and cause the airplane to descend. Similarly, a controlled decrease in airspeed without a change in power setting will create an excess of power and cause the airplane to climb. This fact, coupled with the transient motion of the airplane when the angle of attack is changed rapidly, may lead to the impression that rate of climb and descent can be controlled by changes in angle of attack. While such is true in the region of normal command, for the conditions of *steady* flight, *primary* control of altitude remains the power setting and the *primary* control of airspeed remains the angle of attack. The impressions and habits that can be developed in the region of normal command can bring about disastrous consequences in the region of reversed command.

The characteristics of flight in the region of reversed command are illustrated at point *B* on the second curve of figure 6.2. If the airplane is established in steady, level flight at point *B*, lift is equal to weight and the power available is set equal to the power required. When the airplane is disturbed to some airspeed slightly greater than point *B*, an excess of power exists and, when the airplane is disturbed to some airspeed slightly lower than point *B*, a deficiency of power exists. This relationship is basically unstable because the variation of excess power to either side of point *B* tends to magnify any original disturbance. While the static longitudinal stability of the airplane tends to maintain the original trimmed C_L and airspeed corresponding to that C_L, the phugoid usually has the least satisfactory qualities at the high values of C_L corresponding to low speed flight.

When all factors are considered, flight in the region of reversed command is characterized by a relatively weak tendency of the airplane to maintain the trim speed naturally. In fact it is likely that the airplane will exhibit no inherent tendency to maintain the trim speed in this regime of flight. For this reason, the pilot must give particular attention to precise control of airspeed when operating in the low flight speeds of the region of reversed command.

While flight in the region of normal command may create doubt as to the primary control of airspeed and altitude, operation in the region of reversed command should leave little

doubt about proper flying techniques. For example, if the airplane is established at point *B* in level flight, a controlled increase in airspeed (by reducing angle of attack) without change in power setting will create an excess of power at the higher airspeed and cause the airplane to climb. Also, a controlled decrease in airspeed (by increasing angle of attack) without a change of power setting will create a deficiency of power at the lower airspeed and cause the airplane to descend. This relationship should leave little doubt as to the primary control of airspeed and altitude.

The transient conditions during the changes in airspeed in the region of reversed command are of interest from the standpoint of landing flare characteristics. Suppose the airplane is in steady flight at point *B* and the airplane angle of attack is increased to correspond with the value for the lower airspeed of point *C* (see fig. 6.2). The airplane would not instantaneously develop the lower speed and rate of descent common to point *C* but would approach the conditions of point *C* through some transient process depending on the airplane characteristics. If the airplane characteristics are low wing loading, high L/D, and high lift curve slope, the increase in angle of attack at point *B* will produce a transient motion in which curvature of the flight path demonstrates a definite flare. That is, the increase in angle of attack creates a momentary rate of climb (or reduction of rate of descent) which would be accompanied by a gradual loss of airspeed. Of course, the speed eventually decreases to point *C* and the steady state rate of descent is achieved. If the airplane characteristics are high wing loading, low L/D, and low lift curve slope, the increase in angle of attack at point *B* may produce a transient motion in which the airplane does not flare. That is, the increase in angle of attack may produce such rapid reduction of airspeed and increase in rate of descent that the airplane may be incapable of a flaring flight path without an increase in power setting. Such characteristics may necessitate special landing techniques, particularly in the case of a flameout landing.

Operation in the region of reversed command does not imply that great control difficulty and dangerous conditions will exist. However, flight in the region of reversed command does amplify any errors of basic flying technique. Hence, proper flying technique and precise control of the airplane are most necessary in the region of reversed command.

THE ANGLE OF ATTACK INDICATOR AND THE MIRROR LANDING SYSTEM

The usual errors during the takeoff and landing phases of flight involve improper control of airspeed and altitude along some desired flight path. Any errors of technique are amplified when an adequate visual reference is not available to the pilot. It is necessary to provide the pilot with as complete as possible visual reference field to minimize or eliminate any errors in perception and orientation. The angle of attack indicator and the mirror landing system assist the pilot during the phases of takeoff and landing and allow more consistent, precise control of the airplane.

THE ANGLE OF ATTACK INDICATOR. Many specific aerodynamic conditions exist at particular angles of attack for the airplane. Generally, the conditions of stall, landing approach, takeoff, range, endurance, etc., all occur at specific values of lift coefficient and specific airplane angles of attack. Thus, an instrument to indicate or relate airplane angle of attack would be a valuable reference to aid the pilot.

When the airplane is at high angles of attack it becomes difficult to provide accurate indication of airspeed because of the possibility of large position errors. In fact, for low aspect ratio airplane configurations at high angles of attack, it is possible to provide indications of angle of attack which are more accurate than indications of airspeed. As a result, an angle of attack indicator can be of greatest utility at the high angles of attack.

A particular advantage of an angle of attack indicator is that the indicator is not directly affected by gross weight, bank angle, load factor, velocity, or density altitude. The typical lift curve of figure 6.3 illustrates the variation of lift coefficient, C_L, with angle of attack α. When a particular aerodynamic configuration is in subsonic flight, each angle of attack produces a particular value of lift coefficient. Of course, a point of special interest on the lift curve is the maximum lift coefficient, $C_{L_{max}}$. Angles of attack greater than that for $C_{L_{max}}$ produce a decrease in lift coefficient and constitute the stalled condition of flight. Since $C_{L_{max}}$ occurs at a particular angle of attack, any device to provide a stall warning should be predicated on the function of this critical angle of attack. Under these conditions, stall of the airplane may take place at various airspeeds depending on gross weight, load factor, etc., but always the same angle of attack.

In order to reduce takeoff and landing distances and minimize arresting loads, takeoff and landing will be accomplished at minimum practical speeds. The takeoff and landing speeds must provide sufficient margin above the stall speed (or minimum control speed) and are usually specified at some fixed percentages of the stall speed. As such, takeoff, approach, and landing will be accomplished at specific values of lift coefficient and, thus, particular angles of attack. For example, assume that point *A* on the lift curve is defined as the proper aerodynamic condition for the landing approach. This condition exists as a particular lift coefficient and angle of attack for a specific aerodynamic configuration. When the airplane is flown in a steady flight path at the prescribed angle of attack, the resulting airspeed will be appropriate for the airplane gross weight. Any variation in gross weight will simply alter the airspeed necessary to provide sufficient lift. The use of an angle of attack indicator to maintain the recommended angle of attack will insure that the airplane is operated at the proper approach speed—not too low or too high an airspeed.

In addition to the use of the angle of attack indicator during approach and landing, the instrument may be used as a principal reference during takeoff. The use of the angle of attack indicator to assume the proper takeoff angle of attack will prevent both over-rotation and excess takeoff speed. Also, the angle of attack indicator may be applicable to assist in control of the airplane for conditions of range, endurance, maneuvers, etc.

THE MIRROR LANDING SYSTEM. A well planned, stabilized approach is a fundamental requirement for a good landing. However, one of the more difficult problems of perception and orientation is the positioning of the airplane along a proper flight path during approach to landing. While various devices are possible, the most successful form of glide path indicator applicable to both field and shipboard operations is the mirror landing system. The function of the mirror landing system is to provide the pilot with an accurate visual reference for a selected flight path which has the desired inclination and point of touchdown. Utilization of the mirror system will allow the pilot to position the airplane along the desired glide path and touch down at the desired point. When the proper glide path inclination is set, the pilot can be assured that the rate of descent will not be excessive and a foundation is established for a successful landing.

The combination of the angle of attack indicator and the mirror landing system can provide an excellent reference for a landing technique. The use of the angle of attack indicator will provide the airplane with the proper airspeed while the mirror system reference will provide the desired flight path. When shipboard operations are conducted without the mirror system and angle of attack indicator, the landing signal officer must provide the immediate reference of airspeed and flight path. The LSO must perceive and judge the angle of

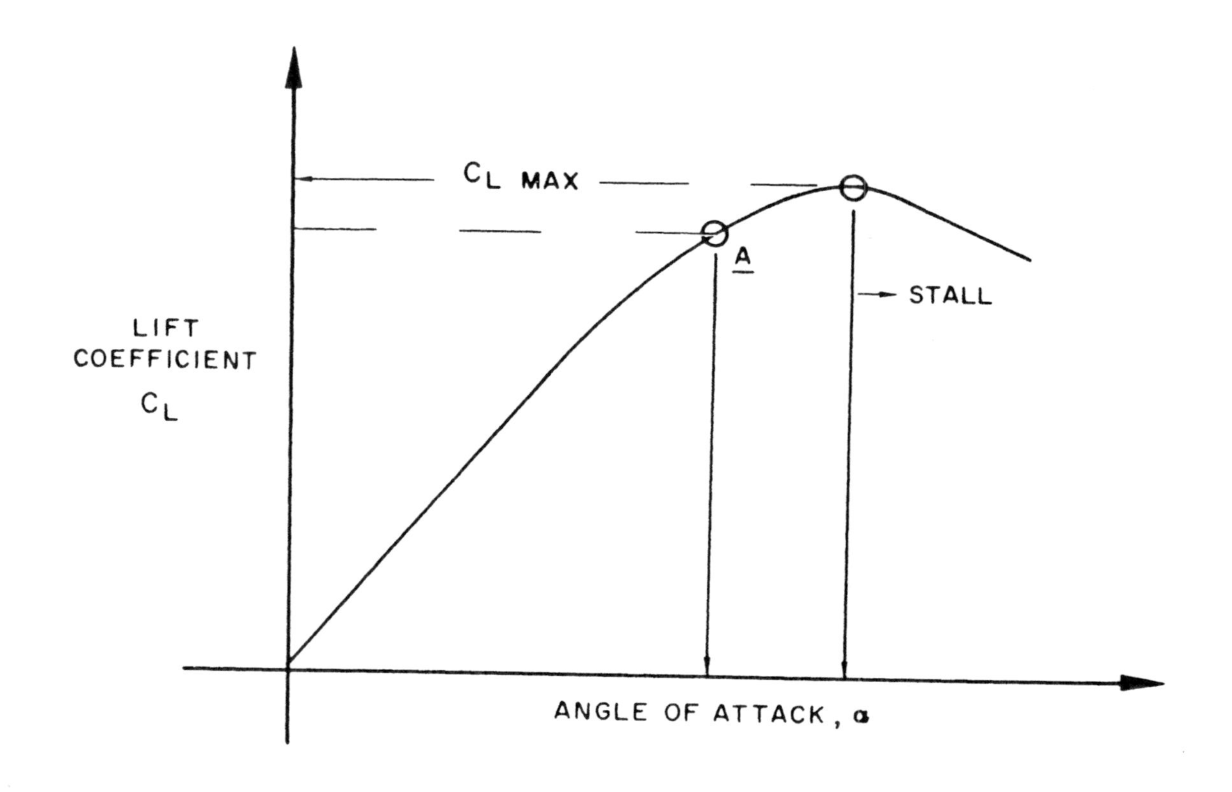

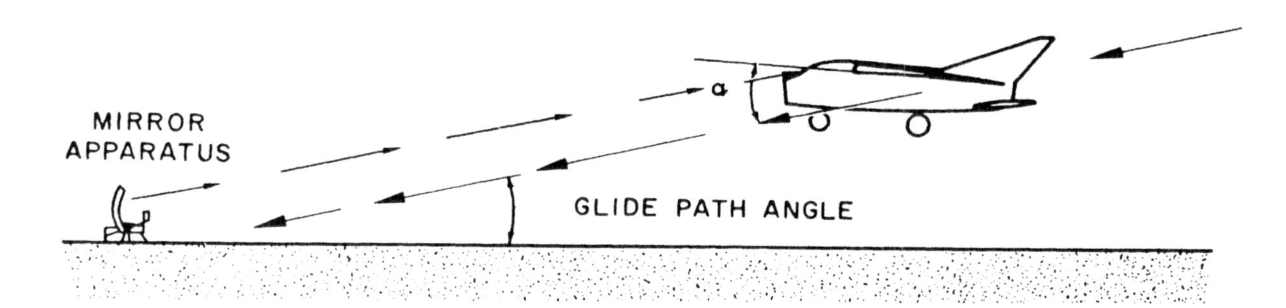

Figure 6.3. Angle of Attack and Glide Path Indication

attack (and, hence, airspeed) and the flight path of the landing aircraft and signal corrections to be made in order to achieve the desired flight path and angle of attack. Because of the field of orientation available to the LSO, he is able to perceive the flight path and angle of attack more accurately than the pilot without an angle of attack indicator and mirror landing system.

THE APPROACH AND LANDING

The specific techniques necessary during the phase of approach and landing may vary considerably between various types of airplanes and various operations. However, regardless of the airplane type or operation, there are certain fundamental principles which will define the basic techniques of flying during approach and landing. The specific procedures recommended for each airplane type must be followed exactly to insure a consistent, safe landing technique.

THE APPROACH. The approach must be conducted to provide a stabilized, steady flight path to the intended point of touchdown. The approach speed specified for an airplane must provide sufficient margin above the stall speed or minimum control speed to allow satisfactory control and adequate maneuverability. On the other hand, the approach speed must not be greatly in excess of the touchdown speed or a large reduction in speed would be necessary prior to ground contact. Generally, the approach speed will be from 10 to 30 percent above the stall speed depending on the airplane type and the particular operation.

During the approach, the pilot must attempt to maintain a smooth flight path and prepare for the touchdown. A smooth, steady approach to landing will minimize the transient items of the flight path and provide the pilot better opportunity to perceive and orientate the airplane along the desired flight path. Steep turns must be avoided at the low speeds of the approach because of the increase in drag and stall speed in the turn. Figure 6.4 illustrates the typical change in thrust required caused by a steep turn. A steep turn may cause the airplane to stall or the large increase in induced drag may create an excessive rate of descent. In either case, there may not be sufficient altitude to effect recovery. If the airplane is not properly lined up on the final approach, it is certainly preferable to take a waveoff and go around rather than "press on regardless" and attempt to salvage a decent landing from a poor approach.

The proper coordination of the controls is an absolute necessity during the approach. In this sense, due respect must be given to the primary control of airspeed and rate of descent for the conditions of the steady approach. Thus, the proper angle of attack will produce the desired approach airspeed; too low an angle of attack will incur an excess speed while an excessive angle of attack will produce a deficiency of speed and may cause stall or control problems. Once the proper airspeed and angle of attack are attained the primary control of rate of descent during the steady approach will be the power setting. For example, if it is realized that the airplane is above the desired glide path, a more nose-down attitude without a decrease in power setting will result in a gain in airspeed. On the other hand, if it is realized that the airplane is below the desired glide path, a more nose-up attitude without an increase in power setting will simply allow the airplane to fly more slowly and—in the region of reversed command—eventually produce a greater rate of descent. For the conditions of steady flight, angle of attack is the primary control of airspeed and power setting is the primary control of rate of climb and descent. This is especially true during the steady approach to landing. Of course, the ability of the powerplant to produce rapid changes in thrust will affect the specific technique to be used. If the powerplant is not capable of producing immediate controlled changes in thrust, the operating technique must account for this

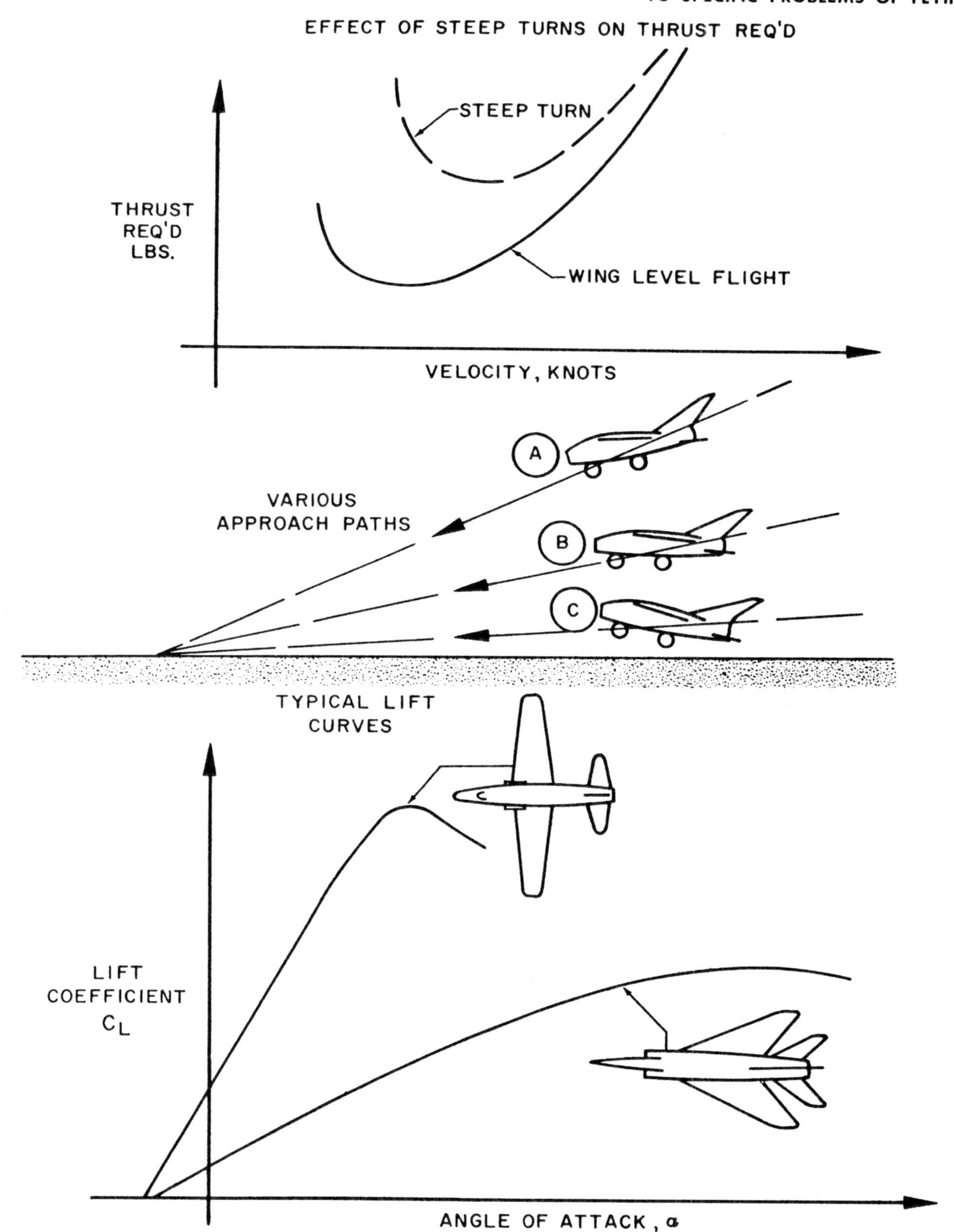

Figure 6.4. The Approach and Landing

deficiency. It is most desirable that the powerplant be capable of effecting rapid changes in thrust to allow precise control of the airplane during approach.

The type of approach path is an important factor since it affects the requirement of the flare, the touchdown rate of descent, and—to some extent—the ability to control the point of touchdown. Approach path *A* of figure 6.4 depicts the steep, low power approach. Such a flight path generally involves a low power setting near idle conditions and a high rate of descent. Precise control of the airplane is difficult and an excess airspeed usually results from an approach path similar to *A*. Waveoff may be difficult because of the required engine acceleration and the high rate of descent. In addition, the steep approach path with high rate of descent requires considerable flare to reduce the rate of descent at touchdown. This extreme flare requirement will be difficult to execute with *consistency* and will generally result in great variation in the speed, rate of descent, and point of touchdown.

Approach path *C* of figure 6.4 typifies the long, shallow approach with too small an inclination of the flight path. Such a flight path requires a relatively high power setting and a deficiency of airspeed is a usual consequence. This extreme of an approach path is not desirable because it is difficult to control the point of touchdown and the low speed may allow the airplane to settle prematurely short of the intended landing touchdown.

Some approach path between the extremes of *A* and *C* must be selected, e.g., flight path *B*. The desirable approach path must not incur excessive speed and rate of descent or require excessive flaring prior to touchdown. Also, some moderate power setting must be required which will allow accurate control of the flight path and provide suitable waveoff characteristics. The approach flight path cannot be too shallow for excessive power setting may be required and it may be difficult to judge and control the point of touchdown. The LSO, mirror landing system, and various approach lighting systems will aid the pilot in achieving the desired approach flight path.

THE LANDING FLARE AND TOUCHDOWN. The specific techniques of landing flare and touchdown will vary considerably between various types of airplanes. In fact, for certain types of airplanes, a flare from a properly executed approach may not be desirable because of the possibility of certain critical dynamic landing loads or because of the necessity for a certain standard of technique when aerodynamic flare characteristics are critical. The landing speed should be the lowest practical speed above the stall or minimum control speed to reduce landing distances and arresting loads. Generally, the landing speed will be from 5 to 25 percent above the stall speed depending on the airplane type and the particular operation.

The technique required for the landing will be determined in great part by the aerodynamic characteristics of the airplane. If the airplane characteristics are low wing loading, high *L/D*, and relatively high lift curve slope, the airplane usually will have good landing flare characteristics. If the airplane characteristics are high wing loading, low *L/D*, and relatively low lift curve slope, the airplane may not possess desirable flare characteristics and landing technique may require a minimum of flare to touchdown. These extremes are illustrated by the lift curves of figure 6.4.

In preparation for the landing, several factors must be accounted for because of their effect on landing distance, landing loads, and arresting loads. These factors are:

(1) Landing *gross weight* must be considered because of its effect on landing speed and landing loads. Since the landing is accomplished at a specific angle of attack or margin above the stall speed, gross weight will define the landing speed. In addition, the gross weight is an important factor in determining the landing distance and energy

dissipating requirements of the brakes. There will be a maximum design landing weight specified for each airplane and this limitation must be respected because of critical landing loads, arresting loads, or brake requirements. Of course, any airplane will have a limiting touchdown rate of descent specified with the maximum landing weight and the principal landing load limitations will be defined by the combination of gross weight and rate of descent at touchdown.

(2) The surface *winds* must be considered because of the large effect of a headwind or tailwind on the landing distance. In the case of the crosswind, the component of wind along the runway will be the effective headwind or tailwind velocity. Also, the crosswind component across the runway will define certain requirements of lateral control power. The airplane which exhibits large dihedral effect at high lift coefficients is quite sensitive to crosswind and a limiting crosswind component will be defined for the configuration.

(3) *Pressure altitude and temperature* will affect the landing distance because of the effect on the true airspeed for landing. Thus, pressure altitude and temperature must be considered to define the density altitude.

(4) The *runway condition* must be considered for its effect on landing distances. Runway slope of ordinary values will ordinarily favor selection of a runway for a favorable headwind at landing. The surface condition of the runway will determine braking effectiveness and ice or water on the runway may produce a considerable increase in the minimum landing distance.

Thus, preparation for the landing must include determination of the landing distance of the airplane and comparison with the runway length available. Use of the angle of attack indicator and the mirror landing system will assist the pilot in effecting touchdown at the desired location with the proper airspeed. Of course, the landing is not completed until the airplane is slowed to turn off the runway. Control of the airplane must be maintained after the touchdown and proper technique must be used to decelerate the airplane.

TYPICAL ERRORS. There are many undesirable consequences when basic principles and specific procedures are not followed during the approach and landing. Some of the typical errors involved in landing accidents are outlined in the following discussion.

The steep, low power approach leads to an *excessive rate of descent* and the possibility of a hard landing. This is particularly the case for the modern, low aspect ratio, swept wing airplane configuration which incurs very large induced drag at low speeds and does not have very conventional flare characteristics. For this type of airplane in a steep, low power approach, an increased angle of attack without a change of power setting may not cause a reduction of rate of descent and may even increase the rate of descent at touchdown. For this reason, a moderate stabilized approach is necessary and the principal changes in rate of descent must be controlled by changes in power setting and principal changes in airspeed must be controlled by changes in angle of attack.

An *excessive angle of attack* during the approach and landing implies that the airplane is being operated at too low an airspeed. Of course, excessive angle of attack may cause the airplane to stall or spin and the low altitude may preclude recovery. Also, the low aspect ratio configuration at an excessively low airspeed will incur very high induced drag and will necessitate a high power setting or otherwise incur an excessive rate of descent. An additional problem is created by an excessive angle of attack for the airplane which exhibits a large dihedral effect at high lift coefficients. In this case, the airplane would be more sensitive to crosswinds and adequate lateral control may not be available to effect a safe landing at a critical value of crosswind.

USNAVY
USNAVY
USNAVY
USNAVY
USNAVY
USNAVY
USNAVY

Excess airspeed at landing is just as undesirable as a deficiency of airspeed. An *excessive airspeed* at landing will produce an undesirable increase in landing distance and the energy to be dissipated by the brakes for the field landing or excessive arresting loads for the shipboard landing. In addition, the excess airspeed is a corollary of too low an angle of attack and the airplane may contact the deck or runway nose wheel first and cause damage to the nose wheel or begin a porpoising of the airplane. During a flare to landing, any excess speed will be difficult to dissipate due to the reduction of drag due to ground effect. Thus, if the airplane is held off with excess airspeed the airplane will "float" with the consequence of a barrier engagement, barricade engagement, bolter, or considerable runway distance used before touchdown.

A fundamental requirement for a good landing is a well planned and executed approach. The possibility of errors during the landing process is minimized when the airplane is brought to the point of touchdown with the proper glide path and airspeed. With the proper approach, there is no need for drastic changes in the flight path, angle of attack, or power setting to accomplish touchdown at the intended point on the deck or runway. Late corrections to line up with the deck or diving for the deck are common errors which eventually result in landing accidents. Accurate control of airspeed and glide path are absolutely necessary and the LSO, angle of attack indicator, and the mirror landing system provide great assistance in accurate control of the airplane.

THE TAKEOFF

As in the case of landing, the specific techniques necessary may vary greatly between various types of airplanes and various operations but certain fundamental principles will be common to all airplanes and all operations. The specific procedures recommended for each airplane type must be followed exactly to insure a consistent, safe takeoff flying technique.

TAKEOFF SPEED AND DISTANCE. The takeoff speed of any airplane is some minimum practical airspeed which allows sufficient margin above stall and provides satisfactory control and initial rate of climb. Depending on the airplane characteristics, the takeoff speed will be some value 5 to 25 percent above the stall or minimum control speed. As such, the takeoff will be accomplished at a certain value of lift coefficient and angle of attack specific to each airplane configuration. As a result, the *takeoff airspeed* (*EAS* or *CAS*) of any specific airplane configuration is a function of the gross weight at takeoff. Too low an airspeed at takeoff may cause stall, lack of adequate control, or poor initial climb performance. An excess of speed at takeoff may provide better control and initial rate of climb but the higher speed requires additional distance and may provide critical conditions for the tires.

The *takeoff distance* of an airplane is affected by many different factors other than technique and, prior to takeoff, the takeoff distance must be determined and compared with the runway length available. The principal factors affecting the takeoff distance are as follows:

(1) The *gross weight* of the airplane has a considerable effect on takeoff distance because it affects both takeoff speed and acceleration during takeoff roll.

(2) The surface *winds* must be considered because of the powerful effect of a headwind or tailwind on the takeoff distance. In the case of the crosswind, the component of wind along the runway will be the effective headwind or tailwind velocity. In addition, the component of wind across the runway will define certain requirements of lateral control power and the limiting component wind must not be exceeded.

(3) *Pressure altitude and temperature* can cause a large effect on takeoff distance, especially in the case of the turbine powered

airplane. Density altitude will determine the true airspeed at takeoff and can affect the takeoff acceleration by altering the powerplant thrust. The effect of temperature alone is important in the case of the turbine powered aircraft since inlet air temperature will affect powerplant thrust. It should be noted that a typical turbojet airplane may be approximately twice as sensitive to density altitude and five to ten times as sensitive to temperature as a representative reciprocating engine powered airplane.

(4) Specific *humidity* must be accounted for in the case of the reciprocating engine powered airplane. A high water vapor content in the air will cause a definite reduction in takeoff power and takeoff acceleration.

(5) The *runway condition* will deserve consideration when the takeoff acceleration is basically low. The runway slope must be compared carefully with the surface winds because ordinary values of runway slope will usually favor choice of the runway with headwind and upslope rather than downslope and tailwind. The surface condition of the runway has little bearing on takeoff distance as long as the runway is a hard surface.

Each of these factors must be accounted for and the takeoff distance properly computed for the existing conditions. Since obstacle clearance distance is generally a function of the same factors which affect takeoff distance, the obstacle clearance distance is usually related as some proportion of the takeoff distance. Of course, the takeoff and obstacle clearance distances related by the handbook data will be obtained by the techniques and procedures outlined in the handbook.

TYPICAL ERRORS. The takeoff distance of an airplane should be computed for each takeoff. A most inexcusable error would be to attempt takeoff from a runway of insufficient length. Familiarity with the airplane handbook performance data and proper accounting of weight, wind, altitude, temperature, etc., are necessary parts of flying. Conditions of high gross weight, high pressure altitude and temperature, and unfavorable winds create the extreme requirements of runway length, especially for the turbine powered airplane. Under these conditions, use of the handbook data is mandatory and no guesswork can be tolerated.

One typical error of takeoff technique is the premature or excess pitch rotation of the airplane. *Premature or excess pitch rotation* of the airplane may seriously reduce the takeoff acceleration and increase the takeoff distance. In addition, when the airplane is placed at an excessive angle of attack during takeoff, the airplane may become airborne at too low a speed and the result may be a stall, lack of adequate control (especially in a crosswind), or poor initial climb performance. In fact, there are certain low aspect ratio configurations of airplanes which, at an excessive angle of attack, will not fly out of ground effect. Thus, over-rotation of the airplane during takeoff may hinder takeoff acceleration or the initial climb. It is quite typical for an airplane to be placed at an excess angle of attack and become airborne prematurely then settle back to the runway. When the proper angle of attack is assumed, the airplane simply accelerates to the takeoff speed and becomes airborne with sufficient initial rate of climb. In this sense, the appropriate rotation and takeoff speeds or an angle of attack indicator must be used.

If the airplane is subject to a *sudden pull-up or steep turn* after becoming airborne, the result may be a stall, spin, or reduction in initial rate of climb. The increased angle of attack may exceed the critical angle of attack or the increase in induced drag may be quite large. For this reason, any clearing turns made immediately after takeoff or deck launch must be slight and well within the capabilities of the airplane.

In order to obviate some of the problems of a deficiency of airspeed at takeoff, usual result can be an excess of airspeed at takeoff. The principal effect of an *excess takeoff airspeed* is the greater takeoff distance which results. The general effect is that each 1 percent excess takeoff velocity incurs approximately 2 percent additional takeoff distance. Thus, excess speed must be compared with the additional runway required to produce the higher speed. In addition, the aircraft tires may be subject to critical loads when the airplane is at very high rolling speeds and speeds in excess of a basically high takeoff speed may produce damage or failure of the tires.

As with the conditions of landing, excess velocity or deficiency of velocity at takeoff is undesirable. The proper takeoff speeds and angle of attack must be utilized to assure satisfactory takeoff performance.

GUSTS AND WIND SHEAR

The variation of wind velocity and direction throughout the atmosphere is important because of its effect on the aerodynamic forces and moments on an airplane. As the airplane traverses this variation of wind velocity and direction during flight, the changes in airflow direction and velocity create changes in the aerodynamic forces and moments and produce a response of the airplane. The variation of airflow velocity along a given direction exists with shear parallel to the flow direction. Hence, the velocity gradients are often referred to as the wind "shear."

The effect of the *vertical gust* has important effects on the airplane at high speed because of the possibility of damaging flight loads. The mechanism of vertical gust is illustrated in figure 6.5 where the vertical gust velocity is added vectorially to the flight velocity to produce some resultant velocity. The principal effect of the vertical gust is to produce a change in airplane angle of attack, e.g., a positive (up) gust causes an increase in angle of attack while a negative (down) gust causes a decrease in angle of attack. Of course, a change in angle of attack will effect a change in lift and, if some critical combination of high gust intensity and high flight speed is encountered, the change in lift may be large enough to cause structural damage.

At low flight speeds during approach, landing, and takeoff, the effect of the vertical gust is due to the same mechanism of the change in angle of attack. However, at these low flight speeds, the problem is one of possible incipient stalling and sinking rather than overstress. When the airplane is at high angle of attack, a further increase in angle of attack due to a gust may exceed the critical angle of attack and cause an incipient stalling of the airplane. Also, a decrease in angle of attack due to a gust will cause a loss of lift and allow the airplane to sink. For this reason, any deficiency of airspeed will be quite critical when operating in gusty conditions.

The effect of the *horizontal gust* differs from the effect of the vertical gust in that the immediate effect is a change of airspeed rather than a change in angle of attack. In this sense, the horizontal gust is of little consequence in the major airplane airloads and strength limitations. Of greater significance is the response of the airplane to horizontal gusts and wind shear when operating at low flight speeds. The possible conditions in which an airplane may encounter horizontal gusts and wind shear are illustrated in figure 6.5. As the airplane traverses a shear of wind direction, a change in headwind component will exist. Also, a climbing or descending airplane may traverse a shear of wind velocity, i.e., a wind profile in which the wind velocity varies with altitude.

The response of an airplane is much dependent upon the airplane characteristics but certain basic effects are common to all airplanes. Suppose that an airplane is established in steady, level flight with lift equal to weight, thrust equal to drag, and trimmed so

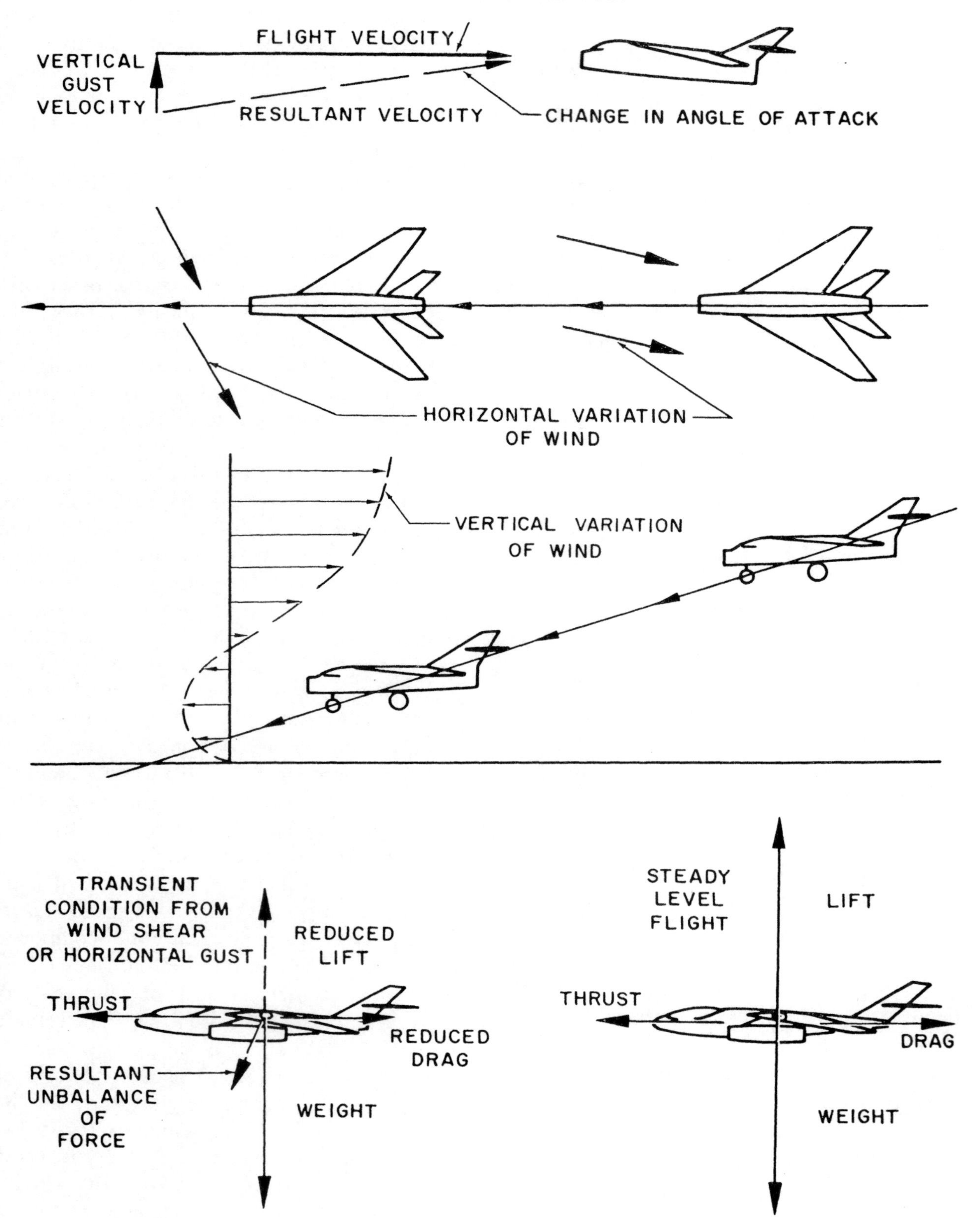

Figure 6.5. Effect of Wind Shear

there is no unbalance of pitching, yawing, or rolling moment. If the airplane traverses a sharp wind shear equivalent to a horizontal gust, the resulting change in airspeed will disturb such an equilibrium. For example, if the airplane encounters a sharp horizontal gust which reduces the airspeed 20 percent, the new airspeed (80 percent of the original value) produces lift and drag at the same angle of attack which are 64 percent of the original value. The change in these aerodynamic forces would cause the airplane to accelerate in the direction of resultant unbalance of force. That is, the airplane would accelerate down and forward until a new equilibrium is achieved. In addition, there would be a change in pitching moment which would produce a response of the airplane in pitch.

The response of the airplane to a horizontal gust will differ according to the gust gradient and airplane characteristics. Generally, if the airplane encounters a sharp wind shear which reduces the airspeed, the airplane tends to sink and incur a loss of altitude before equilibrium conditions are achieved. Similarly, if the airplane encounters a sharp wind shear which increases the airspeed, the airplane tends to float and incur a gain of altitude before equilibrium conditions are achieved.

Significant vertical and horizontal gusts may be due to the terrain or atmospheric conditions. The proximity of an unstable front or thunderstorm activity in the vicinity of the airfield is likely to create significant wind shear and gust activity at low altitude. During gusty conditions every effort must be made for precise control of airspeed and flight path and any changes due to gusts must be corrected by proper control action. Under extreme gusts conditions, it may be advisable to utilize approach, landing, and takeoff speeds slightly greater than normal to provide margin for adequate control.

POWER-OFF GLIDE PERFORMANCE

The gliding performance of an airplane is of special interest for the single-engine airplane in the case of powerplant failure or malfunction. When a powerplant failure or malfunction occurs, it is usually of interest to obtain a gliding flight path which results in the minimum glide angle. The minimum glide angle will produce the greatest proportion of glide distance to altitude loss and will result in maximum glide range or minimum expenditure of altitude for a specific glide distance.

GLIDE ANGLE AND LIFT-DRAG RATIO. In the study of climb performance, the forces acting on the airplane in a steady climb (or glide) produce the following relationship:

$$\sin \gamma = \frac{T-D}{W}$$

where

γ = angle of climb, degrees
T = thrust, lbs.
D = drag, lbs.
W = lbs.

In the case of power-off glide performance, the thrust, T, is zero and the relationship reduces to:

$$\sin \gamma = -\frac{D}{W}$$

By this relationship it is evident that the minimum angle of glide—or minimum negative climb angle—is obtained at the aerodynamic conditions which incur the minimum total drag. Since the airplane lift is essentially equal to the weight, the minimum angle of glide will be obtained when the airplane is operated at maximum lift-drag ratio, $(L/D)_{max}$. When the angle of glide is relatively small, the ratio of glide distance to glide altitude is numerically equal to the airplane lift-drag ratio.

$$\text{glide ratio} = \frac{\text{glide distance, ft.}}{\text{glide altitude, ft.}}$$

$$\text{glide ratio} = (L/D)$$

Figure 6.6 illustrates the forces acting on the airplane in a power-off glide. The equilibrium of the steady glide is obtained when the summation of forces in the vertical and horizontal directions is equal to zero.

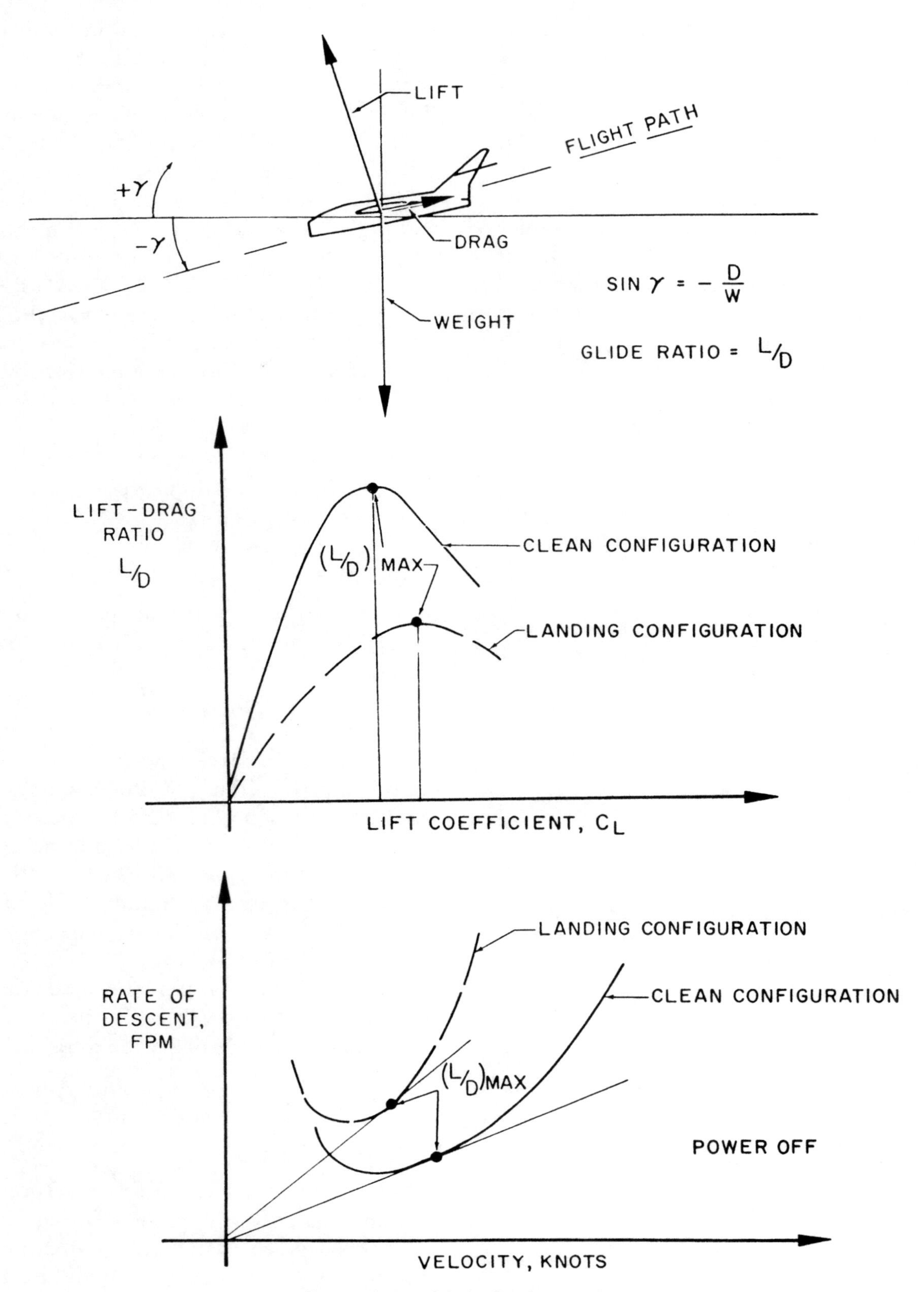

Figure 6.6. Glide Performance

In order to obtain maximum glide ratio, the airplane must be operated at the angle of attack and lift coefficient which provide maximum lift-drag ratio. The illustration of figure 6.6 depicts a variation of lift-drag ratio, L/D, with lift coefficient, C_L, for a typical airplane in the clean and landing configurations. Note that $(L/D)_{max}$ for each configuration will occur at a specific value of lift coefficient and, hence, a specific angle of attack. Thus, the maximum glide performance of a given airplane configuration will be unaffected by gross weight and altitude when the airplane is operated at $(L/D)_{max}$. Of course, an exception occurs at very high altitudes where compressibility effects may alter the aerodynamic characteristics. The highest value of (L/D) will occur with the airplane in the clean configuration. As the airplane is changed to the landing configuration, the added parasite drag reduces $(L/D)_{max}$ and the C_L which produces $(L/D)_{max}$ will be increased. Thus, the best glide speed for the landing configuration generally will be less than the best glide speed for the clean configuration.

The power-off glide performance may be appreciated also by the graph of rate of descent versus velocity shown in figure 6.6. When a straight line is drawn from the origin tangent to the curve, a point is located which produces the maximum proportion of velocity to rate of descent. Obviously, this condition provides maximum glide ratio. Since the rate of descent is proportional to the power required, the points of tangency define the aerodynamic condition of $(L/D)_{max}$.

FACTORS AFFECTING GLIDE PERFORMANCE. In order to obtain the minimum glide angle through the air, the airplane must be operated at $(L/D)_{max}$. The subsonic $(L/D)_{max}$ of a given airplane configuration will occur at a specific value of lift coefficient and angle of attack. However, as can be noted from the curves of figure 6.6, small deviations from the optimum C_L will not cause a drastic reduction of (L/D) and glide ratio. In fact, a 5 percent deviation in speed from the best glide speed will not cause any significant reduction of glide ratio. This is fortunate and allows the specifying of convenient glide speeds which will be appropriate for a range of gross weights at which power-off gliding may be encountered, e.g., small quantities of fuel remaining.

An attempt to *stretch a glide* by flying at speeds above or below the best glide speed will prove futile. As shown by the illustration of figure 6.6, any C_L above or below the optimum will produce a lift-drag ratio less than the maximum. If the airplane angle of attack is increased above the value for $(L/D)_{max}$, a transient reduction in rate of descent will take place but this process must be reserved for the landing phase. Eventually, the steady-state conditions would be achieved and the increased angle of attack would incur a lower airspeed and a reduction in (L/D) and glide ratio.

The effect of *gross weight* on glide performance may be difficult to appreciate. Since $(L/D)_{max}$ of a given airplane configuration will occur at a specific value of C_L, the gross weight of the airplane will not affect the glide ratio if the airplane is operated at the optimum C_L. Thus, two airplanes of identical aerodynamic configuration but different gross weight could glide the same distance from the same altitude. Of course, this fact would be true only if both airplanes are flown at the specific C_L to produce $(L/D)_{max}$. The principal difference would be that the heavier airplane must fly at a higher airspeed to support the greater weight at the optimum C_L. In addition, the heavier airplane flying at the greater speed along the same flight path would develop a greater rate of descent.

The relationship which exists between gross weight and velocity for a particular C_L is as follows:

$$\frac{V_2}{V_1}=\sqrt{\frac{W_2}{W_1}} \quad \text{(constant } C_L\text{)}$$

where

V_1 = best glide speed corresponding to some original gross weight, W_1

V_2 = best glide speed corresponding to some new gross weight, W_2

As a result of this relationship, a 10 percent increase in gross weight would require a 5 percent increase in glide speed to maintain $(L/D)_{max}$. While small variations in gross weight may produce a measurable change in best glide speed, the airplane can tolerate small deviations from the optimum C_L without significant change in (L/D) and glide ratio. For this reason, a standard, single value of glide speed may be specified for a small range of gross weights at which glide performance can be of importance. A gross weight which is considerably different from the normal range will require a modification of best glide speed to maintain the maximum glide ratio.

The *effect of altitude* on glide performance is insignificant if there is no change in $(L/D)_{max}$. Generally, the glide performance of the majority of airplanes is subsonic and there is no noticeable variation of $(L/D)_{max}$ with altitude. Any specific airplane configuration at a particular gross weight will require a specific value of dynamic pressure to sustain flight at the C_L for $(L/D)_{max}$. Thus, the airplane will have a best glide speed which is a specific value of equivalent airspeed (EAS) independent of altitude. For convenience and simplicity, this best glide speed is specified as a specific value of indicated airspeed (IAS) and compressibility and position errors are neglected. The principal effect of altitude is that at high altitude the true airspeed (TAS) and rate of descent along the optimum glide path are increased above the low altitude conditions. However, if $(L/D)_{max}$ is maintained, the glide angle and glide ratio are identical to the low altitude conditions.

The *effect of configuration* has been noted previously in that the addition of parasite drag by flaps, landing gear, speed brakes, external stores, etc. will reduce the maximum lift-drag ratio and cause a reduction of glide ratio. In the case where glide distance is of great importance, the airplane must be maintained in the clean configuration and flown at $(L/D)_{max}$.

The *effect of wind* on gliding performance is similar to the effect of wind on cruising range. That is, a headwind will always reduce the glide range and a tailwind will always increase the glide range. The maximum glide range of the airplane in still air will be obtained by flight at $(L/D)_{max}$. However, when a wind is present, the optimum gliding conditions may not be accomplished by operation at $(L/D)_{max}$. For example, when a headwind is present, the optimum glide speed will be increased to obtain a maximum proportion of *ground* distance to altitude. In this sense, the increased glide speed helps to minimize the detrimental effect of the headwind. In the case of a tailwind, the optimum glide speed will be reduced to maximize the benefit of the tailwind. For ordinary wind conditions, maintaining the glide speed best for zero wind conditions will suffice and the loss or gain in glide distance must be accepted. However, when the wind conditions are extreme and the wind velocity is large in comparison with the glide speed, e.g., wind velocity greater than 25 percent of the glide speed, changes in the glide speed must be made to obtain maximum possible ground distance.

THE FLAMEOUT PATTERN. In the case of failure of the powerplant, every effort should be made to establish a well-planned, stabilized approach if a suitable landing area is available. Generally a 360° overhead approach is specified with the approach beginning from the "high key" point of the flameout pattern. The function of a standardized pattern is to provide a flight path well within the capabilities of the airplane and the abilities of the pilot to judge and control the flight path. The flight handbook will generally specify the particulars of the flameout pattern such as the altitude at the high key, glide speeds, use of flaps, etc. Of course, the particulars of the flameout pattern will be determined by the aerodynamic characteristics of the airplane. A principal factor is the effect of glide ratio, or $(L/D)_{max}$, on the altitude required at the high key point at the beginning of the flameout pattern. The airplane with a low value of $(L/D)_{max}$ will require a high altitude at the high key point.

The most favorable situation during a flameout would be for the airplane to in position to arrive over the intended landing area the altitude for the high key point. In this case, the standard flameout pattern could be utilized. If the airplane does not have sufficient glide range to arrive at the landing area with the altitude for the high key point, it is desirable to fit the approach into the lower portions of the standard flameout approach. If it is not possible to arrive at the intended landing area with sufficient altitude to "play" the approach, serious consideration should be given to ejection while sufficient altitude remains. Deviations from a well-planned approach such as the standard flameout pattern may allow gross errors in judgment. A typical error of a non-standard or poorly executed flameout approach is the use of excessive angles of bank in turns to correct the approach. Because of the great increase in induced drag at large angles of bank, excessive rates of descent will be incurred and there will be further deviations from a desirable flight path.

The power-off gliding characteristics of the airplane can be simulated in power on flight by certain combinations of engine power setting and position of the speed brake or dive flap. This will allow the pilot to become familiar with the power-off glide performance and the flameout landing pattern. In addition, the simulated flameout pattern is useful during a precautionary landing when the powerplant is malfunctioning and there is the possibility of an actual flameout.

The final approach and landing flare will be particularly critical for the airplane which has a low glide ratio but a high best glide speed. These airplane characteristics are typical of the modern configuration of airplane which has low aspect ratio, sweepback, and high wing loading. Since these airplane characteristics also produce marginal flare capability in power-off flight, great care should be taken to follow the procedure recommended for the specific airplane.

As an example of the power-off glide performance of an airplane with low aspect ratio, sweepback, and high wing loading, a best glide speed of 220 knots and a glide ratio of 6 may be typical. In such a case, the rate of descent during the glide at low altitude would be on the order of 3,700 *FPM*. Any deviations from the recommended landing technique cannot be tolerated because of the possibility of an excessive rate of descent. Either premature flare or delayed flare may allow the airplane to touch down at a rate of descent which would cause structural failure. Because of the marginal flare characteristics in power-off flight, the best glide speed recommended for the landing configuration may be well above the speed corresponding to the exact maximum lift-drag ratio. The greater speed reduces induced drag and provides a greater margin for a successful power-off landing flare.

In the extreme case, the power-off glide and landing flare characteristics may be very critical for certain airplane configurations. Thus, a well-planned standard flameout pattern and precise flying technique are necessary and, if very suitable conditions are not available, the recommended alternative is simple: *eject!*

EFFECT OF ICE AND FROST ON AIRPLANE PERFORMANCE

Without exception, the formation of ice or frost on the surfaces of an airplane will cause a detrimental effect on aerodynamic performance. The ice or frost formation on the airplane surfaces will alter the aerodynamic contours and affect the nature of the boundary layer. Of course, the most important surface of the airplane is the wing and the formation of ice or frost can create significant changes in the aerodynamic characteristics.

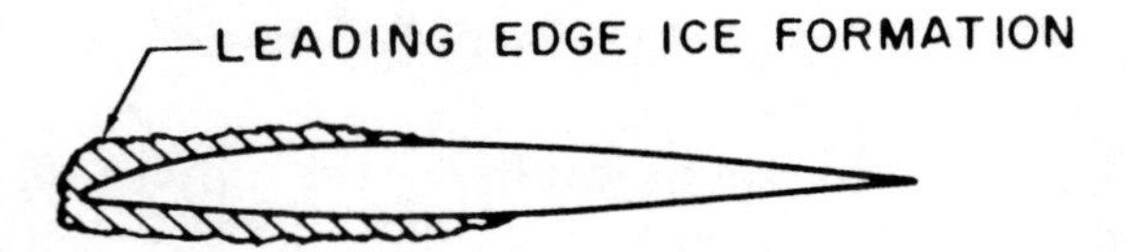

UPPER SURFACE FROST

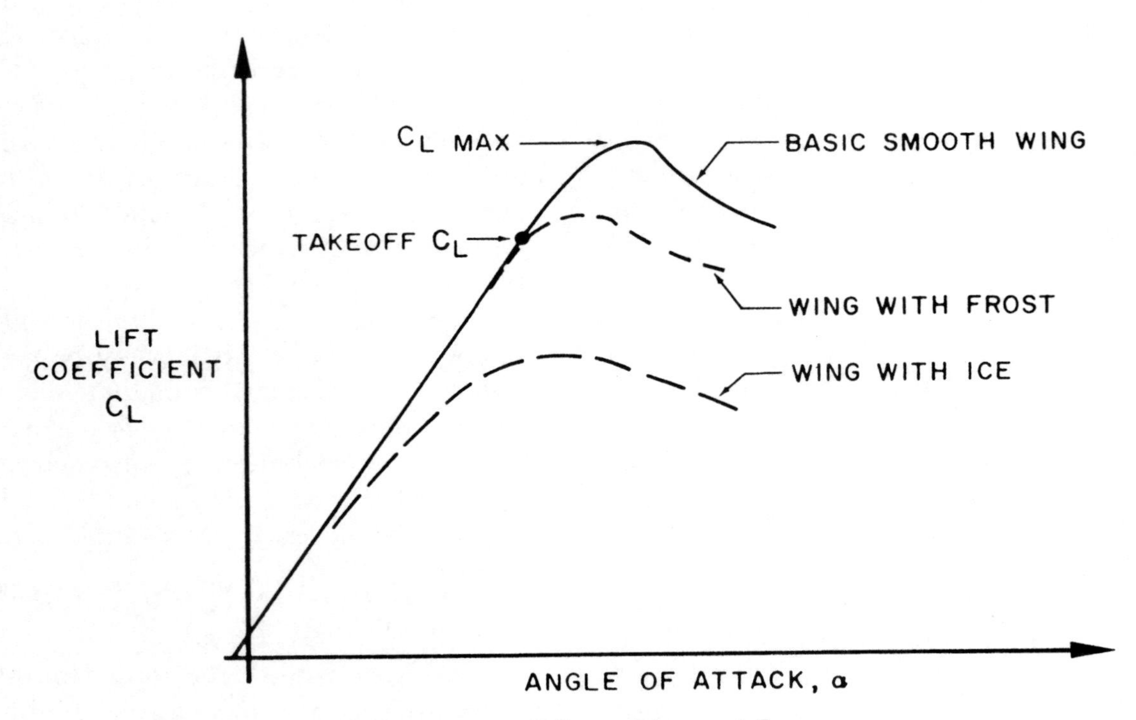

Figure 6.7. Effect of Ice and Frost

A large formation of ice on the leading edge of the wing can produce large changes in the local contours and severe local pressure gradients. The extreme surface roughness common to some forms of ice will cause high surface friction and a considerable reduction of boundary layer energy. As a result of these effects, the ice formation can produce considerable increase in drag and a large reduction in maximum lift coefficient. Thus, the ice formation will cause an increase in power required and stall speed. In addition, the added weight of the ice formation on the airplane will provide an undesirable effect. Because of the detrimental effects of ice formation, recommended anti-icing procedures must be followed to preserve the airplane performance.

The effect of frost is perhaps more subtle than the effect of ice formation on the aerodynamic characteristics of the wing. The accumulation of a hard coat of frost on the wing upper surface will provide a surface texture of considerable roughness. While the basic shape and aerodynamic contour is unchanged, the increase in surface roughness increases skin-friction and reduces the kinetic energy of the boundary layer. As a result, there will be an increase in drag but, of course, the magnitude of drag increase will not compare with the considerable increase due to a severe ice formation. The reduction of boundary layer kinetic energy will cause incipient stalling of the wing, i.e., separation will occur at angles of attack and lift coefficients lower than for the clean, smooth wing. While the reduction in $C_{L_{max}}$ due to frost formation ordinarily is not as great as that due to ice formation, it is usually unexpected because it may be thought that large changes in the aerodynamic shape (such as due to ice) are necessary to reduce $C_{L_{max}}$. However, the kinetic energy of the boundary layer is an important factor influencing separation of the airflow and this energy is reduced by an increase in surface roughness.

The general effects of ice and frost formation on the lift characteristics is typified by the illustration of figure 6.7.

The effect of ice or frost on takeoff and landing performance is of great importance. The effects are so detrimental to the landing and takeoff that no effort should be spared to keep the airplane as free as possible from any accumulation of ice or frost. If any ice remains on the airplane as the landing phase approaches it must be appreciated that the ice formation will have reduced $C_{L_{max}}$ and incurred an increase in stall speed. Thus, the landing speed will be greater. When this effect is coupled with the possibility of poor braking action during the landing roll, a critical situation can exist. It is obvious that great effort must be made to prevent the accumulation of ice during flight.

In no circumstances should a formation of ice or frost be allowed to remain on the airplane wing surfaces prior to takeoff. The undesirable effects of ice are obvious but, as previously mentioned, the effects of frost are more subtle. If a heavy coat of hard frost exists on the wing upper surface, a typical reduction in $C_{L_{max}}$ would cause a 5 to 10 percent increase in the airplane stall speed. Because of this magnitude of effect, the effect of frost on takeoff performance may not be realized until too late. The takeoff speed of an airplane is generally some speed 5 to 25 percent greater than the stall speed, hence the takeoff lift coefficient will be value from 90 to 65 percent of $C_{L_{max}}$. Thus, it is possible that the airplane with frost cannot become airborne at the specified takeoff speed because of premature stalling. Even if the airplane with frost were to become airborne at the specified takeoff speed, the airplane could have insufficient margin of airspeed above stall and turbulence, gusts, turning flight could produce incipient or complete stalling of the airplane.

The increase in drag during takeoff roll due to frost or ice is not considerable and there will not be any significant effect on the initial acceleration during takeoff. Thus, the effect of frost or ice will be most apparent during the later portions of takeoff if the airplane is unable to become airborne or if insufficient margin above stall speed prevents successful initial climb. In no circumstances should a formation of ice or frost be allowed to remain on the airplane wing surfaces prior to takeoff.

ENGINE FAILURE ON THE MULTIENGINE AIRPLANE

In the case of the single-engine airplane, powerplant failure leaves only the alternatives of effecting a successful power-off landing or abandoning the airplane. In the case of the multiengine airplane, the failure of a powerplant does not necessarily constitute a disaster since flight may be continued with the remaining powerplants functioning. However, the performance of the multiengine airplane with a powerplant inoperative may be critical for certain conditions of flight and specific techniques and procedures must be observed to obtain adequate performance.

The effect of a powerplant failure on the multiengine turbojet airplane is illustrated by the first chart of figure 6.8 with the variation of required and available thrust with velocity. If half of the airplane powerplants are inoperative, e.g., single-engine operation of a twin-engine airplane, the maximum thrust available at each velocity is reduced to half that available prior to the engine failure. The variation of thrust required with velocity may be affected by the failure of a powerplant in that there may be significant increases in drag if specific procedures are not followed. The inoperative powerplant may contribute additional drag and the pilot must insure that the additional drag is held to a minimum. In the case of the propeller powered airplane, the propeller must be feathered, cowl flaps closed, etc., as the increased drag will detract considerably from the performance.

The principal effects of the reduced available thrust are pointed out by the illustration of figure 6.8. Of course, the lower available thrust will reduce the maximum level flight speed but of greater importance is the reduction in excess thrust. Since the acceleration and climb performance is a function of the excess thrust and power, the failure of a powerplant will be most immediately appreciated in this area of performance. As illustrated in figure 6.8, loss of one-half the maximum available thrust will reduce the excess thrust to less than half the original value. Since some thrust is required to sustain flight, the excess which remains to accelerate and climb the airplane may be greatly reduced. The most critical conditions will exist when various factors combine to produce a minimum of excess thrust or power when engine failure occurs. Thus, critical conditions will be common to *high gross weight* and *high density altitude* (and high temperatures in the case of the turbine powered airplane) as each of these factors will reduce the excess thrust at any specific flight condition.

The asymmetrical power condition which results when a powerplant fails can provide critical control requirements. First consideration is due the yawing moment produced by the asymmetrical power condition. Adequate directional control will be available only when the airplane speed is greater than the minimum directional control speed. Thus, the pilot must insure that the flight speed never falls below the minimum directional control speed because the application of maximum power on the functioning powerplants will produce an uncontrollable yaw if adequate directional control is unavailable. A second consideration which is due the propeller powered airplane involves the rolling moments caused by the slipstream velocity. Asymmetrical power on the propeller airplane will create a dissymmetry of the slipstream velocities on the wing and create rolling moments which must be controlled. These slipstream induced rolling moments will be greatest at high power and low velocity and the pilot must be sure of adequate lateral control, especially for the crosswind landing.

The effect of an engine failure on the remaining range and endurance is specific to the airplane type and configuration. If an engine fails during optimum cruise of the turbojet airplane, the airplane must descend and experience a loss of range. Since the turbojet airplane is generally overpowered at $(L/D)_{max}$, a loss of a powerplant will not cause a significant change in maximum endurance. If an engine fails during cruise of a reciprocating powered airplane, there will be a significant loss of range only if the maximum range condition cannot be sustained with the remaining powerplants operating within the cruise power rating. If a power greater than the maximum cruise rating is necessary to sustain cruise, the specific fuel consumption increases and causes a reduction of range. Essentially the same relationship exists regarding maximum endurance of the reciprocating powered airplane.

When critical conditions exist due to failure of a powerplant, the pilot must appreciate the reduced excess thrust and operate the airplane within specific limitations. If the engine-out performance of the airplane is marginal, the pilot must be aware of the very detrimental effect of steep turns. Due to the increased load factor in a coordinated turn, there will be an increase in stall speed and—of greater importance to engine-out performance—an increase in induced drag. The following table illustrates the effect of bank angle on stall speed and induced drag.

TABLE 6.1

Bank angle, ϕ, degrees	Load factor	Percent increase in stall speed	Percent increase in induced drag (at constant velocity)
0	1.0000	0	0
5	1.0038	0.2	0.8
10	1.0154	0.7	3.1
15	1.0353	1.7	7.2
20	1.0642	3.2	13.3
25	1.1034	5.0	21.7
30	1.1547	7.5	33.3
35	1.2208	10.5	49.0
40	1.3054	14.3	70.4
45	1.4142	18.9	100.0
60	2.000	41.4	300.0

The previous table of values illustrates the fact that coordinated turns with less than 15° of bank cause no appreciable effect on stall speed or induced drag. However, note that 30° of bank will increase the induced drag by 33.3 percent. Under critical conditions, such an increase in induced drag (and, hence, total drag) would be prohibitive causing the airplane to descend rather than climb. The second graph of figure 6.7 illustrates the case where the steep turn causes such a large increase in required thrust that a deficiency of thrust exists. Whenever engine failure produces critical performance conditions it is wise to limit all turns to 15° of bank wherever possible.

Another factor to consider in turning flight is the effect of sideslip. If the turn is not coordinated to hold sideslip to a minimum, additional drag will be incurred due to the sideslip.

The use of the flaps and landing gear can greatly affect the performance of the multiengine airplane when a powerplant is inoperative. Since the extension of the landing gear and flaps increases the parasite drag, maximum performance of the airplane will be obtained with airplane in the clean configuration. In certain critical conditions, the extension of the landing gear and full flaps may create a deficiency of thrust at any speed and commit the airplane to descend. This condition is illustrated by the second graph of figure 6.8. Thus, judicious use of the flaps and landing gear is necessary in the case of an engine failure.

In the case of engine failure immediately after takeoff, it is important to maintain airspeed in excess of the minimum directional control speed and accelerate to the best climb speed. After the engine failure, it will be favorable to *climb only as necessary to clear obstacles until the airplane reaches the best climb speed.* Of course, the landing gear should be retracted as soon as the airplane is airborne to reduce parasite drag and, in the case of the propeller powered airplane, it is imperative that the wind milling propeller be feathered. The flaps should be retracted only as rapidly as the increase in

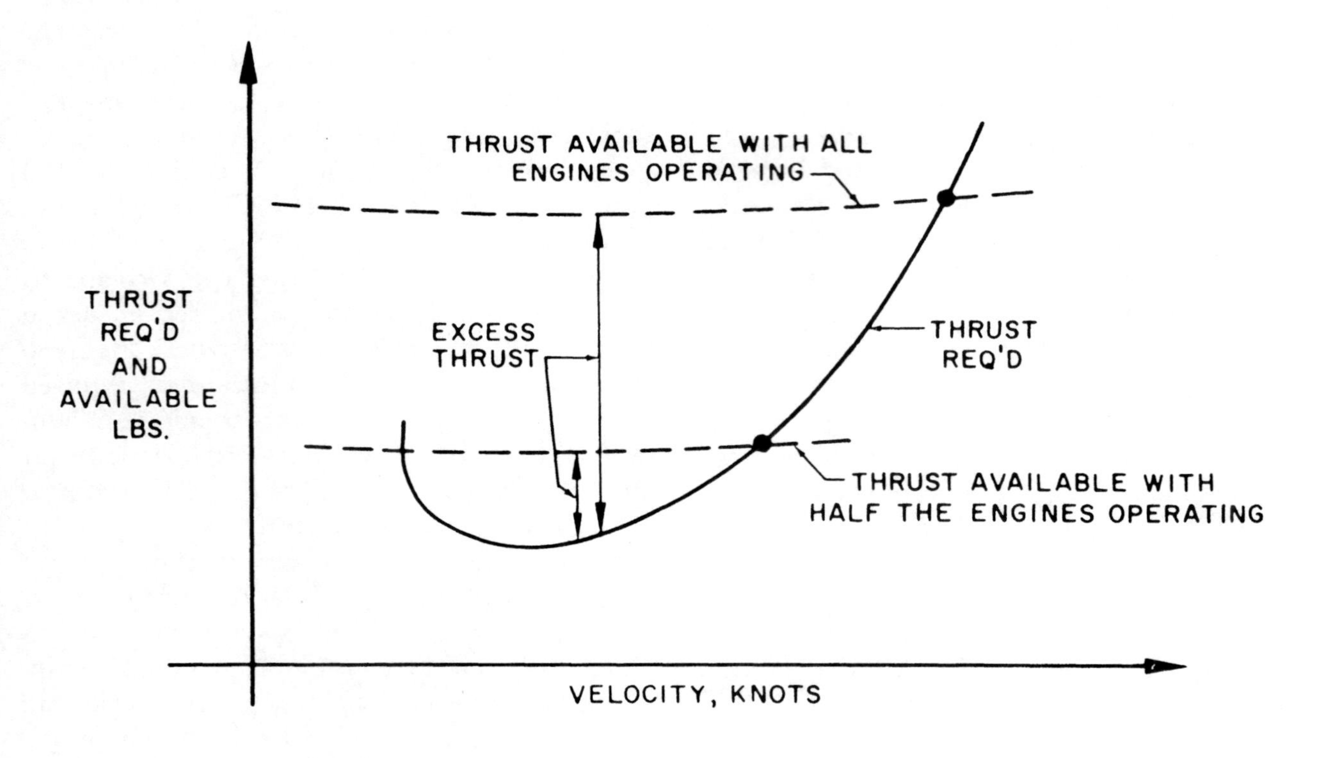

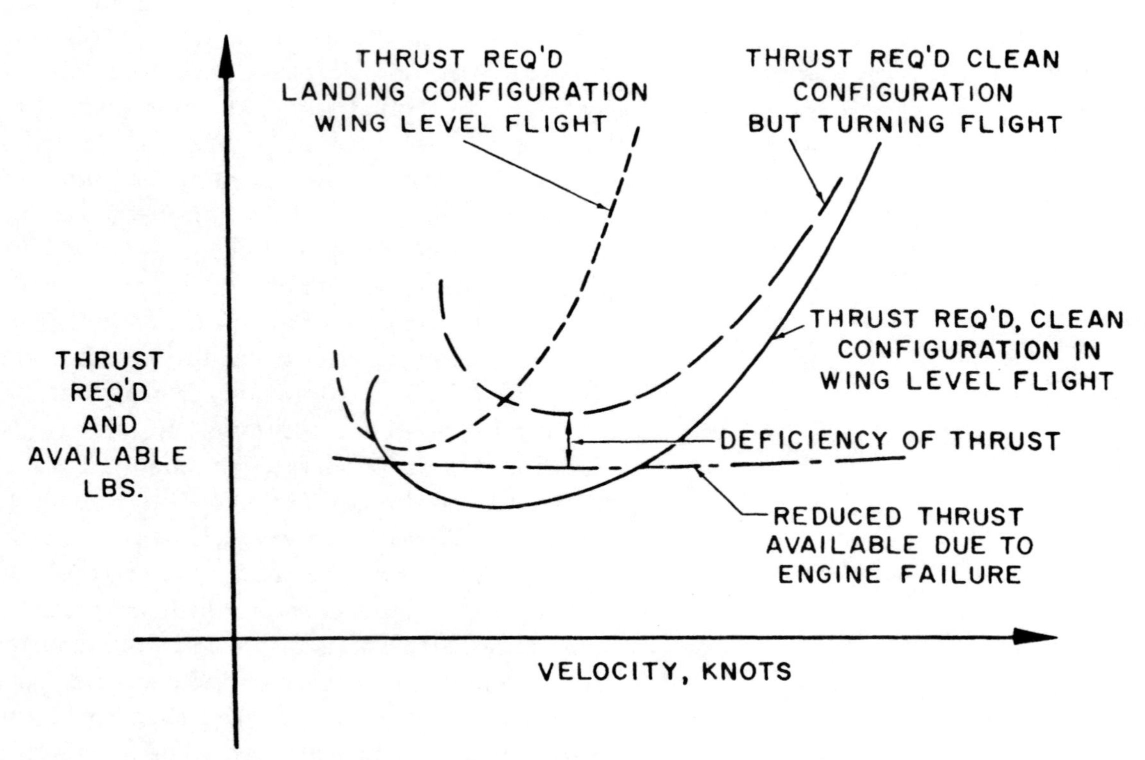

Figure 6.8. Engine Failure on Multi-engine Aircraft

airspeed will allow. If full flap deflection is utilized for takeoff it is important to recall that the last 50 percent of flap deflection creates more than half the total drag increase but less than half the total change in $C_{L_{max}}$. Thus, for some configurations of airplanes, a greater reduction in drag may be accomplished by partial retraction of the flaps rather than retraction of the landing gear. Also, it is important that no steep turns be attempted because of the undesirable increase in induced drag.

During the landing with an engine inoperative, the same fundamental precautions must be observed as during takeoff, i.e., minimum directional control speed must be maintained (or exceeded), no steep turns should be attempted, and the extension of the flaps and landing gear must be well planned. In the case of a critical power condition it may be necessary to delay the extension of the landing gear and full flaps until a successful landing is assured. If a waveoff is necessary, maximum performance will be obtained cleaning up the airplane and accelerating to the best climb speed before attempting any gain in altitude.

At all times during flight with an engine inoperative, the pilot must utilize the proper techniques for control of airspeed and altitude, e.g., for the conditions of steady flight, angle of attack is the primary control of airspeed and excess power is the primary control of rate of climb. For example, if during approach to landing the extension of full flaps and landing gear creates a deficiency of power at all speeds, the airplane will be committed to descend. If the approach is not properly planned and the airplane sinks below the desired glide path, an increase in angle of attack will only allow the airplane to fly more slowly and descend more rapidly. An attempt to hold altitude by increased angle of attack when a power deficiency exists only causes a continued loss of airspeed. Proper procedures and technique are an absolute necessity for safe flight when an engine failure occurs.

GROUND EFFECT

When an airplane in flight nears the ground (or water) surface, a change occurs in the three dimensional flow pattern because the local airflow cannot have a vertical component at the ground plane. Thus, the ground plane will furnish a restriction to the flow and alter the wing upwash, downwash, and tip vortices. These general effects due to the presence of the ground plane are referred to as "ground effect."

AERODYNAMIC INFLUENCE OF GROUND EFFECT. While the aerodynamic characteristics of the tail and fuselage are altered by ground effects, the principal effects due to proximity of the ground plane are the changes in the aerodynamic characteristics of the wing. As the wing encounters ground effect and is maintained at a constant lift coefficient, there is a reduction in the upwash, downwash, and the tip vortices. These effects are illustrated by the sketches of figure 6.9. As a result of the reduced tip vortices, the wing in the presence of ground effect will behave as if it were of a greater aspect ratio. In other words, the induced velocities due to the tip (or trailing) vortices will be reduced and the wing will incur smaller values of induced drag coefficient, C_{D_i}, and induced angle of attack, α_i, for any specific lift coefficient, C_L.

In order for ground effect to be of a significant magnitude, the wing must be quite close to the ground plane. Figure 6.9 illustrates one of the direct results of ground effect by the variation of induced drag coefficient with wing height above the ground plane for a representative unswept wing at constant lift coefficient. Notice that the wing must be quite close to the ground for a noticeable reduction in induced drag. When the wing is at a height equal to the span ($h/b=1.0$), the reduction in induced drag is only 1.4 percent. However, when the wing is at a height equal to one-fourth the span ($h/b=0.25$), the reduction in induced drag is 23.5

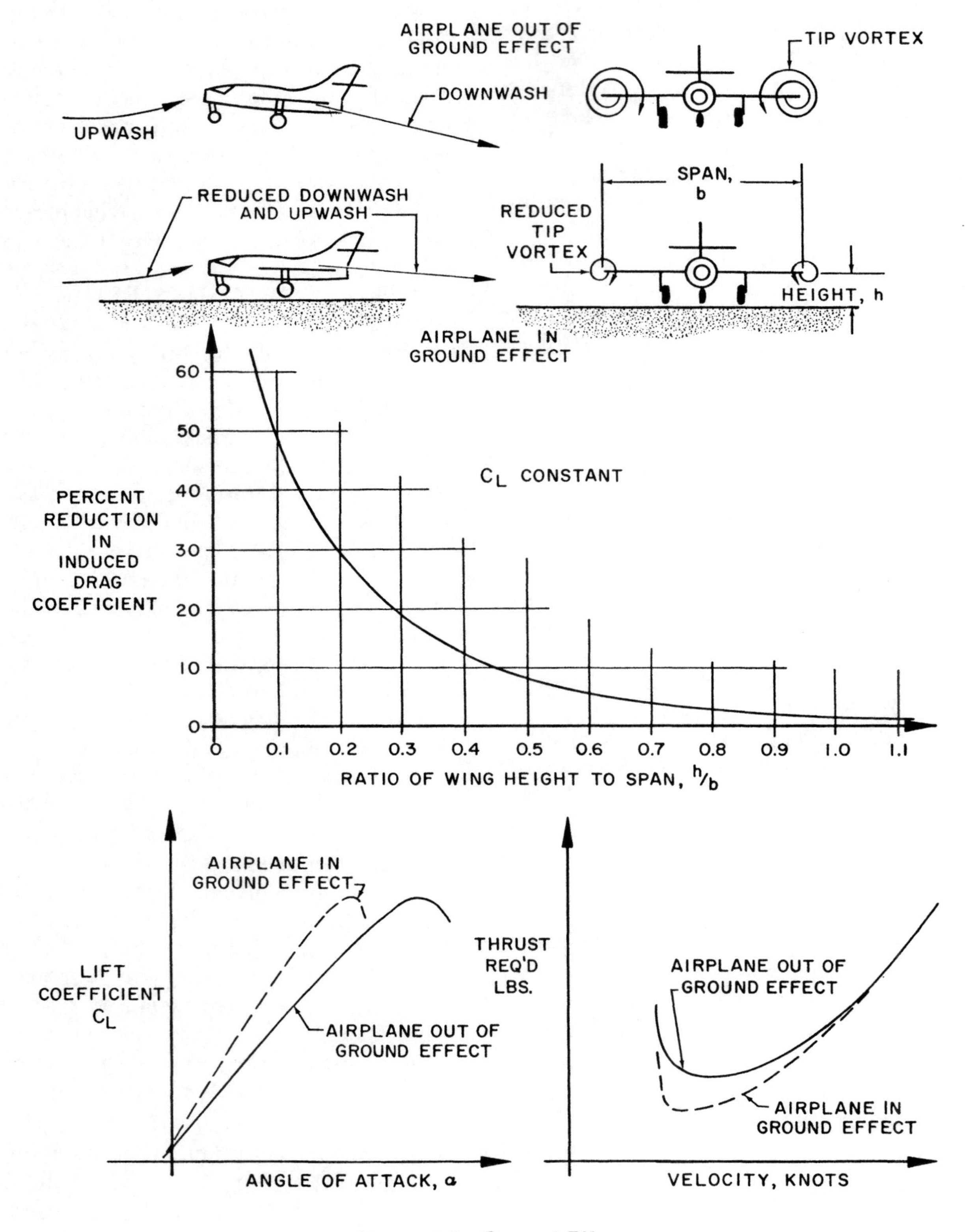

Figure 6.9. Ground Effect

percent and, when the wing is at a height equal to one-tenth the span ($h/b=0.1$), the reduction in induced drag is 47.6 percent. Thus, a large reduction in induced drag will take place only when the wing is very close to the ground. Because of this variation, ground effect is most usually recognized during the liftoff of takeoff or prior to touchdown on landing.

The reduction of the tip or trailing vortices due to ground effect alters the spanwise lift distribution and reduces the induced angle of attack. In this case, the wing will require a lower angle of attack in ground effect to produce the same lift coefficient. This effect is illustrated by the lift curves of figure 6.9 which show that the airplane in ground effect will develop a greater slope of the lift curve. For the wing in ground effect, a lower angle of attack is necessary to produce the same lift coefficient or, if a constant angle of attack is maintained, an increase in lift coefficient will result.

Figure 6.9 illustrates the manner in which ground effect will alter the curve of thrust required versus velocity. Since induced drag predominates at low speeds, the reduction of induced drag due to ground effect will cause the most significant reduction of thrust required (parasite plus induced drag) only at low speeds. At high speeds where parasite drag predominates, the induced drag is but a small part of the total drag and ground effect causes no significant change in thrust required. Because ground effect involves the induced effects of airplane when in close proximity to the ground, its effects are of greatest concern during the takeoff and landing. Ordinarily, these are the only phases of flight in which the airplane would be in close proximity to the ground.

GROUND EFFECT ON SPECIFIC FLIGHT CONDITIONS. The overall influence of ground effect is best realized by assuming that the airplane descends into ground effect while maintaining a constant lift coefficient and, thus, a constant dynamic pressure and equivalent airspeed. As the airplane descends into ground effect, the following effects will take place:

(1) Because of the reduced induced angle of attack and change in lift distribution, a smaller wing angle of attack will be required to produce the same lift coefficient. If a constant pitch attitude is maintained as ground effect is encountered, an increase in lift coefficient will be incurred.

(2) The reduction in induced flow due to ground effect causes a significant reduction in induced drag but causes no direct effect on parasite drag. As a result of the reduction in induced drag, the thrust required at low speeds will be reduced.

(3) The reduction in downwash due to ground effect will produce a change in longitudinal stability and trim. Generally, the reduction in downwash at the horizontal tail increases the contribution to static longitudinal stability. In addition, the reduction of downwash at the tail usually requires a greater up elevator to trim the airplane at a specific lift coefficient. For the conventional airplane configuration, encountering ground effect will produce a nose-down change in pitching moment. Of course, the increase in stability and trim change associated with ground effect provide a critical requirement of adequate longitudinal control power for landing and takeoff.

(4) Due to the change in upwash, downwash, and tip vortices, there will be a change in position error of the airspeed system, associated with ground effect. In the majority of cases, ground effect will cause an increase in the local pressure at the static source and produce a lower indication of airspeed and altitude.

During the *landing phase* of flight, the effect of proximity to the ground plane must be understood and appreciated. If the airplane is brought into ground effect with a constant angle of attack, the airplane will experience an increase in lift coefficient and reduction in thrust required. Hence, a "floating" sensation may be experienced. Because of the reduced drag and power-off deceleration in ground effect, any excess speed at the point of flare may incur a considerable "float" distance. As the airplane nears the point of touchdown on the approach, ground effect will be most realized at altitudes less than the wing span. An exact appreciation of the ground effect may be obtained during a *field* approach with the mirror landing system furnishing an exact reference of the flight path. During the final phases of the field approach as the airplane nears the ground plane, a reduced power setting is necessary or the reduced thrust required would allow the airplane to climb above the desired glide path. During shipboard operations, ground effect will be delayed until the airplane passes the edge of the deck and the reduction in power setting that is common to field operations should not be encountered. Thus, a habit pattern should not be formed during field landings which would prove dangerous during carrier operations.

An additional factor to consider is the aerodynamic drag of the airplane during the landing roll. Because of the reduced induced drag when in ground effect, aerodynamic braking will be of greatest significance only when partial stalling of the wing can be accomplished. The reduced drag when in ground effect accounts for the fact that the brakes are the most effective source of deceleration for the majority of airplane configurations.

During the *takeoff phase* of flight ground effect produces some important relationships. Of course, the airplane leaving ground effect encounters just the reverse of the airplane entering ground effect, i.e., the airplane leaving ground effect will (1) require an increase in angle of attack to maintain the same lift coefficient, (2) experience an increase in induced drag and thrust required, (3) experience a decrease in stability and a nose-up change in moment, and (4) usually a reduction in static source pressure and increase in indicated airspeed. These general effects should point out the possible danger in attempting takeoff prior to achieving the recommended takeoff speed. Due to the reduced drag in ground effect the airplane may *seem* capable of takeoff below the recommended speed. However, as the airplane rises out of ground effect with a deficiency of speed, the greater induced drag may produce marginal initial climb performance. In the extreme conditions such as high gross weight, high density altitude, and high temperature, a deficiency of airspeed at takeoff may permit the airplane to become airborne but be incapable of flying out of ground effect. In this case, the airplane may become airborne initially with a deficiency of speed, but later settle back to the runway. It is imperative that no attempt be made to force the airplane to become airborne with a deficiency of speed; the recommended takeoff speed is necessary to provide adequate initial climb performance. In fact, ground effect can be used to advantage if no obstacles exist by using the reduced drag to improve initial acceleration.

The results of the airplane leaving ground effect can be most easily realized during the deck launch of a heavily loaded airplane. As the airplane moves forward and passes over the edge of the deck, whatever ground effect exists will be lost immediately. Thus, proper rotation of the airplane will be necessary to maintain the same lift coefficient and the increase in induced drag must be expected.

The rotor of the helicopter experiences a similar restraint of induced flow when in proximity to the ground plane. Since the induced rotor power required will predominate at low flight speeds, ground effect will produce a considerable effect on the power required at low speeds. During hovering and flight at low speeds, the elevation of the rotor above the ground plane will be an important factor determining the power required for flight.

The *range* of the reciprocating powered airplane can be augmented by the use of ground effect. When the airplane is close to the ground or water surface the reduction of induced drag increases the maximum lift-drag ratio and causes a corresponding increase in range. Of course, the airplane must be quite close to the surface to obtain a noticeable increase in $(L/D)_{max}$ and range. The difficulty in holding the airplane at the precise altitude without contacting the ground or water will preclude the use of ground effect during ordinary flying operations. The use of ground effect to extend range should be reserved as a final measure in case of emergency. Because of the very detrimental effect of low altitude on the range of the turbojet, ground effect will not be of a particular advantage in an attempt to augment range.

The most outstanding examples of the use of ground effect are shown in the cases of multiengine airplanes with some engines inoperative. When the power loss is quite severe, the airplane may not be capable of sustaining altitude and will descend. As ground effect is encountered, the reduced power required may allow the airplane to sustain flight at extremely low altitude with the remaining powerplants functioning. In ground effect, the reciprocating powered airplane will encounter a greater $(L/D)_{max}$ which occurs at a lower airspeed and power required and the increase in range may be quite important during emergency conditions.

INTERFERENCE BETWEEN AIRPLANES IN FLIGHT

During formation flying and inflight refueling, airplanes in proximity to one another will produce a mutual interference of the flow patterns and alter the aerodynamic characteristics of each airplane. The principal effects of this interference must be appreciated since certain factors due to the mutual interference may enhance the possibility of a collision.

One example of interference between airplanes in flight is shown first in figure 6.10 with the effect of lateral separation of two airplanes flying in line abreast. A plane of symmetry would exist halfway between two identical airplanes and would furnish a boundary of flow across which there would be no lateral components of flow. As the two airplane wing tips are in proximity, the effect is to reduce the strength of the tip or trailing vortices and reduce the induced velocities in the vicinity of wing tip. Thus, each airplane will experience a local increase in the lift distribution as the tip vortices are reduced and a rolling moment is developed which tends to roll each airplane away from the other. This disturbance may provide the possibility of collision if other airplanes are in the vicinity and there is delay in control correction or overcontrol. If the wing tips are displaced in a fore-and-aft direction, the same effect exists but generally it is of a lower magnitude.

The magnitude of the interference effect due to lateral separation of the wing tips depends on the proximity of the wing tips and the extent of induced flow. This implies that the interference would be greatest when the tips are very close and the airplanes are operating at high lift coefficients. An interesting ramification of this effect is that several airplanes in line abreast with the wing tips quite close will experience a reduction in induced drag.

An indirect form of interference can be encountered from the vortex system created by a preceding airplane along the intended flight path. The vortex sheet rolls up a considerable distance behind an airplane and creates considerable turbulence for any closely following airplane. This wake can prove troublesome if airplanes taking off and landing are not provided adequate separation. The rolled-up vortex sheet will be strongest when the preceding airplanes is large, high gross weight, and operating at high lift coefficients. At times this turbulence may be falsely attributed to propwash or jetwash.

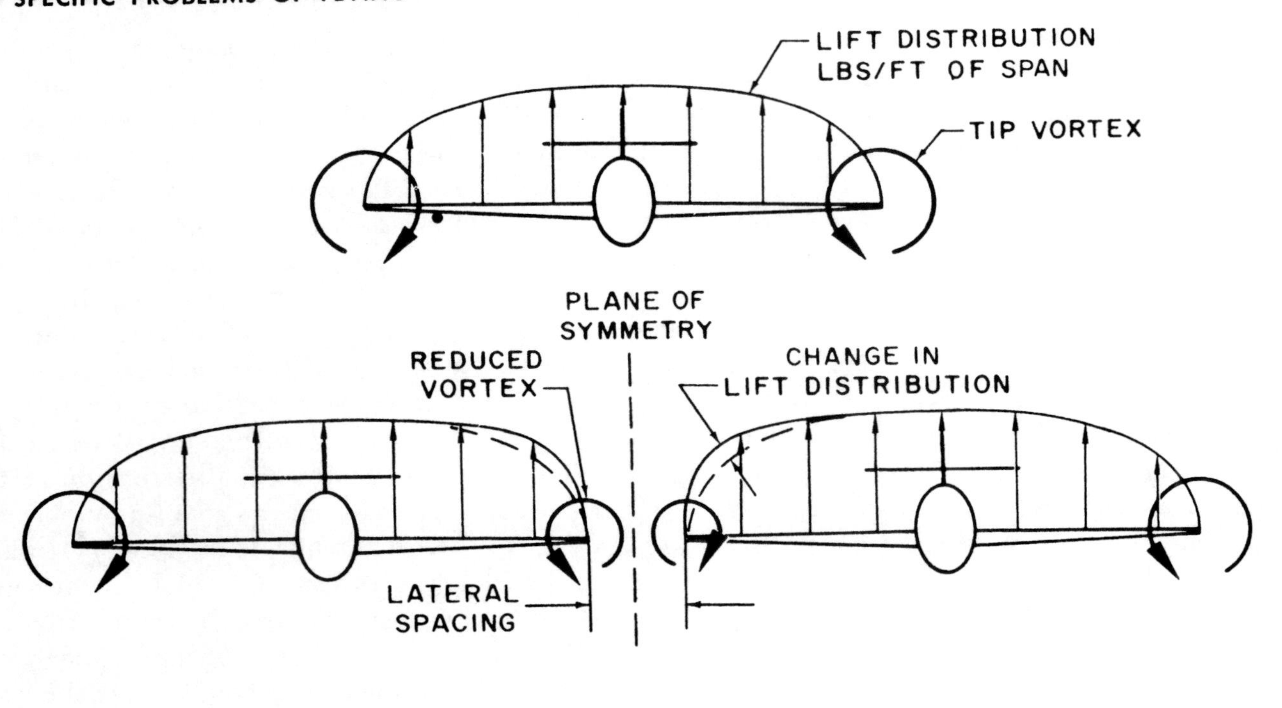

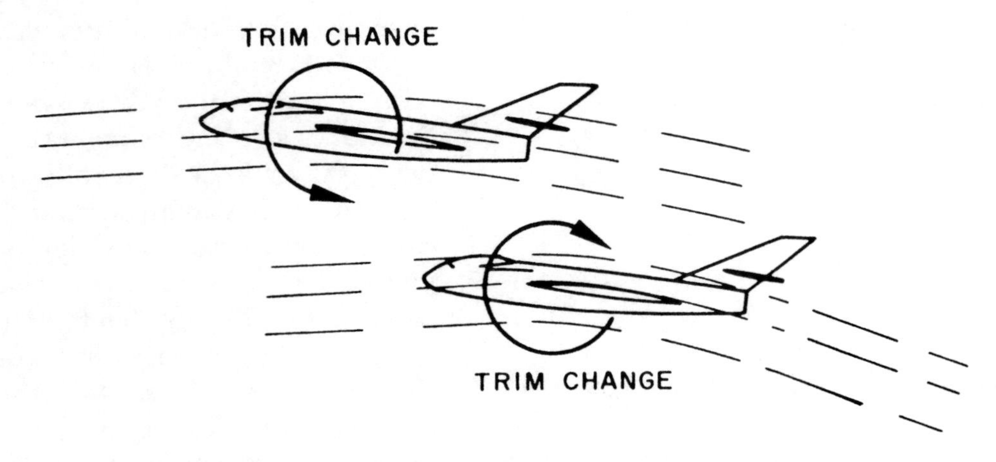

Figure 6.10. Interference Between Airplanes in Flight

Another important form of direct interference is common when the two airplanes are in a trail position and stepped down. As shown in figure 6.10, the single airplane in flight develops upwash ahead of the wing and downwash behind and any restriction accorded the flow can alter the distribution and magnitude of the upwash and downwash. When the trailing airplane is in close proximity aft and below the leading airplane a mutual interference takes place between the two airplanes. The leading airplane above will experience an effect which would be somewhat similar to encountering ground effect, i.e., a reduction in induced drag, a reduction in downwash at the tail, and a change in pitching moment nose down. The trailing airplane below will experience an effect which is generally the opposite of the airplane above. In other words, the airplane below will experience an increase in induced drag, an increase in downwash at the tail, and a change in pitching moment nose up. Thus, when the airplanes are in close proximity, a definite collision possibility exists because of the trim change experienced by each airplane. The magnitude of the trim change is greatest when the airplanes are operating at high lift coefficients, e.g., low speed flight, and when the airplanes are in close proximity.

In formation flying, this sort of interference must be appreciated and *anticipated*. In crossing under another airplane, care must be taken to anticipate the trim change and adequate clearance must be maintained, otherwise a collision may result. The pilot of the leading aircraft will know of the presence of the trailing airplane by the trim change experienced. Obviously, some anticipation is necessary and adequate separation is necessary to prevent a disturbing magnitude of the trim change. In a close diamond formation the leader will be able to "feel" the presence of the slot man even though the airplane is not within view. Obviously, the slot man will have a difficult job during formation maneuvers because of the unstable trim changes and greater power changes required to hold position.

A common collision problem is the case of an airplane with a malfunctioning landing gear. If another airplane is called to inspect the malfunctioning landing gear, great care must be taken to maintain adequate separation and preserve orientation. Many instances such as this have resulted in a collision when the pilot of the trailing airplane became disoriented and did not maintain adequate separation.

During inflight refueling, essentially the same problems of interference exist. As the receiver approaches the tanker from behind and below, the receiver will encounter the downwash from the tanker and require a slight, gradual increase in power and pitch attitude to continue approach to the receiving position. While the receiver may not be visible to the pilot of the tanker, he will anticipate the receiver coming into position by the slight reduction in power required and nose down change in pitching moment. Adequate clearance and proper position must be maintained by the pilot of the receiver for a collision possibility is enhanced by the relative positions of the airplanes. A hazardous condition exists if the pilot of the receiver has excessive speed and runs under the tanker in close proximity. The trim change experienced by both airplanes may be large and unexpected and it may be difficult to avoid a collision.

In addition to the forms of interference previously mentioned, there exists the possibility of strong interference between airplanes in supersonic flight. In this case, the shock waves from one airplane may strongly affect the pressure distribution and rolling, yawing, and pitching moments of an adjacent airplane. It is difficult to express general relationships of the effect except that magnitude of the effects will be greatest when in close proximity at low altitude and high q. Generally, the trailing airplane will be most affected.

BRAKING PERFORMANCE

For the majority of airplane configurations and runways conditions, the airplane brakes furnish the most powerful means of deceleration. While specific techniques of braking are required for specific situations, there are various fundamentals which are common to all conditions.

Solid *friction* is the resistance to relative motion of two surfaces in contact. When relative motion exists between the surfaces, the resistance to relative motion is termed "kinetic" or "sliding" friction; when no relative motion exists between the surfaces, the resistance to the impending relative motion is termed "static" friction. The minute discontinuities of the surfaces in contact are able to mate quite closely when relative motion impends rather than exists, so static friction will generally exceed kinetic friction. The magnitude of the friction force between two surfaces will depend in great part on the types of surfaces in contact and the magnitude of force pressing the surfaces together. A convenient method of relating the friction charactersitics of surfaces in contact is a proportion of the friction force to the normal (or perpendicular) force pressing the surfaces together. This proportion defines the coefficient of friction, μ.

$$\mu = F/N$$

where

μ = coefficient of friction (mu)
F = friction force, lbs.
N = normal force, lbs.

The coefficient of friction of tires on a runway surface is a function of many factors. Runway surface condition, rubber composition, tread, inflation pressure, surface friction shearing stress, relative slip speed, etc., all are factors which affect the coefficient of friction. When the tire is rolling along the runway without the use of brakes, the friction force resulting is simple rolling resistance. The coefficient of rolling friction is of an approximate magnitude of 0.015 to 0.030 for dry, hard runway surface.

The application of brakes supplies a torque to the wheel which tends to retard wheel rotation. However, the initial application of brakes creates a braking torque but the initial retarding torque is balanced by the increase in friction force which produces a driving or rolling torque. Of course, when the braking torque is equal to the rolling torque, the wheel experiences no acceleration in rotation and the equilibrium of a constant rotational speed is maintained. Thus, the application of brake develops a retarding torque and causes an increase in friction force between the tire and runway surface. A common problem of braking technique is application of excessive brake pressure which creates a braking torque greater than the maximum possible rolling torque. In this case, the wheel loses rotational speed and decelerates until the wheel is stationary and the result is a locked wheel with the tire surface subject to a full slip condition.

The relationship of friction force, normal force, braking torque, and rolling torque is illustrated in figure 6.11.

The effect of slip velocity on the coefficient of friction is illustrated by the graph of figure 6.11. The conditions of zero slip corresponds to the rolling wheel without brake application while the condition of full, 100 percent slip corresponds to the locked wheel where the relative velocity between the tire surface and the runway equals the actual velocity. With the application of brakes, the coefficient of friction increases but incurs a small but measurable apparent slip. Continued increase in friction coefficient is obtained until some maximum is achieved then decreases as the slip increases and approaching the 100 percent slip condition. Actually, the peak value of coefficient of friction occurs at an incipient skid condition and the relative slip apparent at this point consists primarily of elastic shearing deflection of the tire structure.

When the runway surface is dry, brush-finished concrete, the maximum value for the coefficient of friction for most aircraft tires is on the order of 0.6 to 0.8. Many factors can determine small differences in this peak value of friction coefficient for dry surface conditions. For example, a soft gum rubber composition can develop a very high value of coefficient of friction but only for low values of surface shearing stress. At high values of surface shearing stress, the soft gum rubber will shear or scrub off before high values of friction coefficient are developed. The higher strength compounds used in the production of aircraft tires produce greater resistance to surface shear and scrubbing but the harder rubber has lower intrinsic friction coefficient. Since the high performance airplane cannot afford the luxury of excessive tire weight or size, the majority of airplane tires will be of relatively hard rubber and will operate at or near the rated load capacities. As a result, there will be little difference between the peak values of friction coefficient for the dry, hard surface runway for the majority of aircraft tires.

If high traction on dry surfaces were the only consideration in the design of tires, the result would be a soft rubber tire of extreme width to create a large footprint and reduce surface shearing stresses, e.g., driving tires on a drag racer. However, such a tire has many other characteristics which are undesirable such as high rolling friction, large size, poor side force characteristics, etc.

When the runway has water or ice on the surface, the maximum value for the coefficient of friction is reduced greatly below the value obtained for the dry runway condition. When water is on the surface, the tread design becomes of greater importance to maintain contact between the rubber and the runway and prevent a film of water from lubricating the surfaces. When the rainfall is light, the peak value for friction coefficient is on the order of 0.5. With heavy rainfall it is more likely that sufficient water will stand to form a liquid film between the tire and the runway. In this case, the peak coefficient of friction rarely exceeds 0.3. In some extreme conditions, the tire may simply plane along the water without contact of the runway and the coefficient of friction is much lower than 0.3. Smooth, clear ice on the runway will cause extremely low values for the coefficient of friction. In such a condition, the peak value for the coefficient of friction may be on the order of 0.2 or 0.15.

Note that immediately past the incipient skidding condition the coefficient of friction decreases with increased slip speed, especially for the wet or icy runway conditions. Thus, once skid begins, a reduction in friction force and rolling torque must be met with a reduction in braking torque, otherwise the wheel will decelerate and lock. This is an important factor to consider in braking technique because the skidding tire surface on the locked wheel produces considerably less retarding force than when at the incipient skid condition which causes the peak coefficient of friction. If the wheel locks from excessive braking, the sliding tire surface produces less than the maximum retarding force and the tires become relatively incapable of developing any significant side force. Stop distance will increase and it may be difficult—if not impossible—to control the airplane when full slip is developed. In addition, at high rolling velocities on the dry surface runway, the immediate problem of a skidding tire is not necessarily the loss of retarding force but the imminence of tire failure. The pilot must insure that the application of brakes does not produce some excessive braking torque which is greater than the maximum rolling torque and particular care must be taken when the runway conditions produce low values of friction coefficient and when the normal force on the braking surfaces is small. When it is difficult to perceive or distinguish a skidding condition, the value of an antiskid or automatic braking system will be appreciated.

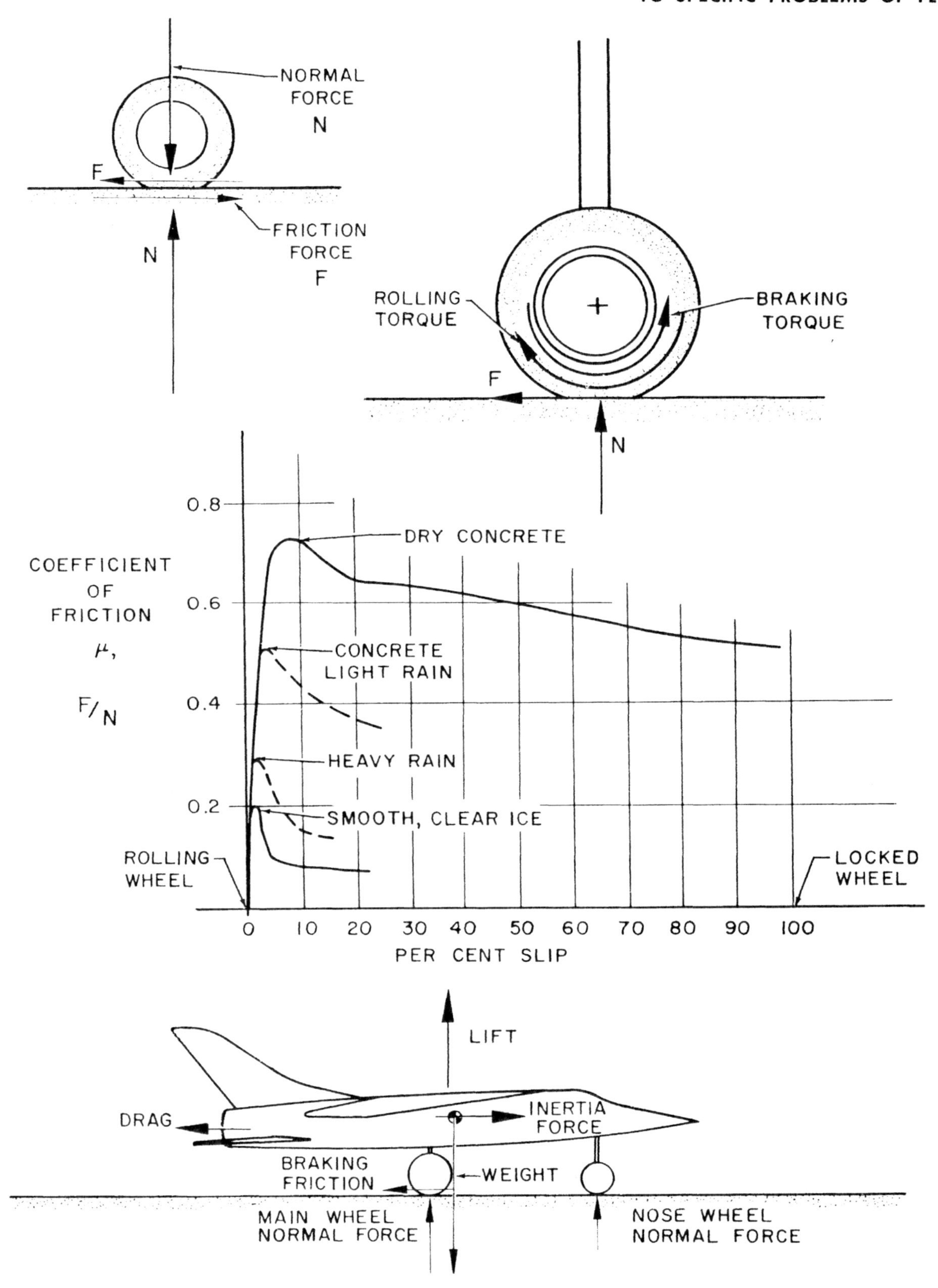

Figure 6.11. Braking Performance

BRAKING TECHNIQUE. It must be clearly distinguished that the techniques for minimum stop distance may differ greatly from the techniques required to minimize wear and tear on the tires and brakes. For the majority of airplane configurations, brakes will provide the most important source of deceleration for all but the most severe of icy runway conditions. Of course, aerodynamic drag is very durable and should be utilized to decelerate the airplane if the runway is long enough and the drag high enough. Aerodynamic drag will be of importance only for the initial 20 to 30 percent of speed reduction from the point of touchdown. At speeds less than 60 to 70 percent of the landing speed, aerodynamic drag is of little consequence and brakes will be the principal source of deceleration regardless of the runway surface. For the conditions of minimum landing distance, aerodynamic drag will be a principal source of deceleration *only* for the initial portion of landing roll for very high drag configurations on very poor runway conditions. These cases are quite limited so considerable importance must be assigned to proper use of the brakes to produce maximum effectiveness.

In order to provide the maximum possible retarding force, effort must be directed to produce the maximum normal force on the braking surfaces. (See figure 6.11.) The pilot will be able to influence the normal force on the braking surfaces during the initial part of the landing roll when dynamic pressure is large and aerodynamic forces and moments are of consequence. During this portion of the landing roll the pilot can control the airplane lift and the distribution of normal force to the landing gears.

First to consider is that any positive lift will support a part of the airplane weight and reduce the normal force on the landing gear. Of course, for the purposes of braking friction, it would be to advantage to create negative lift but this is not the usual capability of the airplane with the tricycle landing gear. Since the airplane lift may be considerable immediately after landing, retraction of flaps or extension of spoilers immediately after touchdown will reduce the wing lift and increase the normal force on the landing gear. With the retraction of flaps, the reduced drag is more than compensated for by the increased braking friction force afforded by the increased normal force on the braking surfaces.

A second possible factor to control braking effectiveness is the distribution of normal force to the landing gear surfaces. The nose wheel of the tricycle landing gear configuration usually has no brakes and any normal force distributed to this wheel is useful only for producing side force for control of the airplane. Under conditions of deceleration, the nose-down pitching moment created by the friction force and the inertia force tends to transfer a significant amount of normal force to nose wheel where it is unavailable to assist in creating friction force. For the instant after landing touchdown, the pilot may control this condition to some extent and regain or increase the normal force on the main wheels. After touchdown, the nose is lowered until the nose wheel contacts the runway then brakes are applied while the stick is eased back without lifting the nose wheel back off the runway. The effect is to minimize the normal force on the nose wheel and increase the normal force on braking surfaces. While the principal effect is to transfer normal force to the main wheels, there may be a significant increase in normal force due to a reduction in net lift, i.e., tail download is noticeable. This reduction in net lift tends to be particular to tailless or short coupled airplane configurations.

The combined effect of flap retraction and aft stick is a significant increase in braking friction force. Of course, the flaps should not be retracted while still airborne and aft stick should be used just enough without lifting the nosewheel off the runway. These techniques are to no avail if proper use of the brakes does not produce the maximum coefficient of friction. The incipient skid condition will produce the maximum coefficient of friction but this peak is difficult to recognize and maintain without an antiskid system. Judicious use of the brakes is necessary to obtain the peak coefficient of friction but not develop a skid or locked wheel which could cause tire failure, loss of control, or considerable reduction in the friction coefficient.

The capacity of the brakes must be sufficient to create adequate braking torque and produce the high coefficient of friction. In addition, the brakes must be capable of withstanding the heat generated without fading or losing effectiveness. The most critical requirements of the brakes occur during landing at the maximum allowable landing weight.

TYPICAL ERRORS OF BRAKING TECHNIQUE. Errors in braking technique are usually coincident with errors of other sorts. For example, if the pilot lands an airplane with excessive airspeed, poor braking technique could accompany the original error to produce an unsafe situation. One common error of of braking technique is the application of braking torque in excess of the maximum possible rolling torque. The result will be that the wheel decelerates and locks and the skid reduces the coefficient of friction, lowers the capability for side force, and enhances the possibility of tire failure. If maximum braking is necessary, caution must be used to modulate the braking torque to prevent locking the wheel and causing a skid. On the other hand, maximum coefficient of friction is obtained at the incipient skidding condition so sufficient brake torque must be applied to produce maximum friction force. Intermittent braking serves no useful purpose when the objective is maximum deceleration because the periods between brake application produce only slight or negligible cooling. Brake should be applied smoothly and braking torque modulated at or near the peak value to insure that skid does not develop.

One of the important factors affecting the landing roll distance is landing touchdown speed. Any excess velocity at landing causes a large increase in the minimum stop distance and it is necessary that the pilot control the landing precisely so to land at the appropriate speed. When landing on the dry, hard surface runway of adequate length, a tendency is to take advantage of any excess runway and allow the airplane to touchdown with excess speed. Of course, such errors in technique cannot be tolerated and the pilot must strive for precision in *all* landings. Immediately after touchdown, the airplane lift may be considerable and the normal force on the braking surfaces quite low. Thus, if excessive braking torque is applied, the wheel may lock easily at high speeds and tire failure may take place suddenly.

Landing on a wet or icy runway requires judicious use of the brakes because of the reduction in the maximum coefficient of friction. Because of reduction in the maximum attainable value of the coefficient of friction, the pilot must anticipate an increase in the minimum landing distance above that applicable for the dry runway conditions. When there is considerable water or ice on the runway, an increase in landing distance on the order of 40 to 100 percent must be expected for similar conditions of gross weight, density altitude, wind, etc. Unfortunately, the conditions likely to produce poor braking action also will cause high idle thrust of the turbojet engine and the extreme case (smooth, glazed ice or heavy rain) may dictate shutting down the engine to effect a reasonable stopping distance.

REFUSAL SPEEDS, LINE SPEEDS, AND CRITICAL FIELD LENGTH

During takeoff, it is necessary to monitor the performance of the airplane and evaluate the acceleration to insure that the airplane will achieve the takeoff speed in the specified distance. If it is apparent that the airplane is not accelerating normally or that the airplane or powerplant is not functioning properly, a decision must be made to refuse or continue takeoff. If the decision to refuse takeoff is made early in the takeoff roll, no problem exists because the airplane has not gained much speed and a large portion of runway distance is unused. However, at speeds near the takeoff speed, the airplane has used a large portion of the takeoff distance and the distance required to stop is appreciable. The problem which exists is to define the highest speed attained during takeoff acceleration from which the airplane may be decelerated to a stop on the runway length remaining, i.e., the "refusal speed."

The refusal speed will be a function of takeoff performance, stopping performance, and the length of available runway. The ideal situation would be to have a runway length which exceeds the total distance required to accelerate to the takeoff speed then decelerate from the takeoff speed. In this case, the refusal speed would exceed the takeoff speed and there would be little concern for the case of refused takeoff. While this may be the case for some instances, the usual case is that the runway length is less than the "accelerate-stop" distance and the refusal speed is less than the takeoff speed. A graphical representation of the refused takeoff condition is illustrated in figure 6.12 by a plot of velocity versus distance. At the beginning of the runway, the airplane starts accelerating and the variation of velocity and distance is defined by the takeoff acceleration profile. The deceleration profile describes the variation of velocity with distance where the airplane is brought to a stop at the end of the runway. The intersection of the acceleration and deceleration profiles then defines the refusal speed and the refusal distance along the runway. Of course, an allowance must be made for the time spent at the refusal speed as the power is reduced and braking action is initiated.

During takeoff, the airplane could be accelerated to any speed up to the refusal speed, then decelerated to a stop on the runway remaining. Once past the refusal speed, the airplane cannot be brought to a stop on the runway remaining and the airplane is committed to an unsafe stop. If takeoff is refused when above the refusal speed, the only hope is for assistance from the arresting gear, runway barrier, or an extensive overrun at the end of the runway. This fact points to the need for planning of the takeoff and the requirement to monitor the takeoff acceleration.

If the refusal speed data are not available, the following equations may be used to approximate the refusal speed and distance:

$$V_r = V_{to} \sqrt{\frac{Ra}{S_{to} + \left[S_L \times \left(\frac{V_{to}}{V_L}\right)^2\right]}}$$

$$S_r = S_{to} \left(\frac{V_r}{V_{to}}\right)^2$$

where

V_r = refusal speed
S_r = refusal distance

and for the appropriate takeoff configuration,

V_{to} = takeoff speed
S_{to} = takeoff distance
V_L = landing speed
S_L = landing distance
R_a = runway length available

These approximate relationships do not account for the time spent at the refusal point and must not be used in lieu of accurate handbook data.

In the case of the single-engine airplane, the pilot must monitor the takeoff performance to recognize malfunctions or lack of adequate acceleration prior to reaching the refusal speed. Obviously, it is to advantage to recognize the

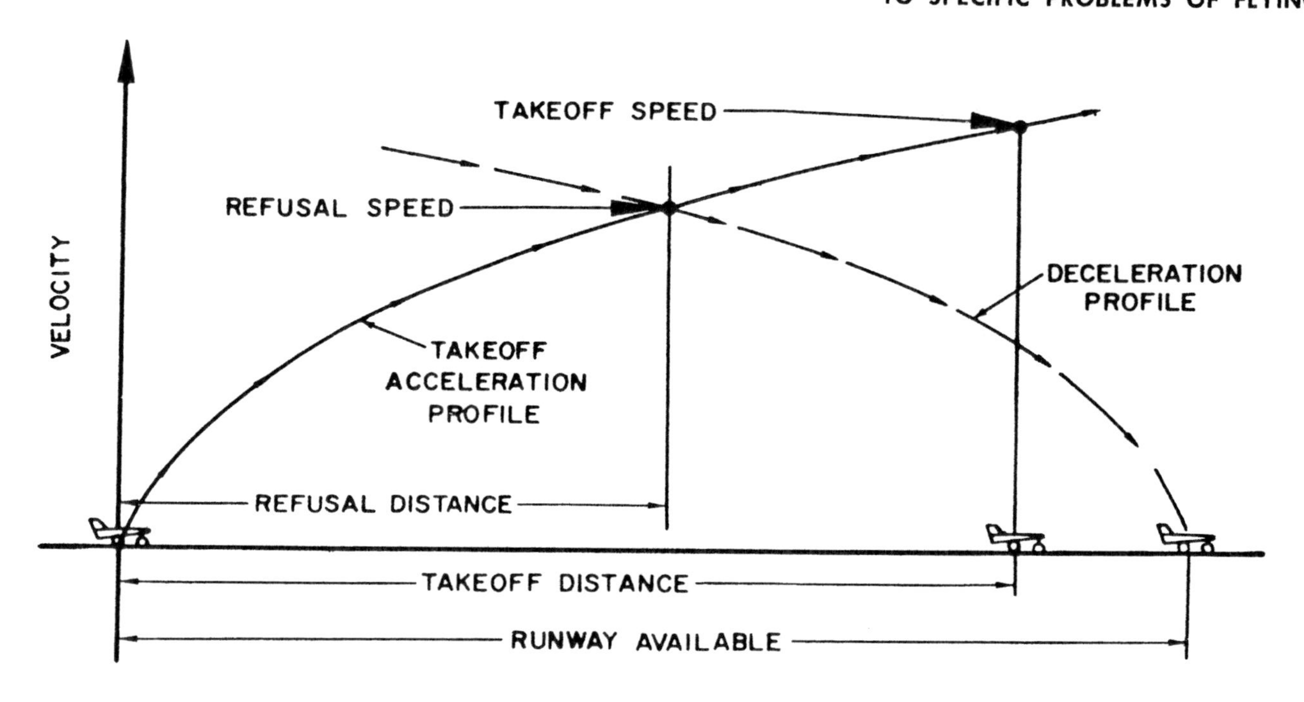

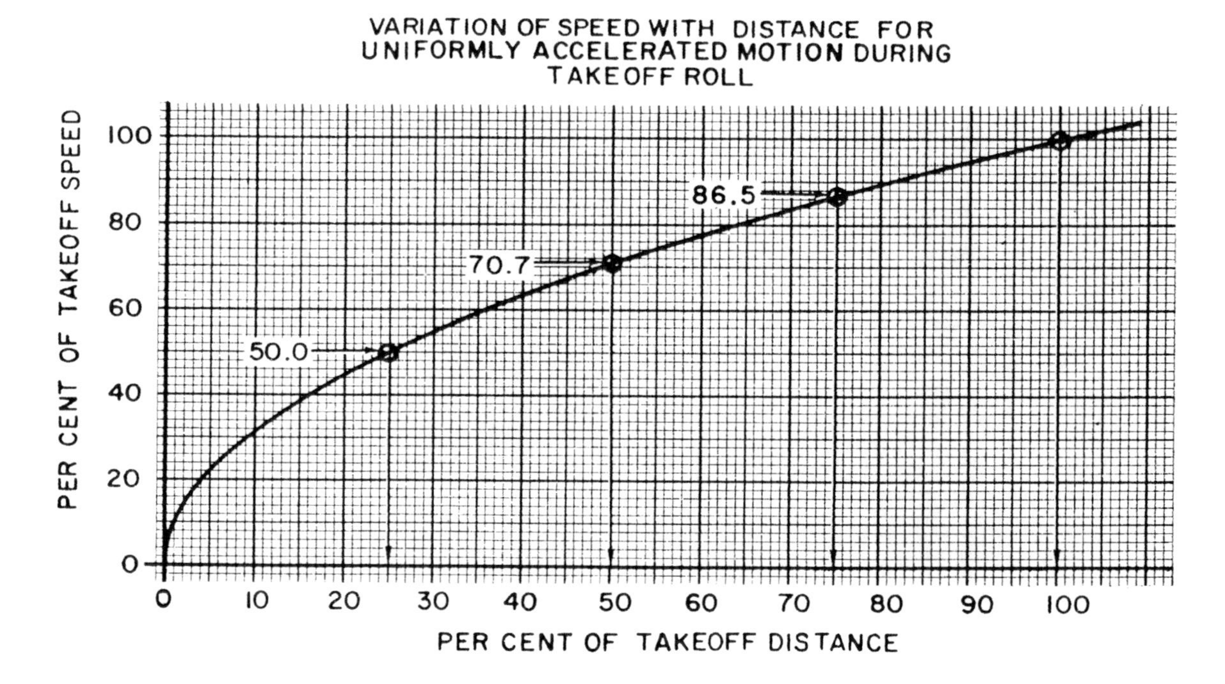

Figure 6.12. Refused Takeoff and Takeoff Velocity Variation

possibility of a refused takeoff before exceeding the refusal speed. To this end, the pilot must carefully evaluate airplane and powerplant performance and judge the acceleration of the airplane by the use of "line speeds." The accelerated motion of the airplane during takeoff roll will define certain relationships between velocity and distance when the acceleration of the airplane is normal. By comparison of predicted and actual speeds at various points along the runway, the pilot can evaluate the acceleration and assess the takeoff performance.

An example of an acceleration profile is shown by the second illustration of figure 6.12, where the variation of velocity and distance is defined for the case of uniformly accelerated motion, i.e., constant acceleration. While the case of uniformly accelerated motion does not correspond exactly to the takeoff performance of all airplanes, it is sufficiently applicable to illustrate the principle of line speeds and acceleration checks. If the takeoff acceleration of the airplane were constant, the airplane would develop specific percentages of the takeoff speed at specific percentages of the takeoff distance. Representative values from figure 6.12 are as follows:

Percent of takeoff distance	*Percent of takeoff velocity*	*Percent of takeoff time*
0	0	0
25	50.0	50.0
50	70.7	70.7
75	86.5	86.5
100	100	100

As an example of this uniformly accelerated motion, the airplane upon reaching the halfway point of takeoff roll would have spent 70.7 percent of the total takeoff time and accelerated to 70.7 percent of the takeoff speed. If the airplane has not reached a specific speed at a specific distance, it is obvious that the acceleration is below the predicted value and the airplane surely will not achieve the takeoff speed in the specified takeoff distance. Therefore, properly computed line speeds at various points along the runway will allow the pilot to monitor the takeoff performance and recognize a deficiency of acceleration. Of course, a deficiency of acceleration must be recognized prior to reaching some point along the runway where takeoff cannot be safely achieved or refused.

The fundamental principles of refusal speeds and line speeds are applicable equally well to single-engine and multiengine airplanes. However, in the case of the multiengine airplane additional consideration must be given to the decision to continue or refuse takeoff when engine failure occurs during the takeoff roll. If failure of one engine occurs prior to reaching the refusal speed, takeoff should be discontinued and the airplane brought to a stop on the remaining runway. If failure of one engine occurs after exceeding the refusal speed, the airplane is committed to continue takeoff with the remaining engines operative or an unsafe refused takeoff. In some cases, the remaining runway may not be sufficient to allow acceleration to the takeoff speed and the airplane can neither takeoff or stop on the runway remaining. To facilitate consideration of this problem, several specific definitions are necessary.

(1) *Takeoff and initial climb speed:* A speed, usually a fixed percentage above the stall speed, at which the airplane will become airborne and best clear obstacles immediately after takeoff. For a particular airplane in the takeoff configuration, this speed (in *EAS* or *CAS*) is a function of gross weight but in no circumstances should it be less than the minimum directional control speed for the critical asymmetrical power condition. Generally, the takeoff and initial climb speed is referred to as the "V_2" speed.

(2) *Critical engine failure speed:* A speed achieved during the takeoff roll at which failure of one engine will require the same distance to continue accelerating with the operative engines to accomplish safe takeoff *or* refuse takeoff and decelerate to a stop utilizing the airplane brakes. At critical engine failure

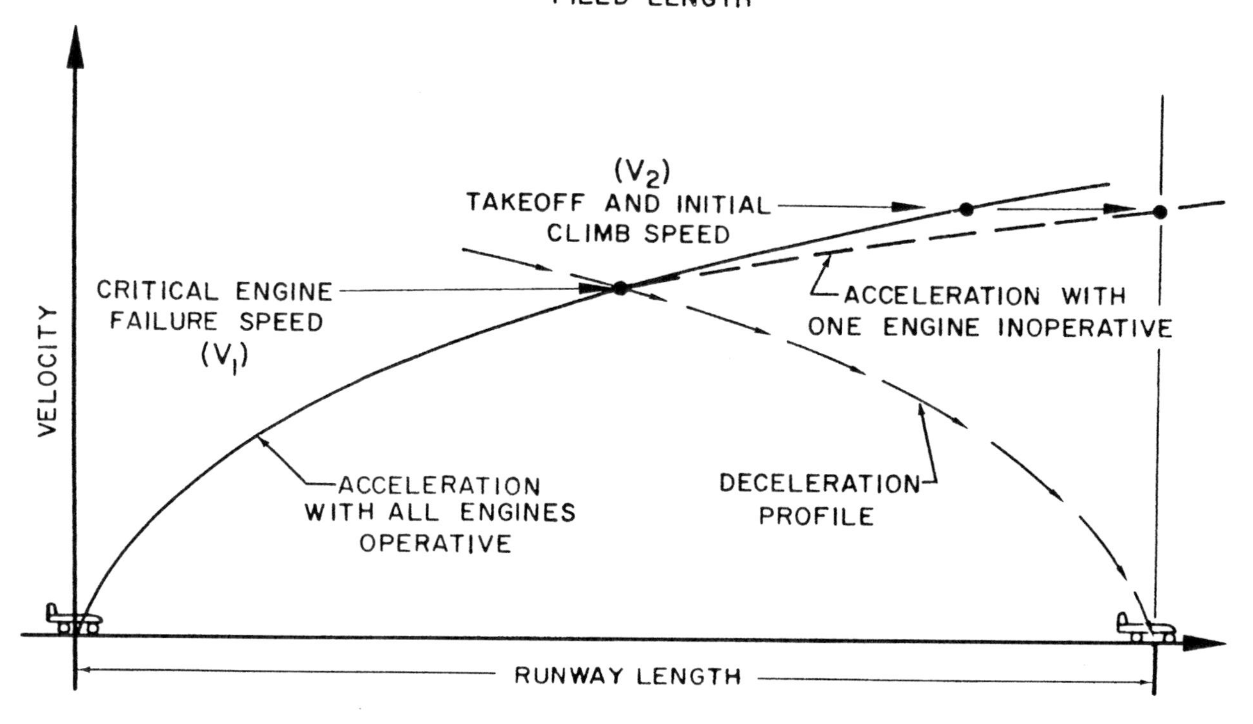

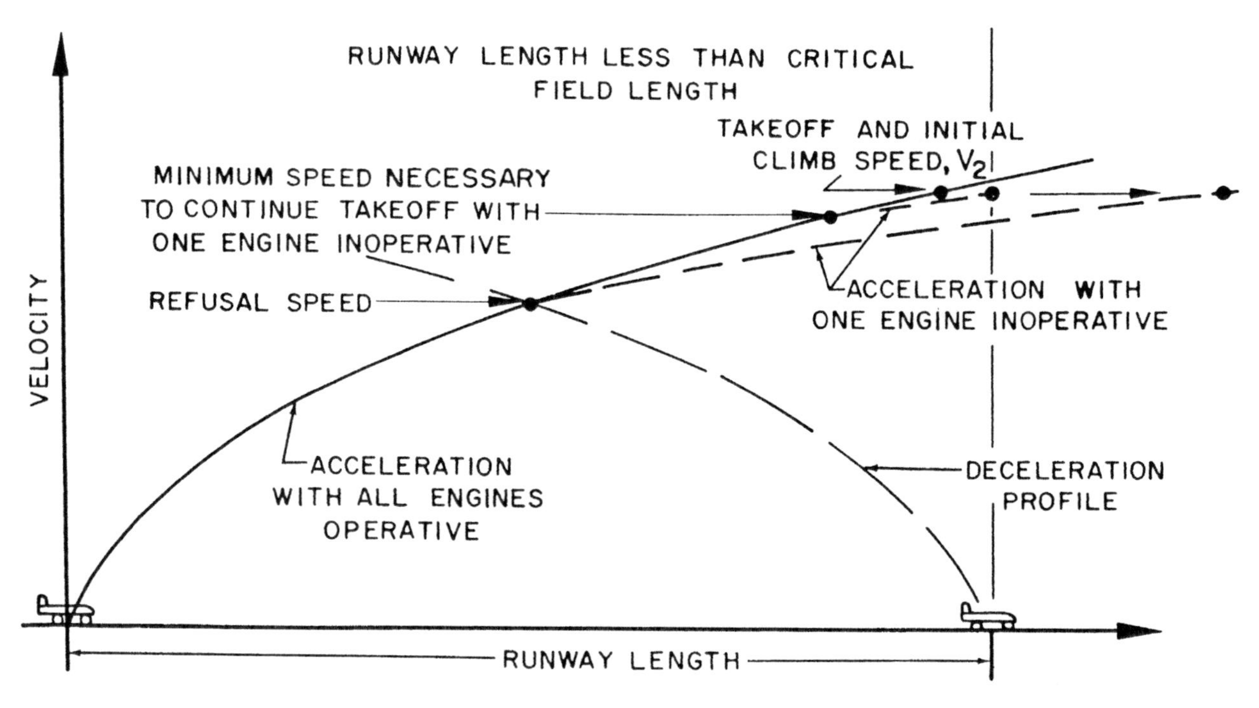

Figure 6.13. Critical Field Length

speed, the distance necessary to continue takeoff with one engine inoperative is equal to the stopping distance. The critical engine failure speed is generally referred to as the "V_1" speed and it is a function of the same factors which determine the takeoff performance, e.g., density altitude, gross weight, temperature, humidity, etc.

(3) *Critical field length:* The runway length necessary to accelerate with all engines operative to the critical engine failure speed (V_1) then continue accelerating to the takeoff and initial climb speed (V_2) with one engine inoperative and achieve safe takeoff or refuse takeoff. By this definition, critical field length describes the minimum length of runway necessary for safe operation of the multiengine airplane. Obviously, the critical field length is a function of the same factors affecting the takeoff distance of the airplane.

The conditions of V_1, V_2, and critical field length are illustrated by figure 6.13. The first illustration of figure 6.13 depicts the case where the *runway length is equal to the critical field length.* In this case, the airplane could accelerate to V_1 with all engines operative then either continue takeoff safely with one engine inoperative or refuse takeoff and decelerate to a stop on the remaining runway. For this condition, an engine failure occurring at *less* than V_1 speed dictates that takeoff must be refused because inadequate distance remains to effect a safe takeoff at V_2 speed. However, at or below V_1 speed, adequate distance remains to bring the airplane to a stop. If engine failure occurs at some speed *greater* than V_1 speed, takeoff should be continued because adequate distance remains to accelerate to V_2 speed and effect a safe takeoff with one engine inoperative. If engine failure occurs beyond V_1 speed, inadequate distance remains to brake the airplane to a stop on the runway.

The second illustration of figure 6.13 depicts the case where the *runway length is less than the critical field length.* In this case, the term of "V_1" speed is not applicable because of inadequate distance and the refusal speed is less than the minimum speed necessary to continue a safe takeoff with one engine inoperative. If engine failure occurs below refusal speed, the takeoff must be refused and adequate distance remains to effect a stop on the runway. If engine failure occurs above refusal speed but below the minimum speed necessary to continue takeoff with one engine inoperative, an accident is inevitable. Within this range of speeds, the airplane cannot effect a safe takeoff at V_2 with one engine inoperative or a safe stop on the remaining runway. For this reason, the pilot must properly plan the takeoff and insure that the runway available is equal to or greater than the critical field length. If the runway available is less than the critical field length, there must be sufficient justification for the particular operation because of the hazardous consequences of engine failure between the refusal speed and the minimum speed necessary to continue takeoff with one engine inoperative. Otherwise, the gross weight of the airplane should be reduced in attempt to decrease the critical field length to equal the available runway.

SONIC "BOOMS"

From the standpoint of public relations and the maintaining of friendly public support for Naval Aviation, great care must be taken to prevent sonic booms in populated areas. While the ordinary sonic boom does not carry any potential of physical damage, the disturbance must be avoided because of the undesirable annoyance and apprehension. As supersonic flight becomes more commonplace and an ordinary consequence of flying operations, the prevention of sonic booms in populated areas becomes a difficult and perplexing job.

When the airplane is in supersonic flight, the local pressure and velocity changes on the airplane surfaces are coincident with the formation of shock waves. The pressure jump through the shock waves in the immediate vicinity of the airplane surfaces is determined by the local flow changes at these surfaces. Of course, the strength of the shock waves and the pressure jump through the wave decreases rapidly with distance away from the airplane. While the pressure jump through the shock wave decreases with distance away from the surface, it does not disappear completely and a measurable—but very small—pressure will exist at a considerable distance from the airplane.

Sound is transmitted through the air as a series of very weak pressure waves. In the ordinary range of audible frequencies, the threshold of audibility for intensity of sound is for pressure waves with an approximate R.M.S. value of pressure as low as 0.0000002 psf. Within this same range of frequencies, the threshold of feeling for intensity of sound is for pressure values with an approximate R.M.S. value of pressure of 0.2 to 0.5 psf. Continuous sound at the threshold of feeling is of the intensity to cause painful hearing. Thus, the shock waves generated by an airplane in supersonic flight are capable of creating audible sound and, in the extreme case, can be of a magnitude to cause considerable disturbance. Pressure jumps of 0.02 to 0.3 psf have been recorded during the passage of an airplane in supersonic flight. As a result, the sonic "booms" are the pressure waves generated by the shock waves formed on the airplane in supersonic flight.

The source of sonic booms is illustrated by figure 6.14. When the airplane is in level supersonic flight, a pattern of shock waves is developed which is much dependent on the configuration and flight Mach number of the airplane. At a considerable distance from the airplane, these shock waves tend to combine along two common fronts and extend away from the airplane in a sort of conical surface. The waves decrease in strength with distance away from the airplane but the pressure jump remains of an audible intensity for a considerable distance from the airplane. If the wave extends to the ground or water surface, it will be reflected and attenuated to some extent depending on the character of the reflecting surface. Of course, if this attached wave form is carried across a populated area at the surface, the population will experience the pressure waves as a sonic boom.

The intensity of the boom will depend on many different factors. The characteristics of the airplane generating the shock waves will be of some importance since a large, high drag, high gross weight airplane in flight at high Mach number will be transferring a greater energy to the air mass. Flight altitude will have an important bearing on boom intensity since at high altitude the pressure jump across a given wave form is much less. In addition, at high altitude a greater distance exists between the generating source of the pressure disturbance and the ground level and the strength of the wave will have a greater distance in which to decay. The ordinary variation of temperature and density plus the natural turbulence of atmosphere will tend to reflect or dissipate the shock wave generated at high altitude. However, in a stable, quiescent atmosphere, the pressure wave from the airplane in high supersonic flight at high altitude may be of an audible magnitude at lateral distances as great as 10 to 30 miles. Thus, supersonic flight over or adjacent to populated areas will produce a sonic boom.

Actually, it is not necessary for an airplane to fly supersonic over or adjacent to a populated area to create a sonic boom. This possibility is shown by the second illustration of figure 6.14 where an airplane decelerates to subsonic from a supersonic dive. As the airplane slows to subsonic from supersonic speed, the airplane will release the leading bow and tail waves which formed as the airplane accelerated from subsonic to supersonic speed. The release of these shock waves is analogous to the case where a surface ship slows to below the wave propagation speed and releases the bow wave which then travels out ahead of

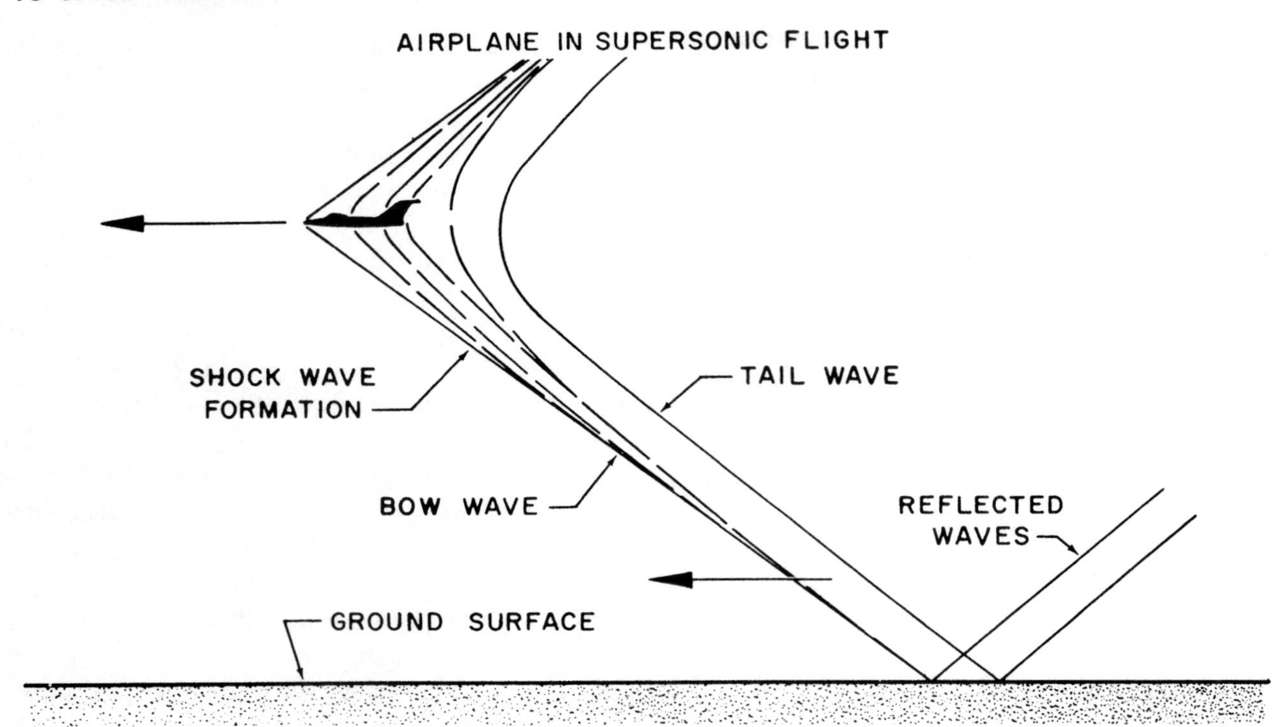

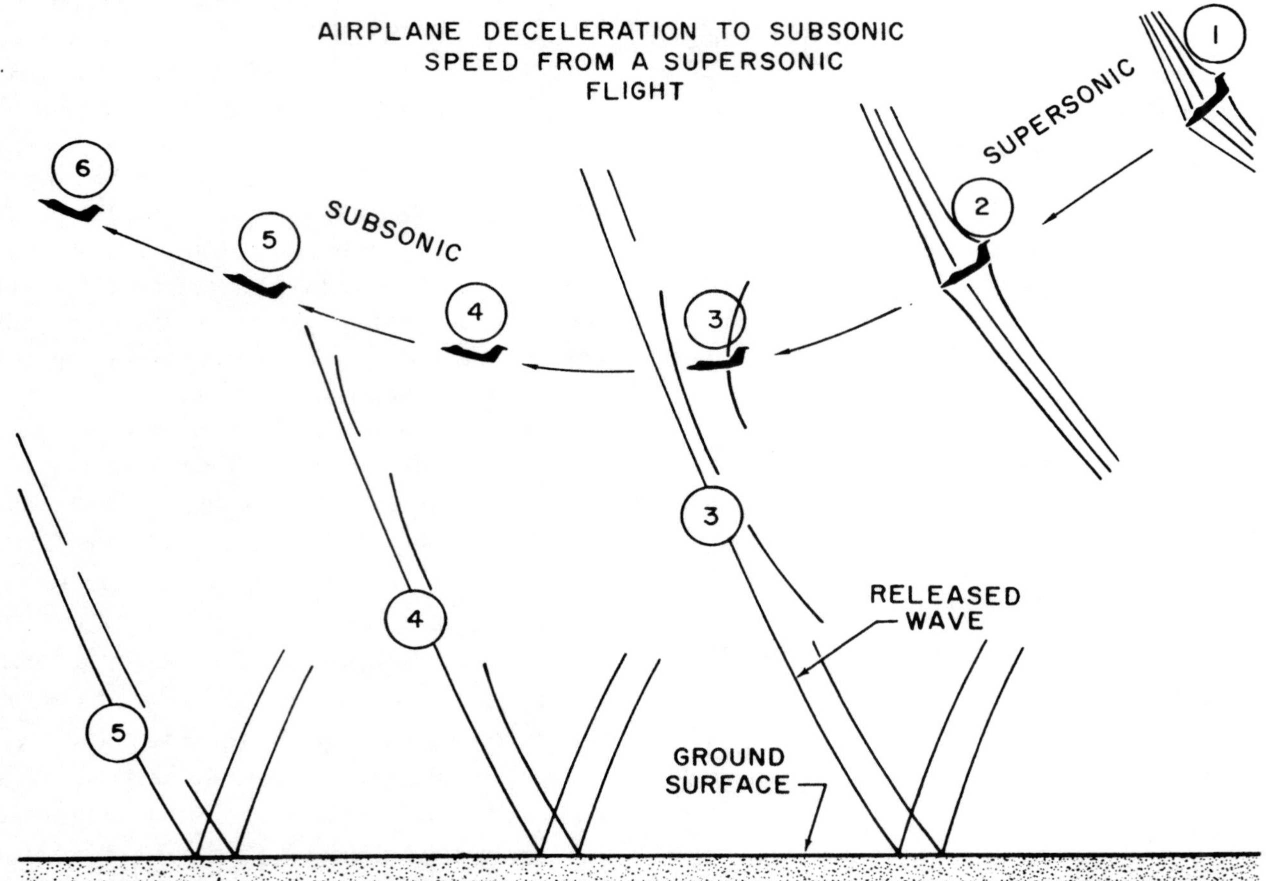

Figure 6.14. Sonic Booms

the ship. When the airplane slows to subsonic, the shock wave travels out ahead of the airplane in a form which is somewhat spherical. Because there are density variations through the shock wave, the shock wave moving ahead of the airplane can cause aberrations in light waves and it may appear to the pilot as if a large sheet of clear cellulose or plastic were in front of the airplane. In addition, the density variation and initial shape of the wave leaving the airplane may cause reflection of sunlight which would appear as a sudden, brilliant "flash" to the pilot.

Of course, the wave released by decelerating to subsonic speed can travel out ahead of the airplane and traverse a populated area to cause a sonic boom. The initial direction of the released wave will be the flight path of the airplane at the instant it decelerates to subsonic speed. To be sure, the released wave should *not* be aimed in the direction of a populated area, even if a considerable distance away. There are instances where a released wave has been of an audible magnitude as far as 30 to 40 miles ahead of the point of release. The released pressure wave will be of greatest intensity when created by a large, high drag configuration at low altitude. Since the wave intensity decreases rapidly with distance away from the source, the boom will be of strongest audibility near the point of release.

It should become apparent that sonic booms are a byproduct of supersonic aviation and, with supersonic flight becoming more commonplace, the problem is more perplexing. The potential of sonic booms is mostly of the audible nature and nuisance of the disturbance. The damage potential of the ordinary sonic boom is quite small and the principal effects are confined to structures which are extremely brittle, low strength, and have characteristic high residual stresses. In other words, only the extremes of pressure waves generated by airplanes in flight could possibly cause cracked plaster and window glass. Such materials are quite prone to sharp dynamic stresses and, when superimposed on the high residual stresses common to the products and building construction, slight but insignificant damage may result. Actually, the most objectionable feature of the sonic boom is the audibility and the anxiety or apprehension caused by the sharp, loud noise which resembles a blast.

The pressure jump through the shock waves in the immediate vicinity of the airplane is much greater than those common to the audible "booms" at ground level. Thus, airplanes in close formation at supersonic speeds may encounter considerable interference between airplanes. In addition, to eliminate even the most remote possibility of structural damage, a high speed airplane should not make a supersonic pass close to a large airplane which may have low limit load factor and be prone to be easily disturbed or damaged by a strong pressure wave.

HELICOPTER PROBLEMS

The main difference between helicopter and an airplane is the main source of lift. The airplane derives its lift from a fixed airfoil surface while the helicopter derives lift from a rotating airfoil called the rotor. Hence, the aircraft will be classified as either "fixed-wing" or "rotating wing." The word "helicopter" is derived from the Greek words meaning "helical wing" or "rotating wing."

Lift generation by a "rotating wing" enables the helicopter to accomplish its unique mission of hovering motionless in the air, taking off and landing in a confined or restricted area, and autorotating to a safe landing following a power failure. Lift generation by "rotating wing" is also responsible for some of the unusual problems the helicopter can encounter. Since the helicopter problems are due to particular nature of the rotor aerodynamics, the basic flow conditions within the rotor must be considered in detail. For simplicity, the initial discussion will consider only the hovering rotor. Although the term hovering usually means remaining over a particular spot on the ground, it shall be considered here as flight at zero airspeed. This is necessary because the aerodynamic characteristics of the rotor depend on its motion with respect to the air and not the ground. Hovering in a 20 knot wind is aerodynamically equivalent to flying at an airspeed of 20 knots in a no-wind condition, and the characteristics will be identical in the two conditions.

The first point to realize is that the rotor is subject to the same physical laws of aerodynamics and motion that govern flight of the fixed-wing airplane. The manner in which the rotor is subject to these laws is much more complicated due to the complex flow conditions.

Rotor lift can be explained by either of two methods. The first method, utilizing simple momentum theory based on Newton's Laws, merely states that lift results from the rotor accelerating a mass of air downward in the same way that the jet engine develops thrust by accelerating a mass of air out the tailpipe. The second method of viewing rotor lift concerns the pressure forces acting on the various sections of the blade from root to tip. The simple momentum theory is useful in determining only lift characteristics while the "blade element" theory gives drag as well as lift characteristics and is useful in giving a picture of the forces at work on the rotor. In the "blade element" theory, the blade is divided up into "blade elements" as shown in figure 6.15. The forces acting on each blade element are analyzed. Then the forces on all elements are summed up to give the characteristics of the whole rotor. The relative wind acting on each segment is the resultant of two velocity components: (1) the velocity due to the rotation of the blades about the hub and (2) the induced velocity, or downwash velocity caused by the rotor. the velocity due to rotation at a particular element is proportional to the rotor speed and the distance of the element from the rotor hub. Thus, the velocity due to rotation varies linearly from zero at the hub to a maximum at the tip. A typical blade section with the forces acting on it is shown in figure 6.15.

A summation of the forces acting perpendicular to the plane of rotation (tip path plane) will determine the rotor thrust (or lift) characteristics while summation of the moments resulting from forces acting in the plane of rotation will determine the rotor torque characteristics. As a result of this analysis, the rotor thrust (or lift) is found to be proportional to the air density, a nondimensional thrust coefficient, and the square of the tip speed, or linear speed of the tip of the blade. The thrust coefficient is a function of the average blade section lift coefficient and the rotor solidity, which is the proportion of blade area to disc area. The lift coefficient is identical to that used in airplane aerodynamics while the solidity is analagous to the aspect ratio in airplane aerodynamics. The rotor torque is found to be proportional to a nondimensional torque coefficient, the air density, the disc area, the square of the tip speed, and the blade radius. The torque coefficient is dependent upon the average profile drag coefficient of the blades, the blade pitch angle, and the average lift coefficient of the blades. The torque can be thought to result from components of profile and induced drag forces acting on the blades, similar to those on an airplane.

As in the airplane, there is one angle of attack or blade pitch condition that will result in the most efficient operation Unfortunately, the typical helicopter rotor operates at a near constant RPM and thus a constant true airspeed and cannot operate at this most efficient condition over a wide range of altitude and gross weight as the fixed-wing airplane. The airplane is able to maintain an efficient angle of attack at various altitudes and gross weights by flying at various airspeeds but the helicopter will operate with a near constant rotor velocity and vary blade angle to contend with variations in altitude and gross weight.

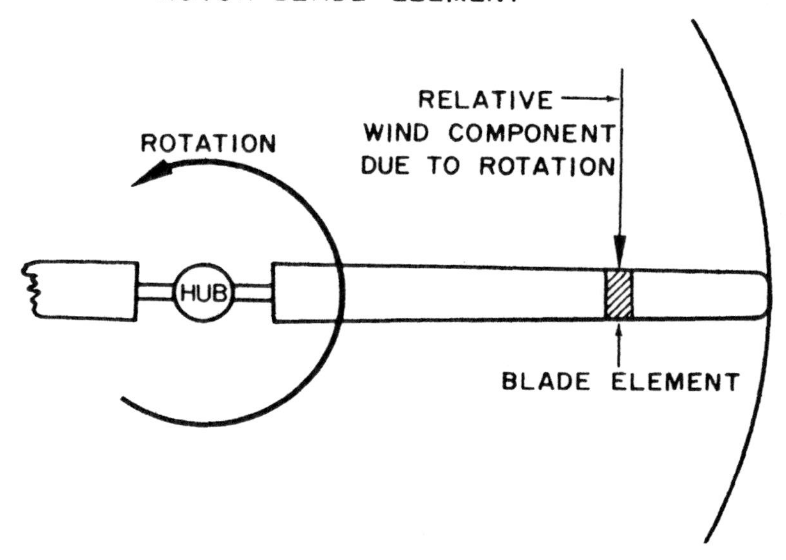

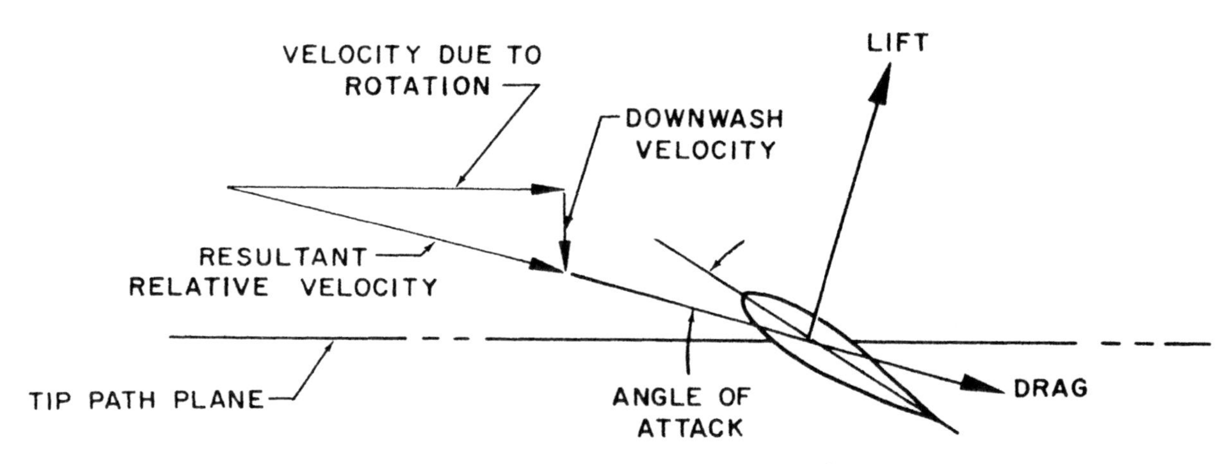

Figure 6.15. Rotor Blade Element Aerodynamics

If the rotor could operate within a wide range of rotor speed, the efficiency and performance could be improved.

With the previous relationships established for the rotor in hovering flight, the effect of forward flight or rotor translation can be considered. With forward flight, a third velocity component, that of the forward velocity of the helicopter, must be considered in determining the relative wind acting on each rotor blade element. Since the entire rotor moves with the helicopter, the velocity of air passing over each of the elements on the advancing blade is increased by the forward speed of the helicopter and the velocity of the air passing over each element of the retreating blade is decreased by the same amount. This is shown in figure 6.16.

If the blade angles of attack on both advancing and retreating blades remained the same as in hovering flight, the higher velocity on the advancing blade would cause a dissymmetry of lift and the helicopter would tend to roll to the left. It was this effect that created great difficulty during many early helicopter and autogiro projects. Juan De La Cierva was the first to realize what caused this effect and he solved the problem by mounting his autogiro blades individually on flapping hinges, thus allowing a flapping action to automatically correct the dissymmetry of lift that resulted from forward flight. This is the method still used in an articulated rotor system today. The see-saw, or semi-rigid, rotor corrects the lift dissymmetry by rocking the entire hub and blades about a gimbal joint. By rocking the entire rotor system forward, the angle of attack on the advancing blade is reduced and the angle of attack on the retreating blade is increased. The rigid rotor must produce cyclic variation of the blade pitch mechanically as the blade rotates to eliminate the lift dissymmetry. Irrespective of the method used to correct the dissymmetry of lift, identical aerodynamic characteristics result. Thus, what is said about rotor aerodynamics is equally valid for all types of rotor systems.

By analyzing the velocity components acting on the rotor blade sections from the blade root to the tip on both advancing and retreating blades, a large variation of blade section angle of attack is found. Figure 6.16 illustrates a typical variation of the local blade angle of attack for various spanwise positions along the advancing and retreating blades of a rotor at high forward speed. There is a region of positive angles of attack resulting in positive lift over the entire advancing blade. Immediately next to the hub of the retreating blade there is an area of reversed flow where the velocity due to the forward motion of the helicopter is greater than the rearward velocity due to the blade rotation. The next area is a negative stall region where, although the flow is in the proper direction relative the blade, the angle of attack exceeds that for negative stall. Progressing out the retreating blade, the blade angle of attack becomes less negative, resulting in an area of negative lift. Then the blade angle becomes positive again, resulting in a positive lift region. The blade angle continues to increase until near the tip of the retreating blade the positive stall angle of attack is exceeded, resulting in stalling of the tip section. This wide variation in blade section angles of attack results in a large variation in blade section lift and drag coefficients. The overall lift force on the left and right sides of the rotor disc are equalized by cyclically varying the blade pitch as explained previously, but the drag variation is not eliminated. This drag variation causes a shaking force on the rotor system and contributes to the vibration of the helicopter.

RETREATING BLADE STALL. Retreating blade stall results whenever the angle of attack of the blade exceeds the stall angle of attack of the blade section. This condition occurs in high speed flight at the tip of the retreating blade since, in order to develop the same lift as the advancing blade, the retreating

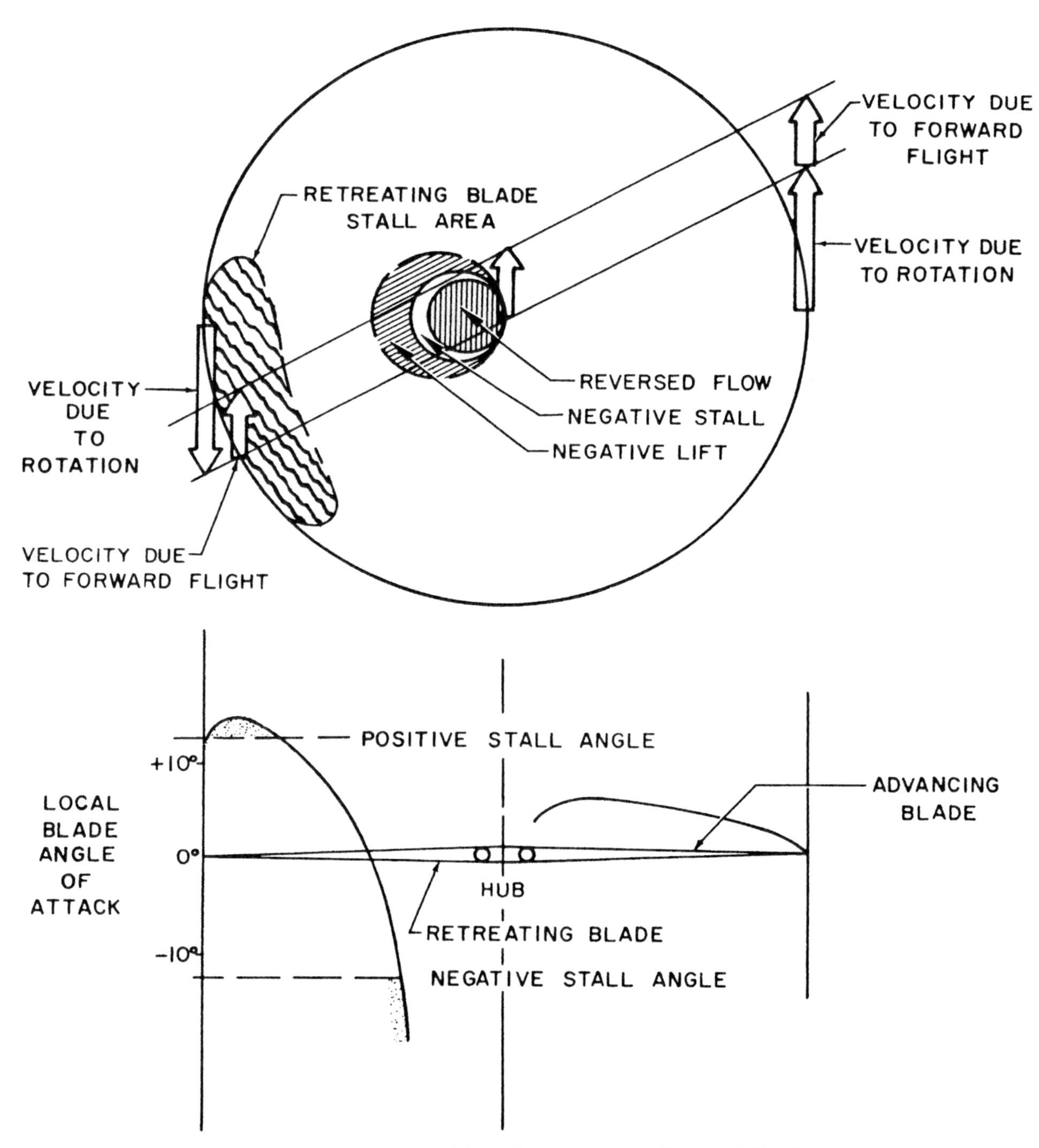

Figure 6.16. Rotor Flow Conditions in Forward Flight

blade must operate at a greater angle of attack. If the blade pitch is increased or the forward speed increased the stalled portion of the rotor disc becomes larger with the stall progressing in toward the hub from the tip of the retreating blade. When approximately 15 percent of the rotor disc is stalled, control of the helicopter will be impossible. Flight tests have determined that control becomes marginal and the stall is considered severe when the outer one-quarter of the retreating blade is stalled. Retreating blade stall can be recognized by rotor roughness, erratic stick forces, a vibration and stick shake with a frequency determined by the number of blades and the rotor speed. Each of the blades of a three-bladed rotor will stall as it passes through the stall region and create a vibration with three beats per rotor revolution. Other evidence of retreating blade stall is partial or complete loss of control or a pitch-up tendency which can be uncontrollable if the stall is severe.

Conditions favorable for the occurrence of retreating blade stall are those conditions that result in high retreating blade angles of attack. Each of the following conditions results in a higher angle of attack on the retreating blade and may contribute to retreating blade stall:

1. High airspeed
2. Low rotor RPM—operation at low rotor RPM necessitates the use of higher blade pitch to get a given thrust from the rotor, thus a higher angle of attack
3. High gross weight
4. High density altitude
5. Accelerated flight, high load factor
6. Flight through turbulent air or gusts—sharp updrafts result in temporary increase in blade angle of attack
7. Excessive or abrupt control deflections during maneuvers

Recovery from a stalled condition can be effected only by decreasing the blade angle of attack below the stall angle. This can be accomplished by one or a combination of the following items depending on severity of the stall:

1. Decrease collective pitch
2. Decrease airspeed
3. Increase rotor RPM
4. Decrease severity of accelerated maneuver or control deflection

If the stall is severe enough to result in pitch-up, forward cyclic to attempt to control pitch-up is ineffective and may aggravate the stall since forward cyclic results in an increase in blade angle of attack on the retreating blade. The helicopter will automatically recover from a severe stall since the airspeed is decreased in the nose high attitude but recovery can be assisted by gradual reduction in collective pitch, increasing RPM, and leveling the helicopter with pedal and cyclic stick.

From the previous discussion, it is apparent that there is some degree of retreating blade stall even at moderate airspeeds. However, the helicopter is able to perform satisfactorily until a sufficiently large area of the rotor disc is stalled. Adequate warning of the impending stall is present when the stall condition is approached slowly. There is inadequate warning of the stall *only* when the blade pitch or blade angle of attack is increased rapidly. Therefore, unintentional severe stall is most likely to occur during abrupt control motions or rapid accelerated maneuvers.

COMPRESSIBILITY EFFECTS. The highest relative velocities occur at the tip of the advancing blade since the speed of the helicopter is added to the speed due to rotation at this point. When the Mach number of the tip section of the advancing blade exceeds the critical Mach number for the rotor blade section, compressibility effects result. The critical Mach number is reduced by thick, highly cambered airfoils and critical Mach number decreases with increased lift coefficient. Most helicopter blades have symmetrical sections and therefore have relatively high critical Mach numbers at low lift coefficients. Since the principal effects of compressibility are the large increase in drag and rearward shift of the airfoil aerodynamic center, compressibility effects on the helicopter increase the power required to maintain rotor RPM and cause rotor roughness, vibration, stick shake, and an undesirable structural twisting of the blade.

Since compressibility effects become more severe at higher lift coefficients (higher blade angles of attack) and higher Mach numbers, the following operating conditions represent the most adverse conditions from the standpoint of compressibility:

1. High airspeed
2. High rotor RPM
3. High gross weight
4. High density altitude
5. Low temperature—the speed of sound is proportional to the square root of the absolute temperature. Therefore, sonic velocity will be more easily obtained at low temperatures when the sonic speed is lower.
6. Turbulent air—sharp gusts momentarily increase the blade angle of attack and thus lower the critical Mach number to the point where compressibility effects may be encountered on the blade.

Compressibility effects will vanish by decreasing the blade pitch. The similarities in the critical conditions for retreating blade stall and compressibility should be noticed but one basic difference must be appreciated—compressibility occurs at HIGH RPM while retreating blade stall occurs at LOW RPM. Recovery technique is identical for both with the exception of RPM control.

AUTOROTATION CHARACTERISTICS. One of the unique characteristics of helicopters is their ability to take part of the energy of the airstream to keep the rotor turning and glide down to a landing with no power. Consideration of the rotor during a vertical autorotation will provide an understanding of why the rotor continues to rotate without power. During autorotation, the flow of air is upward through the rotor disc and there is a vertical velocity component equal to the rate of descent of the helicopter. In addition, there is a velocity component due to rotation of the rotor. The vector sum of these two velocities is the relative wind for the blade element. The forces resulting from the relative wind on each particular blade section will provide the reason why the rotor will continue to operate without power. First, consider a blade element near the tip of the blade as illustrated in figure 6.17. At this point there is a lift force acting perpendicular to the relative wind and a drag force acting parallel to the relative wind through the aerodynamic center. Since the rotation of the rotor is affected only by forces acting in the plane of rotation, the important forces are components of the lift and drag force in the plane of rotation. In this low angle of attack high speed tip section, the net in-plane force is a drag force which would tend to retard the rotor. Next, consider a blade section at about the half-span position as illustrated in figure 6.17. In this case, the same forces are present, but the in-plane component of lift force is greater than the drag force and this results in a net thrust or forward force in the plane of rotation which tends to drive the rotor.

During a steady autorotation, there is a balance of torque from the forces along the blade so that the RPM is maintained in equilibrium at some particular value. The region of the rotor disc where there is a net drag force on the blade is called the "propeller region" and the region of the rotor disc where there is a net in-plane thrust force is called the "autorotation region." These regions are shown for vertical autorotation and forward speed (or normal) autorotation in figure 6.17. Forces acting on the rotor blades in forward flight autorotation are similar to those in vertical autorotation but the difference will consist mainly of shifts of the autorotation region to the left and the addition of reverse

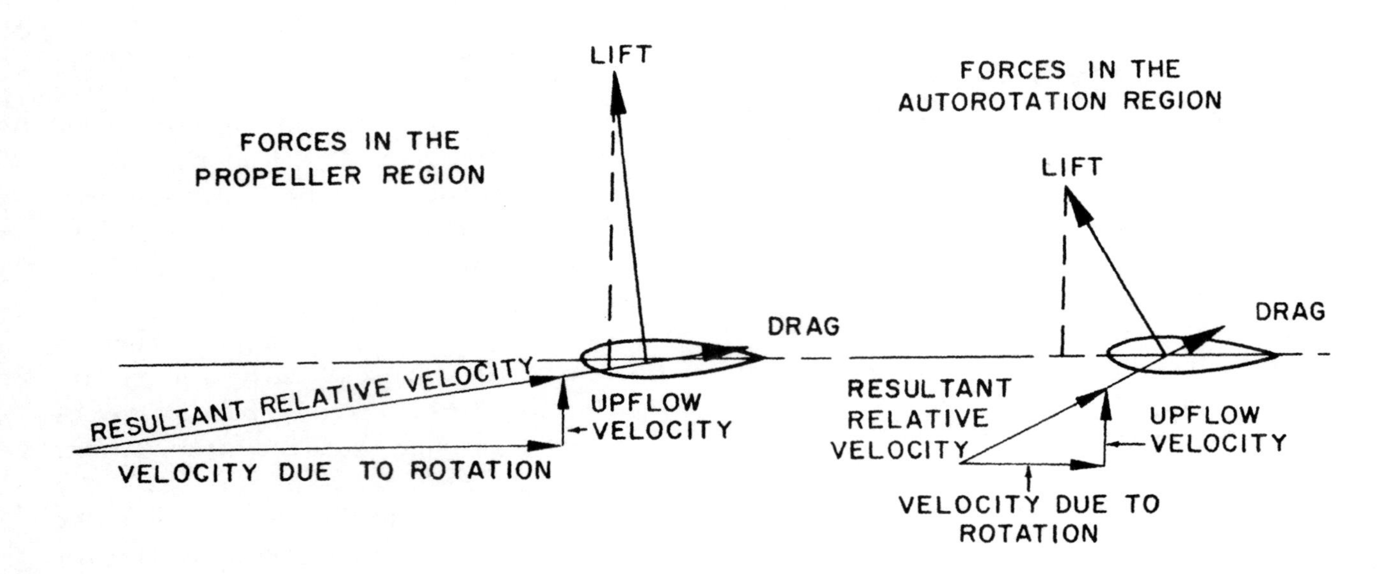

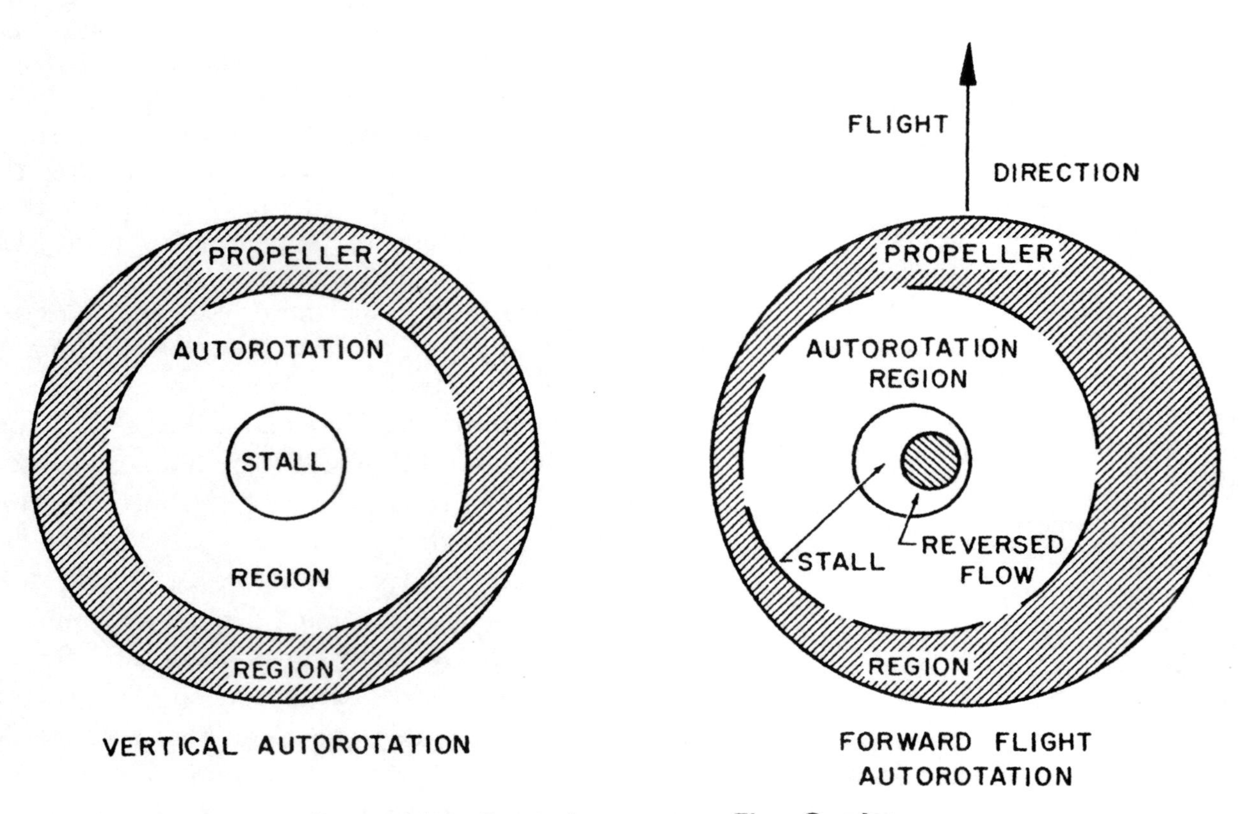

Figure 6.17. Rotor Autorotation Flow Conditions

flow and negative stall regions similar to the powered flight condition.

Autorotation is essentially a stable flight condition. If external disturbances cause the rotor to slow down, the autorotation region of the disc automatically expands to restore the rotor speed to the original equilibrium condition. On the other hand, if an external disturbance causes the rotor to speed up, the propeller region automatically expands and tends to accelerate the rotor to the original equilibrium condition. Actually the stable autorotation condition will exist only when the autorotational speed is within certain limits. If the rotor speed is allowed to slow some excessive amount, then the rotor becomes unstable and the RPM will decrease even further unless the pilot immediately corrects the condition by proper control action.

In case of engine failure, the fixed-wing airplane will be glided at maximum lift-drag ratio to produce maximum glide distance. If minimum rate of descent is desired in power-off flight rather than maximum glide distance, the fixed-wing airplane will be flown at some lower airspeed. Actually, the minimum rate of descent will occur at minimum power required. The helicopter exhibits similar characteristics but ordinarily the best autorotation speed may be considered that speed that results in the minimum rate of descent rather than maximum glide distance. The aerodynamic condition of the rotor which produces minimum rate of descent is:

Maximum ratio of

$$\frac{(\text{Mean blade lift coefficient})^{3/2}}{\text{Mean blade drag coefficient}}$$

It is this ratio which determines the autorotation rate of descent. Figure 6.18 illustrates the variation of autorotation rate of descent with equivalent airspeed for a typical helicopter. Point *A* on this curve defines the point which produces autorotation with minimum rate of descent. Maximum glide distance during autorotation descent would be obtained at the flight condition which produces the greatest proportion between airspeed and rate of descent. Thus, a straight line from the origin tangent to the curve will define the point for maximum autorotative glide distance. This corresponds to Point *B* of figure 6.18. If the helicopter is being glided at the speed for maximum glide distance, a decrease in airspeed would reduce the rate of descent but the glide distance would decrease. If the helicopter is being glided at the speed for minimum rate of descent, the rate of descent (steady state) can not be reduced but the glide distance can be increased by increasing the glide speed to that for maximum distance. Weight and wind affect the glide characteristics of a helicopter the same way an airplane is affected. Ideally, the helicopter autorotates at a higher equivalent airspeed at higher gross weight or when autorotating into a headwind.

In addition to aerodynamic forces which act on the rotor during autorotation, inertia forces are also important. These effects are usually associated with the pilot's response time because the rate a pilot reacts to a power failure is quite critical. The time necessary to reduce collective pitch and enter autorotation becomes critical if the rotor inertia characteristics are such as to allow the rotor to slow down to a dangerous level before the pilot can react. With power on, the blade pitch is relatively high and the engine supplies enough torque to overcome the drag of the blades. At the instant of power failure the blades are at a high pitch with high drag. If there is no engine torque to maintain the RPM, the rotor will decelerate depending on the rotor torque and rotor inertia. If the rotor has high rotational energy the rotor will lose RPM less rapidly, giving the pilot more time to reduce collective pitch and enter autorotation. If the rotor has low rotational energy, the rotor will lose RPM rapidly and the pilot may not be able to react quickly enough to prevent a serious loss of rotor RPM. Once the collective pitch is at the low pitch limit, the rotor RPM can be increased only by a sacrifice in altitude or airspeed. If insufficient altitude is available to exchange for rotor speed, a hard landing is inevitable. Sufficient rotor rotational energy must be available to permit adding collective pitch to reduce the helicopter's rate of descent before final ground contact.

In the case of most small helicopters, at least 300 feet of altitude is necessary for an average pilot to set up a steady autorotation and land the helicopter safely without damage. This minimum becomes 500 to 600 feet for the larger helicopters, and will be even greater for helicopters with increased disc loading. These characteristics are usually presented in the flight handbook in the form of a "dead man's curve" which shows the combinations of airspeed and altitude above the terrain where a successful autorotative landing would be difficult, if not impossible.

A typical "dead man's curve" is shown in figure 6.18. The most critical combinations are due to low altitude and low airspeed illustrated by area *A* of figure 6.18. Less critical conditions exist at higher airspeeds because of the greater energy available to set up a steady autorotation. The lower limit of area *A* is a finite altitude because the helicopter can be landed successfully if collective pitch is held rather than reduced. In this specific case there is not sufficient energy to reach a steady state autorotation. The maximum altitude at which this is possible is approximately ten feet on most helicopters.

Area *B* on the "dead man's curve" of figure 6.18 is critical because of ground contact flight speed or rate of descent, which is based on the strength of the landing gear. The average pilot may have difficulty in successfully flaring the helicopter from a high speed flight condition without allowing the tail rotor to strike the ground or contacting the ground at an excessive airspeed. A less critical zone is sometimes shown on this curve to indicate that higher ground contact speeds can be permitted when the landing surface is smooth. In addition, various stability and control characteristics of a helicopter may produce critical conditions in this area. The critical areas of the "dead man's curve" should be avoided unless such operation is a specific mission requirement.

POWER SETTLING. The term "power settling" has been used to describe a variety of flight conditions of the helicopter. True "power settling" occurs only when the helicopter rotor is operating in a rotary flow condition called the "vortex ring state." The flow through the rotor in the "vortex ring state" is upward near the center of the disc and downward in the outer portion, resulting in a condition of zero net thrust on the rotor. If the rotor thrust is zero, the helicopter is effectively free-falling and extremely high rates of descent can result.

The downwash distribution within the rotor is shown in figure 6.19 for the conditions of normal hovering and power settling. Part *A* of figure 6.19 illustrates the typical downwash distribution for hovering flight. If sufficient power were not available to hover at this condition, the helicopter would begin to settle at some rate of descent depending on the deficiency of power. This rate of descent would effectively decrease the downwash throughout the rotor and result in a redistribution of downwash similar to Part *B* of figure 6.19. At the outer portion of the rotor disc, the local induced downwash velocity is greater than the rate of descent and downflow exists. At the center of the rotor disc, the rate of descent is greater than the local induced downwash velocity and the resultant flow is upward. This flow condition results in the rotary "vortex ring" state. By reference to the basic momentum theory it is apparent that the rotor will produce no thrust in this condition if the net mass flow of air through the rotor is zero. It is important to note that the main lifting part of the rotor is not stalled. The rotor roughness and loss of

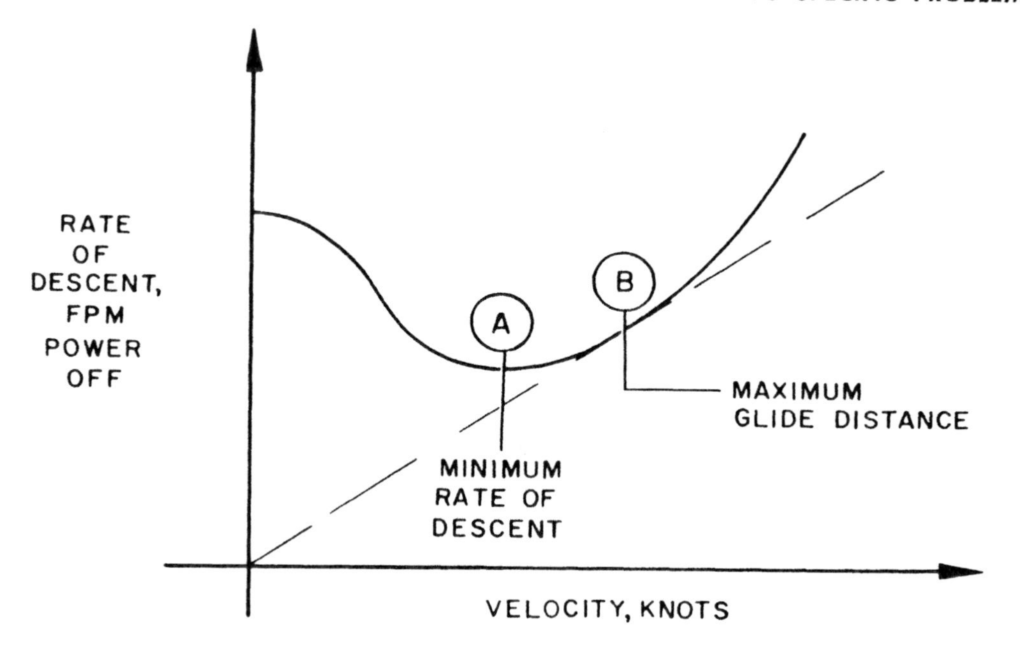

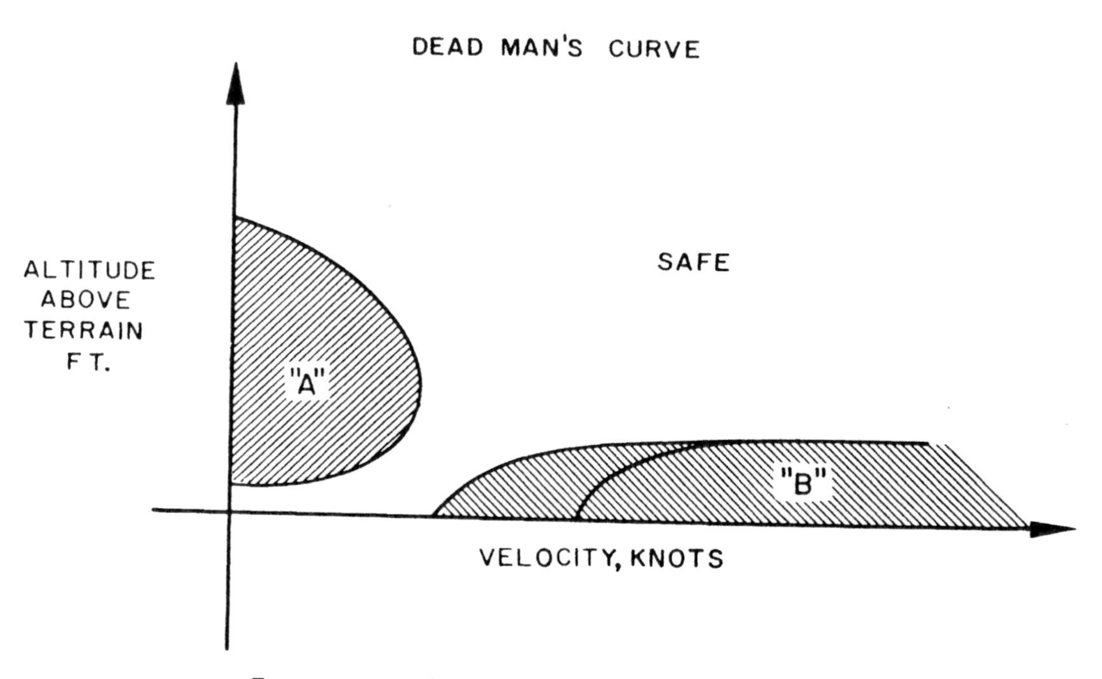

Figure 6.18. Autorotation Characteristics

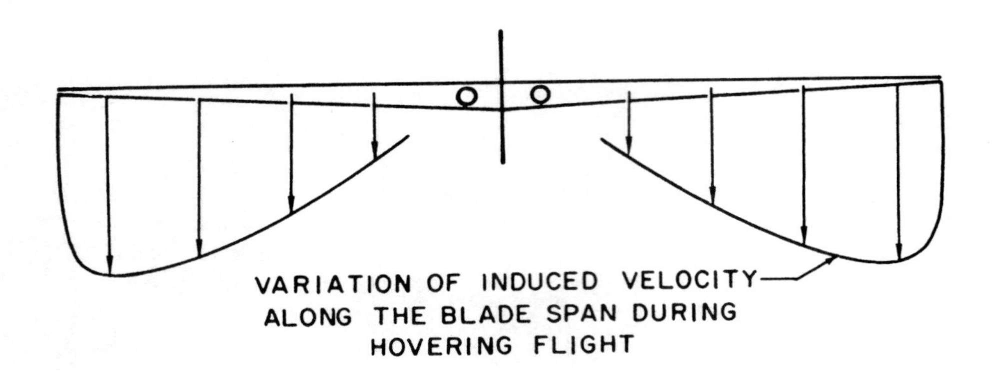

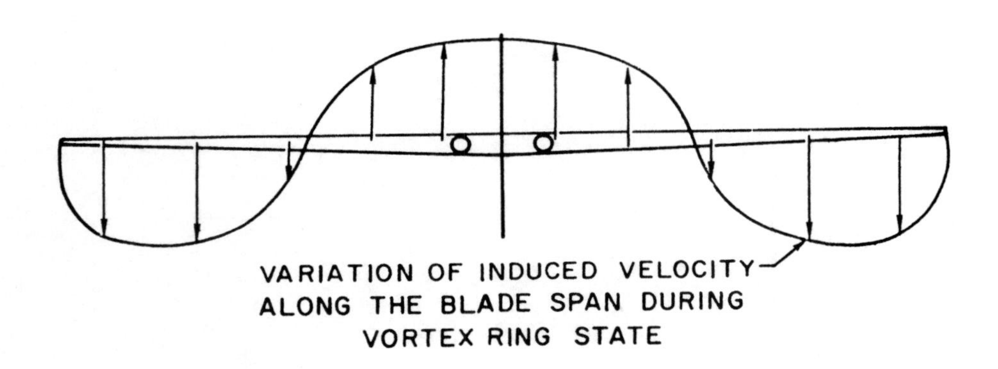

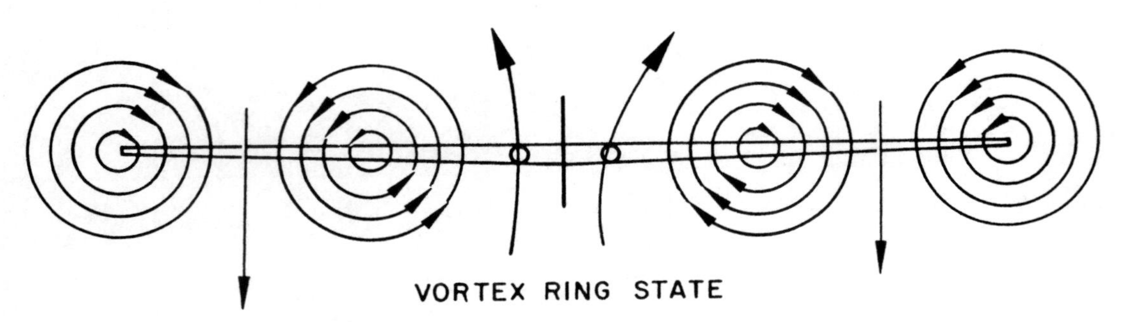

Figure 6.19. Rotor Downwash Distribution

control experienced during "power settling" results from the turbulent rotational flow on the blades and the unsteady shifting of the flow in and out spanwise along the blade. There is an area of positive thrust in the outer portion of the rotor as a result of the mass of air accelerated downward and an area of negative thrust at the center of the rotor as a result of the mass of air flowing upward. The rotor is stalled only near the hub but no important effect is contributed because of the low local velocities.

Operation in the "vortex ring" state is a transient condition and the helicopter will seek equilibrium by descending. As the helicopter descends, a greater upflow through the disc results until eventually the flow is entirely up through the rotor and the rotor enters autorotation where lower rates of descent can be achieved. Unfortunately, considerable altitude will be lost before the autorotative type of flow is achieved and a positive recovery technique *must be applied* to minimize the loss of altitude. "Power settling" can be recognized by rotor roughness, loss of control due to the turbulent rotational flow, and a very high rate of descent (as high as 3,000 fpm). It is most likely to be encountered inadvertently when attempting to hover when sufficient power is not available because of high gross weight or high density altitude.

Recovery from "power settling" can be accomplished by getting the rotor out of the "vortex ring state." If the condition is encountered with low power, rapid application of full power may increase the downwash sufficiently to get the rotor out of the condition. If the condition is encountered at high or maximum power or, if maximum power does not effect a recovery, increasing airspeed by diving will result in recovery with minimum loss of altitude. This type of recovery is most effective but adequate cyclic control must be available. If cyclic control has been lost, recovery must be effected by reducing power and collective pitch and entering autorotation. When normal autorotation has been established, a normal power recovery from the autorotation can be made. While such a recovery technique is effective, considerable altitude may be lost. Hence, diving out of the power settling condition provides the most favorable means of recovery.

Actually, real instances of true "power settling" are quite rare. A condition often described incorrectly as "power settling" is merely a high sink rate as a result of insufficient power to terminate an approach to landing. This situation frequently occurs during high gross weight or high density altitude operation. The flow conditions within the rotor are quite normal and there is merely insufficient power to reduce rate of descent and terminate an approach. Such a situation becomes more critical with a steep approach since the more rapid descent will require more power to terminate the approach.

THE FLIGHT HANDBOOK

For the professional aviator, there are few documents which are as important as the airplane flight handbook. The information and data contained in the various sections of the flight handbook provide the basis for safe and effective operation of the airplane.

Various sections of the flight handbook are devoted to the following subjects:

(1) *Equipment and Systems.* With the mechanical complexity of the modern airplane, it is imperative that the pilot be familiar with every item of the aircraft. Only through exact knowledge of the equipment can the pilot properly operate the airplane and contend with malfunctions.

(2) *Operating Procedures.* Good procedures are mandatory to effect safe operation of the airplane and its equipment. The complexity of modern equipment dictates the use of special and exact procedures of operation and any haphazard or non-standard procedure is an

invitation for trouble of many sorts. The normal and emergency procedures applicable to each specific airplane will insure the proper operation of the equipment.

(3) *Operating Limitations.* The operation of the airplane and powerplant must be conducted within the established limitations. Failure to do so will invite failure or malfunction of the equipment and increase the operating cost or possibly cause an accident.

(4) *Flight Characteristics.* While all aircraft will have certain minimum requirements for flying qualities, the actual peculiarities and special features of specific airplanes will differ. These particular flight characteristics must be well known and understood by the pilot.

(5) *Operating Data.* The performance of each specific airplane defines its application to various uses and missions. The handbook operating data must be available at all times to properly plan and execute the flight of an aircraft. Constant reference to the operating data will insure safe and effective operation of the airplane.

Great time and effort are expended in the preparation of the flight handbook to provide the most exact information, data, and procedures. Diligent *study* and continuous *use* of the flight handbook will ensure that the greatest effectiveness is achieved from the airplane while still operating within the inherent capabilities of the design.

SELECTED REFERENCES

1. Dommasch, Sherby, and Connolly
   "Airplane Aerodynamics"
   Pitman Publishing Co.
   2d Edition, 1957
2. Perkins and Hage
   "Airplane Performance, Stability, and Control"
   John Wiley and Sons
   1949
3. E. A. Bonney
   "Engineering Supersonic Aerodynamics"
   McGraw-Hill Book Co.
   1950
4. Hurt, Vernon, and Martin
   "Aeronautical Engineering, Section I, Manual of Instruction, Aviation Safety Officer Course"
   University of Southern California
   1958
5. Fairchild, Magill, and Brye
   "Principles of Helicopter Engineering"
   University of Southern California
   1959

INDEX